Advances in Pharmaceutical Product Development

Keerti Jain • Awesh K. Yadav
Editors

Advances in Pharmaceutical Product Development

Springer

Editors
Keerti Jain
Drug Delivery and Nanomedicine Research Laboratory
Department of Pharmaceutics
National Institute of Pharmaceutical Education and Research (NIPER)
Lucknow, Uttar Pradesh, India

Awesh K. Yadav
Department of Pharmaceutics
National Institute of Pharmaceutical Education and Research (NIPER)
Lucknow, Uttar Pradesh, India

ISBN 978-981-97-9229-0 ISBN 978-981-97-9230-6 (eBook)
https://doi.org/10.1007/978-981-97-9230-6

This Springer imprint is published by the registered company Springer Nature Singapore Pte Ltd.
The registered company address is: 152 Beach Road, #21-01/04 Gateway East, Singapore 189721, Singapore

Preface

Development of any pharmaceutical product starts with determination of various physicochemical properties of a new chemical entity having potential to be developed as a new drug molecule. These physicochemical properties are determined in the preformulation studies, one of the important stages in the development of pharmaceutical formulation apart from prototype development, scale up studies, and commercialization of pharmaceutical formulation. Preformulation studies assist in identification of possible problems which could be anticipated while developing the pharmaceutical formulation; further it also helps in determining one of the appropriate dosage forms to develop prototype, which could finally lead to a safe, effective, and stable pharmaceutical product for a new drug molecule. Optimization for the scale-up studies facilitates transformation of process to clinical level from preclinical experiments. In this book, we focus on compiling all the stages involved in pharmaceutical product development including the importance, requirement, and effect of each stage and process. This book has covered prototype development for pharmaceutical formulations, scale-up studies, optimization, testing, packaging and commercialization of different dosage forms like tablets, suspensions, emulsions, nanocrystals, inhalational products, sterile products, herbal formulations, etc.

This book provides important and informative topics including quality by design, process analytical technology, perspective related to additives or excipients used in pharmaceutical product development, advances in tablet formulations, pharmaceutical product development of drug nanocrystals, inhalational drug delivery systems, herbal formulations and challenges associated with the development of these formulations, drug repositioning and virtual screening, and pharmacovigilance and pharmacogenomics as well. Chapter 1–3 compiles the information regarding the pharmaceutical formulation development, design, and optimization for materials and pharmaceutical products. Chapter 4 provides a detailed account of formulation additives for pharmaceutical product development. Chapter 5–14 discusses different pharmaceutical products and their developments like tablets and other solid dosage forms, liquid and poly-disperse systems including suspensions and emulsions, sterile products, topical and transdermal drug delivery systems, ophthalmic formulations, inhalational drug delivery systems, and drug nanocrystals. Chapter 15 deals with herbal formulations whereas Chaps. 16 and 17 are focused on drug repurposing and virtual screening, and preclinical, clinical studies, pharmacogenomics, pharmacovigilance, and commercialization of pharmaceutical drug products.

We believe that this book will be a valuable source of information to its audience and will assist in understanding the updates on pharmaceutical formulation development. We are thankful to all the contributors for their worthy manuscripts which helped significantly in the compilation of this book. We are grateful to our institution, NIPER-Raebareli, colleagues, students, and family members for their support and encouragement. We are also thankful to Springer Nature and all the team members who worked tirelessly to publish this book and bring it in the market.

Lucknow, Uttar Pradesh, India Keerti Jain
Lucknow, Uttar Pradesh, India Awesh K. Yadav

Contents

Editors and Contributors

About the Editors

Keerti Jain is presently working as Assistant Professor in Department of Pharmaceutics, National Institute of Pharmaceutical Education and Research (NIPER)—Raebareli. She earned her M. Pharm. and Doctorate degree from Department of Pharmaceutical Sciences, Dr. H. S. Gour Central University, Sagar. Dr. Jain previously worked as Associate Professor in ISF College of Pharmacy, Moga, India and as Assistant Professor in Delhi Pharmaceutical Sciences and Research University (DPSRU), New Delhi. She did her Post Doctorate from M. S. University of Vadodara, India as SERB-National Post-Doctoral Fellow. She has about 15 years of research experience on Nanomedicine based drug delivery systems. Dr. Jain has filed six patents from her innovative research and received funding for research projects from CSIR, ICMR and SERB, New Delhi. She has edited several books and contributed to a few book chapters. She has been awarded with various research fellowships including SERB–NPDF, CSIR-SRF, UGC-JRF and awarded with many awards for her research work including Pharmaceutical Science Alumni Award-2006, most innovative Idea award in LUFTHANSA impact week, SPER Women Forum—Young Scientist Award and prestigious ICMR—Shakuntala Amir Chand Prize for the year 2019. She has been invited to present her research work at several national and seven international conferences and enlisted among World's Top 2% Scientists, consecutively for year 2020, 2021, 2022 and 2023 in the field of Pharmacology and Pharmacy, a list created by Stanford University, USA. Dr. Jain has published more than 70 research papers in many prestigious international journals and also served as a reviewer for several reputed international journals.

Awesh K. Yadav is currently working as Assistant Professor of Pharmaceutics, National Institute of Pharmaceutical Education and Research (NIPER)—Raebareli (A Transit Campus at Lucknow) India, Dr. Yadav has completed his B. Pharm, M. Pharm and Ph.D. degree from Dr. Hari Singh Gour University, Sagar (MP) India. Dr. Yadav served as Principal at Bhagyoday Tirth Pharmacy College Sagar (MP) India. Earlier, he served as Assistant Manager in R&D (Formulation) at Ipca Laboratories Limited Mumbai India. Then, he worked as Research Scientist at Himalaya Drug Company Bangalore, India and as a Research Assistant at Sun Pharm Advanced Research Centre (SPARC), Vadodara (GJ). His research interest is in drug delivery and drug targeting using various drug delivery systems and ligands i.e., Nanoparticles, Nanodiamonds, fullerenes etc. He has been conferred with various prestigious awards notably, Outstanding Teacher award by APTI MP State Branch. He has received various research fellowships including ICMR-SRF, UGC-JRF and awarded for his research work and has served as a reviewer for several international journals. He has also published various research articles in the peer-reviewed international journal and authored or co-authored a few books and book chapters. He is a MEMBER of several international scientific societies and organizations importantly, Association of Pharmaceutical Teachers of India (APTI), Society of ethnopharmacology, Society of Pharmaceutical Education and Research, India.

Contributors

Md Imtiyaz Alam Department of Pharmaceutics, National Institute of Pharmaceutical Education and Research (NIPER) Raebareli, Lucknow, Uttar Pradesh, India

Shristi Arya Department of Pharmaceutics, National Institute of Pharmaceutical Education and Research, Hyderabad, Telangana, India

Deepankar Bahuguna Department of Pharmaceutics, National Institute of Pharmaceutical Education and Research, Hyderabad, Telangana, India

Gopal Bajad Department of Pharmaceutics, National Institute of Pharmaceutical Education and Research, Hyderabad, Telangana, India

Shalini Bajaj School of Pharmaceuticals Sciences, Jaipur National University, Jaipur, Rajasthan, India

Ashish Baldi Pharma Innovation Lab, Department of Pharmaceutical Sciences and Technology, Maharaja Ranjit Singh Punjab Technical University, Bathinda, Punjab, India

B. Behera Delhi Institute of Pharmaceutical Sciences and Research, DPSRU, New Delhi, Delhi, India

Rohit Bhawale Pharmaceutical Nanotechnology Research Laboratory, Department of Pharmaceutics, National Institute of Pharmaceutical Education and Research (NIPER), Ministry of Chemical and Family Welfare, Hyderabad, Telangana, India

Pooja Das Bidla Pharmaceutics Research Projects Laboratory, Department of Pharmaceutical Sciences, Dr. Harisingh Gour Vishwavidyalaya, Sagar, Madhya Pradesh, India

Ishwar Chandra Department of Pharmaceutics, National Institute of Pharmaceutical Education and Research Guwahati, Guwahati, Assam, India

Meenakshi K. Chauhan Delhi Institute of Pharmaceutical Sciences and Research, DPSRU, New Delhi, Delhi, India

Pawan Devangan Department of Pharmaceutics, National Institute of Pharmaceutical Education and Research, Hyderabad, Telangana, India

Amit K. Dubey School of Pharmaceuticals and Population Health Informatics, Faculty of Pharmacy, DIT University, Dehradun, India

Debasis Gantayat School of Pharmaceuticals and Population Health Informatics, Faculty of Pharmacy, DIT University, Dehradun, India

Rohit Garg Department of Pharmaceutics, National Institute of Pharmaceutical Education and Research (NIPER) Raebareli, Lucknow, Uttar Pradesh, India

Arvind Gulbake Department of Pharmaceutics, National Institute of Pharmaceutical Education and Research Guwahati, Guwahati, Assam, India

Madhu Gupta Department of Pharmaceutics, Delhi Pharmaceutical Sciences and Research University, New Delhi, India

Sourabh Jadhav Department of Pharmaceutics, National Institute of Pharmaceutical Education and Research Guwahati, Guwahati, Assam, India

Keerti Jain Drug Delivery and Nanomedicine Research Laboratory, Department of Pharmaceutics, National Institute of Pharmaceutical Education and Research (NIPER), Raebareli, Lucknow, Uttar Pradesh, India

Sanjay K. Jain Pharmaceutics Research Projects Laboratory, Department of Pharmaceutical Sciences, Dr. Harisingh Gour Vishwavidyalaya, Sagar, Madhya Pradesh, India

Vineet K. Jain Department of Pharmaceutics, Delhi Pharmaceutical Sciences and Research University (DPSRU), New Delhi, India

Lalit Lata Jha Department of Pharmaceutics, School of Pharmacy, Faculty of Pharmacy, Parul University, Vadodara, Gujarat, India

Nagphase Nakshatra Jitendra Department of Pharmaceutics, National Institute of Pharmaceutical Education and Research (NIPER) Raebareli, Lucknow, Uttar Pradesh, India

Mit Joshi Department of Pharmacology, Institute of Pharmacy, Nirma University, Ahmedabad, India

Jovita Kanoujia Amity Institute of Pharmacy, Amity University Madhya Pradesh (AUMP), Gwalior, Madhya Pradesh, India

Chantibabu Katta Department of Pharmaceutics, National Institute of Pharmaceutical Education and Research, Hyderabad, Telangana, India

Amanjot Kaur Rayat Institute of Pharmacy, Ropar, Punjab, India

Jaspreet Kaur University Institute of Pharmaceutical Sciences, UGC Centre of Advanced Studies (UGC-CAS), Panjab University, Chandigarh, India

Rayat Institute of Pharmacy, Ropar, Punjab, India

Ankita Kishore Amity Institute of Pharmacy, Amity University Madhya Pradesh (AUMP), Gwalior, Madhya Pradesh, India

Harshita Krishnatreyya Department of Pharmaceutics, National Institute of Pharmaceutical Education and Research Guwahati, Guwahati, Assam, India

Akhilesh Kumar Division of Medicine, ICAR-Indian Veterinary Research Institute, Bareilly, Uttar Pradesh, India

Ankaj Kumar Department of Pharmaceutics, National Institute of Pharmaceutical Education and Research Guwahati, Guwahati, Assam, India

Anoop Kumar Department of Pharmacology, Delhi Institute of Pharmaceutical Sciences and Research (DIPSAR), Delhi Pharmaceutical Sciences and Research University (DPSRU), New Delhi, India

Sumit Kumar Department of Pharmaceutical Sciences, Central University Haryana, Mahendragarh, Haryana, India

Soham Loharkar Department of Pharmaceutics, National Institute of Pharmaceutical Education and Research, Hyderabad, Telangana, India

Jitender Madan Department of Pharmaceutics, National Institute of Pharmaceutical Education and Research, Hyderabad, Telangana, India

Alok Kumar Mahor Institute of Pharmacy, Bundelkhand University, Jhansi, Uttar Pradesh, India

Manju University Institute of Pharmaceutical Sciences, UGC Centre of Advanced Studies (UGC-CAS), Panjab University, Chandigarh, India

Manshi Department of Pharmaceutics and Pharmacy Practice, SGT College of Pharmacy, SGT University, Gurugram, Haryana, India

Neelesh Kumar Mehra Pharmaceutical Nanotechnology Research Laboratory, Department of Pharmaceutics, National Institute of Pharmaceutical Education and Research (NIPER), Ministry of Chemical and Family Welfare, Hyderabad, Telangana, India

Neeraj Mishra Amity Institute of Pharmacy, Amity University Madhya Pradesh (AUMP), Gwalior, Madhya Pradesh, India

Shashank Shekar Mishra School of Pharmaceuticals and Population Health Informatics, Faculty of Pharmacy, DIT University, Dehradun, India

Gijith K. Mohan Drug Delivery and Nanomedicine Research Laboratory, Department of Pharmaceutics, National Institute of Pharmaceutical Sciences and Research (NIPER)—Raebareli, Lucknow, Uttar Pradesh, India

Atul Mourya Department of Pharmaceutics, National Institute of Pharmaceutical Education and Research, Hyderabad, Telangana, India

Kalpana Nagpal Amity Institute of Pharmacy, Amity University, Noida, Uttar Pradesh, India

Pritish Kumar Panda Pharmaceutics Research Projects Laboratory, Department of Pharmaceutical Sciences, Dr. Harisingh Gour Vishwavidyalaya, Sagar, Madhya Pradesh, India

Bhoomika M. Patel National Forensic Sciences University, Gandhinagar, Gujarat, India

Dhruvi Patel Department of Pharmaceutics, Parul Institute of Pharmacy & Research, Faculty of Pharmacy, Parul University, Vadodara, Gujarat, India

Hitarthi Mayur Patel Department of Pharmaceutics, School of Pharmacy, Faculty of Pharmacy, Parul University, Vadodara, Gujarat, India

Manisha Patel Drug Delivery and Nanomedicine Research Laboratory, Department of Pharmaceutics, National Institute of Pharmaceutical Sciences and Research (NIPER)—Raebareli, Lucknow, Uttar Pradesh, India

Meenakshi Patel Department of Pharmaceutics, School of Pharmacy, Faculty of Pharmacy, Parul University, Vadodara, Gujarat, India

Parth Patel Drug Delivery and Nanomedicine Research Laboratory, Department of Pharmaceutics, National Institute of Pharmaceutical Sciences and Research (NIPER)—Raebareli, Lucknow, Uttar Pradesh, India

Anchal Pathak Drug Delivery and Nanomedicine Research Laboratory, Department of Pharmaceutics, National Institute of Pharmaceutical Sciences and Research (NIPER)—Raebareli, Lucknow, Uttar Pradesh, India

T. Pawan Kumar Department of Pharmaceutics, National Institute of Pharmaceutical Education and Research, Hyderabad, Telangana, India

Shivam Ramdas Pawar Drug Delivery and Nanomedicine Research Laboratory, Department of Pharmaceutics, National Institute of Pharmaceutical Education and Research (NIPER), Raebareli, Lucknow, Uttar Pradesh, India

Vibhooti Pippola School of Pharmaceuticals and Population Health Informatics, Faculty of Pharmacy, DIT University, Dehradun, India

Harvinder Popli Department of Pharmaceutics, Delhi Pharmaceutical Sciences and Research University (DPSRU), New Delhi, India

Sarjana Raikwar Pharmaceutics Research Projects Laboratory, Department of Pharmaceutical Sciences, Dr. Harisingh Gour Vishwavidyalaya, Sagar, Madhya Pradesh, India

Neha Raina Department of Pharmaceutics, Delhi Pharmaceutical Sciences and Research University, New Delhi, India

V. Rajesh Department of Pharmaceutics, National Institute of Pharmaceutical Education and Research Guwahati, Guwahati, Assam, India

Anjali Rajora Amity Institute of Pharmacy, Amity University, Noida, Uttar Pradesh, India

Chetan Ram School of Pharmaceuticals and Population Health Informatics, Faculty of Pharmacy, DIT University, Dehradun, India

Sunny Rathee Pharmaceutics Research Projects Laboratory, Department of Pharmaceutical Sciences, Dr. Harisingh Gour Vishwavidyalaya, Sagar, Madhya Pradesh, India

Sweta Roy Department of Pharmacology, Delhi Institute of Pharmaceutical Sciences and Research (DIPSAR), Delhi Pharmaceutical Sciences and Research University (DPSRU), New Delhi, India

P. Saikiran Department of Pharmaceutics, National Institute of Pharmaceutical Education and Research, Hyderabad, Telangana, India

Khyati Saini School of Pharmaceuticals and Population Health Informatics, Faculty of Pharmacy, DIT University, Dehradun, India

Shivani Saraf Pharmaceutics Research Projects Laboratory, Department of Pharmaceutical Sciences, Dr. Harisingh Gour Vishwavidyalaya, Sagar, Madhya Pradesh, India

Sonal Setya Department of Pharmaceutics and Pharmacy Practice, SGT College of Pharmacy, SGT University, Gurugram, Haryana, India

Ruchika Sharma Centre for Precision Medicine and Pharmacy, Delhi Pharmaceutical Sciences and Research University, New Delhi, India

Satish Shilpi School of Pharmaceuticals and Population Health Informatics, Faculty of Pharmacy, DIT University, Dehradun, India

K. Shubham Delhi Institute of Pharmaceutical Sciences and Research, DPSRU, New Delhi, Delhi, India

Prem Prakash Singh Institute of Pharmacy, Bundelkhand University, Jhansi, Uttar Pradesh, India

Vaibhavi Srivastava Pharmaceutical Nanotechnology Research Laboratory, Department of Pharmaceutics, National Institute of Pharmaceutical Education and Research (NIPER), Ministry of Chemical and Family Welfare, Hyderabad, Telangana, India

Sushama Talegaonkar Department of Pharmaceutics, School of Pharma Sciences, DPSRU, New Delhi, Delhi, India

Sofiya Tarannum Drug Delivery and Nanomedicine Research Laboratory, Department of Pharmaceutics, National Institute of Pharmaceutical Education and Research (NIPER)—Raebareli, Lucknow, Uttar Pradesh, India

Darshana Tijare Department of Pharmaceutics, National Institute of Pharmaceutical Education and Research, Hyderabad, Telangana, India

Harithasree Veerabromma Department of Pharmaceutics, National Institute of Pharmaceutical Education and Research, Hyderabad, Telangana, India

Awesh K. Yadav Department of Pharmaceutics, National Institute of Pharmaceutical Education and Research (NIPER) Raebareli, Lucknow, Uttar Pradesh, India

Monu Yadav Amity Institute of Pharmacy, Amity University Gurgaon, Gurgaon, Haryana, India

Advances in Development of Pharmaceutical Products

1

Keerti Jain, Sofiya Tarannum, Gijith K. Mohan, Manisha Patel, Anchal Pathak, Parth Patel, Vineet K. Jain, Ashish Baldi, and Harvinder Popli

Abstract

The process of pharmaceutical product development may be classified into four stages including preformulation, prototype development, scale-up studies and commercialisation of pharmaceutical formulations. Pharmaceutical product development begins with the estimation of physicochemical properties of new chemical entity with druggable properties. These physicochemical properties are determined in the first stage and one of the foremost steps in the development of pharmaceutical formulation, i.e. preformulation studies. Preformulation studies aid in determining the possible obstacles which could be anticipated in pharmaceutical formulation development; further, it also assists in identifying the appropriate dosage form and route of administration to develop a prototype, which could be finally developed in a safe, effective and stable pharmaceutical product for the new drug molecule. In this chapter, we have discussed advances in pharmaceutical product development as well as about all these four important stages, i.e. preformulation, prototype development, scale-up studies and commercialisation of pharmaceutical formulations involved in pharmaceutical product development.

K. Jain (✉) · S. Tarannum · G. K. Mohan · M. Patel · A. Pathak · P. Patel
Drug Delivery and Nanomedicine Research Laboratory, Department of Pharmaceutics, National Institute of Pharmaceutical Education and Research (NIPER)—Raebareli, Lucknow, Uttar Pradesh, India
e-mail: keertijain.02@niperraebareli.edu.in

V. K. Jain · H. Popli
Department of Pharmaceutics, Delhi Pharmaceutical Sciences and Research University (DPSRU), New Delhi, India

A. Baldi
Pharma Innovation Lab, Department of Pharmaceutical Sciences and Technology, Maharaja Ranjit Singh Punjab Technical University, Bathinda, Punjab, India

K. Jain, A. K. Yadav (eds.), *Advances in Pharmaceutical Product Development*, https://doi.org/10.1007/978-981-97-9230-6_1

Keywords

Preformulation · Efficacy · Stability · Prototype · Commercialisation · Pilot plant · Scale-up

1.1 Introduction

Pharmaceutical product development is becoming interdisciplinary day by day with the involvement of techniques of biotechnology, nanotechnology, etc. Therefore, researchers are investing substantial efforts in developing a product in a systematic manner with the involvement of optimisation and evaluation. The pharmaceutical product development process is generally classified into four stages including preformulation, prototype development, scale-up studies and commercialisation of pharmaceutical formulations. Pharmaceutical products must comply with the guidelines issued by regulatory authorities so that its quality, efficacy and safety can be ensured (Amarji et al. 2018). The development of pharmaceutical products is carried out under the supervision of the formulation scientists, who are expert and knowledgeable enough to produce novel pharmaceutical products covering all the steps, starting from the conceptualisation of product to its final packaging that can enter the market and boost the pharmaceutical industry (Destro and Barolo 2022). Pharmaceutical product development is evolving day by day due to inclusion of both the conventional and novel techniques and approaches of several fields like biotechnology, molecular biology and genetics apart from the core pharma, so that unmet medical challenges can be tackled in a better way by using the new therapies (Ojha et al. 2021).

A detailed information and data regarding the properties of active pharmaceutical ingredients (APIs), excipients, the interaction between all the elements of the pharmaceutical products and manufacturing techniques are collected in the initial phases of research that together are called as the preformulation studies. As the novel drug development is time- and resource-consuming process, the generic products and improvised dosage form developments are preferred as their safety as well as efficacy database is already available, contributing to their rapid development along with the exemption of clinical trials (Patravale et al. 2016). Recent advancements in pharmaceutical research and development highlight the development of new analytical technologies that supports the preformulation research like preformulation parameters, stability studies, chiral purity analysis of therapeutic agents, chiral swapping probabilities, preformulation solvent selection and equilibrium and intrinsic solubility analysis of main compounds in suitable solvents. These analytical technologies are important as they give correct findings in no time, which can be explored by the formulation scientists (Bharate and Vishwakarma 2013).

After the preformulation studies, prototype development is the next step in the development of pharmaceutical products. Ideas are needed that are then conceptualised, and as far as the pharmaceutical products are concerned, it basically revolves around the product's efficacy, stability and safety. Like a seed, the concept is then germinated by experimental laboratory work into a prototype (Sathish et al. 2016). The development of prototype is one of the essential steps, needed to be taken while developing a novel product as it includes the basic working model of the product that will be finalised and transformed into the final product. Prototype development is categorised into four phases, i.e. (i) identification of the basic requirements for the product development, (ii) preparation and implementation of the design to meet the basic requirements identified in phase 1, (iii) experiments with the prototype, and (iv) revision of prototype to accomplish basic requirements. Prototype development of all the dosage forms like parenteral, solid oral, topical, transdermal and inhalational is usually carried out so that they can be moulded into their best version without any defects and flaws.

As the final pharmaceutical products at the lab scale are obtained after the preformulation studies and prototype development, the products are needed to be scaled up for fulfilling the demands of the patients (Raval et al. 2018). Large scale success of pharmaceutical products depends on the tactic that how well the small-scale experimentation has been performed using the facility of pilot plant and correct application of scale-up techniques. The pilot plant is a small-scale process plant designed to obtain information that can be used in industrial-scale processes. They allow us to see the feasibility of industrial-scale designs and processes, thus reducing the initial investment cost (Strovel et al. 2016).

Nevertheless, a huge gap exists between small scale and large scale because of the limited suitable scale-up methodologies that can be seen in case of nanoformulations like liposomes, dendrimers, polymeric nanoparticles, etc. Several advancements are taking place in scale-up processes, because of which many nanoformulations came into the market. The scale-up should abide by the rules of current good manufacturing practices (cGMP), which ensures that the necessary measures are taken for production as well as quality control so that production processes and testing are properly defined, validated, reviewed and documented. Along with that, by using the GMP guidelines, officials also inspect the working personnel, sites and materials and make sure that the developed pharmaceutical products are safe and efficacious (WHO 2024). GMP dictates how manufacturers handle the distribution of their products to ensure they maintain quality and reach patients in a safe state. This might involve regulations for storage conditions, transportation protocols and traceability throughout the supply chain. By incorporating the legal components, GMP fosters a system where quality and safety are prioritised at every stage, from raw materials to the final product reaching the patient (FDA, 2024).

Pharmaceutical companies submit a detailed application to the relevant regulatory body. This application contains comprehensive data on the drug, including its chemical composition, manufacturing processes, preclinical animal testing results and, most importantly, clinical trial data from all phases. Regulatory agencies meticulously evaluate the submitted data. The regulatory approval process is often lengthy and intricate. This is because the stakes are incredibly high—a drug that is not safe or effective can have devastating consequences for patients. The extensive documentation and thorough evaluation ensure potential risks are attenuated/minimised and the drug's benefits are well established before it reaches the public (Moini et al. 2023).

Marketing campaigns and sales representatives are the critical part of commercialisation that play crucial role in raising awareness about the pharmaceutical products among healthcare professionals and patients who may get benefit from it.

However, ethical guidelines ensure that these efforts are truthful and balanced, focusing on the product's benefits and potential drawbacks. Strategies are carefully crafted to target the appropriate audience. This might involve medical journals, educational conferences for healthcare providers or patient advocacy groups (Amarji et al. 2018). In an era of online marketing and artificial intelligence (AI), the commercialisation of pharmaceutical products by managing their life cycle in the best possible manner is way easier with better reach as compared to the prior decades (Shanbhogue et al. 2021). Post-marketing surveillance and pharmacovigilance also come into action as this ongoing process involves actively monitoring the drug's safety profile once it is in the market. It includes collecting and analysing data on adverse events experienced by patients. This might involve modifying prescribing guidelines, issuing warnings or even recalling the drug if necessary. AI automates tedious tasks like data extraction, freeing up human reviewers to focus on complex cases or analysis. AI reduces human error in data extraction and analysis, leading to more reliable data for identifying safety concerns. AI can analyse vast amounts of data quickly, allowing for earlier detection of potential safety issues with medications (Schmider et al. 2019).

1.2 Preformulation Studies

Technical understanding of different characteristics of a drug molecule is crucial for designing the appropriate dosage form with optimum therapeutic efficacy and patient compliance. Preformulation, the cornerstone of this process, provides the fundamental scientific information for formulation development. It is the initial step, where parameters are investigated before the pharmaceutical product is formulated into an appropriate dosage form. This is not just a step but a crucial foundation for the entire process. A variety of study sets focused to explore the physical and chemical characteristics of a novel molecule are included in preformulation studies. These physical and chemical characteristics significantly impact the medication's biofate.

Strong proof-of-concept and preclinical studies are necessary to understand the characteristics of the therapeutically active agent and any pertinent additives that may impact the final formulation and drug performance. Scientists are guided in this process by their deep understanding and comprehensive knowledge of the preformulation parameters of the drug (Ahirwar and Shukla 2023). Some of the essential preformulation parameters are discussed in the following subsections:

1.2.1 Solubility

The solubility of therapeutically active agents plays a significant role in developing pharmaceutical products. Poor solubility can also negatively impact the compound's in vivo behaviour and complicate test design. As a result, poor solubility could prevent development of new treatments. Lattice energy, weak bonding forces, co-solvency, pH, additives, dielectric constant, hydrotrophy, complexation, temperature, lipophilicity, ionisation potential, solubilisation by surfactant, pressure and molecular volume are some of the factors that generally affect solubility of drug molecules. These criteria might help to develop the final formulation with reduced possibilities of drug failure if they were thoroughly examined during preformulation studies (Ahirwar and Shukla 2023).

In conclusion, factors such as temperature, pH levels, solvent system, diluents, additives and the physical state of the drug molecules could change solubility profile of drug. Adequate water solubility is necessary for therapeutic efficacy in a normal pH range of 1–8. If the solubility of a therapeutically active agent is not optimal, significant efforts are required to increase it. Poor solubility, defined as 10 mg/mL, can lead to incomplete or irregular absorption under physiological conditions (Bandopadhyay et al. 2018;Ahirwar and Shukla 2023).

1.2.2 Partition Coefficient

The partition coefficient (log *P*) is the ratio of unionised drug distributed between the organic and aqueous phases. The ability of drug molecule to cross the lipidic membrane is provided by the oil-water partition coefficient. For the best possible intestinal absorption of drug molecules, an optimised lipophilic/hydrophilic balance is one of the most important elements as biological membranes are lipidic, but the outer layer is coated with a hydrophilic mucus membrane. Hence, the partition coefficient value significantly impacts how much of a drug will get absorbed and reach the systemic circulation.

Equal distribution of drug in water and partitioning solvent is represented by a log *P* value of 0. Higher lipidic solubility is indicated by a log *P* value larger than 1, while higher water solubility is indicated by a value less than 1. A suitable hydrophilic-lipophilic balance is required for maximum solubility and absorption. Since it is nearly impossible to determine the log *P* value in a biological system, several techniques are available to calculate the partition coefficient of a molecule in vitro. These techniques include the following methods: shake flask, countercurrent/filter probe and software-based computation methods. The shaking flask method, which uses the octanol-water system to assess the partitioning characteristics of drugs, is widely used (Bandopadhyay et al. 2018; Barbosa et al. 2019; Bastin et al. 2000). Octanol is chosen as the partitioning solvent for several reasons, as mentioned below:

- Because it has a polar head and a nonpolar tail, octanol is thought to mimic the lipoidal nature of biological membranes.
- It is an organic compound that is immiscible with water, though some water is expected to be present in the polar head portion.
- Solubility profile of most of the drug molecules in octanol resemble their lipophilicity (Bastin et al. 2000).

1.2.3 Polymorphism and Crystallinity

Drugs and excipients are found in various crystalline or amorphous states based on their chemistry and isolation methods. Molecules can arrange themselves in different geometric configurations during crystallisation, resulting in a structure with distinct packing arrangements or orientations for the created crystals. These multiple states are generally known as polymorphs and this phenomenon is known as polymorphism. The physicochemical properties of each polymorphic form, such as solubility and melting point, can significantly impact stability and bioavailability. Furthermore, polymorphism might impact the compression properties of pharmaceutical products, i.e. the orthorhombic form of paracetamol shows better compressibility than its monolithic form due to the presence of sliding planes in the orthorhombic form (Bastin et al. 2000).

The crystal structures of amorphous solids are typically disorganised; hence, the shift in the physical structure of solids as the temperature rises is not that sharp and provides a broad melting peak in differential scanning calorimetric analysis. The temperature at which solid particles start melting is called glass transition temperature. This is another valuable preformulation feature that needs to be considered while optimising manufacturing procedures (such as temperature selection in heat sterilisation for injectables, wet granulation for tableting, freeze-drying of injectables, etc.) since they may alter the polymorphic structure and therefore the physicochemical and biological properties of the final product (Bandopadhyay et al. 2018).

In order to guide the establishment of a manufacturing process that guarantees the creation and maintenance of the preferred polymorph, it is crucial to determine whether a candidate for development has the propensity to exist in different polymorphic states, the properties of each polymorph (melting point, density, hardness, optical properties, hygroscopicity, solubility, stability, etc.) and the conditions under which each may be formed. Selecting a sufficiently stable polymorph at room temperature is very desirable, as is defining the temperature settings (during manufacture and storage) under which polymorphic change or instability could negatively impact the efficacy, safety and stability of drug molecule (Bandopadhyay et al. 2018; Barbosa et al. 2019; Bastin et al. 2000).

1.2.4 Bulk Properties

Particle size and shape determine the dissolution rate and impact the primary and bulk powder characteristics. Particle size and distribution, flow characteristics and dissolution are important properties that should be researched in dissolution science and when developing new formulations. The physicochemical features of medicinal salts are frequently inextricably linked to their melting point. During formulation development, low melting point APIs often show signs of plastic deformation, leading to aggregation and caking and affecting the flow and compressibility performance. These, in turn, impact several crucial quality characteristics of the drug product (at low doses), including homogeneity of content, friability, rates of disintegration and in vitro dissolution (Bharate and Vishwakarma 2013).

Bastin et al. examined the physicochemical characteristics of many salts of the antiarthritic medication RPR 200765 to highlight the significance of flow properties. Among the four types of salt that they studied—mesylate, hydrochloride, hydrobromide and camphor sulfonate—mesylate salt showed good handling qualities, including good flow, which is necessary for the formulation stage, and nonhygroscopic nature, which made the production of capsules and tablets easier. The flow properties of mesylate salt were superior compared to the other salts investigated (hydrochloride, hydrobromide and camphor sulfonate) (Bastin et al. 2000).

Most of the drugs available in the market are in a salt form, either weak bases or acids. Consequently, it is critical to comprehend the behaviour of molecules during ionisation at particular pH values. Therefore, to comprehend dosage form stability, solubility, bioavailability and efficacy of drug molecules, the effects of drug ionisation, ionic strength, pH and temperature are simultaneously evaluated at the preformulation stage (Bharate and Vishwakarma 2013).

Apart from this, the stability study of active pharmaceutical ingredients, compatibility studies with other excipients, hygroscopicity and degradation profile must also be studied as part of the preformulation study to get regulatory approval for pharmaceutical products. In summary, preformulation is a proactive stage that plays a pivotal role in converting a novel chemical entity into a pharmaceutical product that is stable, safe and effective for human use.

1.3 Prototype Development

Generally, ideas for the development of new products are the descriptive statements that can be written or verbalised, and further these ideas are refined into product concepts that involve benefits of consumers based on the product characteristics. The concepts are then developed into the prototype, which is the working model of the final product. Therefore, it is a crucial stage of the development of new pharmaceutical product, and it involves the development of the working model based on the characteristics of the product obtained during wet lab work for its commercialisation. The ideas in the pharmaceutical industries are protected by very strong intellectual property system such as patents. Intellectual property rights (IPR) are the legal rights granted to the inventors to protect their invention for specific period, and the patentee gets market exclusivity. IPR promotes the economic growth and healthy competition among the scientists from the pharmaceutical research and development field.

There are four phases of prototype development, where phase 1 is the identification of the basic requirements for the product development, phase 2 is the preparation and implementation of the design to meet those requirements, phase 3 involves the experiments with the prototype and phase 4 includes the revision of the prototype and accomplishing those requirements (Singh et al. 2017). To conceptualise a pharmaceutical product from an idea, various parameters should be studied during preformulation studies, including chemical nature and stability of the active pharmaceutical ingredients. Further, the safety and efficacy during preclinical or clinical stages are considered. During preclinical phase of drug development, biopharmaceutical and physiochemical characteristics of drugs are analysed to assess its potential and to reduce the failure rate during the clinical trial as shown in Fig. 1.1.

1.3.1 Experimental Design and Optimisation of Product

While developing pharmaceutical product, it is always desirable to establish optimal and efficient process performance including improved product qualities, enhanced process flexibility and reduced environmental impact. Therefore, finding optimal composition and establishment of processing conditions are the main objectives while developing novel pharmaceutical product of desirable performance. Experimental designs involve the strategic planning to establish optimisation procedures to develop pharmaceutical formulation with desired attributes. While establishing an experimental design, the optimal conditions are established for several factors such as temperature, concentration of excipients, pH, etc., that affect the results or responses of a procedure (Dejaegher and Heyden 2011). During optimisation, one or more factors or independent variables are varied, and response or dependent variables are observed. Finally, experimental domain is obtained from

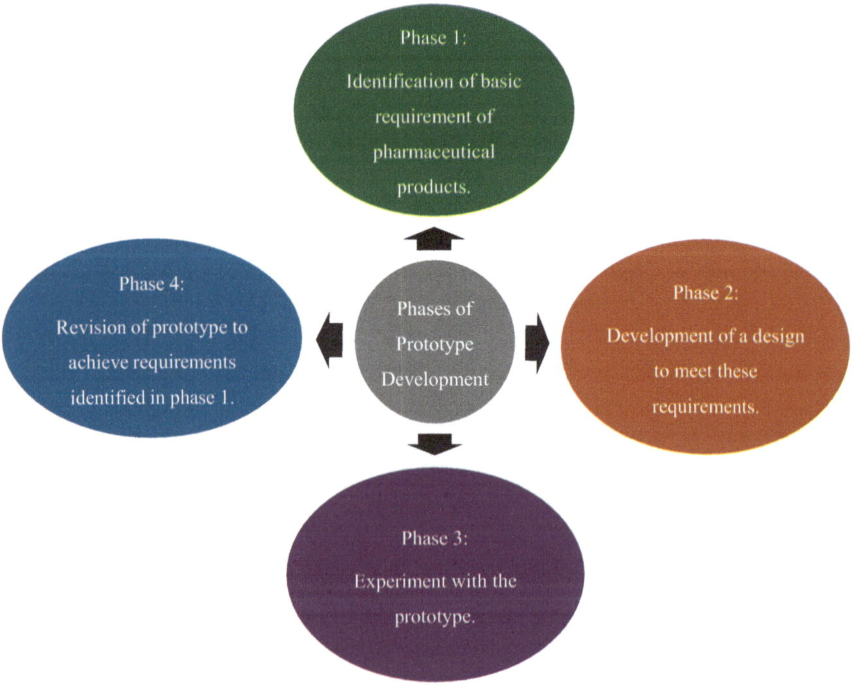

Fig. 1.1 Phases of prototype development

the responses of dependent variables, which shows the optimum values of dependent variabled to get the best possible product with an efficient procedure.

Experimental design involves various steps, where the first step is identifying the rationale of the experimentation and the kind of data desired from the experiment. The second and third steps are the selection of the factors and responses, respectively, while the fourth step is the designing of the experiments. The experiments are replicated multiple times to obtain statistically significant results. In final steps, experiments are conducted, results are analysed and conclusion is drawn. Based on the experimental design, the process is optimised to obtain the pharmaceutical product with the specific attributes, which further complies with the guidelines of the regulatory bodies and meets the quality standards (Armstrong 2006; Gibson 2005).

1.3.2 Parenteral Dosage Forms

Parenteral dosage forms are administered to obtain immediate therapeutic effect, and it includes the administration routes like subcutaneous, intravenous and intramuscular. However, parenteral administrations have several disadvantages such as infection risk at the administration site, and in the case of dosing error, its effect cannot be reversed. However, sometimes it is the preferred route of administration due to higher bioavailability, therapeutic agent acts in seconds due to the direct delivery to the systemic circulation and it can be administered to the unconscious person. There are various cases where parenteral route is the only option such as administration of insulin or paclitaxel or aminoglycoside antibiotics such as gentamicin (Camacho Vieira et al. 2024).

Parenteral dosage forms should be free of foreign particles because they are meant to be injected straight into the systemic circulation. In addition, there are several additional quality-related criteria that should be considered while developing parenteral formulations. The product's sterility is one of the most important quality factors as the parenteral-administered formulation subjected to the systemic circulation directly. The autoclave is commonly used technique to achieve sterility in formulations by heating them under pressure. However, certain drugs, including peptide and protein therapeutics, are heat-sensitive and can be deteriorated by excessive heat or pressure; in this case, sterilisation by filtration can be used to achieve sterility. Another factor that is considered while developing parenteral formulation is the pH, which should be as close as possible to the physiological pH. However, pH can also influence the stability and solubility of the drug in the solution, and in many cases the optimum pH for stability and the solubility does not coincide with the physiological pH. In the case of difference between the physiologic pH and the pH of the formulation, the formulation should be adjusted to a pH value that maintains the stability and adequate solubility of the drug while at the same time not differing considerably from the physiologic pH. For instance, solutions with pH values higher than 9 can cause tissue necrosis.

The solubility of drug molecule in the formulation is another crucial consideration in the optimisation of parenteral dosage forms. Therapeutic agents can be made more soluble by employing various techniques such as adjusting the pH of the solution, incorporating solubilising agents like surfactants, using a co-solvent, formation of solid dispersion and turning the drug into a soluble complex (Volkova et al. 2023; Pardhi and Jain 2021; Pardhi et al. 2024a,b). The right selection of excipients to prevent drug-excipient or excipient-excipient interactions during formulation, as well as the amount of the injectable solution, is another aspect that must be considered. Therefore, development of parenteral dosage forms using optimized protocol, with the aim of creating a safe and effective treatment strategy as well as a potential cure, is of utmost importance.

1.3.3 Oral Dosage Form

Oral solid dosage forms are used to achieve therapeutic effect in the mouth, throat, digestive tract or for systemic action following oral or gastrointestinal absorption. Drug and excipients can be milled, dried, encapsulated, blended, granulated or tableted to provide oral solid dosage forms. Various oral solid dosage forms such as tablets, capsules, lozenges, powders, granules, etc., have been widely used for delivering active pharmaceutical ingredients due to their convenience and consequent patient compliance. Tablets, unit solid dosage forms administered to achieve immediate drug release or modified release, are the most investigated oral solid dosage form (Sohail Arshad et al. 2021). Since powder substances are typically used to formulate solid dosage forms like tablets and capsules, a thorough characterisation and understanding of the powder's properties and effects on various manufacturing processes are mandatory. This is because the properties of the powder not only affect the manufacturing process but also the drug's biopharmaceutical characteristics, like the drug's ability to dissolve in the gastrointestinal tract. The size and form of the powder can affect the flow property of the powder, which is crucial for the tableting process as an uneven flow might result in different tablet fillings, which can vary the content of the tablets. Other processes affected by the particle size and shape are the mixing and the compaction processes (Vashisht et al. 2015).

Physical-mechanical characteristics of tablets, such as disintegration, porosity, density, presence or absence of structural flaws and viscoelastic qualities, affect the performance of pharmaceutical formulations. Early and effective evaluation of physical-mechanical properties of tablets during drug product development and pharmaceutical production is very important. Furthermore, the USFDA established quality by design, a scientific, risk-based framework to encourage creativity and efficiency in pharmaceutical research, manufacturing and quality assurance. To gain a comprehensive understanding of the critical quality attributes of the completed dosage forms, the USFDA released guidance on quality-by-design and process analytical technology tools. Most traditional characterisation techniques cause sample invasiveness and only weakly demonstrate spatial mapping. Furthermore, these methods offer offline/at batch end monitoring, which is not very useful for making subjective real-time operational modifications. Novel noninvasive analytical techniques have emerged because of relatively recent developments in software, technology, instrumentation and design, and they are presently being used to evaluate tablets (Dave et al. 2017).

1.3.4 Transdermal Dosage Form

Pharmaceutical formulations meant to be applied topically—either for systemic therapy (transdermal formulation) or local treatment (topical formulation)—need to have several variables carefully adjusted to ensure appropriate drug penetration and achieve the intended result. The stratum corneum layer, which is the outermost layer of the skin, is followed by the layers of epidermis and dermis and, lastly, the subcutaneous tissue. A barrier that stops foreign objects from penetrating the skin is the stratum corneum. For the transdermal dosage form, drugs must first enter the stratum corneum and then should enter systemic circulation. The concentration of the drug, its capacity to diffuse throughout the stratum corneum and its partition coefficient between the stratum corneum and its carrier are the primary determinants of penetration into the stratum corneum. Biological parameters like age, sex and the metabolic capacity of skin determine the absorption process across the skin. Penetration enhancer, such as urea or phospholipids, can improve absorption by increasing the capacity of drug to permeate the skin. Other enhancers include alcohols and alkanols (ethanol or decanol), terpenes, sulfoxides (dimethyl sulfoxide), glycols, azones (laurocapram) and surfactants (Yang et al. 2024). The field of transdermal drug delivery system is experiencing continuous growth and development due to the swift increase of essential knowledge that drives industrial development. It is anticipated that as transdermal drug delivery technology advances, disease prevention, diagnosis and control will improve, and as a result quality of patient life would increase globally (Alkilani et al. 2015).

1.3.5 Inhalational Formulation

Inhalation of the therapeutic agents directly into the lungs is used to treat respiratory tract illnesses including bronchitis and asthma. Inhalation dosage forms are crucial in this regard, where the drug is delivered directly to the lung, hence minimising the risk of systemic side effects associated with systemic drug administration. The medication delivery mechanism or device is especially of significant importance for inhalational formulations and has a significant influence on the doctor's choice and the patient's acceptance of a certain formulation.

Inhalational formulations are delivered via nebulisers, metered dose inhalers and dry powder inhalers. Different criteria influence the choice of inhalers; for example, paediatric patients have better compliance with nebulisers. The powder's particle size is a crucial component to consider when creating inhalational formulations, as it influences the powder's ability to enter the respiratory tract and its potential to reach the lung. Particles less than 5 μm can enter the lung, whereas larger particles are expelled from the respiratory tract. Generally, particles larger than 3 μm are needed to enter the alveoli, but it should be remembered that particles smaller than 1 μm, for example, are typically exhaled from the respiratory system. Therefore, to create a formulation with the required properties, particle size in inhalation formulations is crucial and needs to be carefully tuned (Gibson 2005).

1.3.6 Nasal Formulations

Both local and systemic treatments can be administered via the nasal route; the former is crucial for treating a variety of nasal ailments, including rhinitis. Intranasal dosage forms offer several advantages over other administration routes, such as bypassing first-pass metabolism, which is crucial for the drugs with low metabolic stability, simple administration method and quick absorption. The nasal route of administration can also be utilised to transport medications to the systemic circulation. Nasal formulations require a proper choice of excipients, administration device and avoidance of any incompatibilities. Other important factors include the selection of proper preservative in the formulation to prevent the microbial growth, i.e. benzalkonium chloride is generally used as a preservative in nasal formulations. In cases where a poor penetration is observed, such as with proteins or peptides, a penetration enhancer may be used to increase the nasal absorption. Ensuring the stability and compatibility of the substances utilised in the formulation is also crucial (Amarji et al. 2018; Suthar et al. 2023; Sahu et al. 2024).

1.3.7 Ophthalmic Dosage Form

Different eye diseases are topically treated using ophthalmic formulations, as a significant number of drugs can reach the site of action via administration of ophthalmic formulations into the eyes. Optimisation of ophthalmic formulations involves the consideration of various factors like adjusting the pH and osmolarity of the solution to ensure the drug's solubility and the formulation's sterility. There is a narrow range of permissible pH values that are close to the physiologic pH; in general, the pH of the solution needs to be as close to the physiologic pH as possible. To ensure that the solution is isotonic, the osmolarity of the solution must also be adjusted to prevent irritation. The sterility of the ocular formulations is also of prime importance; generally, preservatives are employed to inhibit microbial growth in the ophthalmic dosage forms (Rowe et al. 2017).

The selection of preservative relies on formulation pH, presence of surfactant, type of API/excipients, etc. Besides pH-adjusting agents, tonicity modifiers, viscosity modifiers, buffers and preservatives are also used in the formulation of ophthalmic dosage forms. The viscosity modifier prolongs the retention of drug on the ocular surface, thereby increasing bioavailability. Certain formulations might contain penetration enhancers and/or solubilisers such cremophor and polysorbates (Zhang et al. 2022). These constituents enhance the molecule's solubility in aqueous media and could augment its penetration, thereby reducing the dosage necessary to attain therapeutic efficacy and augmenting bioavailability (Amarji et al. 2018).

1.3.8 Stability Analysis of Pharmaceutical Products

In the early stage of the dosage form development, formulation and active drug stability are assessed, where guidelines of the International Council for Harmonisation of Technical Requirements for Pharmaceuticals for Human Use (ICH) or other present regulatory criteria are followed when conducting a stability analysis. Understanding of environmental factors like temperature, moisture and light, retesting intervals and storage settings, which affect shelf life, are of utmost importance. Stability testing objectives vary depending on the stage of the drug development and discovery process. The initial stage of product development seeks to ascertain the drug's inherent rigidity; potential interaction of excipients and drug; and effect of acidity, alkalinity, humidity, oxygen and light on drug stability (Chaurasia 2016). Data on the physicochemical stability of the pharmaceutical product is obtained via accelerated stability analysis, which aids in determining the most likely pathway for their breakdown. The determination of a stable formulation is the main goal of preclinical research. Confirming the stability and data integrity for the formulation batches is the purpose of stability testing in clinical trials (Smrity et al. 2016).

The main goal of a stability study is to determine the variation in the quality of a formulation over time under different environmental conditions. Stability study comprises the determination of test parameters to determine the factors that can influence the potency, microbiological stability and physical stability of drug and formulation. Physicochemical characteristics of active ingredients; presence of excipients; process used to produce drug and its product; type of dosage form; packaging used; exposure of drug to environmental factors during transportation, storage and handling; and susceptibility of ingredients to oxidation, reduction, hydrolysis or racemisation are just a few of the factors affecting the stability of the pharmaceutical product (Blessy et al. 2014).

To anticipate the possibility of potential degradation products, regulatory guidelines advise conducting forced degradation studies using acidic/alkali hydrolysis, oxidation, thermolysis and photolysis. Nevertheless, none of the recommendations provided precise instructions for the type, level, quantity or duration of stressors to be used. Furthermore, a product's stability testing technique may vary depending on where in the world it is intended to be distributed or marketed. The ICH suggested conditions for various climatic zones must be the basis for setting up circumstances for real-time, long-term and rapid stability tests. A few unique specifications can be used to construct the stability testing technique for special products like phytopharmaceuticals or biotechnological products. Therefore, while creating the protocol for stability research, a wide range of relevant aspects must be considered (Sengupta et al. 2018).

1.4 Scale-Up Studies

All the industries, including the pharmaceutical industry, need to ensure that the product is meeting all the quality attributes before commercialisation. It must be noted that, unlike the other common products, pharmaceuticals must meet regulatory requirements during manufacturing as well before and after marketing of the products (Raval et al. 2018). It is not advisable to manufacture products on a large scale and proceed towards quality control in terms of economic aspects. Pharmaceutical plants use pilot plant scale-up to establish an efficient manufacturing method, so that commercial products can be produced on a large scale from a lab scale formula. Pilot comprises the elements, material, man, method and machine, which all contribute to the manufacturing process. Scale-up refers to the utilisation of the information from a pilot plant model to make a prototype (Bandarapalle et al. 2024). Alternatively, it can be referred as the technology transfer from research to the production unit to increase the production of a pharmaceutical product.

1.4.1 Pilot Plant

A pilot plant is defined as the process of developing a reliable, feasible manufacturing process from a lab scale to a marketable product. A pilot plant is a small industrial system that is run to provide data about the system's behaviour for use in the larger facility designed for large-scale production. It provide the necessary guidelines to make the manufacturing process sustainable, bringing more economic benefits to the organisation. Pilot plant studies are used for formula standardisation, evaluating various relevant processing equipment, and optimising and controlling target production rates (Rane et al. 2023). It is a pre-commercial system that produces new pharmaceutical products on a small scale as on experimental aspects. This batch of products is used to check the feasibility of large-scale manufacturing and meet government regulatory requirements (Dhobale et al. 2018). The main objectives of pilot plant studies are listed below:

- To identify the potential challenges and overcome those during the scale-up
- Production and process control guideline preparation
- To study the crucial elements of the process
- Preparation and supply of master manufacturing formula for manufacturing
- Evaluation and validation of process and equipment
- Evaluation of the formula to produce a stable batch
- To produce physically and chemically stable therapeutic dosage forms
- Review of the processing equipment
- To avoid the scale-up problems
- To try the process on a model of the proposed plant before committing a large sum of money to a production unit

These objectives are critical outcomes from a pilot plant and are very important requirements for the efficient functioning.

1.4.1.1 Requirements for a Pilot Plant Scale-Up

Before considering setting up a pilot plant, a few requirements are required to be met:

- The formula should be thoroughly verified to determine its ability to withstand large-scale production.
- The correct identification of the processing equipment, which is most economical, simple and reliable in the production.
- Ensuring the availability of raw materials, which is required to produce the product.
- Evaluation, validation and finalization of production and process controls.
- Provide necessary records and reports to support good manufacturing practices (GMPs) and record of the formulation process, equipment utilised and specifications (Kerone 2024).

1.4.2 Current Good Manufacturing Practices (cGMP)

In the healthcare industry, maintenance of high-quality standards is an important factor, as it affects the safety and efficacy of medicines (Sharma et al. 2023). The cGMP is an assortment of different policies, guidelines, codes and directives that have been developed and issued by international organisations and institutions in cooperation with the pharmaceutical industry and several national regulatory agencies (Patel and Chotai 2008). The cGMP aims at the proper design, monitoring and control of manufacturing processes and facilities. It mandates that pharmaceutical manufacturers maintain appropriate control over their manufacturing processes, ensuring the identity, strength, quality and purity of their products. It also provides the essential guidelines for the production, distribution and supply of drugs, medical devices and other healthcare products (Sharma et al. 2023; Gouveia et al. 2015). The components of the cGMP can be summarised as described in Fig. 1.2, which includes quality control of finished product, quality control of packaging, written procedure and other relevant documentation, manufacturing of premises with good design, protective clothing, and properly designed and maintained equipment (Shukla et al. 2016).

Implementation of cGMP in the production and testing practice ensures the quality of the product. cGMP guidelines should be followed by the quality assurance (QA) system with the most care to achieve the goals (Shukla et al. 2016). The major development of the GMP in association with the continuous efforts to make the pharmaceutical manufacturing safer and more efficient is enlisted in Table 1.1.

During the manufacturing process, there may be many instances where the pro duction must be regulated in accordance with the demand of the product as well as the stability concerns of the product.

Fig. 1.2 Schematic presentation of various components of cGMP

Table 1.1 Timeline of the development of GMP (FDA, 2004)

Date	Milestone
1906	The Bureau of Chemistry passed the 1906 Pure Food and Drugs Act, prohibiting interstate commerce in misbranded and adulterated foods, drinks and drugs
1933	FDA recommended revising the 1906 Pure Food and Drugs Act
1938	FDA passed the 1938 Federal Food, Drugs, and Cosmetics Act, which provides identity and quality standards for food
Mid-1960s	FDA decided to clarify the FDCA through GMP regulations
1968	FDA proposed food GMP regulations
1969	FDA finalised food GMP regulations
Early 1970s	FDA considered promulgating industry-specific regulations
Late 1970s	FDA decided to revise the general GMPs rather than adopting industry-specific GMPs
1986	FDA published revised food GMPs
2002	FDA formed Food GMP Modernization Working Group
2004	FDA announced effort to modernise food GMPs

1.4.3 Regulatory Approval

FDA's Scale-Up and Post Approval Changes (SUPAC) guidelines refers to the process of scaling up and the modifications made in the manufacturing process, equipment, composition and site after approval. It is necessary that the raw materials,

Table 1.2 SUPAC guidelines for the changes made after the approval in batch size

SUPAC guidelines		
Level 1	Definition of level	Change in batch size, up to and including a factor of ten times the size of the pilot/biobatch
	Test documentation	Chemistry, dissolution and bioequivalence documentation
	Filing documentation	Annual report (all information including long-term stability data)
Level 2	Definition of level	Changes in batch size beyond a factor of ten times the size of the pilot/biobatch
	Test documentation	Chemistry, dissolution and bioequivalence documentation
	Filing documentation	• Changes Being Effected supplement (all information including accelerated stability data) • Annual report (long-term stability data)

process, equipment or manufacturing site and batch size can be changed in due course of time, so that quality aspects of finished product can be maintained. The SUPAC guideline evaluates the change based on:

- Levels of change.
- Recommended chemistry, manufacturing and controls tests for each level of change.
- In vitro dissolution tests and/or in vivo bioequivalence tests for each level of change.
- Documentation that should support the change. For those changes filed in a "changes being effected supplement" [21 CFR 314.70 (c)], the FDA may, after a review of the supplemental information, decide that whether the changes are or are not approvable. SUPAC guidelines for the changes made after the approval in batch size are summarised in Table 1.2.

SUPAC guidelines necessitates that the entire scale-up process be validated every time the process is scaled up by a factor of at least 10. It has a lot of advantages in the rapid increase in the batch size of drugs and a reduction in the unmarketable drugs due to stability issues. As per FDA guidelines, a quality-by-design (QbD) approach is an efficient and quick approach to produce high-quality pharmaceutical products. According to FDA, QbD is a systematic approach used for product and process design and development. QbD, as defined by the International Conference on Harmonization guidelines (ICH Q8: pharmaceutical development), directs the scale-up process to produce quality products (Jang et al. 2020). Because there are broader ranges and limits based on product and process understanding, a QbD concept has less regulatory burden. Modifications made within these parameters do not need permission in advance.

1.5 Commercialisation

The commercialisation of pharmaceutical products is a complex and multifaceted process that encompasses various stages from drug discovery to sending the stock to distributors. It requires strategic planning, rigorous regulatory amenability and ongoing management to ensure the products' success in the market. Implementing strong pharmacovigilance systems, adhering to regulatory criteria like SUPAC and life cycle management are crucial components of commercialisation (Bugusu et al. 2011; Ahmed and Vveinhardt 2014).

1.5.1 Life Cycle Management of Pharmaceutical Product

Product life cycle management is a strategic process that helps a company to manage its products efficiently from production to the market. Product life cycle management can improve the product development process of a company and allows to make better business decisions (Bugusu et al. 2011). The healthcare industry has been looking for useful tools and protocols to help with decision-making to improve the quality of care, save costs and reduce inefficient use of resources. Product life cycle management is essential in the pharmaceutical industry to manage the dynamic healthcare environment, which includes various stakeholders, rapid market changes and regulatory challenges. Effective product life cycle management strategies enhance innovation, optimise resources and improve patient outcomes. Product life cycle management is vital for overcoming competitive pressures and government price regulations, helping companies to increase market share and avoid the decline phase by identifying and addressing key factors influencing the product life cycle (Ahmed and Vveinhardt 2014). Product life cycle management includes several parameters, which are described below:

1.5.1.1 Product Development and Launch

The first stage of life cycle management includes discovery, development and approval of the drug. This phase is characterised by significant investment in research and clinical trials to ensure the safety and efficacy of the product. When a product gets approval, it is introduced to the market and advertised to build awareness and acceptability among patients and medical professionals.

1.5.1.2 Increase Market Access and Product Reach

Post-launch, the focus shifts to increasing market penetration and expanding the product reach. It may involve obtaining approvals for additional indications, new formulations or extended-release versions of the drug. These efforts help to broaden the patient base and extend the product market life.

1.5.1.3 Product Optimisation and Enhancement

Throughout its life cycle, a pharmaceutical product may undergo various optimisations to enhance its performance, safety or patient adherence. It can include reformulations to improve stability or bioavailability, packaging improvements or the development of combination therapies. This not only improves patient outcomes but also helps in maintaining the product competitive edge.

1.5.1.4 Regulatory Compliance and Post-marketing Surveillance

Continuous monitoring of product safety and efficacy through post-marketing surveillance is crucial. This involves collecting and analysing data on adverse events, ensuring compliance with regulatory requirements and making necessary updates to product labelling or usage guidelines. Effective pharmacovigilance systems play a critical role in this aspect of life cycle management.

1.5.1.5 Life Cycle Extension Strategies

As a product approaches the end of its patent life, companies often employ various strategies to extend its market exclusivity. This can include developing new indications, obtaining orphan drug status or creating over the counter (OTC) versions. Such strategies help to prolong the market presence of the product and revenue generation (Blue et al. 2022; Ahmed and Vveinhardt 2014; Mousavi et al. 2022).

1.6 Role of SUPAC Guidelines (Scale-Up and Post Approval Changes)

SUPAC guidelines are regulatory framework established by the Food and Drug Administration (FDA) to provide guidance to pharmaceutical sponsors in regard to making changes to drug products, specifically focusing on modified-release solid oral dosage forms. SUPAC outlines recommendations for manufacturers regarding changes in components, composition, site of manufacturing, scale-up or scale-down of production and manufacturing processes and equipment during the post-approval period. By following the SUPAC guidelines, manufacturers can ensure that any modifications made to modified-release solid oral dosage forms maintain product quality, safety and efficacy while meeting regulatory requirements. The guideline offers clear instructions on conducting chemistry, manufacturing, control tests, in vitro dissolution testing and in vivo bioequivalence documentation for different levels of changes, facilitating decision-making processes and enhancing communication between manufacturers and regulatory authorities. Overall, SUPAC serves as a valuable resource for manufacturers seeking to implement changes to drug products in a compliant and effective manner (Flynn et al. 1999). The SUPAC guideline categorises changes into different levels based on their impact. The levels of changes are as follows:

Level 1 change: This level typically involves minor changes that have minimal impact on the product. Examples include changes in container closure systems, batch size or manufacturing equipment within the same facility.

Level 2 change: Level 2 changes consist of site changes within a contiguous campus or between facilities in adjacent city blocks where certain conditions are met, such as using the same equipment, SOPs, environmental conditions and personnel common to both manufacturing sites.

Level 3 change: This level involves more significant changes that may impact product quality, safety or efficacy. Examples include changes in critical manufacturing processes, equipment or components that could affect product performance (FDA 1995, 1997).

By categorising changes into different levels, the SUPAC guidelines provide manufacturers with a structured approach to assess and implement modifications to modified-release solid oral dosage forms while considering the potential impact on product quality and regulatory compliance. The SUPAC guideline holds significant importance in the pharmaceutical industry by providing clear and structured recommendations for manufacturers seeking to make changes to modified-release solid oral dosage forms. By following the guidelines outlined in SUPAC, manufacturers can ensure the maintenance of product quality, safety and efficacy throughout the post-approval period. Compliance with the SUPAC guideline not only helps to mitigate risks associated with changes but also facilitates regulatory compliance by guiding manufacturers on the necessary tests and documentation required for different levels of changes. Moreover, SUPAC plays a crucial role in enhancing communication between manufacturers and regulatory authorities, fostering a common understanding of expectations and requirements for post-approval changes in drug products. Overall, the SUPAC guidelines serve as a valuable resource for manufacturers, supporting informed decision-making processes and ensuring that modifications to drug products are implemented effectively while upholding regulatory standards and patient safety (Malinowski et al. 1997; Flynn et al. 1999; Yelvigi et al. 2013).

The implementation of the SUPAC guideline carries significant implications on a global scale within the pharmaceutical industry. Manufacturers worldwide can benefit from adopting the structured approach outlined in SUPAC to ensure the quality, safety and efficacy of modified-release solid oral dosage forms during post-approval changes. By adhering to the recommendations outlined in the guideline, companies can navigate regulatory requirements more effectively, regardless of their geographical location. The harmonised approach to managing changes in drug products not only facilitates compliance with regulatory standards but also promotes consistency in product quality and performance across different markets. Furthermore, the global adoption of the SUPAC guideline fosters collaboration and communication between manufacturers, regulatory bodies and healthcare authorities on an international level, contributing to the overall advancement of pharmaceutical standards and patient care worldwide (Malinowski et al. 1997; FDA 1995, 1997).

1.7 Role of Pharmacovigilance

Pharmacovigilance is a critical component of healthcare systems worldwide, focusing on the detection, assessment, understanding and prevention of adverse effects or any other drug-related problems. It plays a crucial role in ensuring the safety and efficacy of medications throughout its life cycle, from pre-marketing clinical trials to post-marketing surveillance.

The main aim of pharmacovigilance is to detect, assess and prevent adverse effects and other drug-related issues to ensure the safe and effective use of drugs. By monitoring drug safety throughout the life cycle of drugs, pharmacovigilance aims to minimise risks and optimise the benefit-risk profile of medications for patient care. Additionally, pharmacovigilance plays a crucial role in promoting public health by enhancing the understanding of drug safety and facilitating informed decision-making in healthcare (Najafi 2018). Through early detection of adverse events, pharmacovigilance contributes to identifying potential risks during preclinical and clinical trials, enabling proactive risk management strategies. Continuously monitoring and analysing safety data, pharmacovigilance assesses and manages the risks associated with pharmaceutical products, ensuring compliance with regulatory standards and facilitating the approval process of new drugs. Moreover, pharmacovigilance enhances the overall safety profile of medications by collecting and analysing real-world data to address safety concerns, drug interactions and long-term effects (Najafi 2018; Pitts et al. 2016). Some key roles of pharmacovigilance in the development of pharmaceutical products include:

- **Early detection of adverse events**: Pharmacovigilance helps in the early detection of adverse drug reactions (ADRs) during preclinical and clinical trials, allowing for timely intervention and risk mitigation strategies.
- **Risk assessment and management**: By continuously monitoring and analysing safety data, pharmacovigilance assesses the risks associated with pharmaceutical products and implements risk management plans to minimise potential harm to patients.
- **Regulatory compliance**: It ensures compliance with regulatory requirements by reporting adverse events to regulatory authorities, contributing to the approval process of new drugs and providing post-marketing surveillance to monitor the safety of pharmaceutical products in real-world situations.
- **Improving patient care**: The safe and effective use of medications and pharmacovigilance ultimately aim to improve patient care outcomes, enhance public health and build trust in the pharmaceutical industry (Talbot and Nilsson 1998; Nikookalam et al. 2024).

Overall, pharmacovigilance plays a critical role in the development of pharmaceutical products by ensuring their safety, efficacy and quality throughout the entire drug development process and post-marketing phase.

In conclusion, the commercialisation of pharmaceutical products is a complex process that requires strategic life cycle management, adherence to regulatory guidelines like SUPAC and robust pharmacovigilance systems. These elements are crucial for ensuring the safety, efficacy and market success of pharmaceutical products. By effectively managing these aspects, pharmaceutical companies can maximise the value of their products and contribute to improved patient outcomes and public health.

1.8 Conclusion

Pharmaceutical product development occurs in several stages. The first stage is preformulation which is crucial, laying the groundwork for further development. It involves three main areas of study: bulk characterisation, solubility and stability analysis. Further, prototype development, scale-up and regulatory approval are very important steps which are advancing day by day in one way or another so that the patient's need of safe and effective medicines can be fulfilled. Emerging trends in pharmaceutical development include pharmacovigilance, pharmacogenomics and many more. Pharmacovigilance is the ongoing process of monitoring a drug's safety profile after it reaches the market. AI technology is playing an increasingly important role in analysing vast amounts of data on adverse events to identify potential safety concerns early. Pharmacogenomics explores the link between a person's genes and their response to medications. This personalised medicine approach aims to develop drugs tailored to individual patients' genetic makeup, potentially leading to more effective and safer treatments. The focus on patient safety and personalised medicine has indeed brought new challenges to pharmaceutical research and development. Advancements in technology and a growing focus on patient-centric healthcare are driving innovation in pharmaceutical research and development. The future holds promise for more effective, personalised and safer medications for patients.

Acknowledgements The authors (KJ, ST, GKM, MP, AP and PP) are grateful to the Department of Pharmaceuticals, Ministry of Chemicals and Fertilizers, Government of India, for providing the facilities to write this manuscript. The NIPER-R communication number for this manuscript is NIPER-R/Communication/583.

References

Ahirwar K, Shukla R (2023) Preformulation studies: a versatile tool in formulation design. In: Drug formulation design. IntechOpen, pp 1–28

Ahmed RR, Vveinhardt J (2014) Product development & commercialization in pharmaceutical industry. Pharma Innov 3(9, Part A):40

Alkilani AZ, McCrudden MT, Donnelly RF (2015) Transdermal drug delivery: innovative pharmaceutical developments based on disruption of the barrier properties of the stratum corneum. Pharmaceutics 7(4):438–470

Amarji B, Abed SN, Bairagil U, Deb PK, Al-Attraqchi O, Choudhury AA, Tekade RK (2018) Four stages of pharmaceutical product development: preformulation, prototype development and scale-up, biological aspects, and commercialization. Elsevier Inc., pp 637–668. https://doi.org/10.1016/B978-0-12-814423-7.00018-6

Armstrong NA (2006) Pharmaceutical experimental design and interpretation, 2nd edn. CRC Press

Bandarapalle K, Kumarachari RK, Sri KT, Neeraja B, Likhitha C, Chaitanya K et al (2024) A comprehensive review on pilot plant scale up and platform technology. Future J Pharm Health Sci 4(1):14–25

Bandopadhyay S, Bandyopadhyay N, Deb PK, Singh C, Tekade RK (2018) Preformulation studies of drug substances, protein, and peptides: role in drug discovery and pharmaceutical product development. In: Dosage form design considerations. Academic, pp 401–433

Barbosa JS, Nolasco MM, Ribeiro-Claro P, Paz FA, Braga SS (2019) Preformulation studies of the γ-cyclodextrin and montelukast inclusion compound prepared by comilling. J Pharm Sci 108(5):1837–1847

Bastin RJ, Bowker MJ, Slater BJ (2000) Salt selection and optimisation procedures for pharmaceutical new chemical entities. Org Process Res Dev 4:427–435

Bharate SS, Vishwakarma RA (2013) Impact of preformulation on drug development. Expert Opin Drug Deliv 10(9):1239–1257

Blessy M, Patel RD, Prajapati PN, Agrawal YK (2014) Development of forced degradation and stability indicating studies of drugs—a review. J Pharm Anal 4(3):159–165

Blue LE, Guan X, Joubert MK, Kuhns ST, Moore S, Semin DJ, Wikström M, Wypych J, Goudar CT (2022) State-of-the-art and emerging trends in analytical approaches to pharmaceutical-product commercialization. Curr Opin Biotechnol 78:102800

Bugusu B, Lay Ma UV, Floros JD (2011) Products and their commercialization. In: Nanotechnology in the agri-food sector: implications for the future. Wiley, pp 149–170

Camacho Vieira C, Peltonen L, Karttunen AP, Ribeiro AJ (2024) Is it advantageous to use quality by design (QbD) to develop nanoparticle-based dosage forms for parenteral drug administration? Int J Pharm 657:124163

Chaurasia G (2016) A review on pharmaceutical preformulation studies in formulation. Int J Pharm Sci Res 7(6):2313–2320

Dave VS, Shahin HI, Youngren-Ortiz SR, Chougule MB, Haware RV (2017) Emerging technologies for the non-invasive characterization of physical-mechanical properties of tablets. Int J Pharm 532(1):299–312

Dejaegher B, Heyden YV (2011) Experimental designs and their recent advances in set-up, data interpretation, and analytical applications. J Pharm Biomed Anal 56(2):141–158

Destro F, Barolo M (2022) A review on the modernization of pharmaceutical development and manufacturing—trends, perspectives, and the role of mathematical modeling. Int J Pharm 620:121715

Dhobale AV, Mahale AM, Shirsat M, Pethkar S, Chakote V (2018) Recent advances in pilot plant scale up techniques - a review. Indo Am J Pharm Res 8(4):1060–1068

FDA (1995) Guidance for industry immediate release solid oral dosage forms scale-up and post-approval changes: chemistry, manufacturing, and controls, in vitro dissolution testing, and in vivo bioequivalence documentation. U.S. Department of Health and Human Services Food and Drug Administration, Centre for Drug Evaluation and Research (CDER), November 1995. pp 1–30. https://www.fda.gov/media/70949/download. Accessed 15 Jul 2024

FDA (1997) Guidance for industry SUPAC-MR: modified release solid oral dosage forms scale-up and post-approval changes: chemistry, manufacturing, and controls; in vitro dissolution testing and in vivo bioequivalence documentation. U.S. Department of Health and Human Services Food and Drug Administration, Centre for Drug Evaluation and Research (CDER), September 1997. pp 1–52. https://www.fda.gov/media/70956/download. Accessed 15 Jul 2024

FDA (2004) Good Manufacturing Practices for the 21st Century for Food Processing (2004 Study) Section 1: Current Food Good Manufacturing Practices. https://www.fda.gov/food/current-good-manufacturing-practicescgmpsfood-and-dietary-supplements/good-manufacturing-

practices-21st-century-food-processing-2004-study-section-1-current-food-good. Accessed on 5.12.2024
FDA (2024) Facts about the current good manufacturing practice (CGMP). FDA. https://www.fda.gov/drugs/pharmaceutical-quality-resources/facts-about-current-good-manufacturing-practice-cgmp. Accessed 6 Aug 2024
Flynn GL, Shah VP, Tenjarla SN, Corbo M, DeMagistris D, Feldman TG, Franz TJ, Miran DR, Pearce DM, Sequeira JA, Swarbrick J, Wang JC, Yacobi A, Zatz JL (1999) Assessment of value and applications of *in vitro* testing of topical dermatological drug products. Pharm Res 16(9):1325–1330
Gibson M (2005) Pharmaceutical preformulation and formulation: a practical guide from candidate drug selection to commercial dosage form, 2nd edn. CRC Press. ISBN 9781138031975
Gouveia B, Rijo P, Gonçalo T, Reis C (2015) Good manufacturing practices for medicinal products for human use. J Pharm Bioallied Sci 7(2):87
Jang EH, Park YS, Kim MS, Choi DH (2020) Model-based scale-up methodologies for pharmaceutical granulation. Pharmaceutics 12(5):453
Kerone (2024) Objectives of pilot plant scale up techniques. https://kerone.com/blog/objectives-of-pilot-plant-scale-up-techniques. Accessed 6 Aug 2024
Malinowski H, Marroum P, Uppoor VR, Gillespie W, Ahn HY, Lockwood P, Henderson J, Baweja R, Hossain M, Fleischer N, Tillman L (1997) Draft guidance for industry extended-release solid oral dosage forms: development, evaluation and application of in vitro-in vivo correlations. In: In Vitro-in vivo correlations. Springer US, Boston, MA, pp 269–288
Moini J, Logalbo A, Schnellmann JG (2023) Drug approval and drug regulations. In: Neuropsychopharmacology, pp 13–28. eBook ISBN: 9780323959735
Mousavi A, Mohammadzadeh M, Zare H (2022) Developing a system dynamic model for product life cycle management of generic pharmaceutical products: its relation with open innovation. J Open Innov Technol Mark Complex 8(1):14
Najafi S (2018) Importance of pharmacovigilance and the role of healthcare professionals. J Pharmacovigil 6(01):1–2
Nikookalam M, Hosseini A, Moghaddasi H, Asadi F, Zarghi A (2024) A framework of national pharmacovigilance system for developing countries. Iran J Pharm Sci 20(2):169–181
Ojha B, Jain VK, Mehra NK, Jain K (2021) Nanotechnology: introduction and basic concepts. Dendrimers in nanomedicine, pp 1–17
Pardhi VP, Jain K (2021) Impact of binary/ternary solid dispersion utilizing poloxamer 188 and TPGS to improve pharmaceutical attributes of bedaquiline fumarate. J Drug Del Sci Tech 62:102349
Pardhi VP, Pathak A, Jain K (2024a) Solid dispersions of bedaquiline fumarate to improve its pharmaceutical attributes: a comparative study between PEG and PVP. J Drug Del Sci Tech 94:105461
Pardhi VP, Patel P, Vaish A, Jain K (2024b) Inclusion complexes of bedaquiline fumarate with β-cyclodextrin and its derivatives: In silico, in vitro and in vivo evaluation. J Drug Del Sci Tech 91:105253
Patel KT, Chotai NP (2008) Pharmaceutical GMP: past, present, and future—a review. Pharmazie 63(4):251–255
Patravale VB, Disouza JI, Rustomjee M (eds) (2016) Pharmaceutical product development: insights into pharmaceutical processes, management and regulatory affairs, 1st edn, pp 1–456. https://doi.org/10.1201/b19579
Pitts PJ, Le Louet H, Moride Y, Conti RM (2016) 21st century pharmacovigilance: efforts, roles, and responsibilities. Lancet Oncol 17(11):e486–e492
Rane NU, More PK, Salunkhe AA, Rane KA (2023) A review on pilot plant scale-up considerations for solid orals. Int J Res Publ Rev 4(7):1806–1815
Raval N, Tambe V, Maheshwari R, Deb PK, Tekade RK (2018) Scale-up studies in pharmaceutical products development. In: Dosage form design considerations. Elsevier, pp 669–700. https://doi.org/10.1016/B978-0-12-814423-7.00019-8

Rowe RC, Sheskey PJ, Quinn ME (2017) Handbook of pharmaceutical excipients, 8th edn. Pharmaceutical Press
Sahu RC, Suthar T, Kumar D, Singh P, Datusalia AK, Jain K (2024) Novel ligand conjugated poly (propylene imine) dendrimers for brain targeted delivery of tacrine hydrochloride. J Drug Del Sci Tech 92:105336
Sathish KV, Sudharshan RD, Saibaba S, Rajkamal B (2016) Regulatory aspects of product development stages. Int J Inst Pharm Life Sci 6(3):2249–6807
Schmider J, Kumar K, LaForest C, Swankoski B, Naim K, Caubel PM (2019) Innovation in pharmacovigilance: use of artificial intelligence in adverse event case processing. Clin Pharmacol Ther 105(4):954–961
Sengupta P, Chatterjee B, Tekade RK (2018) Current regulatory requirements and practical approaches for stability analysis of pharmaceutical products: a comprehensive review. Int J Pharm 543(1–2):328–344
Shanbhogue MH, Thirumaleshwar S, Tegginamath PK, Somareddy HK (2021) Artificial intelligence in pharmaceutical field—a critical review. Curr Drug Deliv 18(10):1456–1466
Sharma A, Gamta V, Luthra G (2023) The importance of good manufacturing practices (GMP) in the healthcare industry. J Pharm Res Int 35(18):75–90
Shukla A, Vishnoi G, Das DR (2016) Current good manufacturing guidelines for medicinal product. J Drug Deliv Therap 6(2):57–61
Singh H, Khurana LK, Singh R (2017) Pharmaceutical development. In: Pharmaceutical medicine and translational clinical research. Elsevier. https://doi.org/10.1016/B978-0-12-802103-3.00003-1
Smrity SZ, Saifuddin AHM, Sultana S (2016) Stability analysis of formulated emulsion containing black cumin (Nigella sativa) oil. Am J Biomed Life Sci 4(3):49–53
Sohail Arshad M, Zafar S, Yousef B, Alyassin Y, Ali R, AlAsiri A, Chang MW, Ahmad Z, Ali Elkordy A, Faheem A, Pitt K (2021) A review of emerging technologies enabling improved solid oral dosage form manufacturing and processing. Adv Drug Deliv Rev 178:113840
Suthar T, Patel P, Singh P, Datusalia AK, Yadav AK, Jain K (2023) Hesperidin microemulsion: formulation optimization, characterization, and in vitro evaluation. J Drug Del Sci Tech 80:104166
Strovel J, Sittampalam S, Coussens NP, Hughes M, Inglese J, Kurtz A, Andalibi A, Patton L, Austin C, Baltezor M, Beckloff M, Weingarten M, Weir S (2016) Early drug discovery and development guidelines: for academic researchers, collaborators, and start-up companies. In: Markossian S, Grossman A, Arkin M, Auld D, Austin C, Baell J, Brimacombe K, Chung TDY, Coussens NP, Dahlin JL, Devanarayan V, Foley TL, Glicksman M, Gorshkov K, Haas JV, Hall MD, Hoare S, Inglese J, Iversen PW, Lal-Nag M, Li Z, Manro JR, McGee J, McManus O, Pearson M, Riss T, Saradjian P, Sittampalam GS, Tarselli M, Trask OJ Jr, Weidner JR, Wildey MJ, Wilson K, Xia M, Xu X (eds) Assay guidance manual. Eli Lilly & Company and the National Center for Advancing Translational Sciences, Bethesda, MD
Talbot JC, Nilsson BS (1998) Pharmacovigilance in the pharmaceutical industry. Br J Clin Pharmacol 45(5):427
Vashisht V, Jain K, Kaur S, Mehra NK (2015) Recent advances in Granulation technologies. Int. J. Pharm. Sci. 5:1144–1154
Volkova T, Simonova O, Perlovich G (2023) Revisiting the solubility-permeability relationship with hydrophobic drug umifenovir in pluronic solutions: impact of pH and co-solvent. Pharmaceutics 15(2):422. https://doi.org/10.3390/pharmaceutics15020422
WHO (2024) Health products policy and standards. https://www.who.int/teams/health-product-policy-and-standards/standards-and-specifications/gmp. Accessed 6 Aug 2024
Yang C, Wang W, Xue L, Wang Y (2024) Current status and prospects of nanosuspension for improved transdermal targeted drug delivery. J Drug Deliv Sci Technol 97(105751):1–13
Yelvigi M, Chambliss WG, Sam T (2013) Scale-up and post approval changes (SUPAC). In: Encyclopedia of pharmaceutical science and technology, vol 6. CRC Press, pp 3052–3055
Zhang R, Yang J, Luo Q, Shi J, Xu H, Zhang J (2022) Preparation and in vitro and in vivo evaluation of an isoliquiritigeninloaded ophthalmic nanoemulsion for the treatment of corneal neovascularization. Drug Deliv 2022(1):2217–2233. https://doi.org/10.1080/10717544.2022.2096714

Design of Materials and Product Specifications for Pharmaceutical Dosage Forms

2

Ankita Kishore, Neeraj Mishra, Jovita Kanoujia, Prem Prakash Singh, and Alok Kumar Mahor

Abstract

The supply of standard-quality pharmaceutical products to consumers is the prime accountability of manufacturers. To sustain this, pharmaceutical companies must comply with all explicit norms and guidelines, producing conventional to novel drug delivery systems to make them marketable. Reproducibility is the mandatory criterion in the successful fabrication of dosage forms; therefore, many new inputs are being carried out, including a quality-by-design (QbD) approach from the starting material up to the FP. The ICH Q6A guideline explains specifications in terms of analytical procedures, various tests, references and suitable acceptance criteria, and materials and DP must conform to these specifications. Three types of specifications have been set for various developmental stages of a dosage form: in-process specification, release specification and shelf life specification. In the ICH Q8 guideline, QbD is well explained. ICH Q6A, ICH Q6B guidelines and QbD help design the FP within specifications. Identification of critical quality attributes (CQAs), quality target product profile (QTPP) and necessary process parameters (CPPs) are required for QbD implementation. It works on identifying risks, executing the design of experiments (DOE), analysing risks and, finally, defining the design space. A control approach is used throughout the process to ensure that goods have a constant and predetermined quality. Different guidelines of regulatory authorities also must be followed by pharmaceutical companies to get their products approved for marketing.

A. Kishore · N. Mishra · J. Kanoujia
Amity Institute of Pharmacy, Amity University Madhya Pradesh (AUMP), Gwalior, Madhya Pradesh, India

P. P. Singh · A. K. Mahor (✉)
Institute of Pharmacy, Bundelkhand University, Jhansi, Uttar Pradesh, India
e-mail: alokmahor522@bujhansi.ac.in

K. Jain, A. K. Yadav (eds.), *Advances in Pharmaceutical Product Development*,
https://doi.org/10.1007/978-981-97-9230-6_2

Keywords

ICH guidelines · Quality-based design · Specifications · Pharmaceutical dosage forms · Quality target product profile

2.1 Introduction

The production of quality products is the prime focus of the pharmaceutical industry to supply effective and safe medicines to patients Haleem et al. (2015). A quality product can be designed by considering the quality of raw material (RM) used, process, equipment, technical knowledge and packaging material, as well as the guidance documents, regulation and compliance programme Altria (1998). The quality system model is an essential tool for manufacturing industries to implement for a better-quality product. When implementing the quality system model, the design of guidelines, development, strict implementation and promotion are required to adhere to the standards. These guidelines are a minimum standard and are reevaluated and updated as new technology advancements occur Balagué & Saarti (2011).

Designing specifications for the RM and FP is a regulatory measure to ensure the superiority of the pharmaceutical product. The International Conference on Harmonization of Technical Requirements for the Registration of Pharmaceuticals for Human Use (ICH) with WHO was established. The officials from the United States, Japan and the European Union (EU) felt the need to harmonise the requirements of pharmaceutical products during the International Conference of Drug Regulatory Authorities. ICH took another initiative in the form of ICHQ6 specifications (quality guideline of ICH), which gives the quality standards for tests and must be followed by manufacturers for drug substances (DS).

2.1.1 The Objective of Design of Materials and Product Specifications

The pharmacopeial monograph must be followed by a DS or the product manufactured from it. If the monograph is not available, it should nevertheless adhere to the NDA's (new drug application) provisions. Disparities in technical requirements were reduced by ICH standards to harmonise the NDAs. Regulatory agencies of some potential pharmaceutical markets, including the United States, Japan and the EU, have taken this initiative. The responses from both regulatory representatives and the industry were considered for harmonisation. ICHQ6B focuses on developing specifications for biologicals; on the other hand, ICH6A is an ICH guideline established to provide the specifications of new drug molecules and DPs. The single set of international specifications for new drug substance (NDS) and their products was also issued by the Food and Drug Administration (FDA) (ICH Official website 2023b; Center for Drug Evaluation, Research 2023).

ICHQ6 guidelines focus on determining the essential quality characteristics that affect the efficacy and safety of a DS or DP and provide references for analytical procedures. The testing procedures and criteria for accepting the result are considered in ICHQ6 guidelines. In addition, QbD methodologies and clinical experience are tools used in setting the specifications, and regulatory bodies like the FDA now request this information (Europa.eu 2023).

All manufacturers must submit specifications for approval to regulatory bodies along with a justification of their acceptance criteria. Systematic characterisation of RM and FP is the prime requirement for manufacturers and the key to maintaining quality. As a result, creating specifications has become an essential stage in developing pharmaceutical products (Kumar and Palmieri Jr. 2010).

2.1.2 Product Specification

Product specifications serve as a quality benchmark and set the best strategies for achieving the manufacturer's objective. The maker must adhere to these requirements to meet quality standards. Pharmacopeial monographs, regarded as authorised specifications, are a significant source of verified information (Kumar and Palmieri Jr. 2010). ICH guidelines provide specific information on multiple aspects of the product development process, as shown in Table 2.1.

The upper and lower specifications impact the shelf life of the product. The applicant must provide suitable analytical validation data in the marketing authorisation application (MAA) (Flynn et al. 2022). The specification was evaluated in terms of quality during the development phase, and the validation of the

Table 2.1 Different guidelines are given by ICH

S. No.	ICH guideline name	Specifications related to	References
1	ICHQ1	Stabilities	EMA (2018a)
2	ICHQ2	Validity of analytical procedure	EMA (2018b)
3	ICHQ3	Impurities in NDS and NDP	EMA (2018c, d)
4	ICHQ4	Pharmacopeia	EMA (2018e)
5	ICHQ5	Quality of biotechnological product	ICH Official website (2023a)
6	ICHQ6	Specifications	
7	ICHQ7	Good manufacturing practice	
8	ICHQ8	Pharmaceutical development	
9	ICHQ9	Quality risk management	
10	ICHQ10	Pharmaceutical quality system	
11	ICHQ11	Development and manufacture of drug substance	
12	ICHQ12	Life cycle management	
13	ICHQ13	Continuous manufacturing of drug substances and drug products	
14	ICHQ14	Analytical procedure development	

manufacturing process should be established. The upper and lower specification boundaries impact the shelf life of the product. The likelihood of a batch being rejected increases with tighter specification limits, whereas the likelihood of accepting a pharmaceutical product of poor quality increases with broader specs (Riley et al. 2013).

The regulatory bodies are aware of that. However, only a small quantity of data might be accessible during filing time. Therefore, amended acceptance criteria can be proposed, and as a result, the industry has the chance to adjust its specifications and provide justification for each process and associated acceptance criterion (Center for Drug Evaluation, Research 2023).

The specifications applied to DP and DS during the development phase are used to evaluate their quality; for this reason, specifications must be carefully designed with the necessary justification. The RM certificate of analysis (CoA) must include a detailed description of these specifications and a citation to the most recent individual monograph from the pharmacopoeia if available. RM manufacturers must be frequently tested following pharmacopeial guidelines. If standardised specifications are lacking, they typically test their products following all major pharmacopoeias to guarantee widespread acceptance. End-users first assess their RM suppliers by conducting complete testing on the first lots of the products they purchase; subsequently, they only regularly conduct identification testing, accepting the certificates of analysis (CoA) of their suppliers, or they only conduct complete testing on a limited number of batches under a documented standard operating procedure (Certificate of Analysis 2023).

2.1.3 Types of Product Specification

2.1.3.1 In-Process Specification

Testing standards that must be passed before a drug ingredient or DP is approved for sale. However, it must be verified that the product's qualities remain the same from the in-process to the released state (Center for Drug Evaluation, Research 2023).

2.1.3.2 Release Specification

These specifications are the final requirements of the DS and DP based on the decision to release or reject the batch. Setting stringent parameters for the product's release is more demanding than specifying shelf life. The DP manufacturer sets these internal requirements to ensure that the DP continues to meet its shelf life requirements (Kumar and Palmieri Jr. 2010).

2.1.3.3 Shelf Life Specification

The regulatory acceptance standard valid for the pharmaceutical product's shelf life. Typically, regulatory authorities accept internal or pharmacopeial specifications for shelf life, to which the product must adhere throughout its entire shelf life (Center for Drug Evaluation, Research 2023).

2.1.4 Specification Design

During the drug development process, specifications are first created for RM, in-process materials, intermediates, packaging components and the final product. Some of the ICH guidelines help in designing the specifications related to the DS or DP. Updating specifications is a continuous process as the manufacturer gains experience and knowledge about the product and process, and revisions to specifications emerge. The minimum tests required for any DS specification include a description test, identification, assay and impurity test. Limits for the aforementioned tests must be justified based on the product's development data or prior knowledge. The finalised specifications have been approved by the regulatory body (Pharmaguideline 2019).

2.1.5 Specification Justification

Any specification is incomplete without proper justification. Justify each procedure when suggesting a specification (ICH Harmonised Tripartite Guideline 2023).

1. Final specifications are met by meeting in-process specifications considering the operational variability.
2. Each specification is verified/tested by validated analytical methods to control analytical variations. The technique should demonstrate specificity.
3. The same established specifications were shown to be utilised for analysing stability samples, scale-up samples and validation samples. However, additional analysis can be performed during validation to validate/support the manufacturing process.
4. Clinical and preclinical studies directly link with product characteristics/specifications.

2.1.6 Revision of Specification

Specifications can be updated and revised based on the manufacturer's experience with the product and process. However, this must not impact the DP's safety, efficacy and identity. Specifications may alter due to pharmacopeial changes or revisions. Such modifications must be evaluated for their effect on the drug result. Any specification updates must follow a change control process, and the necessary regulatory clearance must be obtained before implementation (Geigert 2019a; Patel et al. 2019).

2.2 ICH Guidelines for the Product Specifications

ICH has developed ICH guidelines, which are exceptional in that they bring together different agencies of regulations and the industries involved in pharmaceutical manufacturing to discuss and create ICH guidelines (Khagga et al. 2019). A rising number of regulatory bodies have started implementing these ICH principles since the organisation's founding in 1990. Since then, ICH has steadily adapted to address increasingly global changes in the pharmaceutical industry (ICH Official website 2023b).

2.2.1 The Mission of ICH

The goal of the ICH is to promote greater global harmonisation so that high-quality, secure and effective medicines can be produced, licensed and kept in the most resource-effective way possible while still fulfilling strict requirements (EMA 2018f; Turner 2019).

2.2.2 Aims of the ICH

ICH guidelines aim to lay consistent framework for technical specifications for human-use pharmaceuticals (Bhavna and Bhargava 2022).

ICH guidelines aim to guarantee that medicines can be manufactured and licenced quickly, safely and of superior quality (Harron 2013).

The following ICH guidelines are used explicitly in the final product specifications (Elder 2017):

1. **ICH Q6A**
2. **ICH Q6B**
3. **ICH Q8 R2**

Their details are as follows:

2.2.3 ICH Q6A Guideline

This guideline is about the universal and specific test procedures and acceptable ranges for their results set for New Drugs and NDP: Q6A: Chemical Substances (ICH Official website 2023a).

2.2.3.1 Objective

This recommendation aims to contribute as much as possible to developing a singular collection of international standards for NDP and novel drugs (ICH Official website 2023a). This guideline deals with specifications, like those processes, tests and approval standards, that are crucial to guarantee the standards of NDS and NDP both at the time of release and up to their expiration date (Knight 2014).

The ICHQ6A guideline includes NDP along with combination products; if necessary, it can also include a new DS (Kestur et al. 2023). It does not consider any DP or substances that are in any kind of clinical research developmental stage. Depending on the situation, this guideline can apply to low molecular weight artificial peptides and semisynthetic or synthetic antibiotics (ICH Official website 2023a;

Knight 2014). The dosage forms involved in this guideline are small-volume parenteral and large-volume parenteral (Ain et al. 2020; Bergren and Subbarao 2019), solid (JBGSR.MS.ID.00087.Text—biogeneric science and research 2023) and liquid oral dosage forms (Jîtcă et al. 2023). Sometimes, the extension can also be done in topical formulations and inhalational and transdermal drug delivery systems (Riley and Yang 2020). Some general concepts should be clearly understood as they are essential in setting harmonised specifications. Based on different situations, these concepts are taken into consideration. This guideline (ICHQ6A) demonstrates its importance in providing an understanding of the circumstances under which the following concepts are necessary to be considered (Table 2.2):

Under this guideline, NDP and NDS are defined as follows:

Table 2.2 Concepts to be considered for setting harmonisation specifications

S. No.	Concept	Explanation	References
1	Periodic or skip testing	• Under this concept, instead of being performed batch by batch, specific tests are conducted on previously selected batches at the time of product release and at particular time intervals • All skipped batches must still meet all requirements for the approval of the product • Applicability may be on oral solid dosage forms for microbiological testing and residual solvent testing	Morris (2016)
2	Acceptance criterion for release vs. shelf life	• This idea is only specific to DS • It usually involves setting up more stringent regulations for a drug's release than those that should be regulated throughout its shelf life	Kajiwara and Shikano (2020)
3	In-process tests	• These tests are carried out while a DS or DP is being produced • These tests are only used to make changes in the processing parameters within a specific range • Examples such as, this test will apply to the friability and hardness of core tablets, which later are going to be coated, but it will not apply to later after coating on individual weights of a tablet. Further these coated tablets would be subjected to final release testing.	Ng (2022)
4	Design and development considerations	• Information collected when developing an NDP or substance can be utilised to make a base for specification set. On this base, one can propose excluding or replacing some tests. For example: – Testing for microbes for medicinal substances and solid pharmaceutical dosage forms has been shown to resist microbial growth throughout development – Depending on its relation to product performance, tests for particle size come under this class and can be performed during developmental stages or as a release test	ICH Official website (2023a)

(Continued)

Table 2.2 (continued)

S. No.	Concept	Explanation	References
5	Limited data available at filing	• The process of establishing acceptance criteria may be impacted because only a limited amount of information may be easily accessible at the moment of filing • At the time of filing, safety and effectiveness should always be the primary considerations for the acceptance criteria • When there is a lack of information, the tests and acceptance criteria that were initially authorised should be evaluated when new data is gathered to consider any potential revisions	Chen et al. (2009)
6	Parametric release	• In this, the regulatory authority has permitted to perform parametric release for some DP in place of regular release tests • One example is sterility testing for DP, which has been terminally sterilised	Pandit et al. (2019)
7	Alternate method	• Alternate methods are the tests that can be utilised for measuring an aspect when they have a comparable level of or greater level of quality control over the DP or DS than the official method • For example, tablets that are in stable form at all times of manufacturing can be released based on spectrophotometric testing rather than a chromatographic procedure that is officially approved	Friedel et al. (2018)
8	Pharmacopoeial tests and acceptance criteria	• If any one test is mentioned in different pharmacopoeias of different regions and this test is having differences from each other, then one harmonisation is needed. This harmonisation is only possible when all regional regulatory agencies approve the established processes and acceptance criteria	De Jong (2016)
9	Evolving technologies	• The development of both brand-new analytic technologies and improvements to already-existing technologies is ongoing. Such technologies need to be employed when they are thought to provide extra quality assurance or when their usage is otherwise warranted	Schenck et al. (2020)
10	Effect of DS on pharmaceutical product specifications	• Generally speaking, it shouldn't be necessary to test the DP for characteristics that are unique to the medicinal ingredient in the manufacture • For example, testing for synthesis contaminants regulated in the medicinal ingredient and not products of degradation is typically not considered essential	ICH Official website (2023a)
11	Reference standard (RS)	• RS is frequently described and assessed for its intended application using methods apart from those employed in ordinary testing • For novel DS, RS intended for tests, a quantitative technique should evaluate the purity, and the impurities should be adequately recognised and monitored	Singh et al. (2018)

2.2.3.2 New Drug Product

A pharmaceutical product type, such as a solution, capsule, tablet, cream, etc., that has not been registered before in a member state or region and that typically, but not always, comprises a drug component with excipients.

2.2.3.3 New Drug Substance

The specified therapeutic agent (a new chemical or novel molecular entity) has not yet been approved in a specific region or member state. It may be a salt, simple or complex ester of a drug that has already been granted authorisation.

Two types of tests are to be performed when testing these NDPs or new drug substances (NDS): (1) universal tests and (2) specific tests.

Specific tests for NDP have further been categorised for tablets, oral liquids and parenteral DP (ICH Official website 2023a).

2.2.3.4 Universal Tests

Some standard tests (Table 2.3) must be performed for all NDS and NDP, known as universal tests (Elder 2013).

2.2.3.5 Specific Tests

In addition to performing universal testing, several specific tests based on the drug substance and DP must also be carried out. In some circumstances or when new information becomes available, additional tests can be required in addition to those that are described in Table 2.4 for NDS and Table 2.5 for NDP (ICH Official website 2023a; ICH Q 2000).

Table 2.3 Universal tests for new drug substances and new drug products

S. No.	Test name	Description	References
1	Description	• It tells about the state and colour of NDS and size, shape and colour of NDP • Colour should not change during manufacturing and storage	ICH Official website (2023a)
2	Identification	• Specific for recognition of any new drug substance and should be capable of discriminating even those which have close structures to others • Identifies NDS in NDP by a different set of testing	ICH Official website (2023a)
3	Assay	• A technique that indicates stability should be used to determine the composition of the new therapeutic drug. It can also be used to quantify impurities • In a new drug, product assay determines the new drug substance's composition	Rignall (2017)
4	Contamination	• Residual solvents and inorganic and organic impurities are considered impurities • Organic impurities can also be present due to the degradation of NDS in NDP manufacturing • The ICH publications covering contaminants in NDS and solvent leftovers in pharmaceuticals provide detailed information for both NDS and NDP	Parente (2020)

Table 2.4 Specific tests for new drug substances (ICH Official website 2023a; Knight 2014; Pandit et al. 2019)

S. No.	Test name	Description	References
1	Physicochemical properties	• Properties include pH, refractive index, melting point, etc. procedures for these tests are specific	ICH Official website (2023a)
2	Size of particles	• Particle size determination becomes a pivotal factor to consider if any NDS is intended for solids or in suspension	ICH Official website (2023a)
3	Polymorphic forms	• Some techniques for polymorphic studies include the infrared test, differential scanning calorimetry, thermogravimetric analysis, X-ray diffraction, optical spectroscopy and Raman spectroscopy	ICH Official website (2023a)
4	Chiral new drug substance tests	• When one enantiomeric form predominantly works as a new drug substance, its other enantiomeric form must be removed from the study and considered an impurity. This test can be considered as written in the ICH guideline for contaminants in NDS and NDP	Knight (2014)
5	Water content	• This test is crucial when the new drug substance is known to be hygroscopic, susceptible to moisture degradation or a stoichiometric hydrate. Moisture content acceptance criteria should be set through specified test procedures, including loss on drying and moisture presence detected through Karl Fischer titration	Knight (2014)
6	Inorganic impurities	• The appropriate test should be selected for testing inorganic impurities (e.g. catalysts) in NDS during development • Acceptable ranges for residue on the ignition or sulphated ash should be according to pharmacopoeial criteria	Knight (2014)
7	Microbial limits	• The overall number of aerobic microbes, the complete number of moulds and yeasts and the lack of any particular undesirable bacteria (such as *Staphylococcus aureus*, *Pseudomonas aeruginosa*, *Salmonella* and *Escherichia coli*) may need to be specified. Under pharmacopoeial guidelines, they should be established	Pandit et al. (2019)

Table 2.5 Specific tests for new drug products (solid oral dosage forms)

Types of drug products	Tests	Description	References
Solid oral dosage forms: 1. Tablets (coated/uncoated) 2. Hard gelatine capsules 3. Soft capsules or granules	(a) Dissolution	• It includes measurement of drug substance release from DP (immediate release and extended-release DP) • For immediate-release products, tests should be able to exclude products with altered dissolution rates that can affect bioavailability • IVIVC (in vitro/in vivo correlation), which is used to establish acceptability criteria for bioavailability, is used for extended-release dosage forms	Bhavna and Bhargava (2022), ICH Official website (2023a), Kestur et al. (2023), Nickerson et al. (2018), Chen et al. (2017a)
	(b) Disintegration	• Disintegration may be used instead of dissolution for products that dissolve quickly (within 15 min, and are pH dependent. At pH values of 1.2, 4 and 6.8, the dissolution value is less than 80%) and involves products with high solubility across the physiological region (>250 mL of dose or solubility at pH ranges of 1.2–6.8)	ICH Official website (2023a), Nickerson et al. (2018)
	(c) Hardness/friability	• Used for in-process testing • In the case of chewable or dispersible tablets, a specific acceptable range should be included	ICH Official website (2023a), Pandit et al. (2019)
	(d) Uniformity of dosage units	• This test takes into account both the weight of the dosage form and the quantity of the active ingredient it contains; a pharmacopoeial protocol should be followed	ICH Official website (2023a), Chen et al. (2017b)
	(e) Water content	• The effect of moisture absorption on DP should be assessed • The acceptable range can be decided by loss on drying, and moisture detection can be done by Karl Fischer titration	ICH Official website (2023a), Pandit et al. (2019)
	(f) Microbial limits	• The overall number of aerobic microbes, the total number of moulds and yeasts and the absence of any specific harmful bacteria (such as *Staphylococcus aureus*, *Pseudomonas aeruginosa*, *Salmonella* and *Escherichia coli*) may need to be specified in accordance with pharmacopoeial guidelines	ICH Official website (2023a), Roesti (2019)

Specific tests and acceptable ranges that must be applied to NDP have been divided according to the dosage forms, i.e. (ICH Official website 2023a):

1. Solid oral drug products/dosage forms (Table 2.5)
2. Liquid oral drug products/dosage forms (Table 2.6)
3. Small volume and large volume parenteral (Table 2.7)

2.2.4 ICH Q6B Guideline

Test Procedures and Acceptance Criteria for Biotechnological/Biological Products

According to the broad guidelines provided in this document, a unified set of worldwide specifications for biotechnological and biological products should be defined and justified to a practical degree.

2.2.4.1 Scope

- The ideas described in this recommendation apply to proteins and protein chains (polypeptides), as well as to the derivatives and end products (such as conjugates) made from them; this also covers proteins and polypeptides extracted from body fluids and tissues.
- This guideline does not include synthetic proteins and polypeptides, cell metabolites, vitamins, heparins, allergenic extracts, DNA products, traditional vaccines, whole blood, cells and cellular components of blood.
- This guideline is devoid of regulation of clinical/preclinical research material.

2.2.4.2 Specifications

Specific testing is chosen for each product and incorporated into the specifications. Explaining the reasoning behind the acceptable range of acceptance criteria is essential. Data from lots used in preclinical and clinical investigations, data from stability studies, data from lots used to demonstrate manufacturing consistency and pertinent development data should be used to construct and justify acceptance criteria (Pokar et al. 2020; ICH Harmonised Tripartite Guideline 2023; Pallagi et al. 2018).

These specifications have been described under two heads:

1. **Specification for drug substances**
2. **Specifications for drug products**

2.2.4.2.1 Specifications for Drug Substance

All drug substances must pass the examinations and meet the acceptance standards indicated in Table 2.8. The drug substances should undergo pharmacopeial studies (such as endotoxin detection) where required.

2.2.4.2.2 Specifications for Drug Products

All drug products must pass the following test tabulated in Table 2.9.

Table 2.6 Specific tests for new drug products (liquid oral dosage forms)

Types of drug products	Tests	Description	References
1. Oral liquids 2. Powders intended for reconstitution as oral liquids	(a) Dosage unit uniformity	• This test refers to the dosage form's weight and the amount of the active ingredient it contains; a pharmacopeial protocol should be followed • Applicable on both single- and multi-dose packaging	ICH Official website (2023a), Pandit et al. (2019)
	(b) pH	• Strictly followed wherever needed, and the range should be justified	ICH Official website (2023a)
	(c) Microbial limits	• The overall number of aerobic microorganisms, the total number of moulds and yeasts and the absence of any undesirable bacteria (such as *Pseudomonas aeruginosa*, *Escherichia coli*, *Salmonella* and *Staphylococcus aureus*) may need to be specified. Following pharmacopeial guidelines, these should be determined	ICH Official website (2023a), Roesti (2019)
	(d) Antimicrobial preservative content	• An acceptable range should be set • Generally, testing for antimicrobial preservative concentration is done at the time of release	De Jong (2016), Ola et al. (2018)
	(e) Content of antioxidant preservatives	• Antioxidant preservative quantity is tested at the time of release	ICH Official website (2023a), Knight (2014), Pharmacopeial Forum 750 of the USPC or the USP (2009)
	(f) Extractables	• Test for the content extracted and released in oral liquids from containers and closures (e.g. cap liner, rubber stopper, plastic bottles, etc.)	Pandit et al. (2019), Pharmaceutical Technology Editorial Advisory Board (2020)
	(g) Alcohol content	• Testing for alcohol content can be done by assay to match the quantity labelled on the label	ICH Official website (2023a), Knight (2014)
	(h) Dissolution	• Dissolution testing is included in the case of suspensions and powders for reconstituted in the form of suspension • Pharmacopoeial tests are followed. Other tests have to be justified	Knight (2014), Hiwale et al. (2008)

(Continued)

Table 2.6 (continued)

Types of drug products	Tests	Description	References
	(i) Particle size distribution	• Applicable on oral suspensions • Generally performed at the time of release or otherwise justified • This test can also replace dissolution testing if justified	ICH Official website (2023a), Pandit et al. (2019), Ola et al. (2018)
	(j) Re-dispersibility	• Applicable to oral suspension to check re-dispersibility by shaking method	Ola et al. (2018)
	(k) Rheological properties (viscosity/specific gravity)	• Applicable to viscous solutions or suspensions • The acceptance range and test method should be specified	Chan (2014)
	(l) Reconstitution time	• Reconstitution timing should be stated for dry powders intended for reconstitution • Diluent selection should be justified	ICH Official website (2023a), Knight (2014)
	(m) Water content	• When suitable, a water content test and acceptability criterion for oral products that need reconstitution should be recommended	Pandit et al. (2019), Lin (2022)

Table 2.7 Specific tests for NDPs (parenteral dosage forms)

Types of drug products	Tests	Remarks	References
Parenteral dosage forms	(a) Dosage unit uniformity	• This test refers to both the dosage form's weight and the amount of the active ingredient it contains; a pharmacopeial protocol should be followed • Applicable on powders for reconstitution • Applicable on both single- and multi-dose packaging	ICH Official website (2023a)
	(b) pH	• Whenever necessary, this rule must be followed appropriately, and the range must be validated	ICH Official website (2023a)
	(c) Test for sterility	• Every item intended for parenteral use needs a test procedure and an acceptable standard to determine its sterility. This approach may be recommended for drug items that have undergone terminal sterilisation, with data collected throughout development and validation supporting parametric release	Nema and Ludwig (2016), Avis (2018)
	(d) Endotoxins/pyrogens	• Specifying the acceptance range and test procedure for endotoxins, such as the limulus amoebocyte lysate (LAL) test, is essential • In some cases, it may be appropriate to check for the presence of pyrogen instead of endotoxins	Akers (2016)
	(e) Particulate matter	• This test includes acceptable ranges for visible and sub-visible particulates	Beg et al. (2020)
	(f) Water content	• Applicable to products intended to be reconstituted and to nonaqueous parenteral • Loss on drying test and Karl Fischer titration is used depending upon set protocols	Radhakrishnan et al. (2019)
	(g) Antimicrobial preservative content	• An acceptable range should be set • Generally, testing for antimicrobial preservative concentration is done at the time of release	ICH (1999)
	(h) Content of antioxidant preservatives	• Antioxidant preservative quantity is tested at the time of release	ICH (1999)

(Continued)

Table 2.7 (continued)

Types of drug products	Tests	Remarks	References
	(i) Extractables	• Test for the content extracted and released in oral liquids from containers and closures (e.g. cap liner, rubber stopper, plastic bottles, etc.) • More strictly followed as compared to oral liquids	Houston et al. (2022)
	(j) Functionality testing of delivery systems	• Delivery systems are those in which parenteral preparations are generally packaged as auto-injector cartridges or prefilled syringes; functionality tests should be conducted according to established protocols • Generally performed in-process	ICH (1999)
	(k) Osmolarity	• The product's osmolarity should be maintained according to the set tonicity of the product	Sangeetha et al. (2022)
	(l) Particle size distribution	• Applicable on injectable suspensions • Generally performed at the time of release or otherwise justified • This test can also replace dissolution testing if justified	Geigert (2019b)
	(m) Re-dispersibility	• Applicable to injectable suspensions to check re-dispersibility by shaking method	Elder (2017)
	(n) Reconstitution time	• Reconstitution timing should be stated for all parenteral intended reconstitution • Diluent selection should be justified	Elder (2017)

Table 2.8 Specific tests for drug substances

S. No.	Test name	Description	References
1	Appearance and description	• The physical appearance (solid/liquid) and colour of drug substances should be specified	Knight (2014), Pokar et al. (2020)
2	Identity	• Tests for identity must be exact and based on the chemical structure or other distinctive characteristics • Several test methods (physicochemical, immunochemical and biological) may be necessary to prove identity	
3	Purity and impurities	• Results for purity and impurities are method-dependent. A combination of methods is used to determine purity. Analytical techniques can be improved by separating the intended product from impurities and related chemicals • Impurities are categorised into two parts: (1) process and (2) product related; separation techniques should be optimised accordingly	
4	Potency	• Specifications should include a pertinent, verified potency assay	
5	Quantity	• An appropriate assay should be used to calculate the drug substance's quantity, which is often based on protein content (mass) • It might not be necessary to make a different quantity determination in circumstances where product manufacturing is based on potency	

Table 2.9 Specific tests for drug products

S. No.	Test name	Description	References
1	Appearance and description	• The physical appearance (solid/liquid), colour and transparency of the DP should be specified	Knight (2014)
2	Identity	• Tests for identity must be exact and based on the chemical structure or other distinctive characteristics • Several test methods (physicochemical, immunochemical and biological) may be necessary to prove identity • Evaluation tests for a DP may also be implemented for the firm establishment of identity	Knight (2014)
3	Purity and impurities	• During the DP's production, storage or both, impurities may be created or enhanced • Impurities may be process related (if same as drug substance, the examination is not mandatory) or deterioration product while manufacturing or storage (impurities should be quantified and acceptability range should be decided) • Separation techniques should be optimised according to the desired product, deterioration product from drug substance and excipients also	Pokar et al. (2020)

(Continued)

Table 2.9 (continued)

S. No.	Test name	Description	References
4	Potency	• Specifications should include a pertinent, verified potency assay	Pokar et al. (2020)
5	Quantity	• An appropriate assay should be used to calculate the drug substance's quantity, which is often based on protein content (mass) • It might not be necessary to make a different quantity determination in circumstances where product manufacturing is based on potency	Pokar et al. (2020)
6	General tests	• For the examination of the DP's functions, physical description and the evaluation of other quality parameters are necessary. These tests may include, for instance, pH and osmolarity testing	Davis (2013)
7	Additional testing for unique dosage forms	• Some particular dosage forms may require testing in addition to those listed above	Davis (2013)

2.2.5 ICH Q8R2 Guidelines for the Product Specification

Pharmaceutical development aims to create high-quality products through manufacturing to deliver the products' intended performance consistently. Quality should be incorporated into products from the outset because they cannot be tested into them. Data from pharmaceutical development studies can provide a foundation for quality risk management. The development section aims to give reviewers and inspectors a thorough grasp of the product and manufacturing procedure. The minimum approach and enhanced knowledge approaches are the best keys in pharmaceutical development studies. Critical formulation characteristics and process variables are typically determined by evaluating the degree to which their change can affect the efficacy of the DP.

The applicant also has the option of conducting pharmaceutical development studies, which may result in a better understanding of how well a product performs across a broader range of material characteristics, process parameters and processing options; this information, in turn, can be converted into scientific understanding that can further be used to establish more expanded design spaces. The design space ensures the quality of the finished product, a multidimensional combination of the properties of the raw material and process parameters. More adaptable regulatory methods must be created to enable this. Regulatory decisions are predicated on risk via review or inspections:

- Decrease in post-approval submissions
- Enhancement in the manufacturing process
- Real-time quality control

The foundation for science-based submissions and the regulatory assessment of those submissions is the degree of knowledge attained, not the volume of data (Narayan 2011; Mishra et al. 2018).

2.2.5.1 Q8(R2): Structure—Parent Guideline (Knight 2014)

1. Pharmaceutical Development: Introduction
2. Components of the Drug Product
   (a) Drug Substance
   (b) Excipients
3. Drug Product
   (a) Formulation Development
   (b) Overages
   (c) Physicochemical and Biological Properties
4. Manufacturing Process Development
5. Container Closure System
6. Microbiological Attributes
7. Compatibility

2.2.5.1.1 Pharmaceutical Development: Introduction

Pharmaceutical development aims to create a high-quality product and its production method to consistently achieve the product's desired performance. Scientific understanding assists in formulating the design space, specifications and manufacturing controls based on information and expertise gathered through pharmaceutical development research and manufacturing experience. Pharmaceutical development study data can establish a foundation for effective risk management. It should be noted that directly checking the quality of the final product is impractical. However, quality maintenance is crucial throughout each phase of product development.

This section should provide the concepts through which one can understand whether the selected dosage form and proposed formulation are correct for the intended dosage form or not. At the very least, it is necessary to identify and support control techniques for those elements of drug ingredients, excipients, container closing systems and manufacturing methods that are crucial to product quality. Critical formulation characteristics and process variables are often determined by evaluating the degree to which their modification can affect the efficacy of the medicinal product (Maguire et al. 2017; Fahmy et al. 2013).

2.2.5.1.2 Components of Drug Product

Drug Substances

Any DP comprises drug substances and excipients in the form of the formulation. Drug substances are a vital part of DPs, highly affecting the product performance due to their biological and physicochemical properties. Solubility, particle size, water content, biological activity, crystal properties and permeability are some examples of physicochemical and biological properties that may need investigation. These qualities might be connected, so it may be necessary to consider them all. ICH guidelines Q6A and Q6B provide detailed information about the NDS and DP. The drug-excipient interaction study also plays an essential role in pharmaceutical development (PDA 2023).

Excipients

An excipient is another prominent product ingredient that can alter bioavailability, stability and manufacturability. When relevant, the compatibility of excipients with each other should be assessed. Excipients must demonstrate the ability to function throughout the designated DP shelf life and fulfil their intended purpose. Examples of such excipients include disintegrants, antioxidants, release-controlling agents and penetration enhancers. The safety profile of each excipient should be examined in accordance with regulatory guidelines (Maguire et al. 2017).

2.2.5.1.3 Drug Product

Formulation Development

A descriptive summary of the formulation's development process should be provided, along with a list of essential qualities for the product's quality. Consider the intended use and route of administration. Information from formal experimental designs can be used to identify critical or interacting variables and ensure the DP's quality. The formulation design, from the initial idea to the final design, should be highlighted in the summary. The same summary should justify the selection of product components, manufacturing processes and excipient ranges based on experience gained. The information related to bioavailability and bioequivalence with the in vitro dissolution profile should be presented. In vitro/in vivo correlation should be established and justified with valid data wherever applicable (Narayan 2011).

Overages

Generally, it is not recommended to use an excess of a drug substance to make up for manufacturing defects, prolong a product's shelf life or both. The batch formula should include information related to the reason for overages, the amount of overages and justification for the amount (D'Hondt et al. 2014).

Physicochemical and Biological Properties

Identifying and analysing the physicochemical and biological characteristics crucial for the DP's manufacturing, performance and safety is essential. This involves considering the physiological effects of the medication substance and formulation characteristics (DeGrazio and Paskiet 2020).

2.2.5.1.4 Manufacturing Process Development

The selection of the manufacturing process for production should be mentioned in the reports. The controls and improvements in the process should also reflect the proper description. The selection of particular equipment and implementation of the process improvement programme should be part of the documentation. Any crucial process parameters that need to be monitored or controlled for quality should be identified by the manufacturing process development project or process improvement scheme. The data should, for instance, include the following: (1) the manufacturing location, (2) the batch size, (3) the identity of the batches produced (e.g.

batch number) and (4) any significant equipment differences (e.g. different design, operating principle, size) (PDA 2023).

2.2.5.1.5 Container Closure System

It is essential to discuss the decision and justification behind choosing the container closure method for the commercial product. The selection of leading packaging materials, primary packaging materials and secondary packaging materials should be justified. When suitable, the storage and shipping container for bulk DPs should also be considered, as well as the intended use of the DP and the suitability of the container closure system for storage and transportation (shipping) (Gandhi and Roy 2016).

2.2.5.1.6 Microbiological Attributes

Justify performing microbiological tests for sterile products and not for non-sterile products. Selecting and implementing preservative systems in antimicrobial products, including those with inherent antimicrobial properties and those that rely on added preservatives (Chavan et al. 2016).

2.2.5.1.7 Compatibility

To provide pertinent and helpful information for the labelling, the compatibility of the medicinal product with reconstitution diluents (such as precipitation and stability) should be addressed. The recommended in-use shelf life at the recommended storage temperature should be covered with an emphasis on changes in shelf life at extreme concentrations (Csóka et al. 2018).

2.2.5.2 Q8(R2): Structure—Annex

1. Introduction
2. Elements of Pharmaceutical Development
   (a) Quality Target Product Profile
   (b) Critical Quality Attributes
   (c) Risk Assessment: Linking Material Attributes and Process Parameters to Drug Product CQAs
   (d) Design Space
   (e) Control Strategy
   (f) Product Life Cycle Management and Continual Improvement
3. Submission CTD-Q
   (a) Quality Risk Management and Product and Process Development
   (b) Design Space
   (c) Control Strategy
   (d) Drug Substance-Related Information

2.2.5.2.1 Introduction

This annex, part of ICH Q8 Pharmaceutical Development annex, further clarifies the core guideline's essential concepts. Its focus is to explain the concept of QbD and describe how Q8 guidelines will be practised in pharmaceutical development.

2.2.5.2.2 The Elements of Pharmaceutical Development

It should include:

- Defined quality target product profile (QTPP)
- Recognising the important critical quality attributes (CQAs) of drug substances, excipients and drug products
- Choosing a suitable manufacturing method
- Making a control plan (Zhang and Mao 2017)

Quality Target Product Profile

Some factors to consider for the quality goal of the product include delivery systems, dosage strength, container closure system, dosage form, use in a clinical setting and route of administration. The factors affecting pharmacokinetic data as per the finished product (FP) should also be considered, including sterility, stability, purity and drug release (Debevec et al. 2018).

Critical Quality Attributes (CQA)

A CQA is a property that should fall within a suitable limit to guarantee the intended product quality. It can be chemical, physical and biological. It is applicable for excipients, DS, FP and in-process materials. For example, CQAs of solid oral dosage forms are usually elements that affect the product strength, stability, drug release and purity. Some other attributes can be added depending on the delivery system, such as sterility, a critical quality attribute for parenteral. When the dosage form and manufacturing process are chosen, the possible CQAs list may change (Holm et al. 2017).

Risk Assessment: Linking Material Attributes and Process Parameters to Drug Product CQAs

Early in developing a pharmaceutical, risk assessment is usually carried out to recognise which material characteristics and process parameters influence the CQAs of a product. This tool helps rank the potential parameters depending on their impact on product quality with past knowledge and experimental data. The list of parameters can be refined based on experimental data to understand the variables and their interaction using statistical tools (Holm et al. 2017).

Design Space

The design space explains the connection between process inputs, such as material characteristics, process variables and the CQAs. Selecting potential variables with their defined ranges is essential to achieving a quality product. Material properties and process parameters are crucial parts of the design space. Proper justification should be provided for the selection of a particular variable and its range. Process parameters, material attributes or complex mathematical relationships can define a design space. The applicant should discuss potential risks in the scale-up operation and justify the importance of creating a design space from a small or pilot scale to the planned production-scale manufacturing process (Debevec et al. 2018).

Control Strategy

A control strategy is created to guarantee that a product will be consistently produced at the required standard. The control system should be based on dosage form, material properties and process. While real-time release testing can replace final product testing, it cannot replace the batch review and quality control procedures required by GMP. There should be control over material properties (excipients, drug substance, packaging material), unit operations, product specifications, in-process testing and monitoring programmes (Holm et al. 2017).

Product Life Cycle Management and Continual Improvement

Companies can assess novel methods to raise product quality throughout the product life cycle. As more experience is acquired through routine manufacturing, this monitoring could include a trend analysis of the manufacturing process. With more process information, the design space might need to be expanded, shrunk or redefined (Ramnarine and O'Donnell 2018).

2.2.5.2.3 Submission of Pharmaceutical Development and Related Information in Common Technical Documents (CTD) Format

The applicant must provide complete information with proper justification related to product life cycle management.

Quality Risk Management and Product and Process Development

Quality risk management can be applied at various phases when developing products, processes and starting a manufacturing operation. For instance, risk analyses and functional relationships can relate material qualities and process parameters to product CQAs.

Design Space

Design spaces should be explained with essential critical controls and intermediates. Its relationship with the control strategy can also be mentioned.

Control Strategy

In the relevant CTD format sections, comprehensive details about input material controls and process controls should be provided.

Drug Substance-Related Information

Some discussions on drug substance CQAs may be appropriate in pharmaceutical development and should be part of the application if drug substance CQAs can potentially influence the CQAs or manufacturing process of the DP (Mishal and Rathod 2014).

2.3 Conclusion

The pharmaceutical industry's primary duty is to give society access to high-quality medicines. The fundamentals determining whether pharmaceutical items are of high quality are guiding documents, regulations and compliance programmes. One control strategy for maintaining the quality of pharmaceutical products is setting specifications for the raw ingredients to the FP. ICH guidelines are crucial documents that help preserve the quality of pharmaceutical products from start to finish, which manufacturers must abide by to register their products. This document includes many acceptance criteria that each product must meet to be considered. The specific instructions offered by ICH for preserving the quality of products during and after manufacturing are known as ICHQ6A, ICHQ6B and ICHQ8 R2. For NDS and NDP, ICHQ6A administers universal and specific tests, while ICHQ6B administers only specific tests for biotechnological and biological goods. The opportunities provided by ICHQ8R2 are to understand better how well a product performs across a broad range of material characteristics, process parameters and processing options, which can be converted into scientific understanding that establishes a more expansive design space. The QTPP, product and process design and knowledge, scale-up, control strategy and continuous improvement are some of the essential components of pharmaceutical quality by design. Utilising all these quality concepts in design makes a reproducible product easy to manufacture.

References

Ain QU, Farooq MA, Caliskan B, Ahsan A, Aquib M, Hussain Z et al (2020) Stability studies of parenteral products. In: Drug stability and chemical kinetics. Springer Singapore, Singapore, pp 247–263

Akers MJ (2016) Sterile drug products: formulation, packaging, manufacturing and quality. CRC Press, Boca Raton, FL. https://books.google.at/books?id=C2_LBQAAQBAJ

Altria KD (1998) Pharmaceutical raw materials and excipients analysis. In: Analysis of pharmaceuticals by capillary electrophoresis. Vieweg+Teubner Verlag, Wiesbaden, pp 133–152

Avis KE (ed) (2018) Pharmaceutical dosage forms: parenteral medications, 2nd edn. Routledge, London, England. https://books.google.at/books?id=FFMPEAAAQBAJ

Balagué N, Saarti J (2011) Quality management system. In: Managing your library and its quality. Elsevier, pp 71–87

Beg S, Rahman M, Al Robaian M, Sarim S, Alruwaili N, Panda SK (eds) (2020) Pharmaceutical drug product development and process optimization: effective use of quality by design. Taylor & Francis, London, England. https://books.google.at/books?id=FEQPEAAAQBAJ

Bergren M, Subbarao N (2019) Parenteral product specifications and stability. In: Parenteral medications, 4th edn. CRC Press, Boca Raton, FL, pp 977–996

Bhavna OA, Bhargava S (2022) International Council for Harmonisation (ICH) guidelines. In: Ali J, Baboota S (eds) Regulatory affairs in the pharmaceutical industry. Elsevier, San Diego, CA, pp 47–74

Center for Drug Evaluation, Research (2023) Q6A specifications: test procedures and acceptance criteria for new drug substances and new drug products: chemical substances. U.S. Food and Drug Administration, FDA. https://www.fda.gov/regulatory-information/search-fda-guidance-documents/q6a-specifications-test-procedures-and-acceptance-criteria-new-drug-substances-and-new-drug-products

Certificate of Analysis (2023). Nutraceuticalsworld.com. https://www.nutraceuticalsworld.com/knowledge-center/market-segments/testing/certificate-of-analysis/
Chan CC (2014) Specification setting and stability studies in the development of therapeutic delivery solution. In: Therapeutic delivery solutions. Wiley, Hoboken, NJ, pp 269–313
Chavan SD, Pimpodkar NV, Kadam AS, Gaikwad PS (2016) Quality by design. Asian J Res Pharm Sci 6(1):45–50. https://www.indianjournals.com/ijor.aspx?target=ijor:ajrps&volume=6&issue=1&article=007
Chen W, Stithit S, Zheng JY, Hwang R (2009) Specification setting and manufacturing process control for solid oral drug products. In: Qiu Y, Chen Y, Zhang GGZ, Liu L, Porter WR (eds) Developing solid oral dosage forms. Elsevier, San Diego, CA, pp 599–614
Chen Y, Gao Z, Duan JZ (2017a) Dissolution testing of solid products. In: Qiu Y, Chen Y, Zhang GGZ, Yu L, Mantri RV (eds) Developing solid oral dosage forms. Elsevier, San Diego, CA, pp 355–380
Chen W, Stithit S, Zheng JY (2017b) Specification setting and manufacturing process control for solid oral drug products. In: Qiu Y, Chen Y, Zhang GGZ, Yu L, Mantri RV (eds) Developing solid oral dosage forms. Elsevier, San Diego, CA, pp 677–693
Csóka I, Pallagi E, Paál TL (2018) Extension of quality-by-design concept to the early development phase of pharmaceutical R&D processes. Drug Discov Today 23(7):1340–1343. https://www.sciencedirect.com/science/article/pii/S1359644617305688
D'Hondt M, Bracke N, Taevernier L, Gevaert B, Verbeke F, Wynendaele E et al (2014) Related impurities in peptide medicines. J Pharm Biomed Anal 101:2–30. https://www.sciencedirect.com/science/article/pii/S0731708514003008
Davis BS (2013) Quality by design. In: Risk management applications in pharmaceutical and biopharmaceutical manufacturing. Wiley, pp 89–100. https://doi.org/10.1002/9781118514399.ch5
De Jong HJ (2016) Control tests on the finished product. In: International Pharmaceutical Product Registration, 2nd edn. CRC Press, Boca Raton, FL, pp 264–279
Debevec V, Srčič S, Horvat M (2018) Scientific, statistical, practical, and regulatory considerations in design space development. Drug Dev Ind Pharm 44(3):349–364. https://doi.org/10.1080/03639045.2017.1409755
DeGrazio F, Paskiet D (2020) Injectable combination product development: facilitating risk-based assessments for efficiency and patient centric outcomes. J Pharm Sci 109(7):2101–2115. https://www.sciencedirect.com/science/article/pii/S0022354920301799
Elder D (2013) ICH Q6A—specifications: test procedures and acceptance criteria. European Pharmaceutical Review. https://www.europeanpharmaceuticalreview.com/article/23393/ich-q6a-specifications-test-procedures-and-acceptance-criteria/
Elder D (2017) ICH Q6A specifications: test procedures and acceptance criteria for new drug substances and new drug products: chemical substances. In: ICH quality guidelines. Wiley, Hoboken, NJ, pp 433–466
EMA (2018a) ICH Q1A(R2) Stability testing of new drug substances and drug products—scientific guideline. European Medicines Agency. https://www.ema.europa.eu/en/ich-q1a-r2-stability-testing-new-drug-substances-drug-products-scientific-guideline
EMA (2018b) ICH Q2(R2) Validation of analytical procedures—scientific guideline. European Medicines Agency. https://www.ema.europa.eu/en/ich-q2r2-validation-analytical-procedures-scientific-guideline
EMA (2018c) ICH Q3A(R2) Impurities in new drug substances—scientific guideline. European Medicines Agency. https://www.ema.europa.eu/en/ich-q3a-r2-impurities-new-drug-substances-scientific-guideline
EMA (2018d) ICH Q3B(R2) Impurities in new drug products—scientific guideline. European Medicines Agency. https://www.ema.europa.eu/en/ich-q3b-r2-impurities-new-drug-products-scientific-guideline
EMA (2018e) ICH Q4B Evaluation and recommendation of pharmacopoeial texts for use in the ICH regions—scientific guideline. European Medicines Agency. https://www.ema.europa.eu/en/ich-q4b-evaluation-recommendation-pharmacopoeial-texts-use-ich-regions-scientific-guideline

EMA (2018f) International Council on Harmonisation of Technical Requirements for registration of pharmaceuticals for human use (ICH). European Medicines Agency. https://www.ema.europa.eu/en/partners-networks/international-activities/multilateral-coalitions-initiatives/international-council-harmonisation-technical-requirements-registration-pharmaceuticals-human-use

Europa.eu (2023). https://www.ema.europa.eu/en/documents/scientific-guideline/ich-q-6-test-procedures-acceptance-criteria-new-drug-substances-new-drug-products-chemical_en.pdf

Fahmy R, Danielson D, Martinez MN (2013) Quality by design and the development of solid oral dosage forms. In: Advances in delivery science and technology. Springer US, New York, pp 107–129

Flynn R, Plueschke K, Quinten C, Strassmann V, Duijnhoven RG, Gordillo-Marañon M et al (2022) Marketing authorization applications made to the European Medicines Agency in 2018–2019: what was the contribution of real-world evidence? Clin Pharmacol Ther 111(1):90–97. https://doi.org/10.1002/cpt.2461

Friedel HD, Brown CK, Barker AR, Buhse LF, Keitel S, Kraemer J et al (2018) FIP guidelines for dissolution testing of solid oral products. J Pharm Sci 107(12):2995–3002. https://www.sciencedirect.com/science/article/pii/S0022354918305069

Gandhi A, Roy C (2016) Quality by Design (QbD) in pharmaceutical industry: tools, perspectives and challenges. Pharmatutor 4(11):12–20. http://www.pharmatutorjournal.com/index.php/pt/article/view/387

Geigert J (2019a) The art of setting specifications. In: The challenge of CMC regulatory compliance for biopharmaceuticals. Springer International Publishing, Cham, pp 355–380

Geigert J (2019b) Biopharmaceuticals are not chemical drugs. In: The challenge of CMC regulatory compliance for biopharmaceuticals. Springer International Publishing, Cham, pp 33–51

Haleem RM, Salem MY, Fatahallah FA, Abdelfattah LE (2015) Quality in the pharmaceutical industry—a literature review. Saudi Pharm J 23(5):463–469. https://doi.org/10.1016/j.jsps.2013.11.004

Harron DWG (2013) Technical requirements for registration of pharmaceuticals for human use: the ICH process. In: The textbook of pharmaceutical medicine. Blackwell Publishing Ltd., Oxford, UK, pp 447–460

Hiwale P, Amin A, Kumar L, Bansal AK (2008) Variables affecting reconstitution time of dry powder for injection. PharmTech. https://www.pharmtech.com/view/variables-affecting-reconstitution-time-dry-powder-injection

Holm P, Allesø M, Bryder MC, Holm R (2017) Q8(R2): pharmaceutical development. In: ICH quality guidelines. Wiley, Hoboken, NJ, pp 535–577

Houston CT, Rodrigues AD, Smith BB, Wang T, Richardson M (2022) Principles for management of extractables and leachables in ophthalmic drug products. PDA J Pharm Sci Technol 76(3):278–294. http://journal.pda.org/content/76/3/278.abstract

ICH (1999) International conference on harmonisation of technical requirements for registration of pharmaceuticals for human use ICH Harmonised Tripartite Guideline specifications: test procedures and acceptance criteria for new drug substances and new drug products: chemical substances q6a. Ich.org. https://database.ich.org/sites/default/files/Q6A%20Guideline.pdf

ICH Harmonised Tripartite Guideline (2023) International conference on harmonisation of technical requirements for registration of pharmaceuticals for human use. Ich.org. https://database.ich.org/sites/default/files/Q8_R2_Guideline.pdf

ICH Official website (2023a) ICH. Ich.org. https://www.ich.org/page/quality-guidelines

ICH Official website (2023b) ICH. Ich.org. https://www.ich.org/page/mission

ICH Q (2000) Specifications: test procedures and acceptance criteria for new drug substances and new drug products: chemical substances. International Conference on Harmonization, Geneva

JBGSR.MS.ID.00087.Text—biogeneric science and research (2023). Biogenericpublishers.com. https://biogenericpublishers.com/jbgsr.ms.id.00087.text/

Jîtcă C-M, Jîtcă G, Ősz B-E, Puşcaş A, Imre S (2023) Stability of oral liquid dosage forms in pediatric cardiology: a prerequisite for patient's safety—a narrative review. Pharmaceutics 15(4):1306. https://doi.org/10.3390/pharmaceutics15041306

Kajiwara E, Shikano M (2020) Considerations and regulatory challenges for innovative medicines in expedited approval programs: breakthrough therapy and Sakigake designation. Ther Innov Regul Sci 54(4):814–820. https://doi.org/10.1007/s43441-019-00019-z
Kestur U, Patel A, Badawy S, Mathias N, Zhang L (2023) Strategies for managing solid form transformation risk in drug product. J Pharm Sci 112(4):909–921. https://www.sciencedirect.com/science/article/pii/S0022354922005652
Khagga B, Kaitha MV, Dammu R, Mogili S (2019) ICH guidelines—"Q" series (quality guidelines)—a review. GSC Biol Pharm Sci 6(3):089–106. https://doi.org/10.30574/gscbps.2019.6.3.0034
Knight J (2014) Specifications of drug substances and products: development and validation of analytical methods. Org Process Res Dev 18(9):1154–1154. https://doi.org/10.1021/op5002879
Kumar R, Palmieri MJ Jr (2010) Points to consider when establishing drug product specifications for parenteral microspheres. AAPS J 12(1):27–32. https://doi.org/10.1208/s12248-009-9156-6
Lin J (2022) Control strategies for pharmaceutical development. In: Analytical testing for the pharmaceutical GMP laboratory. Wiley, pp 109–142. https://doi.org/10.1002/9781119680475.ch4
Maguire JA, Jiang W, Yu LX (2017) Modern pharmaceutical regulations. In: Qiu Y, Chen Y, Zhang GGZ, Yu L, Mantri RV (eds) Developing solid oral dosage forms. Elsevier, San Diego, CA, pp 1095–1126
Mishal A, Rathod S (2014) Quality by design: a new era of development of quality in pharmaceuticals. Res J Pharm Technol 7(5):581–591. https://www.indianjournals.com/ijor.aspx?target=ijor:rjpt&volume=7&issue=5&article=016
Mishra V, Thakur S, Patil A, Shukla A (2018) Quality by design (QbD) approaches in current pharmaceutical set-up. Expert Opin Drug Deliv 15(8):737–758. https://doi.org/10.1080/17425247.2018.1504768
Morris J (2016) Control of drug substance. In: International Pharmaceutical Product Registration, 2nd edn. CRC Press, Boca Raton, FL, pp 170–184
Narayan P (2011) Overview of drug product development. Curr Protoc Pharmacol 55(1). https://doi.org/10.1002/0471141755.ph0703s55
Nema S, Ludwig JD (2016) Pharmaceutical dosage forms—parenteral medications. In: Nema S, Ludwig JD (eds) Regulations, validation and the future, vol 3, 3rd edn. CRC Press, Boca Raton, FL. https://books.google.at/books?id=Q0m3BgAAQBAJ
Ng LL (2022) Development and validation of analytical procedures. In: Analytical testing for the pharmaceutical GMP laboratory. Wiley, pp 143–167. https://doi.org/10.1002/9781119680475.ch5
Nickerson B, Kong A, Gerst P, Kao S (2018) Correlation of dissolution and disintegration results for an immediate-release tablet. J Pharm Biomed Anal 150:333–340. https://www.sciencedirect.com/science/article/pii/S0731708517322768
Ola M, Bhaskar R, Patil P (2018) Dry syrup: an overview. Indian J Pharm Biol Res 6(03):30–38. http://www.ijpbr.in/index.php/IJPBR/article/view/709
Pallagi E, Ismail R, Paál TL, Csóka I (2018) Initial risk assessment as part of the Quality by Design in peptide drug containing formulation development. Eur J Pharm Sci 122:160–169. https://www.sciencedirect.com/science/article/pii/S0928098718303014
Pandit D, Deb PK, Chadha R (2019) Design of materials and product specifications. In: Tekade RK (ed) Biomaterials and bionanotechnology. Elsevier, San Diego, CA, pp 1–17
Parente E (2020) Description and identification. In: Riley CM, Rosanske TW, Reid G (eds) Specification of drug substances and products. Elsevier, pp 177–194
Patel P, Badjatya JK, Hinge M (2019) Comparative study of regulatory requirements of drug products in emerging market. Int J Drug Regul Aff 7(3):48–62. https://www.semanticscholar.org/paper/4050fe9bb6ba517f71d42f5c946296cd4422e3c8
PDA (2023) Injectable combination product integrated development: a PDA live eLearning course. Pda.org. https://www.pda.org/pda-letter-portal/home/full-article/injectable-combination-product-integrated-development-a-pda-live-elearning-course
Pharmaceutical Technology Editorial Advisory Board (2020) PharmTech. https://www.pharmtech.com/view/pharmaceutical-technology-editorial-advisory-board

Pharmacopeial Forum 750 of the USPC or the USP (2009) Council of Experts. 35
Pharmaguideline (2019) Specifications for new drug substances and new drug products (quality control). Pharmaceutical Guidelines. Pharmaguideline.
Pokar D, Rajput N, Sengupta P (2020) Industrial approaches and consideration of clinical relevance in setting impurity level specification for drug substances and drug products. Int J Pharm 576:119018. https://www.sciencedirect.com/science/article/pii/S0378517319310804
Radhakrishnan V, Banerjee A, Khan MA (2019) Quality by design and control strategies for parenteral dosage forms. In: Parenteral medications, 4th edn. CRC Press, Boca Raton, FL, pp 809–828
Ramnarine E, O'Donnell K (2018) Demonstrating pharmaceutical quality system effectiveness and driving continual improvement: evidence-based risk reduction. PDA J Pharm Sci Technol 72(3):338–345. http://journal.pda.org/content/72/3/338.abstract
Rignall A (2017) ICHQ1A(R2) stability testing of new drug substance and product and ICHQ1C stability testing of new dosage forms. In: ICH quality guidelines. Wiley, Hoboken, NJ, pp 3–44
Riley CM, Yang H (2020) General principles and regulatory considerations: specifications and shelf life setting. In: Riley CM, Rosanske TW, Reid G (eds) Specification of drug substances and products. Elsevier, pp 9–41
Riley CM, Rosanske TW, Riley SR (2013) Specification of drug substances and products: development and validation of analytical methods. Elsevier
Roesti D (2019) Microbiological examination of non-sterile final dosage forms and raw material including acceptance criteria and testing frequency. In: Pharmaceutical microbiological quality assurance and control. Wiley, pp 105–151. https://doi.org/10.1002/9781119356196.ch5
Sangeetha VP, Remya NS, Mohanan PV (2022) Regulatory issues in biological products. In: Biomedical product and materials evaluation. Elsevier, pp 309–327
Schenck L, Erdemir D, Saunders Gorka L, Merritt JM, Marziano I, Ho R et al (2020) Recent advances in co-processed APIs and proposals for enabling commercialization of these transformative technologies. Mol Pharm 17(7):2232–2244. https://doi.org/10.1021/acs.molpharmaceut.0c00198
Singh DK, Sahu A, Kumar S, Singh S (2018) Critical review on establishment and availability of impurity and degradation product reference standards, challenges faced by the users, recent developments, and trends. Trends Analyt Chem 101:85–107. https://www.sciencedirect.com/science/article/pii/S0165993617303710
Turner JR (2019) The international council for harmonisation and a case study in regulatory science. Ther Innov Regul Sci 53(5):561–563. https://doi.org/10.1177/2168479019869740
Zhang L, Mao S (2017) Application of quality by design in the current drug development. Asian J Pharm Sci 12(1):1–8. https://www.sciencedirect.com/science/article/pii/S1818087616300575

3 Optimization Techniques for the Development of Pharmaceutical Products

Sunny Rathee, Shivani Saraf, Pritish Kumar Panda, Sarjana Raikwar, Pooja Das Bidla, and Sanjay K. Jain

Abstract

Quality by design (QbD) is a systematic approach for the development of a product that begins with predefined objectives and emphasizes product, process understanding, and process control, based on sound science and quality risk management (QRM). Quality by design (QbD) has the potential to enhance performance, offer regulatory relief and adaptability, and bring about significant financial gains throughout the product life cycle. Due to its numerous advantages, the quality by design (QbD) method is now widely employed in the pharmaceutical industry. Prior to the development, this method provides a greater understanding of the process and the result, which eventually helps in the development of a higher-quality product. Formulation-based design (FbD) is under extensive investigation by budding scientists for better know-how of the product and process development for an unequivocal universal acceptance. Nowadays, the development of pharmaceutical products, biotechnological products, and new chemical entities using QbD and FbD techniques has resulted in the filing of

S. Rathee
Pharmaceutics Research Projects Laboratory, Department of Pharmaceutical Sciences, Dr. Harisingh Gour Vishwavidyalaya, Sagar, Madhya Pradesh, India

College of Pharmaceutical Sciences (COPS), School of Health Sciences, Dayananda Sagar University, Kanakapura Road, Ramanagara Dt., Bengaluru, Karnataka, India

S. Saraf
Pharmaceutics Research Projects Laboratory, Department of Pharmaceutical Sciences, Dr. Harisingh Gour Vishwavidyalaya, Sagar, Madhya Pradesh, India

Babulal Tarabai Institute of Pharmaceutical Sciences, Sagar, Madhya Pradesh, India

P. K. Panda · S. Raikwar · P. D. Bidla · S. K. Jain (✉)
Pharmaceutics Research Projects Laboratory, Department of Pharmaceutical Sciences, Dr. Harisingh Gour Vishwavidyalaya, Sagar, Madhya Pradesh, India

K. Jain, A. K. Yadav (eds.), *Advances in Pharmaceutical Product Development*, https://doi.org/10.1007/978-981-97-9230-6_3

various new drug applications and abbreviated new drug applications. This chapter focuses on the emergence of QbD and FbD as quality tools, applications of QbD and FbD in different fields and various regulatory guidelines, and hurdles in the implementation of QbD and FbD.

Keywords

Quality by design · Formulation by design · New drug application · Quality risk management

3.1 Introduction

The traditional method used in pharmaceutical product development, which relies on quality by testing, is no longer in use. These methods ensured the quality of products by regulating the raw materials and manufacturing technology. The finished products must pass all quality control tests, and when products fail the quality control tests, the manufacturer must restart the process and identify the error or reason for failure. Therefore, the traditional methods (quality by testing) are very expensive and can lead to variances that reduce the safety of the finished pharmaceutical products (Cunha et al. 2020; Zhang and Mao 2017).

To solve these problems, the quality by design (QbD) strategy was investigated to improve manufacturing processes and ensure the safety and efficacy of finished products. The QbD strategy was developed to establish rigorous specifications and guidelines for the approval of pharmaceutical products. Dr. Joseph M. Juran described the approach QbD first time and utilized it to ensure the quality of the product from planning or design not from evaluation and control. The QbD was approved by regulatory agencies such as FDA (Food and Drug Administration) and EMA (European Medicines Agency) after the publication of various guidelines by ICH, i.e., Q8, Q9, Q10, and Q11. The ICH Q8 guideline described the QbD approach and defined the main objectives of the process, facilitating knowledge and control and controlling risk to final product quality (Bastogne 2017). The concept used in QbD is that quality can be "designed in" to processes through the systemic optimization technique. In optimization, we determined the response of a given variable to the quality of the product was determined, by the implementation of systematic optimization (Jagtap et al. 2022). QbD is a scientific approach that helps in designing pharmaceutical formulations having the intended product qualities (Jain 2014). QbD approach aims to identify the critical parameter that affects the quality of a product. The parameters involved in product variability were identified and eliminated or controlled in a predetermined design space to ensure product quality (Adam et al. 2011). In nanotechnology, statistical tests like Bonferroni, t-tests, and others can be used to compare various treatments and determine differences among them, but QbD approaches enable the optimization of evaluation results (dependent variables) and the analysis of study factors (independent variables). Additionally, the QbD approach also determined the level of an impact on responses in a linear, quadratic, and full quadratic assessment along with the p-value of input variables. The design of the expert (DoE) tool is a statistical method that determined the influence of independent variables on dependent variables. The most commonly used DoE methods are screening design (Plackett–Burman, fractional factorial design, and full factorial design) and response surface methodology (RSM). In the pharmaceutical field, central composite design (CCD) and the Box–Behnken design (BBD) RSM are the most widely used methods for optimization. Doehlert design is also as effective as CCD and BBD, but it was less studied for the nanotechnology-based products (Luiz et al. 2021).

This chapter discussed the QbD approach for pharmaceutical product development, the role of QbD and FbD in product development, various experimental designs employed in the optimization of drug delivery systems, key elements of QbD, various tools used in FbD optimization, FbD Opti-tactics for drug delivery systems, software employed for QbD optimization, and applications of QbD and FbD in different fields.

3.2 Role of QbD and FbD in the Development of Pharmaceutical Products

The conventional optimization method is studying the effect of one factor at a time (OFAT) while maintaining the other parameters constant. This method provides the solution for a specific problematic property but does not assure the true optimum composition or process. This may be attributed to the existence of interactions or the impact of one or more factors on other factors. After optimization, the final product is satisfactory but mostly suboptimal because a better formulation may still exist for the studied conditions. Thus, the traditional OFAT method has various problems such as being time-consuming, expensive, and ineffective at revealing the interactions. Further, this method only provides "just satisfactory" results because detailed studies of all factors are not possible in this approach. This approach cannot study the simultaneous effects of all variables on responses and may result in inconsistent products. To overcome these problems, recently, DoE techniques were investigated for pharmaceutical product development. These methods include the application of appropriate experimental design along with mathematical equations and graphic outcomes that depict a clear picture of variation of the response(s) as a function of the factor(s). These methods provided the best possible solution and need fewer experimental runs to obtain an optimum formulation as compared to traditional methods. Additionally, the screening methods used in the DoE aid in identifying the most critical input variables. Using model equations, one can simulate the behavior of the product or process, saving time, effort, materials, and money. The comparison between conventional and QbD system is depicted in Fig. 3.1.

The most remarkable features of formulation by design (FbD) are that it can predict the response of formulation and may detect and measure the possible interaction and synergistic effects between the variables. The FbD method helps in the understanding of the formulation and can identify the problems and eliminate them

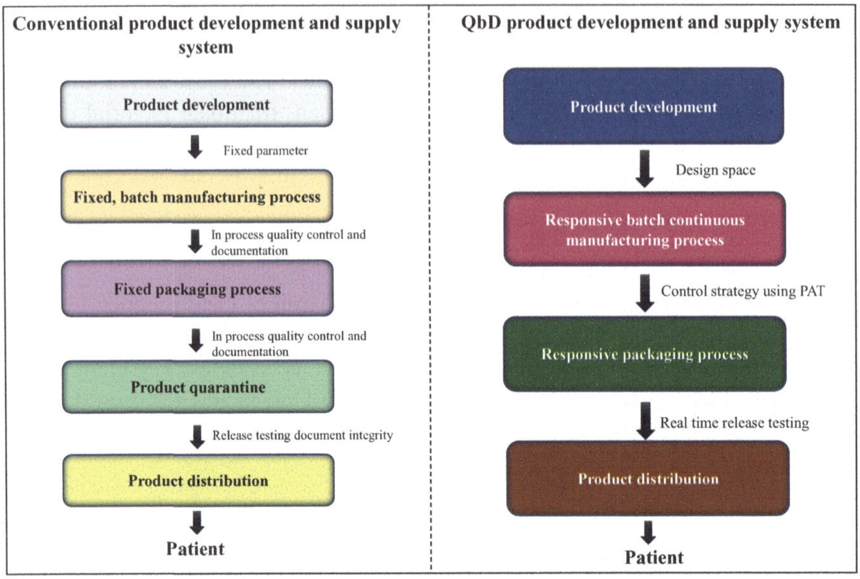

Fig. 3.1 Comparison between conventional and QbD system

more easily. The FbD techniques provide high-quality drug products at an affordable price and ultimately achieved the QbD objectives (Namjoshi et al. 2020).

The use of QbD approaches is to provide quality products and help industry to speed up the regulatory approval process. The development of a successful product needs an understanding of QbD principles and the tools utilized in QbD methods. Various tools such as the design of experiments, risk assessment tools, and PAT (process analytical technology) are employed for the establishment of QbD principles. After the implementation of QbD, changes to the product and process can be managed more effectively.

3.3 Key Experimental Designs Employed for Optimization of Drug Delivery Systems

The designing of an ideal pharmaceutical product requires multiple objectives. In the case of complex drug delivery systems (DDSs), a variety of variables such as drugs, excipients, polymers, and processes are involved (Singh et al. 2020). For the last many years, this task has been carried out via trial and error, accompanied by previous data, experience, and understanding of the formulator (Singh et al. 2005a). The conventional approach involves studying the influence of the corresponding composition and process variable by varying one variable at a time (OVAT) while keeping other factors constant. This technique, at times, is also denoted as one factor at a time (OFAT) or “shotgun” approach (Singh et al. 2005a). In OVAT studies, the first variable is fixed at a favorable value, and the next is observed until no further improvement is achieved in the response variables. This approach can accomplish the solution of a specific problem, but the fulfillment of the true optimum composition or process is never certain. It may be attributed to the presence of interaction(s) such as the synergistic or antagonistic effect of one or more variable(s) on others.

The execution of DoE optimization methods invariably includes the use of experimental designs and the generation of calculated equations and graphic outcomes. This leads to producing a complete picture of the variation of the product/process response(s) as a function of the input variable(s). The use of different types of combination strategies for formulation variables, along with Design of Experiments (DoE), fits experimental data into statistical equations, utilizes these models to predict formulation performance, and optimizes critical responses (Rao and Srinivas 2011).

The general scientific methodology is determined via the conduct of an experiment and following interpretation of its experimental result. There are the twin essential characteristics of both the experimental and outcomes, which lead to generating the scientific methodology (Singh et al. 2006). Different runs or trials are the experiments conducted according to the selected experimental design (ED). Several kinds of EDs are employed for the optimization of different drug delivery systems (Singh et al. 2005a). Commonly used EDs for response surface methodology (RSM), screening, and factor-affecting studies during the pharmaceutical product/process development include the following:

3.3.1 Factorial Designs (FD)

It is the most commonly used ED and involves all levels of a given factor combined with all levels of every other factor in the experiment. Full factorial design (FFD) involves studying the effect of all the factors (k) at various levels (x). It includes various interactions among them, with the total number of experiments which is denoted as xk. There are two types of FD: it can be either "symmetric" if the number of levels is the same for each factor or "asymmetric" in cases of a different number of levels for different factors (Singh and Ahuja 2000).

3.3.2 Fractional Factorial Designs (FFDs)

The fractional factorial design (FFD) involves a finite fraction (1/xr) of a complete or full FD, where r is the degree of fractionation and xk-r is the total number of experiments needed (Doornbos and Haan 1995). When the number of required experiments exceeds the adaptable levels due to an increase in the number of factors or factor levels, then this design is particularly preferred over FD (Singh et al. 2005a). An FFD with $r = 1$, on the other hand, will require only 23-1, i.e., four experiments, estimating a total of three effects are estimated (Singh et al. 2005b).

3.3.3 Plackett–Burman Designs (PBDs)

The Plackett–Burman designs (PBDs) are specialized two-level FFDs used commonly for screening of K, i.e., $N - 1$ factors, where N is a multiple of 4 (Plackett and Burman 1946). It is also known as Hadamard designs or symmetrically reduced 2k-r FDs; the designs can easily be created by using a minimum number of trials (Loukas 2001). It is also known as a nongeometric design because this design cannot be represented as cubes. The PBDs are relatively employed during the screening processes.

3.3.4 Central Composite Designs (CCD)

The central composite design is also known as the Box–Wilson design; the central composite design (CCD) is most widely used for quadratic models (Singh and Ahuja 2000). The design consists of a combination of two-level factorial points ($2n$), axial or star points ($2n$), and a central point (Box and Wilson 1992). Hence, $2n + 2n + 1$ was used to denote the total number of factors combined in CCD.

3.3.5 Box–Behnken Designs (BBD)

Box–Behnken design (BBD) is specially designed to optimize various product and process variables. BBD requires only three levels (−1, 0, 1) (Box and Behnken 1960). It is used to overcome the limitations associated with the CCD, where each factor has been considered at five levels. Hence, increasing the number of factors leads to an increase in the number of experiments. BBD is considered an economical alternative to CCD (Singh and Ahuja 2000).

3.3.6 Equiradial Designs

The equiradial designs consist of N points on a circle about the center of interest in the form of a regular polygon (Singh et al. 2005a). Some limitations are associated with this design, i.e., only a limited number of parameters ($p = 3$) have been selected in this model. This design has special characteristics; it may be rotated by any angle and still have the same properties.

3.3.7 Mixture Designs

In a drug delivery system (DDS), simultaneous variation of all factors at all levels may not be possible under many circumstances. The characteristics of a drug delivery system (DDS) formulated with different excipients depend not only on the proportions of the substances but also on the absolute quantities of each component present in the formulation. In this study, the total sum of the excipients is unity, and none of the fractions can be negative. Hence, the levels of different components can be changed with the limit that the total should not exceed one. In previous studies, the mixture design is most widely used (Singh et al. 2005b). The design region for mixture proportions is simple. The method consists of first generating data from the $n + 1$ experiment, where n is the number of factors (Araujo and Brereton 1996). Based on the $n + 1$ response and predetermined rules, one outcome is eliminated and a new study is performed.

The mixture design (SMD) is a type of mixture design and is denoted as Scheffe's design. It can either be a centroid or lattice design (SLD). The design points are homogeneously distributed over the factor space and form the lattice (Lachman et al. 1976).

3.3.8 Taguchi Designs

The Taguchi design was invented by a Japanese engineer, who proposed several strategies for experimental design that are known as the "Taguchi method." This design utilized two-, three-, and mixed-level fractional factorial designs. Taguchi is denoted as an ED as "offline quality control," as it is a method to confirm good performance in the design stage of products or processes. The objective is to make a product or process less variable (i.e., more robust) in the face of variation over which we have little control. In the Taguchi data analysis, the response variable is not the common raw response or quality characteristics but the signal-to-noise ratio (S/N ratio) (Jain 2014).

3.3.9 Optimal Designs

It is a type of computer-aided design particularly useful when classical designs do not apply. Generally, custom designs are produced based on specific optimality criteria, such as D-, A-, G-, I-, and V-optimality criteria (Singh et al. 2005b). The most widely used criterion in custom designs is D-optimality. This optimality criterion results in minimizing the generalized variance of the parameter estimates for a prespecified model. The D-optimal designs are based on the principle of minimization of parameter variance and covariance.

3.4 Elements of QbD

3.4.1 Quality Target Product Profile (QTPP)

The ICH guideline Q8 defines the QbD as "A methodical strategy for development that starts with predetermined goals and emphasizes knowledge of the product and process as well as process control, based on sound science and quality risk management." This rational aspect of product designing involves the first step to defining a list of quality requirements which is termed as quality target product profile (QTPP). QTPP is defined by ICH Q8 as "A prospective summary of the quality parameters of a pharmaceutical drug product that ideally will be achieved to ensure the desired quality while considering the safety and efficacy of the drug product." These specifications for quality are known as quality attributes, and it is crucial to identify the most critical quality attributes (CQAs) of a formulation to accurately characterize the various QTPP components, or physicochemical properties (Namjoshi et al. 2020).

3.4.2 Critical Quality Attributes (CQAs)

As per FDA, the definition of CQA comprises any physical, chemical, biological, or microbiological property or characteristics that should be within a suitable range or distribution to assure the desired quality of the developed product such as product purity, strength, stability, and drug release (Jain 2014). CQA is having a direct effect on different parameters such as quality, safety, or efficacy of the product. These parameters are associated with drug substances, excipients, and in-process materials, i.e., intermediates and products (Sangshetti et al. 2017). The dosage form and delivery system greatly affect the CQAs.

An Ishikawa fishbone diagram is used to create the "cause-and-effect" relationship among multiple input factors and drug product CQAs (Singh et al. 2005a). The Ishikawa diagram showed the cause–effect relationship for CQAs. The ranking exercise aims to define statistically significant variables among material attributes (MAs) and critical process parameters (CPPs), which greatly affect various CQAs. The risk estimation matrix (REM) is a method to determine the most prevalent risk factor. In this method, the MAs and process parameters (PPs) are assigned varied degrees of risk, viz., high, medium, and low. The degree of risk is based on risk severity, frequency of incidence, and, at times, its detectability (Singh et al. 2005a, 2017). The prioritization strategy is a factor screening, which aids in choosing statically significant CMAs and CPPs while ignoring the "idler" ones. It is useful in subsequent optimization studies which used the DoE approach (Singh et al. 2006, 2017).

3.4.3 Risk Management

Risk assessment is a crucial component of quality by design (QbD) and is an integral part of the quality risk management (QRM) process (Fig. 3.2). During risk assessment, the focus is on identifying the critical quality attributes (CQAs) and their role in achieving the predetermined target. The safety and efficacy of the product are highly dependent on the CQAs, making them of utmost importance. Additionally, the critical material attributes (CMA) and critical process parameters (CPP) are linked to the CQAs in this process (Patil and Pethe 2013).

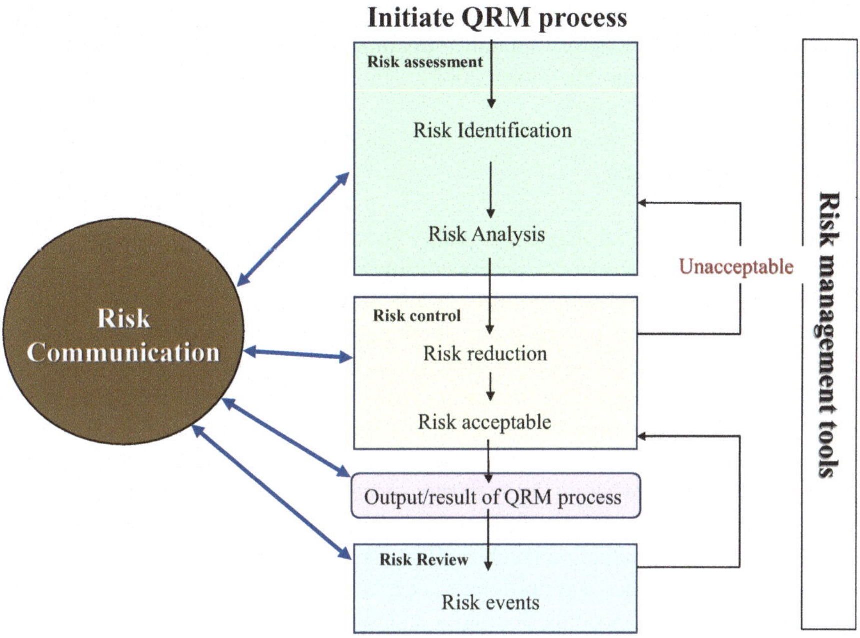

Fig. 3.2 Details of the quality risk management process

There are various qualitative and quantitative tools available for risk assessment. Brainstorming is an effective technique used in the risk assessment process. Other tools include fishbone/Ishikawa diagrams, road maps, preliminary hazard analysis (PHA), fault tree analysis (FTA), RRMA (risk ranking and filtering), FMEA (failure mode and effects analysis), hazard and operability analysis, hazard analysis and critical control points, etc. Among these tools, the fishbone/Ishikawa diagram, FTA, and road maps are used for qualitative risk identification, while FMEA and PHA are used as quantitative tools to categorize the risk (Kalyane et al. 2020).

PHA is a semiquantitative method used for preliminary risk analysis, in which a rank/score is given according to the severity and probability of hazards (Wang et al. 2014). FMEA is a vital technique for risk analysis, involving the ranking of CQA based on the multiplication of severity, occurrence, and detectability in terms of the risk priority number (Lipol and Haq 2011). RRMA is a qualitative technique for risk analysis that categorizes risk as high, low, or medium.

In addition to risk identification and risk analysis, risk evaluation plays a critical role in qualitative risk management. Bar charts and Pareto charts are commonly employed tools used for risk evaluation. These charts provide information regarding the level of significance of the risk (Borgonovo and Smith 2011). Another tool used for risk evaluation is the as low as reasonably practicable (ALARP) approach, which evaluates risk based on the cost-to-benefit ratio. Risks are categorized as unacceptable, broadly acceptable, or tolerable. Unacceptable risks need to be eliminated or reduced to a broadly acceptable or tolerable level. Broadly acceptable risks are well controlled because of their insignificant lower risk, and a tolerable risk criterion is determined through regulatory guidance documents, history, standards, and company policies (Moore and Nicholson 2005).

3.4.4 Design Space

Design space is a crucial aspect of quality by design (QbD) that establishes the range of control for a product or process during manufacturing. QbD offers several benefits that can only be realized through the use of a design space. ICH Q8 (R2) guidelines define the design space as a combination of input variables and process parameters that interact in multiple dimensions to ensure product quality (Mishra et al. 2018). Performing within the design space is not considered a change, while moving outside of it is deemed a change that would typically trigger a regulatory post-approval change process. The design space can be an invariant, experimentally verified design space, where a single design space applies to all production batches, or it can be a fully model-dependent dynamic design space that calculates different areas for individual production batches (Chatterjee et al. 2017).

Dynamic design space is a portion of acceptable spaces for an individual batch, which provides more operating flexibility and can be adjusted based on predefined process parameters. However, the lack of standards for predictive model verification creates uncertainty in developing these types of design spaces. Additionally, creating and maintaining dynamic design space requires more resources, and the business aspect needs to be considered for its flexibility regarding investment. The majority of case studies are available in the literature about the design space generation of the final product formulation or active pharmaceutical ingredient (API) development, with some related to biopharmaceuticals. The number of experiments required for design space generation is based on the complexity of the process. For less complex processes like dry granulation or tablet complexation, fewer experiments are required for the generation of design space compared to more complex processes like biological or API crystallization, which involve multiple factors. Thorough quality risk assessment and thoughtful experimental setup are necessary for the generation of quality design space. Blind experiments can increase the cost of product or process development and take more time than usual.

3.4.5 Control Strategy

The control strategy is a fundamental element of quality by design (QbD), a systematic approach to pharmaceutical development that emphasizes a science-based and risk-based approach to product and process design. The control strategy is a comprehensive plan that identifies and controls the critical quality attributes (CQAs) of a product and process to ensure consistent quality. The control strategy includes all the measures that are used to ensure that the product meets the required specifications and quality standards. It includes the critical process parameters (CPPs), the critical quality attributes (CQAs), the methods used to monitor and control the process, and the specifications for the finished product. The control strategy also takes into account the various risks associated with the product and process and includes measures to mitigate those risks. The control strategy is developed through a process of iterative experimentation and analysis. The initial design of the control strategy is based on the existing knowledge and understanding of the product and process. This initial design is then refined and optimized through experimentation and analysis, using tools such as the design of experiments (DOE) and statistical process control (SPC). The control strategy is implemented throughout the product life cycle, from development to commercialization. The control strategy is continuously monitored and updated to ensure that it remains effective and relevant as the product and process evolve. The control strategy is a planned set of measures that are derived from the existing understanding of the product and the process. These measures are implemented to ensure consistent product quality by controlling various parameters and attributes related to drug substance, drug product materials, facility, and equipment operating conditions. The control strategy can include in-process controls, finished product specifications, associated methods, and the frequency of monitoring and control. This approach is implemented in the pharmaceutical development field to maintain product quality (Bhoop 2014).

A proper understanding of the process helps to minimize the variability of input materials, allowing for tighter process control and more consistent end products. The understanding of product performance can also justify alternative approaches to determine the quality of input materials, such as using disintegration instead of dissolution for quick-disintegrating solid dosage forms. In-process control measures, such as dose uniformity, coupled with a near-infrared assay, can enable real-time release testing, providing a higher level of quality assurance than conventional end-product testing. The control strategy includes controlling the input material attributes based on their impact on processability or product quality, as well as the control of unit operations that affect downstream processing or product quality. It also includes in-process real-time release testing and a monitoring program to verify multivariate prediction models (Pawar et al. 2016). Overall, the control strategy is a key element of QbD, ensuring that the product and process are designed and controlled to consistently meet the required quality standards. It provides a systematic approach to product and process design that is grounded in science and helped to minimize the risks associated with pharmaceutical development (Jiang et al. 2017).

3.4.6 Product Life Cycle Management and Continual Improvement

Product life cycle management and continual improvement are important elements of quality by design (QbD), a systematic approach to pharmaceutical development that emphasizes a science- and risk-based approach to product and process design. The product life cycle begins with product development and continues through commercialization and discontinuation. Throughout this cycle, it is important to manage the product and process to ensure that they meet the required quality standards. This involves monitoring the product and process performance and making adjustments as necessary to maintain consistent quality. The ICH Q10 guidelines provide a framework for product life cycle management and continual improvement. This framework involves establishing a quality management system that incorporates quality planning, quality control, and quality assurance. The quality management system should also include processes for monitoring and analyzing the product and process performance to identify opportunities for improvement. Continual improvement is a key component of the product life cycle management process. This involves using the knowledge gained from monitoring the product and process to identify areas for improvement and implementing changes to improve the quality and efficiency of the product and process. Throughout the product life cycle, there are always opportunities for companies to enhance the quality of their products. The ICH Q10 guidelines emphasize the management of the product life cycle and the continuous improvement of product quality. To achieve this, it is important to monitor whether the product or process performance is meeting the specified benefits as outlined in the design space. This monitoring process could involve analyzing the manufacturing process trends during routine production or using a mathematical model and periodic maintenance. The company's internal quality system manages the maintenance of the model while keeping the design space unchanged (Åsberg et al. 2016). As additional product or process knowledge is gained, there may be a need for the expansion, reduction, or redefinition of the design space. This would enable companies to continue to improve the quality of their products throughout the product life cycle.

Overall, product life cycle management and continual improvement are essential elements of QbD. They provide a systematic approach to product and process design and help to ensure that the product meets the required quality standards throughout its life cycle. By continuously monitoring and improving the product and process, companies can ensure that they remain competitive in the marketplace and meet the evolving needs of their customers. The critical components of quality by design (QbD) and their roles across product development phases are depicted in Fig. 3.3.

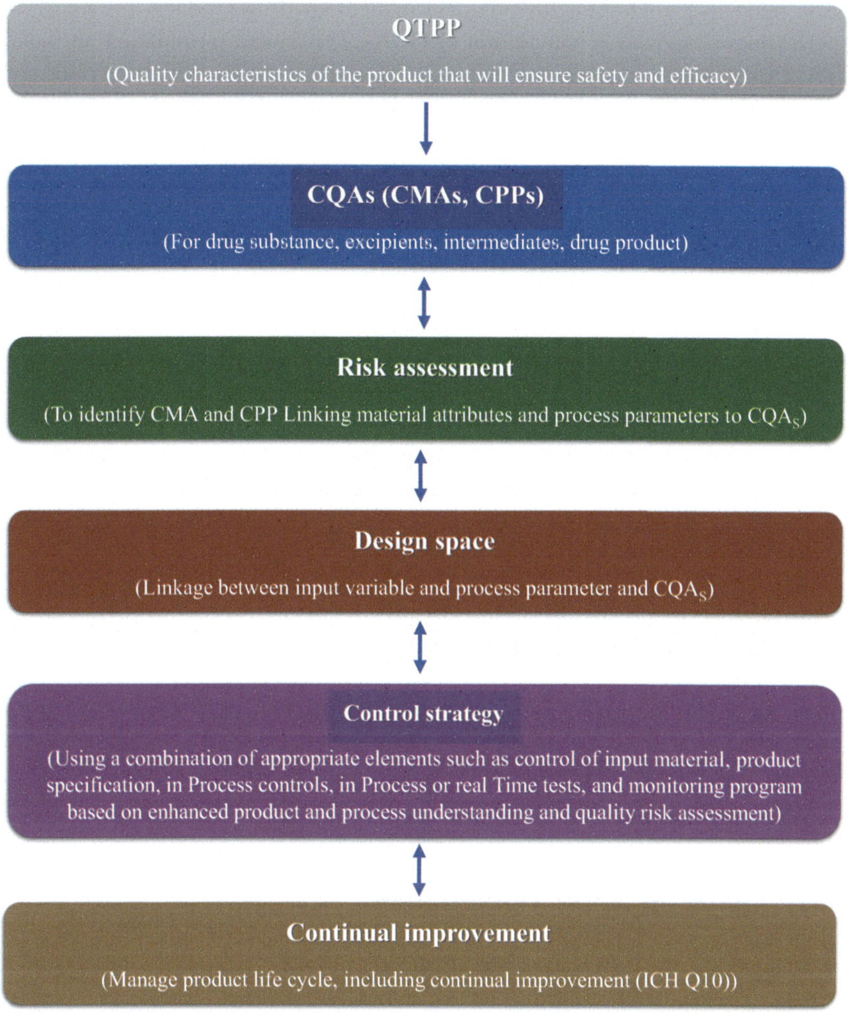

Fig. 3.3 "Critical components of quality by design (QbD) and their roles across product development phases"

3.5 Various Approaches Employed in QbD Optimization

Modern QbD approaches for developing drug products include desired state of manufacturing, end-product testing for validation, testing balances with the design, quality is always accomplished, complements well with Federal QbR, reduced expenditure of resources, and wider operating ranges (Bhoop 2014). QbD success is due to main five vital strengths, viz., (1) meticulous drug product, apt choice of experimental designs, (2) recognition of CQAs, (3) critical formulation attributes (CFAs), (4) critical process parameters (CPPs), and (5) precise definition of design and control space and precise computer-aided optimization (Singh et al. 2005a, 2011). The QbD optimization approaches are more advantageous because they require less number of experiments to attain an optimum formulation; locating the main problem and rectification is easier and showed the interactions between drugs, excipients, and processes lead to stimulating the product performance (Ghaemmaghamian et al. 2022).

3.6 Various Approaches Employed in FbD Optimization

Formulation-based design (FbD) optimization involves the systematic optimization of the formulation of a product to achieve specific desired properties while meeting constraints such as cost, availability of ingredients, and regulatory requirements. Here are some approaches that can be employed in FbD optimization of formulation without plagiarism:

3.6.1 Design of Experiments (DoE)

In this approach, the formulation space is divided into smaller subspaces, and a subset of the subspaces is selected for experimentation. The results of the experiments are then used to build a predictive model of the formulation properties, which can be optimized using FbD techniques.

3.6.2 Constraint-Based Optimization

In this approach, the formulation is optimized subject to a set of constraints such as the cost of the ingredients, the availability of the ingredients, and the desired properties of the final product. The optimization can be performed using techniques such as linear programming or mixed-integer programming.

3.6.3 Multi-objective Optimization

Multi-objective optimization involves optimizing multiple formulation properties simultaneously, which can be conflicting. This approach can be used to find the optimal trade-off between different properties, such as taste, texture, and shelf life.

Surrogate modeling: Surrogate modeling can be used in FbD optimization to construct a surrogate model of the formulation properties that can be optimized using standard optimization techniques.

3.6.4 Expert Systems

Expert systems can be used to incorporate domain knowledge into the optimization process. This approach involves creating a knowledge base of formulation rules and using an inference engine to select the optimal formulation based on the desired properties.

3.6.5 Evolutionary Algorithms

Evolutionary algorithms such as genetic algorithms can be used to optimize the formulation by selecting the optimal combination of ingredients. This approach works well when the formulation space is large and complex.

It is important to ensure that the FbD optimization process for formulation does not involve plagiarism by using appropriate attribution methods and ensuring that the final optimized formulation is unique and not copied from existing formulations.

3.7 Tools Used in Formulation-Based Design (FbD) Optimization

Formulation-based design (FbD) optimization involves breaking down a complex function or formulation into smaller sub-functions or sub-formulations, which can be optimized independently. Here are some tools that can be used in FbD optimization:

(A) Design of experiments (DoE) software: DoE software such as JMP, Minitab, and Design-Expert can be used to plan and execute experiments in FbD optimization. These tools can help to identify significant factors and interactions between factors in the formulation or function.
(B) Optimization software: Optimization software such as MATLAB Optimization Toolbox, Gurobi, and CPLEX can be used to solve the optimization problem in FbD optimization. These tools can help to find the optimal combination of sub-functions or sub-formulations that satisfy the optimization objective and constraints.
(C) Statistical modeling software: Statistical modeling software such as R, SAS, and SPSS can be used to build predictive models of the sub-functions or sub-formulations in FbD optimization. These tools can help to identify significant factors and interactions between factors and predict the response of the sub-functions or sub-formulations to changes in the input variables.
(D) Surrogate modeling software: Surrogate modeling software such as Surrogate Model Toolbox, DACE, and MOE can be used to construct surrogate models of the sub-functions or sub-formulations in FbD optimization. These tools can

help to reduce the computational cost of the optimization by replacing the expensive sub-functions or sub-formulations with a cheaper surrogate model.

(E) Expert systems software: Expert systems software such as Expert System Shell, Drools, and CLIPS can be used to create and deploy expert systems in FbD optimization. These tools can help to incorporate domain knowledge into the optimization process and make intelligent decisions based on the desired properties of the product.

(F) Machine learning software: Machine learning software such as TensorFlow, PyTorch, and scikit-learn can be used to build predictive models of the sub-functions or sub-formulations in formulation-based design (FbD) optimization. These tools can help to identify significant factors and interactions between factors and predict the response of the sub-functions or sub-formulations to change in the input variables.

It is important to choose the appropriate tools for formulation-based design (FbD) optimization based on the problem at hand and the available resources. The use of these tools can help to accelerate the optimization process and achieve optimal results.

3.8 FbD Opti-tactics for Drug Delivery Systems

Liposomes are spherical structures composed of a double-layered lipid membrane that closely resembles natural biological membranes. They are formed by using phospholipids with mixed lipid chains (such as egg PE) or other surfactants via a process known as sonication. Liposomes are being widely investigated for their potential as drug delivery systems for a variety of diseases. To develop liposomes suitable for use, it is crucial to optimize their *in vivo* kinetics, ensure high-throughput production conditions, and design specific liposome compositions. One promising approach for identifying appropriate liposome compositions is to use high-throughput screening methodologies and apply the principles of quality by design (QbD) to the vital elements (factors) of liposome production. In terms of drug development, there are two main ways to achieve selectivity for the drug: discovering new compounds with high selectivity and innovating novel drug administration methods with high selectivity through integration and harmonization of various techniques. Takahara et al. (1997) proposed the "3H" formula for innovative drug administration, which involves combining hybrid, high-quality, and husbandry approaches. In addition, an overview of the optimization studies conducted on various topical mucosal adhesive dosage forms and the parenteral administration of peptide drugs, including insulin and erythropoietin, is also discussed. In 1988, Gregoriadis reported on the initial optimization studies for liposomes.

3.9 Available Software Employed for QbD Optimization

There are several software tools available for quality by design (QbD) optimization, including the following:

3.9.1 Design-Expert

A statistical software package that allows users to design experiments and analyze data. It provides a graphical interface to visualize results and can be used for optimization of process parameters.

3.9.2 SIMCA

It is a multivariate data analysis software that is commonly used for process modeling, monitoring, and optimization. It can be used to analyze large data sets and identify key process variables that impact quality.

3.9.3 Minitab

A statistical software package that provides a wide range of statistical tools for data analysis, design of experiments, and quality control. It also provides graphical tools for visualization and communication of results.

3.9.4 JMP

A data analysis and visualization software that allows users to explore and model data, design experiments, and optimize processes. It provides a user-friendly interface and can be used for both statistical analysis and machine learning.

3.9.5 MATLAB

A programming language and software environment that is widely used for data analysis, modeling, and simulation. It provides a wide range of tools for optimization, including linear programming, nonlinear programming, and genetic algorithms.

3.9.6 Aspen Plus

A process simulation software that is commonly used in the chemical and process industries. It can be used to simulate and optimize chemical processes, including process parameters, equipment design, and product quality.

3.9.7 AutoCAD

A computer-aided design software that can be used for designing and optimizing processes and equipment. It provides a graphical interface to create 2D and 3D designs and can be used to simulate and optimize process flow and equipment layout.

3.10 Applications of QbD and FbD in Different Fields

Quality by design (QbD) and formulation by design (FbD) are systematic approaches to product and process development that are widely applicable in various fields. Here are some examples of their applications:

3.10.1 Pharmaceutical Industry

QbD and FbD have become an essential tool for product development in various scientific fields. It is widely used for analytical method development utilizing different techniques such as HPLC, ultrahigh-performance liquid chromatography, capillary electrophoresis, mass spectrometry, and near-infrared spectroscopy. QbD and FbD are extensively used in the pharmaceutical industry to ensure the quality and consistency of drug products. By designing the product and process based on scientific principles and risk assessment, companies can minimize the variability in drug quality and increase process efficiency. QbD and FbD also help to optimize drug formulation, enhance drug delivery, and improve the safety profile of the drug. In the domain of pharmaceutical formulation development, QbD is employed to optimize the processes of various formulations, including sterile manufacturing, solid dosage forms, modified-release products, gel manufacturing, tableting processes, ANDAs, and API and excipient analysis (Hasnain et al. 2016; Wen and Park 2011; Lionberger et al. 2008; Karmarkar et al. 2011; Orlandini et al. 2014; Musters et al. 2013; Basalious et al. 2011).

Moreover, biopharmaceutical applications of QbD include protein manufacturing, purification of proteins using different chromatographic techniques, and production and characterization of monoclonal antibodies. QbD can also be applied in the fields of genetics and clinical applications (Rouiller et al. 2012; Finkler and Krummen 2016).

Pharmaceutical unit operations and dosage forms can be optimized using the QbD concept, which involves input material, process attributes, and quality

Table 3.1 Applications of QbD in the manufacturing of different dosage forms by using different pharmaceutical unit operations

S. No.	Dosage form	Pharmaceutical unit operations	Model drug	Design of experiment (DoE)	Critical material attributes (CMA)	Critical process parameters (CPP)	Critical quality attributes (CQA)	References
1.	Solid nanocrystalline dry powder	Spray drying	Indomethacin	Full factorial design	Not mentioned	Inlet temperature, flow rate, and aspiration rate	Particle size, moisture content, percent yield, and crystallinity	Kumar et al. (2014)
2.	Controlled-release tablets	Physical mixture, solvent evaporation	Felodipine	Box–Behnken design	Amount of polymer HPMC, amount of polymeric surfactants, amount of Pluronic F127	Preparation technique	Maximum solubility after 30 min, equilibrium solubility after 24 h, dissolution efficiency	Basalious et al. (2011)
3.	Tablets	Roller compaction	Not mentioned	Fractional factorial statistical design	API composition, API excipient ratio	API flow rate, lubricant flow rate, pre-compression pressure	Tablet weight, tablet dissolution, hardness, ribbon density	Singh et al. (2012)

(continued)

Table 3.1 (continued)

S. No.	Dosage form	Pharmaceutical unit operations	Model drug	Design of experiment (DoE)	Critical material attributes (CMA)	Critical process parameters (CPP)	Critical quality attributes (CQA)	References
4.	Tablets	Fluid bed granulation	Not mentioned	Fractional factorial design (screening) central composite design (optimization)	Viscosity, temperature, and concentration of the binder aqueous dispersion	Inlet air temperature, binder spray rate, and air flow rate	Particle size distribution (PSD), bulk and tapped densities, flowability, and angle of repose	Lourenço et al. (2012)
5.	Coated tablets	Film coating	Placebo tablets	Central composite—face centered—response surface design	Solid percent of the coating dispersion	Inlet air temperature, air flow rate, solid level, coating pan speed, spray rate	Appearance (coating defects, gloss, and color uniformity), disintegration time (dissolution of the film coating)	Teckoe et al. (2013)
6.	Orodispersible films	Homogenate membrane method	Theophylline	Central composite design	Percentage of HPMC, percentage of glycerol	Drying temperature	Tensile strength, elongation at break, Young's modulus, disintegration time	Zhang and Mao (2017)

S. No.	Dosage form	Pharmaceutical unit operations	Model drug	Design of experiment (DoE)	Critical material attributes (CMA)	Critical process parameters (CPP)	Critical quality attributes (CQA)	References
7.	Solid lipid nanoparticles (SLN)	Hot-melt extrusion (HME)	Fenofibrate (FBT)	Plackett–Burman (PB) screening design	Lipid concentration, surfactant concentration	Screw speed, barrel temperature, zone of liquid addition	Particle size, polydispersibility index, zeta potential, entrapment efficiency	Patil et al. (2015)
8.	Nanoparticles	Homogenization	Paclitaxel	Box–Behnken design	Surfactant concentration in aqueous phase (%)	Homogenization rate	Average particle size, zeta potential, encapsulation efficiency	Yerlikaya et al. (2013)
9.	Solid lipid nanoparticle (SLN)	Homogenization	Rivastigmine	Factorial design	Drug/lipid ratio, surfactant concentration	Homogenization time	Size, PDI, entrapment efficiency	Shah et al. (2015)
10.	Nanoparticles	O/W emulsification–solvent evaporation	Cyclosporine A (CyA)	Plackett–Burman (PB) design	Type of solvent organic-to-aqueous phase ratio, drug concentration, polymer concentration, surfactant concentration, O/W ratio	Stirring rate	Encapsulation efficiency, particle size, zeta potential, burst release, and dissolution efficiency	Rahman et al. (2010)

attributes such as during fluid bed granulation, roller compaction, film coating, spray drying, hot-melt extrusion, and homogenization. These aspects are summarized in Table 1.2 according to Zhang and Mao (2017).

The use of QbD in nanopharmaceuticals has grown rapidly since 2007, with numerous papers published in peer-reviewed journals discussing its applications for process optimization and nanoparticle product development. Javed et al. (2019) reported that QbD has been used to develop various nanoparticles for drug delivery, with most of the applications originating from Asia (particularly India, 48%) and the United States (33%). Europe accounted for only 11% of the applications, while Africa accounted for 9%. These QbD applications have focused on formulated-related issues (62%), formulation and manufacturing (29%), and identifying critical factors for the development process (9%). These findings are based on studies conducted by Bastogne (2017) and Beg et al. (2018). The applications of QbD in the manufacturing of different dosage forms by using different pharmaceutical unit operations are discussed in Table 3.1.

3.10.2 Food Industry

QbD and FbD are also applied in the food industry to ensure the quality and safety of food products. By designing the process based on scientific principles and risk assessment, companies can minimize the variability in food quality and increase process efficiency. QbD and FbD also help to optimize food formulation; enhance food flavor, texture, and nutritional value; and improve food safety.

3.10.3 Chemical Industry

QbD and FbD are used in the chemical industry to optimize the production of chemicals and materials. By designing the process based on scientific principles and risk assessment, companies can minimize the variability in product quality and increase process efficiency. QbD and FbD also help to optimize product formulation, enhance product performance, and improve the environmental profile of the product.

3.10.4 Biotechnology Industry

QbD and FbD are applied in the biotechnology industry to optimize the production of biologics and biosimilars. By designing the process based on scientific principles and risk assessment, companies can minimize the variability in product quality and increase process efficiency. QbD and FbD also help to optimize product formulation, enhance product performance, and improve the safety profile of the product. It is important to note that implementing QbD in biotechnological product development can help to minimize failures in the final product. The use of different tools for protein characterization is also vital in this approach. Obtaining information related to the composition of amino acid sequence variants, secondary or tertiary structure, and posttranslational modifications is essential in biotechnological product development. Other important factors that contribute significantly to the biological properties of the product include cell line screening, designing expression systems, and process development. By applying QbD tools at the development stage, biotechnological products can be optimized for their intended use, ensuring their safety, quality, and efficacy.

QbD and PAT are becoming increasingly important in the biotechnological industry as they help to minimize failures in the final product, improve process performance, and reduce costs. Many companies such as Sandoz Pharmaceuticals and Boehringer Ingelheim Pharma are implementing these concepts in their biotech product development and process control. By using multivariate analysis tools and spectroscopic tools, they can improve the quality of their products and reduce regulatory burden. Overall, QbD and PAT are important tools that can help the biotechnological industry to develop high-quality products more efficiently and cost-effectively.

3.10.5 QbD in Analytical Method Development

In the analytical QbD approach, critical method attributes (CMA) and critical method variables (CMV) are identified and varied systematically to optimize the method. This helps in achieving the desired performance of the analytical method and also helps in reducing the variability of the method. The use of a quality by design approach in analytical method development can lead to a more efficient and effective method, which can reduce the overall cost and time required for method development and validation (Garg et al. 2015). Moreover, it also helps in meeting the regulatory requirements and improving the quality of the final product. Hence, the analytical QbD approach has become an integral part of the analytical method development process in the pharmaceutical industry. The ICH Q2 (R1) guidelines provide the criteria for the method validation parameters such as linearity, accuracy, precision, and repeatability. But these robustness parameters are analyzed at the final stage of the method development stage, during method validation, and this eventually sometimes leads to undesired, inaccurate results. This necessitates entire method development and validation repetition.

The ICH Q8 (R2) guidelines give space to method development and method validation in a systemic way which will help in developing the rugged and robust analytical method. Analytical QbD (AQbD) involves designing of analytical target profile (ATP) and optimizing the design space for critical performance attributes by varying critical method attributes and critical method variables. Nowadays, the analytical QbD approach is widely used in method development specifically in HPLC to increase robustness and ruggedness (Thakur et al. 2017).

By applying QbD principles in analytical method development, the quality of analytical results can be improved while reducing the overall cost and time required for method development and validation. This approach also helps to identify and control the critical parameters that affect the performance of the analytical method, leading to increased consistency and reliability of the results. Various applications of QbD in the manufacturing of different dosage forms by using different pharmaceutical unit operations are summarized in Table 3.1.

3.11 Conclusion

Quality by design (QbD) is a systematic approach that emphasizes product and process understanding, based on sound science and quality risk management. The use of QbD in the pharmaceutical industry has the potential to enhance performance, offer regulatory relief and adaptability, and bring about significant financial gains throughout the product life cycle. Formulation by design (FbD) is a key part of QbD and is being extensively investigated by budding scientists for better know-how of the product and process development for an unequivocal universal acceptance. QbD and FbD techniques are being increasingly used in the development of pharmaceutical products, biotechnological products, and new chemical entities, leading to the filing of various new drug applications and abbreviated new drug applications. The QbD approach provides a greater understanding of the process and the result, which helps in the development of higher-quality products. However, the implementation of QbD and FbD approaches faces several challenges that need to be addressed. Overall, the emergence of QbD and FbD as quality tools, their applications in different fields, and various regulatory guidelines have revolutionized the pharmaceutical industry and improved the safety and efficacy of finished products.

Acknowledgments We are highly thankful for financial support obtained from ICMR-SRF to Miss Shivani Saraf, Mr. Pritish K. Panda, Mrs. Pooja Das Bidla, and Miss Sarjana Raikwar. Authors report no potential conflict of interest.

Declaration of Competing Interests The authors declare no competing interests.

References

Adam S, Suzzi D, Radeke C, Khinast JG (2011) An integrated quality by design (QbD) approach towards design space definition of a blending unit operation by discrete element method (DEM) simulation. Eur J Pharm Sci 42(1–2):106–115

Araujo P, Brereton R (1996) Experiment design: I. Screening. Anal Chem 15:63–70

Åsberg D, Nilsson M, Olsson S, Samuelsson J, Svensson O, Klick S et al (2016) A quality control method enhancement concept—continual improvement of regulatory approved QC methods. J Pharm Biomed Anal 129:273–281

Basalious EB, El-Sebaie W, El-Gazayerly O (2011) Application of pharmaceutical QbD for enhancement of the solubility and dissolution of a class II BCS drug using polymeric surfactants and crystallization inhibitors: development of controlled-release tablets. AAPS PharmSciTech 12:799–810

Bastogne T (2017) Quality-by-design of nanopharmaceuticals–a state of the art. Nanomedicine 13(7):2151–2157

Beg S, Saini S, Bandopadhyay S, Katare OP, Singh B (2018) QbD-driven development and evaluation of nanostructured lipid carriers (NLCs) of Olmesartan medoxomil employing multivariate statistical techniques. Drug development and industrial pharmacy. 44(3):407–20.

Bhoop BS (2014) Quality by design (QbD) for holistic pharma excellence and regulatory compliance. Pharm Times 46(8):26–33

Borgonovo E, Smith CL (2011) A study of interactions in the risk assessment of complex engineering systems: an application to space PSA. Oper Res 59(6):1461–1476

Box GE, Behnken DW (1960) Some new three level designs for the study of quantitative variables. Technometrics 2(4):455–475

Box GE, Wilson KB (1992) On the experimental attainment of optimum conditions. Breakthroughs in statistics: methodology and distribution. Springer, pp 270–310

Chatterjee S, Moore CM, Nasr MM (2017) An overview of the role of mathematical models in implementation of quality by design paradigm for drug development and manufacture. In: Reklaitis GV, Seymour C, García-Munoz S (eds) Comprehensive quality by design for pharmaceutical product development and manufacture. American Institute of Chemical Engineers, Inc., pp 9–24

Cunha S, Costa CP, Moreira JN, Lobo JMS, Silva AC (2020) Using the quality by design (QbD) approach to optimize formulations of lipid nanoparticles and nanoemulsions: a review. Nanomedicine 28:102206

Doornbos D, Haan P (1995) Optimization techniques in formulation and processing. Encyclopedia Pharm Technol 11:75–160

Finkler C, Krummen L (2016) Introduction to the application of QbD principles for the development of monoclonal antibodies. Biologicals 44(5):282–290

Garg NK, Sharma G, Singh B, Nirbhavane P, Katare OP (2015) Quality by design (QbD)-based development and optimization of a simple, robust RP-HPLC method for the estimation of methotrexate. J Liq Chromatogr Relat Technol 38(17):1629–1637

Ghaemmaghamian Z, Zarghami R, Walker G, O'Reilly E, Ziaee A (2022) Stabilizing vaccines via drying: quality by design considerations. Adv Drug Deliv Rev 187:114313

Hasnain MS, Siddiqui S, Rao S, Mohanty P, Jahan Ara T, Beg S (2016) QbD-driven development and validation of a bioanalytical LC–MS method for quantification of fluoxetine in human plasma. J Chromatogr Sci 54(5):736–743

Jagtap K, Chaudhari B, Redasani V (2022) Quality by design (QbD) concept review in pharmaceuticals. Asian J Res Chem 15:303

Javed MN, Alam MS, Waziri A, Pottoo FH, Yadav AK, Hasnain MS, Almalki FA (2019) QbD applications for the development of nanopharmaceutical products. In Pharmaceutical quality by design 229–253. Academic Press.

Jain S (2014) Quality by design (QBD): a comprehensive understanding of implementation and challenges in pharmaceuticals development. Int J Pharm Pharm Sci 6:29–35

Jiang M, Severson KA, Love JC, Madden H, Swann P, Zang L et al (2017) Opportunities and challenges of real-time release testing in biopharmaceutical manufacturing. Biotechnol Bioeng 114(11):2445–2456

Kalyane D, Raval N, Polaka S, Tekade RK (2020) Quality by design as an emerging concept in the development of pharmaceuticals. The future of pharmaceutical product development and research. Elsevier, pp 1–25

Karmarkar S, Garber R, Genchanok Y, George S, Yang X, Hammond R (2011) Quality by design (QbD) based development of a stability indicating HPLC method for drug and impurities. J Chromatogr Sci 49(6):439–446

Kumar S, Gokhale R, Burgess DJ (2014) Quality by design approach to spray drying processing of crystalline nanosuspensions. Int J Pharm 464(1–2):234–242

Lachman L, Lieberman HA, Kanig JL (1976) The theory and practice of industrial pharmacy. Lea & Febiger Philadelphia

Lionberger RA, Lee SL, Lee L, Raw A, Yu LX (2008) Quality by design: concepts for ANDAs. AAPS J 10:268–276

Lipol LS, Haq J (2011) Risk analysis method: FMEA/FMECA in the organizations. Int J Basic Appl Sci 11(5):74–82

Loukas YL (2001) A Plackett–Burnam screening design directs the efficient formulation of multi-component DRV liposomes. J Pharm Biomed Anal 26(2):255–263

Lourenço V, Lochmann D, Reich G, Menezes JC, Herdling T, Schewitz J (2012) A quality by design study applied to an industrial pharmaceutical fluid bed granulation. Eur J Pharm Biopharm 81(2):438–447

Luiz MT, Viegas JSR, Abriata JP, Viegas F, de Carvalho Vicentini FTM, Bentley MVLB et al (2021) Design of experiments (DoE) to develop and to optimize nanoparticles as drug delivery systems. Eur J Pharm Biopharm 165:127–148

Mishra V, Thakur S, Patil A, Shukla A (2018) Quality by design (QbD) approaches in current pharmaceutical set-up. Expert Opin Drug Deliv 15(8):737–758

Moore K, Nicholson M (2005) The move to performance based standards-managing. Uncertainty University of York Safety Critical

Musters J, van den Bos L, Kellenbach E (2013) Applying QbD principles to develop a generic UHPLC method which facilitates continual improvement and innovation throughout the product lifecycle for a commercial API. Org Process Res Dev 17(1):87–96

Namjoshi S, Dabbaghi M, Roberts MS, Grice JE, Mohammed Y (2020) Quality by design: development of the quality target product profile (QTPP) for semisolid topical products. Pharmaceutics 12(3):287

Orlandini S, Pasquini B, Stocchero M, Pinzauti S, Furlanetto S (2014) An integrated quality by design and mixture-process variable approach in the development of a capillary electrophoresis method for the analysis of almotriptan and its impurities. J Chromatogr A 1339:200–209

Patil AS, Pethe AM (2013) Quality by design (QbD): a new concept for development of quality pharmaceuticals. Int J Pharm Quality Assur 4(2):13–19

Patil H, Feng X, Ye X, Majumdar S, Repka MA (2015) Continuous production of fenofibrate solid lipid nanoparticles by hot-melt extrusion technology: a systematic study based on a quality by design approach. AAPS J 17:194–205

Pawar P, Wang Y, Keyvan G, Callegari G, Cuitino A, Muzzio F (2016) Enabling real time release testing by NIR prediction of dissolution of tablets made by continuous direct compression (CDC). Int J Pharm 512(1):96–107

Plackett RL, Burman JP (1946) The design of optimum multifactorial experiments. Biometrika 33(4):305–325

Rahman Z, Zidan AS, Habib MJ, Khan MA (2010) Understanding the quality of protein loaded PLGA nanoparticles variability by Plackett–Burman design. Int J Pharm 389(1–2):186–194

Rao VS, Srinivas K (2011) Modern drug discovery process: an in silico approach. J Bioinform Seq Anal 2(5):89–94

Rouiller Y, Solacroup T, Deparis V, Barbafieri M, Gleixner R, Broly H et al (2012) Application of Quality by Design to the characterization of the cell culture process of an Fc-Fusion protein. Eur J Pharm Biopharm 81(2):426–437

Sangshetti JN, Deshpande M, Zaheer Z, Shinde DB, Arote R (2017) Quality by design approach: regulatory need. Arab J Chem 10:S3412–S3S25

Shah B, Khunt D, Bhatt H, Misra M, Padh H (2015) Application of quality by design approach for intranasal delivery of rivastigmine loaded solid lipid nanoparticles: effect on formulation and characterization parameters. Eur J Pharm Sci 78:54–66

Singh B, Ahuja N (2000) Pharmaceutical experimental design (Drugs and the pharmaceutical sciences, Vol. 92), Edited by GA Lewis, D. Mathieu and R. Phan-Tan-Luu, Marcel Dekker, New York, 1999. vi+ 498 pp., 23.5 x15. 5 cm., 1.9 lb., Hardcover, ISBN 0-8247-9860-0, Price $175.00. Int J Pharm 1(195):247–248

Singh B, Dahiya M, Saharan V, Ahuja N (2005a) Optimizing drug delivery systems using systematic "design of experiments." Part II: retrospect and prospects. Crit Rev Ther Drug Carrier Syst 22(3):215

Singh B, Kumar R, Ahuja N (2005b) Optimizing drug delivery systems using systematic "design of experiments." Part I: fundamental aspects. Crit Rev Ther Drug Carrier Syst 22(1):27

Singh B, Gupta R, Ahuja N (2006) Computer-assisted optimization of pharmaceutical formulations and processes. In: Jain NK (ed) Pharmaceutical product development. CBS Publishers, New Delhi, pp 273–318

Singh B, Bhatowa R, Tripathi CB, Kapil R (2011) Developing micro-/nanoparticulate drug delivery systems using "design of experiments". Int J Pharm Invest 1(2):75

Singh R, Ierapetritou M, Ramachandran R (2012) An engineering study on the enhanced control and operation of continuous manufacturing of pharmaceutical tablets via roller compaction. Int J Pharm 438(1–2):307–326

Singh B, Saini S, Lohan S, Beg S (2017) Systematic development of nanocarriers employing quality by design paradigms. Elsevier, Nanotechnology-based approaches for targeting and delivery of drugs and genes, pp 110–148

Singh B, Sharma T, Saini S, Kaur R, Jain A, Raza K et al (2020) Systematic development of drug nanocargos using formulation by design (FbD): an updated overview. Crit Rev Ther Drug Carrier Syst 37(3):229

Takahara J, Takayama K, Nagai T (1997) Multi-objective simultaneous optimization technique based on an artificial neural network in sustained release formulations. J control release 49(1):11–20.

Teckoe J, Mascaro T, Farrell TP, Rajabi-Siahboomi AR (2013) Process optimization of a novel immediate release film coating system using QbD principles. AAPS PharmSciTech 14(2):531–540

Thakur D, Kaur A, Sharma S (2017) Application of QbD based approach in method development of RP-HPLC for simultaneous estimation of antidiabetic drugs in pharmaceutical dosage form. J Pharm Investig 47(3):229–239

Wang S, Sun Y, He A, Zheng C, Zheng X (2014) Predictive value of phosphorylated mammalian target of rapamycin for disease-free survival in breast cancer patients receiving neoadjuvant chemotherapy. Oncol Lett 8(6):2642–2648

Wen H, Park K (2011) Oral controlled release formulation design and drug delivery: theory to practice. Wiley

Yerlikaya F, Ozgen A, Vural I, Guven O, Karaagaoglu E, Khan MA et al (2013) Development and evaluation of paclitaxel nanoparticles using a quality-by-design approach. J Pharm Sci 102(10):3748–3761

Zhang L, Mao S (2017) Application of quality by design in the current drug development. Asian J Pharm Sci 12(1):1–8

Pharmaceutical Product Development: Formulation Additives

4

Meenakshi Patel, Dhruvi Patel, Hitarthi Mayur Patel, and Lalit Lata Jha

Abstract

This chapter offers a complete assessment of the key role played by formulation additives in the development of pharmaceutical dosage forms. Beginning with an insightful introduction to formulation additives, the chapter emphasises on their significance in achieving desired product characteristics. A systematic breakdown follows, categorising formulation additives based on the types of dosage forms and providing detailed insights into the additives used in each category, shedding light on their functions and applications. The importance of understanding drug-additive interactions during formulation development has also been discussed, covering physical, chemical and therapeutic incompatibilities, along with analytical techniques to characterise such interactions. Recent advances in additive science, such as functional and co-processed additives and novel materials, are also explored along with international patented excipients. Additionally, the chapter provides insights into related regulatory perspectives, including generally recognised as safe (GRAS), international impurity guidelines (IIG) and International Pharmaceutical Excipients Council (IPEC), to ensure compliance and safety in pharmaceutical formulations. This chapter serves as a necessary source for those engaged in pharmaceutical product devel-

M. Patel (✉)
Department of Pharmaceutics, School of Pharmacy, Faculty of Pharmacy, Parul University, Vadodara, Gujarat, India

D. Patel
Department of Pharmaceutics, Parul Institute of Pharmacy & Research, Faculty of Pharmacy, Parul University, Vadodara, Gujarat, India

H. M. Patel · L. L. Jha
Department of Pharmaceutics, School of Pharmacy, Faculty of Pharmacy, Parul University, Vadodara, Gujarat, India

K. Jain, A. K. Yadav (eds.), *Advances in Pharmaceutical Product Development*, https://doi.org/10.1007/978-981-97-9230-6_4

opment, providing a comprehensive and up-to-date perspective on formulation additives.

Keywords

Formulation additives · Excipients · Drug-excipient compatibility · Co-processed additives

4.1 Introduction to Formulation Additives

Formulation additives in pharmaceuticals refer to the substances that are added during the manufacturing process to enhance the physical properties, stability or performance of the final product. They are pharmacologically inactive organic or inorganic substances obtained from natural, semisynthetic or synthetic source. The selection and quantity of additives incorporated into the formulation depends on the characteristics of drug, type of formulation and route of administration. The additives chosen for the preparation of any kind of dosage form affect the safety, efficacy and pharmacokinetics of the drug and also decide the formulation aspects of the whole formulation development process.

Additives like binders, fillers, disintegrants and lubricants assist in the manufacturing of tablets and capsules, ensuring their physical integrity and facilitating proper dissolution in the body. Similarly, solubilisers, surfactants and complexing agents aid in enhancing the solubility of poorly soluble drugs, thereby improving their absorption and therapeutic efficacy. Additionally, preservatives, antioxidants and stabilisers play a crucial role in preventing degradation and maintaining the potency of medications throughout their shelf life. The careful selection and precise integration of these formulation additives are critical in ensuring the safety, efficacy and quality of pharmaceutical products that ultimately benefit patients worldwide.

Initially considered as inert substances, additives were ignored in safety evaluations due to the belief that they remained inactive within formulations. However, this assumption has changed with time as research has unveiled their active roles and deep impact on the safety and efficacy of medicinal substances. Additives can exhibit physical, chemical and biopharmaceutical interaction with drugs. The various examples in the past have proved that overlooking the safety evaluation of these additives could compromise the overall safety profile of medications. The devastating 1937 sulphanilamide incident, which led to the tragic deaths of numerous children, happened due to the presence of diethylene glycol as solvent in the formulation (Ballentine 1981). This incidence urged the enactment of the Federal Food, Drug, and Cosmetic Act of 1938, leading to transformation in regulatory frameworks. It highlighted the necessity of strict testing and evaluation, recognising that these substances could interact with the body, the drug itself or other additives, ultimately precipitating the toxicity and diminishing the drug efficacy. This chapter shows the significance of formulation additives, explaining their impact on product development, from enhancing bioavailability to ensuring optimum drug delivery systems including the regulatory aspects.

4.2 Importance of Additives in Development of Pharmaceutical Dosage Form

Additives are considered as essential constituents without whom the preparation of pharmaceutical dosage forms would be virtually impossible. Their importance ranges from improving the physical appearance to controlling the release profile of drug from the formulation. Rather than just being an inactive support, additives now play a more important role as they also aid in delivering, safeguarding, stabilising, noticing, elevating and improving patient compliance as well as elegance. Not only the additive but their variability in terms of grade also has a deep impact on the functionality of dosage form (Dave et al. 2015; Zarmpi et al. 2017). "Functionality" refers to the necessary activity of the specified additive, which in other words is an intrinsic quality of the additive in the pharmaceutical formulation (Narang et al. 2017). This functionality is determined by the technique, type and rate of incorporation of the additive in the dosage form.

Additives are added in the formulation according to their properties, to preserve physiochemical property of drug substance and dosage form. Figure 4.1 shows the ideal properties of additives used in the development of pharmaceutical formulation. The importance of ideal properties in formulation additives lies in their ability to optimise drug delivery, ensure product quality and facilitate efficient manufacturing while prioritising patient safety and therapeutic efficacy.

Studies have proved that choosing a right type of additive is essential for ensuring the stability of the formulation (Veronica et al. 2022). One of the primary roles of additives lies in their ability to provide structural integrity and stability to dosage forms. These components, ranging from binders, fillers and disintegrants to preservatives and stabilisers, form the backbone of formulations, ensuring the stability, cohesiveness and robustness of tablets and capsules. Beyond the physical

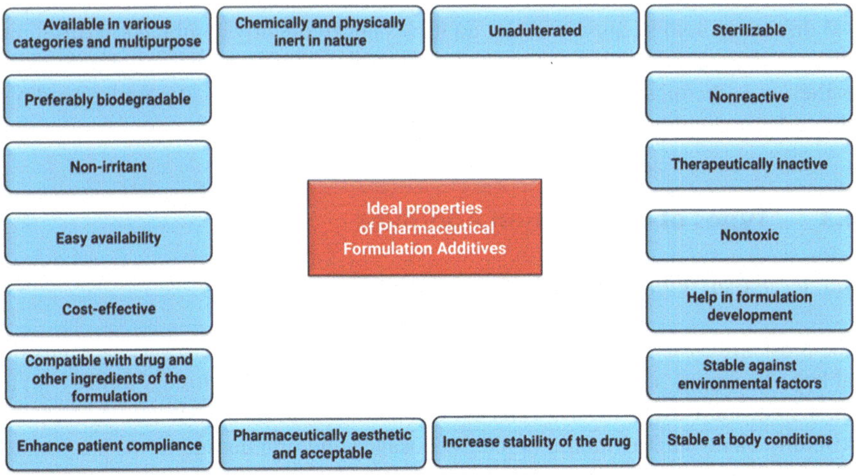

Fig. 4.1 Ideal properties of pharmaceutical formulation additives

construction, additives contribute significantly to the functionality and performance of pharmaceutical formulations (Bodratti and Alexandridis 2018). Additives such as solubilisers, surfactants and complexing agents play a significant role in enhancing the solubility of poorly soluble drugs, thereby improving their absorption and therapeutic efficacy (Vadlamudi and Dhanaraj 2017). They influence the release kinetics, ensuring controlled and sustained delivery of medications to achieve desired therapeutic outcomes while minimising adverse effects (Rosiaux et al. 2014). This indicates that additives not only influence the physical appearance but also affect the dissolution rates, bioavailability and pharmacokinetics of active pharmaceutical ingredients (API) (Li et al. 2023). They may be added in the formulation to improve the drug solubility and permeability consequently increasing the bioavailability (Panakanti and Narang 2015). Additionally, these additives might be introduced to prolong shelf life, maintain consistent product quality or streamline the manufacturing process itself. Additionally, these additives are also necessary to guard against microbial contamination and formulation deterioration (Kusuma et al. 2020). Furthermore, additives contribute to the enhancement of patient compliance by influencing various attributes of the formulation such as taste, appearance and ease of administration (Adamkiewicz and Szeleszczuk 2023; Malaquias et al. 2018). Masking unpleasant tastes, improving palatability and ensuring good visual appearance of the formulation ensure the patient acceptability and adherence to prescribed regimens, hence achieving better treatment of the disease or disorder.

While using the additives in the preparation of dosage form, a thorough understanding of additive science is required to prevent any formulation errors. The additive used in the formulation needs to be compatible both chemically and physically with the active pharmaceutical ingredient and the other ingredients, including the packing material (Abrantes et al. 2016). Therefore, evaluation of drug and additives is required, along with choosing the concurrent use and potential contradiction. Additionally, conventional additives need to be innovated in additives are required to fit in them as per the use of the formulation. There is a gap in the additive design and discovery field because existing additives are frequently combined in different ratios to change their action. They can be categorised based on various criteria such as their use, chemical composition, route of administration, etc., for understanding their roles and applications in drug formulations (Fig. 4.2).

4.3 Types of Formulation Additives

4.3.1 Additives for Oral Solid Dosage Forms

The most widely used types of oral solid medication products are various kinds of tablets and capsules. Generally, capsules and tablets are solid dosage forms with a fairly long shelf life that release the API immediately if its release rate is not purposefully changed. The modified release dosage solid oral dosage forms are generally prepared to sustain the release of drug. In order to attain desired dosage form parameters like drug content, hardness or crushing strength, disintegration time,

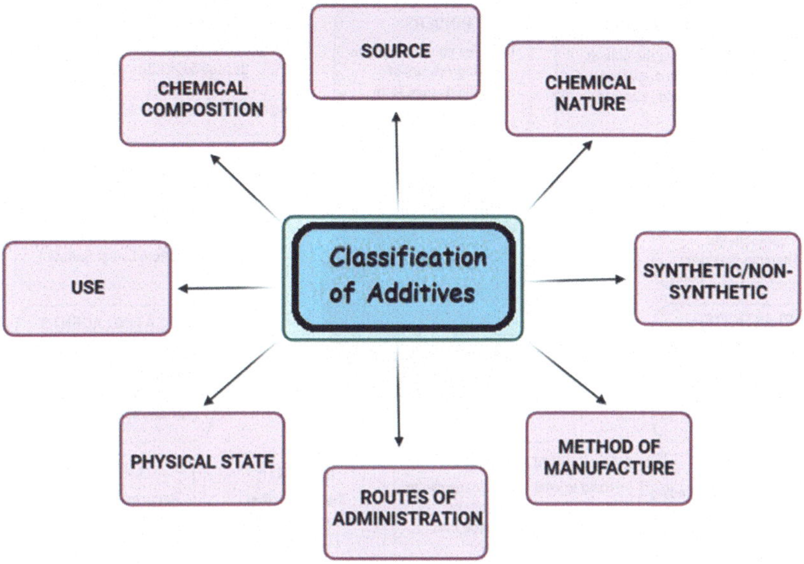

Fig. 4.2 Various criteria for classifying additives

friability, tensile strength and dissolution time, polymeric additives are combined with an API and processed using various methods. Primarily, they aid in ensuring the stability and uniformity of the dosage form throughout its shelf life, preventing degradation and maintaining the drug's efficacy. Additives also contribute to enhancing the physical characteristics of the formulation, such as improving flow properties, compressibility and dissolution rates. Additionally, they assist in masking unpleasant tastes, controlling the release profile of the drug and facilitating easier manufacturing processes. Moreover, certain additives act as fillers, binders, disintegrants or lubricants, optimising the formulation's performance and ensuring consistent drug delivery to the patient (Fig. 4.3). The strategic incorporation of additives is essential to guarantee the quality, safety and effectiveness of oral solid dosage forms.

Various additives utilised in solid dosage form are as follows:

4.3.1.1 Fillers/Diluents

These additives add bulk to the formulation, facilitating the handling and manufacturing processes when the dose is small. Examples include lactose, mannitol, microcrystalline cellulose, calcium phosphates, sorbitol, sucrose, hydrolysed starch, immediately compressible starch and various cellulose derivatives.

4.3.1.2 Binders

Added at a certain point in wet granulation, these can be liquid or dry powders that help to form granules or give cohesive force between particles during direct compression for mechanical strength. Binders such as starch, cellulose derivatives (like

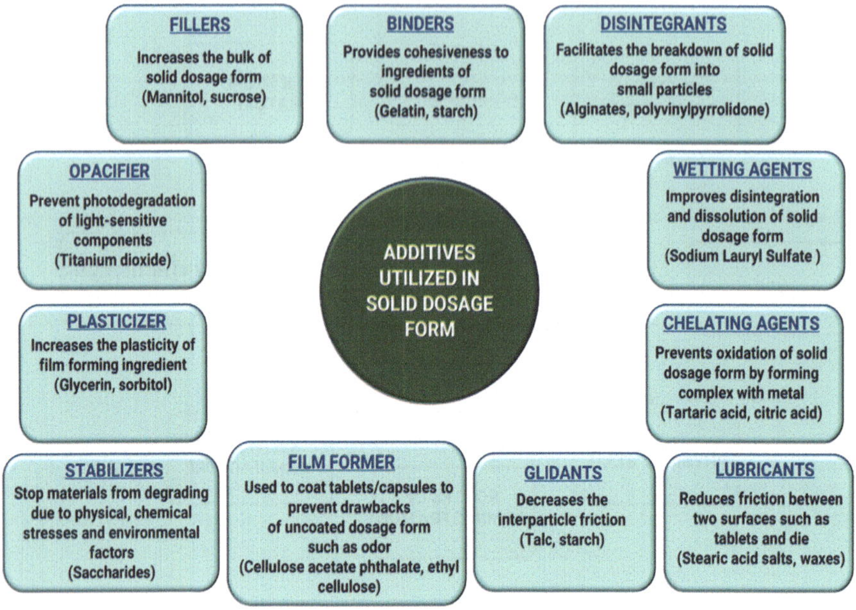

Fig. 4.3 Types of additives used in solid dosage forms

methylcellulose or hydroxypropyl cellulose) and polyvinylpyrrolidone (PVP) are used to hold the active ingredients and other additives together, forming a cohesive mass.

4.3.1.3 Disintegrants

Disintegrants are mixed into the formulation to help break down the dosage form in the gastrointestinal tract, which increases the surface area of particles, aiding in drug dissolution and absorption. Examples include cross-linked sodium carboxymethylcellulose, cellulose, alginates, polyvinylpyrrolidone, starch, starch derivatives, clay, etc. Substances like croscarmellose sodium, crospovidone, sodium starch glycolate and modified cellulose act as superdisintegrants which are used to prepare the fast dispersible oral formulations (Indurkhya et al. 2023).

4.3.1.4 Glidants and Lubricants

Glidants (like colloidal silicon dioxide, Syloid® 244, talc and starch) improve powder flowability (Parekh et al. 2023), while lubricants (such as magnesium stearate or stearic acid) prevent tablet adhesion to the manufacturing equipment and aid in tablet ejection.

4.3.1.5 Coating Agents

Coating agents like cellulose derivatives (cellulose acetate phthalate and ethyl cellulose), shellac or polymethacrylates are used for film coating or enteric coating to modify drug release, protect the drug from degradation or mask taste.

4.3.1.6 Sweeteners, Flavourings and Colouring Agents

These additives improve the palatability and appearance of the dosage form, enhancing patient compliance. Titanium dioxide which is white in colour is employed as an opacifier to create an opaque shell which can prevent photodegradation of light-sensitive components present in the formulation (Blundell et al. 2022).

4.3.1.7 Preservatives and Antioxidants

Preservatives like parabens and antioxidants such as ascorbic acid or tocopherols are used to prolong the shelf life and stability of the dosage form by preventing microbial growth or drug degradation. These additives are essential for transforming an unstable medicine or formulation into one that is well tolerated. For instance, saccharides help to maintain the formulation of lyophilised proteins.

4.3.1.8 Solubilisers

They are not commonly used in solid dosage forms, but certain solubilising agents are incorporated to improve drug solubility and bioavailability, especially for poorly water-soluble drugs. Surfactants like polysorbates (such as polysorbate 80) or sodium lauryl sulphate can aid in improving the solubility of hydrophobic drugs. Cyclodextrins (e.g. hydroxypropyl beta-cyclodextrin) are cyclic oligosaccharides that can form inclusion complexes with poorly soluble drugs, thereby improving their solubility and dissolution rates (Mandadapu et al. 2022).

4.3.2 Additives for Oral Liquid Dosage Forms

The pharmaceutical industry has made extensive use of oral liquid formulations because of their rapid onset of action, high degree of dosage flexibility and ease of swallowing. They are usually divided into two broad categories: monophasic and biphasic formulations, which include a variety of dosage forms. In monophasic liquids, the drug dissolves entirely in the vehicle; in biphasic liquids, the API is suspended or dispersed in the vehicle, creating two phases instead of a complete dissolve. The type of API and the intended therapeutic effect are thought to be the two main determining factors when selecting a liquid dosage form. Various types of additives used in the formulation of liquid dosage forms are given in Fig. 4.4. Additives incorporated into liquid dosage forms play important role in enhancing stability, palatability and overall formulation characteristics. Here are some common additives used in oral liquid formulations:

4.3.2.1 Vehicles

Vehicles acts as primary base that disperses or dissolves drug along with other additives, enabling them to achieve a liquid state. Vehicles are further classified into aqueous and oily vehicles. Aqueous vehicles like glycerine, ethyl alcohol, propylene glycol, and water, as well as oily vehicles such as emulsified bases, organic oily bases, mineral oils, and vegetable oils are commonly used.

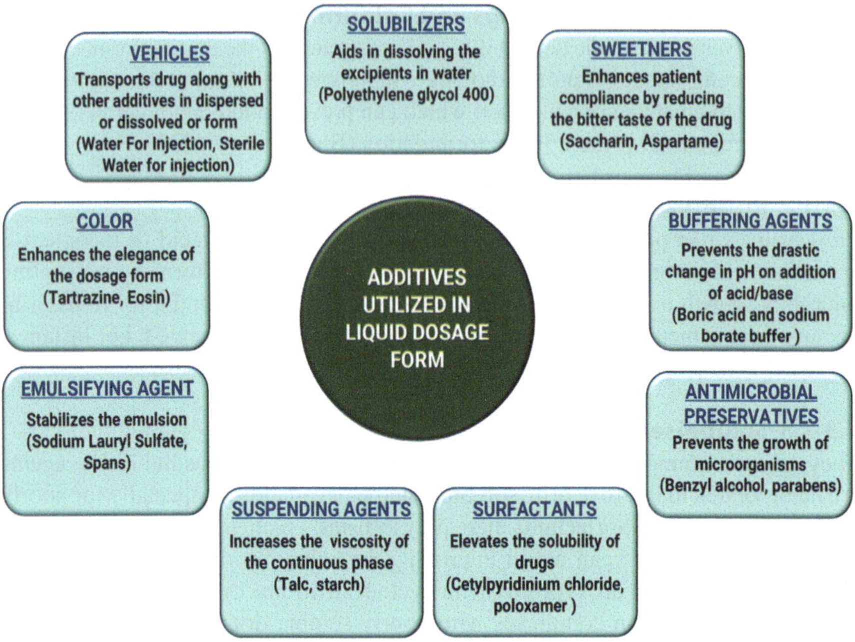

Fig. 4.4 Types of additives used in liquid dosage forms

4.3.2.2 Solubilisers

Hydrogen bond between the particles is broken down by the solubilisers causing them to dissolve in water. Modification in solubility can be done by the use of surface active agents, pH change, co-solvent addition or forming the complex. Ingredients like ethanol, propylene glycol or polyethylene glycol are used to solubilise poorly soluble drugs or aid in the formulation process.

4.3.2.3 Sweeteners

Sweeteners like sucrose and sorbitol or artificial sweeteners such as saccharin are utilised to improve the taste profile, effectively masking bitter or unpleasant tastes and enhancing overall palatability of oral liquid formulations. They are typically added to the formulations at concentrations ranging from 30% to 50%, with the exception of cold syrups, which have 80% sweetener content.

4.3.2.4 pH Adjusters

Substances such as citric acid or sodium citrate are used to regulate and maintain the pH of the liquid formulation within a suitable range for stability and optimal solubility of the API.

4.3.2.5 Preservatives

Preservatives are chemical substances that are added to prevent microbial growth and ensure the stability and safety of the liquid formulation during its shelf life.

Some examples of preservatives are benzyl alcohol (2%), parabens (0.015% to 0.2% w/v), chlorocresol phenol, etc. (Al-Rubaye 2022).

4.3.2.6 Surfactant

They are added in liquid formulations for aiding the solubilisation, emulsification and stabilisation of the formulation. Water-insoluble drugs can be dispersed as a colloidal dispersion using surfactants, which are surface active agents that tend to reduce the surface of interfacial tension between two liquids. Cationic surfactants like cetylpyridinium chloride and anion surfactants like polysorbate, poloxamer, etc., are used in liquid formulations (Gonçalves et al. 2023).

4.3.2.7 Suspending Agent

Suspending agents lower the rate at which suspensions settle by raising the viscosity of the continuous phase. Some of the suspending agents utilised in liquid dosage forms include carbopol, gelatin, kaolin, methylcellulose and hydroxypropyl methyl cellulose (Kumar and Verma 2023).

4.3.2.8 Emulsifying Agent

The role of an emulsifying agent is to reduce the interfacial tension or form a film surrounding the globules that are dispersed in an emulsion emulsifying agents. Sodium lauryl sulphate, sodium dioctyl sulfosuccinate, polymers (Spans and Tweens) and tragacanth are examples of commonly used emulsifying agents (Kamba et al. 2013).

4.3.2.9 Colorants

These additives, both natural and synthetic (e.g. FD&C dyes), serve to provide visual identification, enhance appearance and distinguish between various formulations.

4.3.2.10 Viscosity Modifiers

Substances such as carboxymethylcellulose (CMC), xanthan gum or glycerin are employed to adjust the viscosity of the liquid, improving pourability, suspension stability and ease of administration.

4.3.3 Additives for Transdermal Dosage Forms

Transdermal drug delivery system (TDDS) is a system which is used to deliver the drug into the systemic system through skin. An appropriate selection of additives and procedures is critical in TDDS to regulate the release of the drug. Each additive in TDDS serves a specific purpose, contributing to the effectiveness of formulation, its stability and patient compliance. The selection and combination of these additives are carefully considered to optimise drug delivery through the skin while ensuring safety and efficacy. The classification of TDDS is dependent upon the type of additive utilised, particularly the polymers incorporated during preparation. For instance, to create matrix-based TDDS, where the drug is dispersed within the polymer matrix, polyvinylpyrrolidone, ethyl cellulose or polyethylene glycol is used as polymer, and for formulating reservoir-type TDDS (Rajabalaya et al. 2012), where drug is surrounded by polymeric membrane, ethylene-vinyl acetate, polyethylene or isotactic polypropylene is used as the rate-controlling polymer (Kim et al. 2001).

Some other important additives found in transdermal formulations include:

4.3.3.1 Penetration Enhancers

Penetration enhancers are important additive of TDDS used to overcome the barrier property of stratum corneum (uppermost layer of skin) and enhance drug permeation. These additives improve drug permeation through the skin by altering the skin barrier properties. Examples include fatty acids (oleic acid), alcohols (ethanol), surfactants (Tween, Span) or terpenes (limonene). Although several surface active chemicals and lipophilic solvents have been shown to be useful penetration enhancers, their use is constrained by their sensitivity and irritation. Novel additives like *N*-methyl-2-pyrrolidone, Labrasol® and Transcutol® have proved to be better penetration enhancer than the conventional ones (Bácskay et al. 2023; Nakmode et al. 2022). Recent studies revealed that the type and concentration of the penetration enhancers play an important role in penetrability and characteristics of the formulation (Sakdiset et al. 2023).

4.3.3.2 Solvents/Solubilisers

Substances like propylene glycol, dimethyl sulphoxide (DMSO) or ethanol are used to solubilise the drug and enhance its permeation across the skin. They prevent drug precipitation and aid in maintaining a homogenous and stable pharmaceutical preparation. The careful selection and combination of solvents and solubilisers are critical in designing transdermal drug delivery systems to ensure optimal drug delivery.

4.3.3.3 Adhesives

Acrylic adhesives or silicone adhesives are used to adhere the patch to the skin and maintain its integrity during application (Banerjee et al. 2014). These adhesives must strike a balance between secure attachment and skin compatibility to ensure both effective drug delivery and patient comfort. Typically, they consist of pressure-sensitive materials that adhere to the skin upon application of light pressure. Silicone-based adhesives are commonly used due to their biocompatibility and hypoallergenic properties.

4.3.3.4 Backings and Release Liners

Backing materials provide structural support to the patch and restrict the release of the drug leaking from the membrane on top of the skin patch. Release liners on the other hand protect the adhesive surface of TDDS and are removed before application of the formulation on skin. The material used for both the purposes should be compatible with skin and should not cause any irritation to the skin. Backing layer should have a low modulus, a high rate of moisture-vapour transfer, and high flexibility. Polyethylene terephthalate, ethylene-vinyl acetate, polyethylene, polyurethane, etc., are few examples of polymers used as backing layers.

4.3.3.5 Plasticisers

Plasticisers like glycerin, dibutyl phthalate, polyethylene glycol, propylene glycol and triacetin maintain the flexibility of the patch. Additionally, plasticisers aid in modulating the drug release kinetics by influencing the diffusion characteristics of the active pharmaceutical ingredient through the skin (Pichayakorn et al. 2012).

4.3.3.6 Stabilisers and Antioxidants

To prevent degradation, stabilisers and antioxidants (ascorbic acid, tocopherols) are added to maintain the stability of the formulation. These additives inhibit the formation of free radicals that could potentially compromise the therapeutic properties of the drugs.

4.3.4 Additives for Parenteral Dosage Forms

In order to obtain the required product profile (stability and effectiveness), additives are a crucial component of the creation of parenteral dosage forms. The physicochemical characteristics of the additives utilised in the parenteral formulations are either lyophilised, suspended or are in a solution form. Even with appropriate additive selection, it is essential to employ them carefully during the formulation and manufacturing process in order to prevent unfavourable outcomes including decreased drug solubility, stability and activity. There are varieties of additives used for preparing the desired formulation for parenteral use. For instance, buffers like citrate or phosphate salts help to maintain the desired pH, ensuring the compatibility of the formulation inside the body. Preservatives such as benzyl alcohol or phenol prevent microbial growth in multi-dose vials, extending their shelf life and reducing the risk of contamination. Additionally, tonicity-adjusting agents like sodium chloride or mannitol ensure isotonicity, preventing adverse reactions and uneasiness at the injection site. Co-solvents like propylene glycol or ethanol aid in solubilising poorly water-soluble drugs, facilitating their administration.

4.3.4.1 Additive Used in Lyophilisation

Lyophilisation, a widely used process in the development of parenteral formulations, holds a significantly important role in preserving the stability and efficacy of injectable medications (Pramanick et al. 2013). This process initiates with freezing of the solution followed by the removal of water by sublimation technique, termed as primary drying, and finally water removal by desorption process, referred to as secondary drying. The extensive reduction in moisture content is helpful in preventing microbial growth and chemical reactions in the parenteral formulations. Therefore, it is important to choose an appropriate additive used for this process. Here are the additives used for the lyophilisation process.

4.3.4.1.1 Bulking Agents

These are typically employed for high-potency, low-dose medications that lack the bulk to sustain their own structure. These agents become even more important when the total solid content is less than 2% (Rayaprolu et al. 2018). The bulking agents help in the formation of proper lyophilised cake which ensures the better porosity and subsequent sublimation process. Bulking agents also have the capability to behave as stabilisers and tonicity-adjusting agents. Mannitol is the additive that is most frequently and extensively utilised in lyophilised products. After crystallisation, mannitol has an extremely high eutectic melting point (−1.4 °C) and is well processed during lyophilisation. However, in some situations, the bulking agent's crystallisation could negatively impact the product's physical stability; in these cases, an amorphous bulking agent is recommended. Other substance like lactose, sucrose, polyethylene glycol, glycine, etc., can also be used as bulking agents. Combination of these bulking agents with mannitol is commonly practised to address their individual limitations across various conditions. For instance, Nutropin® lyophilised injection incorporates mannitol along with glycine, while Enbrel® uses sucrose and mannitol as bulking agents.

4.3.4.1.2 Lyoprotectants

Lyoprotection involves stabilising and protecting the molecules from degradation during freeze-drying and subsequent storage. Sucrose and trehalose are frequently utilised for this purpose because they are not prone to crystallisation during lyophilisation, and therefore they do not phase separate from the protein. Trehalose appears to be a more advantageous lyoprotectant compared to sucrose due to its lower hygroscopicity, minimal chemical reactivity and notably higher glass transition temperature (Bjelošević et al. 2020).

4.3.4.1.3 Antioxidants

They are added to prevent the oxidation. Examples of marketed lyophilised injectable biopharmaceuticals containing antioxidants are Vivotif Berna® (live typhoid vaccine) with ascorbic acid and ACTHREL® (corticorelin ovine) containing cysteine hydrochloride and ascorbic acid (Mehmood and Farooq 2015).

4.3.4.1.4 Buffering Agents

In order to prevent drug deterioration during processing, storage and reconstitution, pH control is essential. As a result, a buffering agent must be added to the lyophilised formulation. Since the medication must be reconstituted and kept for a while before it can be given to the patient, the pH stability profile of the active ingredient determines which buffer is best. The drug's optimal stability pH needs to be determined and maintained for this reason. For sensitive compounds, choosing an appropriate buffer and its concentration is crucial.

4.3.4.2 Additives Used in Liquid Injection

Additives used in liquid injections serve various purposes, such as stabilising the API, adjusting the pH, enhancing solubility or ensuring the injection's safety and efficacy. Common additives found in liquid injections include the following:

4.3.4.2.1 Buffers

These maintain the injection's pH within a suitable range to prevent degradation of the API. Common buffers include sodium phosphate, citrate, acetate or glycine.

4.3.4.2.2 Preservatives

Preservatives in the form of antioxidants, antimicrobial and chelating agents are added in the liquid injections to prevent microbial growth in multi-dose vials. The antioxidants are utilised to reduce or inhibit the oxidation reaction of the drug or additives throughout the product's shelf life, while antimicrobial agents are utilised to inhibit the growth of microbes in the dosage form. Monothioglycerol, butylated hydroxytoluene, ascorbic acid, acetylcysteine and sulfurous acid salts (bisulphite, metabisulfite) are the most often utilised antioxidants in formulations that are sterile. In the category of chelating agents, substances like disodium edetate, edetate calcium disodium and diethylenetriamine pentaacetate are used (Rayaprolu et al. 2018).

4.3.4.2.3 Tonicity Adjusters

There are certain substances used to adjust the tonicity of the solution, making it isotonic with body fluids to prevent tissue injury. Dextrose is most preferred tonicity adjuster than glycerol and sodium chloride. Glycerine and mannitol are two more tonicity-adjusting substances that are frequently employed.

4.3.4.2.4 Solvent System

Aqueous and nonaqueous vehicles are the two types of solvent systems utilised in parenteral suspension. Both water-miscible and water-immiscible vehicles fall under the category of nonaqueous vehicles. Usually, the best solvent system for injection is water. However, to increase the solubility and stability in parenteral preparation, nonaqueous water-miscible substances are utilised as co-solvents with water for injection. Propylene glycol, glycerine, *N*-lactamide and ethanol are a few examples of nonaqueous water-miscible vehicles.

4.3.4.2.5 Solubilisers

Surfactants and co-solvents are two main categories into which the solubilising agents can be divided. Surfactants work by lowering the drug components' surface tension, increasing dissolution, while co-solvents are defined as solvents that, when combined with another solvent, can dissolve a solute. Ethanol, glycerol, various grades of polyethylene glycol, etc., can be used as co-solvents (Nema and Brendel 2011). Sorbitan monooleate (Span 80), polyoxyethylene sorbitan monolaurate (Tween 20), polyoxyethylene sorbitan monooleate (Tween 80) are some examples of surfactants, which are used in liquid injections. D-α-Tocopherol polyethylene glycol 1000 succinate is a novel nonionic surfactant which is considered potential emulsifying agent or surfactant for injectables and to inhibit the multi-drug resistance (Rayaprolu and Strom 2013).

4.3.4.2.6 Complexing and Dispersing Agents

Sometimes, complexation is used to increase a drug's solubility in a solvent, particularly water. Cyclodextrins have shown to be incredibly successful additives for solubilising medications that are hydrophobic. Modified cyclodextrins, like hydroxypropyl-β-cyclodextrin and sulfobutylether-β-cyclodextrin, have been found to solubilise and stabilise a variety of injectable medications, such as interleukin-2, oestradiol, dexamethasone and other proteins and peptides, in the parenteral dosage form without seeming to cause any compatibility issues.

4.3.4.2.7 Flocculating/Suspending Agents (For Pharmaceutical Injectable Suspension)

The controlled flocculation method uses one or more flocculating agents to create flocs, or loosely bonded aggregate, that settle quickly but readily redisperse when agitated. A suitable quantity of flocculating agent is used to achieve the maximum volume of sedimentation and inhibit the development of cakes. Traditionally, flocculating agents have included hydrophilic colloids, surfactants and electrolytes. The formation of flocs is dependent on the surface charge of the particles and is facilitated by surfactants and electrolytes, which minimise the electrical forces of repulsion between the particles. Examples of some of the electrolytes utilised in parenteral suspensions are salt/potassium citrate, salt/potassium chloride and sodium/potassium acetate.

4.4 Drug-Additive Interaction Studies in Development of Pharmaceutical Formulations

Pharmaceutical additives are generally considered pharmacologically inert substances, but studies have revealed that they can actively interact physically, chemically and physiologically with drugs to compromise the stability of the finished product and the therapeutic function of the drug component (Gorain et al. 2018). The interaction depends on the amount of the drug, its chemical makeup and the characteristics of the additives used. While drug-additive interactions may have some advantageous effects on occasions, in the majority of cases, they tend to delay drug release and diminish bioavailability. For instance, during a study on the compatibility of superdisintegrants with various model drugs of differing aqueous solubility, it was found that ionic interactions between cationic drugs and anionic superdisintegrants lead to delayed drug release and poor dissolution (Balasubramaniam et al. 2008). A study revealed the positive impact of a drug-additive complex, wherein the presence of cyclodextrin exhibited improved solubility, bioavailability and stability (Liu et al. 2023). Predicting the precise mechanism of such an interaction remains challenging, despite the records of several probable reasons in the literature. These interactions can be categorised as the following:

4.4.1 Physical Incompatibilities

Physical interactions such as van der Waals forces, hydrogen bonding or electrostatic forces between drugs and additives have the potential to influence the properties of drug. Such interactions can have either positive or negative effect on the performance of formulation including its dissolution rate and solid-state characteristics. The undesirable insoluble complex formation between drug and additive causes a slower rate of drug dissolution and absorption like tetracycline and calcium carbonate complex. On the other hand, drug cyclodextrin complex enhances the solubility of poorly water-soluble drugs (Saokham et al. 2018). A capsule formulation prepared using stearic acid with povidone showed decreased dissolution stability and hence indicated that this combination of additives should not be used for the formulation of immediate release formulations (Desai et al. 2008). In another study, the bioavailability of ganciclovir, a BCS class III, drug was enhanced when its solid dispersion was prepared using cyclodextrin and shellac polymers (Gaber et al. 2022). Furthermore, in the course of the study, it was observed that the incorporation of polyvinylpyrrolidone K30 into the binary solid dispersion (comprising indomethacin and kaolin) not only enhanced the physical stabilisation of amorphous indomethacin but also addressed the solubility challenges arising from the presence of kaolin (Bejaoui et al. 2021). There are many such examples showing the significance of physical interaction of additives with the drug molecule.

4.4.2 Chemical Incompatibilities

Another type of interaction that can arise between drug and additives is chemical interaction, and it is imperative to study this for complete pharmaceutical research and avoid incompatibility. These interactions can give rise to covalent or noncovalent bonds, profoundly influencing the drug's stability, solubility and bioavailability. As compared to physical interaction, chemical interactions may lead to adverse impacts on the product, resulting in the formation of a different compound. The studies have proved the complexity of chemical interactions which emphasises their significance in formulation development. One of the examples is the chemical incompatibility observed between magnesium stearate and aspirin which gives rise to several potentially undesirable products, including salicyl salicylic acid, salicylic acid and acetyl salicyl salicylic acid (Mitrevej and Hollenbeck 1983). Whether facilitating or impeding the drug's performance, these interactions necessitate careful consideration during the design and optimisation of pharmaceutical formulations. Among the different types of chemical reactions, leading to incompatibility, hydrolysis stands out as the most common in pharmaceutical formulations (Gabrič et al. 2022). Hydrolysis is a common degradation process, particularly in the presence of water, affecting esters, amides, lactones or lactams. Oxidation, the second-most likely degradation process, involves alcohols, aldehydes, alkaloids, phenols and unsaturated fatty substances. Although less common, other potential reactions include isomerisation, photolysis and polymerisation. These reactions have the potential to reduce the concentration of the API and generate hazardous impurities. Additionally, ionisable drugs may undergo reactions with ionised, soluble additives, resulting in the formation of insoluble products. Studies have shown that the existence of residual reactive impurities in additives contributes to the degradation of the drug product. Specifically, the oxidative degradation of polyethylene glycol and polysorbate additives can lead to the initial formation of aldehyde impurities. These aldehyde impurities have the potential to undergo further oxidation, ultimately resulting in the generation of both formic and acetic acids (Sharma 2019). These chemical transformations underscore the importance of carefully considering the choice of solvent and additives in pharmaceutical formulations to ensure stability and prevent the formation of undesirable by-products.

4.4.3 Therapeutic or Physiological Incompatibilities

Therapeutic incompatibilities arise when a patient's reaction to one or more medications differs from the doctor's expected course of action or degree of effect. These interactions occur within the biological system in the presence of an aqueous environment. This type of interaction significantly influences the absorption of API in the biosystem, thereby affecting their pharmacological activity. The pH of gastric fluid plays a crucial role in drug absorption from the gastrointestinal (GI) environment, influencing the solubility and ionisation of pharmaceutical drugs. Alterations in gastric transit time, influenced by coadministered drugs or additives, directly impact the bioavailability of consumed drugs. Furthermore, GI motility, affected by various formulations such as lipid-based dosage forms, can contribute to variability in drug absorption (Soni et al. 2016). Additionally, alterations in drug metabolism in the liver, mediated by enzymes like cytochrome P-450, can affect drug concentration, leading to potential toxicities or loss of pharmacological activity. The involvement of P-glycoprotein efflux transporter in multidrug-resistant cells, and the incorporation of specific additives in pharmaceutical formulations to inhibit P-gp, demonstrates the complicated relationship between physiological interactions and drug availability, particularly in cancer treatment (Choudhury et al. 2017). Addressing therapeutic incompatibility is crucial in pharmaceutical development to ensure that the final product delivers the intended therapeutic benefits while maintaining stability and safety. Researchers and formulators strive to design formulations that minimise the risk of incompatibility, enhancing the overall quality and effectiveness of pharmaceutical products.

4.4.4 Analytical Techniques to Characterise Drug-Additive Incompatibility

In order to reduce or mitigate the unfavourable responses brought on by the incompatibility, scientists use a variety of analytical techniques for its detection. Thermal methods like hot stage microscopy, isothermal microcalorimetry, thermogravimetric analysis, differential scanning calorimetry and X-ray diffraction assess thermal events, crystallinity and polymorphism, which are vital in performing compatibility screening studies (Rojek and Wesolowski 2023). Technique like high-performance liquid chromatography enables the separation and quantification of individual components, aiding in the identification of degradation products and monitoring changes in composition (Al-Rifai et al. 2022). Furthermore, Fourier transform infrared spectroscopy elucidates molecular structures by measuring infrared radiation absorption, offering insights into chemical changes and functional groups (Rojek et al. 2023). Nuclear magnetic resonance spectroscopy provides detailed information on molecular conformation and interactions between drug and additives (Velip et al. 2023). Mass spectrometry and UV-visible spectroscopy offer quantitative analysis and identification of chemical entities, while electrophoretic techniques can reveal charge-related interactions (Pan et al. 2023; Mishra et al. 2023). In recent times, spectroscopic techniques including near-infrared and solid-state nuclear magnetic resonance spectroscopy have been employed to examine moisture present in the drug or interactions between drug and additive, which could lead to the instability of active principles. These methods vary based on their theory, sample machine and temperature stress, required testing duration, sample number, sensitivity to small changes and demand for either internal or external requirements. Real-time stability studies and particle size analysis contribute further to a comprehensive understanding of drug-additive compatibility, ensuring the quality and safety of pharmaceutical formulations. Finally, studies reveal that the combination of thermal and nonthermal methods proves to be an effective approach for identifying incompatibilities between drugs and additives (Xie et al. 2023).

4.5 Recent Advances in Additive Science

Recent advances in pharmaceutical formulation additive science have revolutionised drug delivery, efficacy and patient experience. One notable breakthrough involves the development of multifunctional additives including functional and co-processed additives. These additives serve multiple purposes, such as improving drug solubility, stability and bioavailability. Co-crystallisation techniques, where two or more compounds are combined to form a new crystalline structure, have been employed to enhance the physical properties of drugs. An example is the co-crystallisation of the antiretroviral drug ritonavir with saccharin, which improves its solubility and bioavailability (Varshosaz et al. 2018). Polymeric additives, such as copovidone and hypromellose acetate succinate, are also gaining prominence (Vadlamudi and Dhanaraj 2017; Mašková et al. 2020). These polymers not only serve as binders and fillers but also contribute to the controlled release of drugs. They enhance the stability of formulations and assist in achieving the desired drug release profiles. Co-processing techniques, where multiple additives are combined to create a single multifunctional excipient, represent another notable advancement. For instance, the co-processing of lactose with microcrystalline cellulose results in a multifunctional excipient that improves tablet hardness, disintegration and drug release. Moreover, the integration of smart polymers and hydrogels into pharmaceutical formulations has opened avenues for controlled drug release by responding to external stimuli. These advancements not only contribute to the development of more effective and patient-specific medications but also prompt a surge in patent filings to protect these innovative technologies globally. Some of the internationally patented additives are given in Table 4.1. Despite the regulatory authorities' recommendations, the firms that manufacture drug products restrict the use of innovative additives since they are required to provide clinical safety data to the agencies before the additive may be used in a product (Pérez et al. 2020).

4.5.1 Functional and Co-processed Additives

To address the needs of formulation experts regarding production costs, improved additive functionality and tablet quality, formulation scientists decided to focus their efforts on developing multifunctional additives with improved performance. Due to their extensive utilisation across various purposes, additives are now considered functional ingredients rather than remaining inactive components. Single-component additives might sometimes lack the performance needed to achieve the desired functionality for formulating or manufacturing specific drugs. Hence, co-processing has become pivotal in designing stable additives with multiple functionalities. These additives stand as innovative blends in pharmaceutical formulations, combining the strengths of both natural and synthetic components. Natural co-processed additives often merge plant-based materials or polysaccharides with synthetic polymers to create stable, multifunctional additives. The blending of natural or synthetic polymers has extensively been studied to create innovative, stable co-processed additives. Plant-based component co-processing has emerged as a significant advantage for numerous pharmaceutical industries globally in recent times (Bhatia et al. 2022). These combinations harness the inherent advantages of natural sources, such as biocompatibility and sustainability, while leveraging the precision and consistency offered by synthetic compounds. Synthetic co-processed additives, on the other hand, result from particular engineering to fuse specific functionalities, enhancing flow properties, compressibility or disintegration rates. Their tailored composition allows for modification of drug formulations to meet targeted release profiles, stability requirements and manufacturing efficiency, exemplifying the synergy achieved through the combination of natural and synthetic elements in pharmaceutical development.

The characteristics of additives can be readily modified extensively through co-processing or particle engineering methods. During co-processing method, additives interact at a sub-particle level to enhance the combined beneficial properties of both additives. Simultaneously, the aim is to hide any undesirable characteristics inherent in each additive. Typically, this is accomplished through co-drying and co-precipitation methods. Both additives are dispersed within a solvent and then dried, yielding a physical mixture adjusted to a targeted size range. Particle size, shape, surface area, porosity, density and other basic solid-state characteristics affect

Table 4.1 List of international patented additives

Patent number	Title	Additives	Uses	References
US10555912B2	Use of polymeric excipients for lyophilisation or freezing of particles	Polyvinyl alcohols, optionally in conjunction with sugars	Cryoprotectant	Foss and Shinde (2018)
US20180055939A1	Excipients for use in adeno-associated virus pharmaceutical formulations and pharmaceutical formulations made therewith	Sorbitol and Tween 20 and Tween 80	Diluent	Sista and Espinoza (2017)
EP3441074A1	Novel vehicles for the transfection of miRNAs	Span and Tween	Nonionic surfactants	Sanchez et al. (2017)
US20120289536A1	Pharmaceutical compositions comprising colloidal silicon dioxide	Colloidal silicon dioxide and cross-linked polyvinylpyrrolidone	Disintegrants	Haeberlin and Kramer (2012)
US20130177649A1	Co-processed tablet excipient composition, its preparation and use	Co-processed additive comprising 40% to 70% lactose, 20% to 50% MCC, 1% to 10% cross-linked sodium starch glycolate and 0.2% to 1% lubricant	Filler, binder, disintegrant and lubricant, respectively	Van Gessel (2010)
US8513329B2	Chemical additives to make polymeric materials biodegradable	Furanone and carrier resin	For enhancing biodegradation of polymeric material	Lake and Adams (2008)

(continued)

Table 4.1 (continued)

Patent number	Title	Additives	Uses	References
US8932629B2	Co-processed microcrystalline cellulose and sugar alcohol as an excipient for tablet formulations	Co-processed microcrystalline cellulose and mannitol	Improving compressibility, lubricant affectability and ejection profile	Li et al. (2007)
US20040228932A1	Pharmaceutical excipient	Multifunctional fibre-rich fraction (FRF)	Binder, disintegrant, filler, dispersing agent, coating agent, film-forming agent, thickener, etc.	Pilgaonkar et al. (2003)
US5989589A	Cross-linked cellulose as some tablet excipients	Cross-linked cellulose	Binder and disintegrant	Cartilier and Chebli (1997)
US5585115A	Pharmaceutical excipient having improved compressibility	Combination of MCC particles and silicon dioxide particles	Improves Compressibility, regardless of whether used in direct compression, dry granulation or wet granulation formulations	Sherwood et al. (1995)
US5128143A	Sustained release excipient and tablet formulation	Xanthan gum and galactomannan gum	Cross-linking agents at gastric pH	Baichwal and Staniforth (1990)
US5004601A	Low-melting mouldable pharmaceutical excipient and dosage forms prepared therewith	Low molecular weight PEG, medium to high molecular weight PEG long-chain saturated carboxylic acid polyethylene oxide and colloidal silica	Mouldable pharmaceutical excipients having low melting point	Snipes (1988)
US4762857A	Trehalose as stabiliser and tableting excipients	Trehalose	Stabiliser	Ernest Bollin and Fletcher (1987)

additive capabilities such as compatibility, flowability, dilution potential, disintegration potential and lubricating potential. Therefore, designing a particle that will yield the required capabilities must come first when developing a novel additive.

4.5.2 Novel Materials and Multi-Materials

The field of pharmaceutical formulations has witnessed a transformative shift with the incorporation of novel materials and multi-material additives, revolutionising drug delivery and therapeutic efficacy. Nanotechnology has emerged as a cornerstone, with lipid nanoparticles (LNPs) and polymeric nanoparticles at the forefront. LNPs exhibit exceptional capabilities in enhancing drug solubility and stability, significantly improving bioavailability. Meanwhile, polymeric nanoparticles enable controlled drug release and targeted delivery, paving the way for personalised medicine. The discovery of 3D printing and multi-material additive manufacturing has further propelled innovation by allowing the creation of intricate dosage forms with distinct material properties (Lee et al. 2017). Co-crystals, formed by combining API with co-crystal formers, contribute to improved drug solubility and stability. Hydrogels, particularly smart hydrogels responsive to environmental stimuli, provide dynamic drug release profiles. The utilisation of polymeric additives, such as biodegradable polymers, has become integral to encapsulating drugs for sustained release. Graphene, carbon nanotubes and metal-organic frameworks offer unique properties for drug delivery, while traditional components like cyclodextrins continue to play a crucial role in enhancing drug solubility. This amalgamation of novel materials and multi-material additives not only addresses long-standing challenges in drug formulation but also opens avenues for the development of advanced, patient-centric pharmaceutical products with improved therapeutic outcomes.

4.6 Related Regulatory Perspectives

The regulatory aspect of pharmaceutical additives is an essential aspect for ensuring the safety and efficacy of medicinal products. Any component of a drug product other than the active ingredient is considered an inactive ingredient, i.e. additive, as defined by 21 CFR 210.3(b)(8). Regulatory bodies, including the U.S. Food and Drug Administration (FDA) and the European Medicines Agency (EMA), closely monitor the use of pharmaceutical additives through stringent guidelines. The regulatory guidelines governing pharmaceutical additives include several key perspectives, including those outlined by the International Pharmaceutical Excipients Council (IPEC), generally recognised as safe (GRAS) status and the Inactive Ingredient Guide (IIG).

4.6.1 GRAS

In the United States, the GRAS status holds particular significance in the regulatory perspectives of pharmaceutical additives. The GRAS designation, overseen by the FDA, signifies that a substance is considered safe for consumption based on scientific evidence. While GRAS primarily applies to food ingredients, it is relevant to pharmaceutical additives that may find dual utility in both drug formulations and food products (Manchanda et al. 2018). Sections 201(s) and 409 of the Federal Food, Drug, and Cosmetic Act, which governs how the FDA reviews and approves certain compounds, serve as the foundation for GRAS. Generally speaking, a material can be added to the GRAS components list if it has completed the necessary safety evaluation conducted by a group of professionals and found not to be harmful when used as intended. The GRAS list's primary goals are to ensure consumers that the goods they are eating are healthy and safe while also accelerating the approval and launch of new items onto the market. A manufacturer may add an ingredient to human food items without first obtaining FDA permission, even though the FDA is responsible for premarket approval of food additives. This is as long as the component is deemed GRAS for the intended application. Imagine if every time a new product was introduced to the market, regulatory bodies like the US FDA had to evaluate every component from scratch. In addition to placing a burden on regulatory bodies' human resources, this would also force new product makers to resubmit data on these components. Alternatively, by simply adding these components to their goods without needing to file for new approvals, producers and scientists may easily generate new products and fulfil new needs, thanks to a pre-approved list of additives. Presently, slightly more than 370 compounds have been granted GRAS status; many of them are also authorised additives that have monographs in the European Pharmacopoeia, Japanese Pharmacopoeia or United States Pharmacopeia—National Formulary. Additives that are listed as GRAS often consist of natural, synthetic or semisynthetic components. When an additive is designated as GRAS, it basically indicates that an expert scientific committee has decided their intended use, and the FDA will not examine it prior to market release. By mandating producers to give information required by authorities and consumers on the safety of substances introduced into consumer products, the GRAS framework offers an extra safeguard for the welfare of the public.

4.6.2 IIG

IIG serves as a comprehensive reference for pharmaceutical manufacturers, providing detailed information on permissible inactive ingredients in drug formulations. It outlines acceptable additives and their respective specifications, offering guidance to industry professionals to ensure compliance with regulatory standards. The pharmaceutical industry can utilise this information to help in the development of new drugs. When an inactive component surfaces in an authorised drug product for a specific route of administration, it is no longer deemed novel for the goal of developing novel drugs, and the next time it appears in a new drug product, it could need a less thorough assessment. For instance, a sponsor may believe that an inactive component is safe to employ in a comparable way for a comparable kind of product if it has been approved in a certain dosage form at a specific concentration.

The IIG, published by the US Food and Drug Administration (FDA), serves as a reference for acceptable inactive ingredients in pharmaceutical products. The Chemical Abstracts Service registry (CAS) is a section of the American Chemical Society that offers extensive electronic services for chemical information. It provides unique numerical identifiers for chemical substances, helping to accurately identify and track these substances. Additionally, the Unique Ingredient Identifier (UNII) system, managed by the FDA, assigns a unique alphanumeric code to each substance, enhancing precision in ingredient identification. In practical terms, when a pharmaceutical manufacturer refers to the IIG for guidance on suitable additives, the listed ingredients are often identified by their CAS numbers. The UNII system, with its unique identifiers, adds an additional layer of specificity in tracking and communication. Collectively, these regulatory tools work together to ensure precision, consistency and compliance in the use of pharmaceutical additives, facilitating regulatory approvals and the development of safe and effective medicines. Together, the IIG, CAS and UNII contribute to a robust regulatory framework, facilitating the development and approval of pharmaceutical products with a focus on safety, efficacy and quality.

4.6.3 IPEC

IPEC is a global organisation committed to establishing and promoting standards for additives used in pharmaceutical products (DeMerlis et al. 2016). The council provides guidelines and best practices for the development, manufacturing and quality control of pharmaceutical additives, contributing to the overall safety and efficacy of drug formulations. By offering a harmonised approach to excipient standards, IPEC facilitates international collaboration and ensures a consistent regulatory framework across different regions. The federation has recommended guidelines for the safety assessment of new additives and good manufacturing practices for bulk pharmaceutical additives, ensuring comprehensive information for the safe use of these substances. Pharmacopoeial harmonisation, managed by the Pharmacopoeial Discussion Group (PDG), has played a significant role in streamlining regulatory processes for additives, contributing to their quality, safety and efficacy. As of 2017, significant progress has been made in harmonising chapters and monographs across different pharmacopoeias (IPEC Federation 2023). Presently, IPEC Europe holds the presidency, while IPEC-Americas and IPEC Japan serve as vice president and treasurer, respectively. Manufacturers adhering to IPEC guidelines demonstrate a commitment to producing high-quality pharmaceuticals, contributing to the confidence of regulatory agencies and fostering global acceptance of their products.

4.7 Conclusion

This chapter offers a comprehensive survey of the multifaceted world of formulation additives in the pharmaceutical industry. The elucidation begins with a thorough introduction, setting the stage for a detailed examination of the importance of additives in shaping the development of pharmaceutical dosage forms. By elucidating the importance of additives in achieving desired characteristics and functionalities, the chapter highlights their essential role as formulation aids. There are various types of formulation additives that are tailored to meet the diverse requirements of oral solids, liquids, transdermal and parenteral dosage forms. However, prior to incorporating any particular additive into formulation development, it is crucial to conduct thorough drug-additive interaction studies. The interaction studies provide important insights into the complexities inherent in formulation development, emphasising the need for thorough understanding and mitigation of physical, chemical and therapeutic incompatibilities. Despite the growing demand for new additives due to advancements in drug product development, it remains imperative to carefully consider the regulatory aspects surrounding their usage. Notably, there is no singular universal regulatory guideline governing pharmaceutical additives, necessitating compliance with a combination of international, regional and national standards and regulations. It can be concluded that there are huge possibilities and future directions in the development of additives, suggesting potential avenues for further advancement in pharmaceutical formulation technology. Since formulations cannot be created without additives, this chapter stands as an essential resource for individuals involved in the development of pharmaceutical products.

References

Abrantes CG, Duarte D, Reis CP (2016) An overview of pharmaceutical excipients: safe or not safe? J Pharm Sci 105(7):2019–2026. https://doi.org/10.1016/j.xphs.2016.03.019

Adamkiewicz L, Szeleszczuk Ł (2023) Review of applications of cyclodextrins as taste-masking excipients for pharmaceutical purposes. Molecules 28(19):6964. https://doi.org/10.3390/molecules28196964

Al-Rifai N, Alshishani A, Saad B, Rasras A, Zahra J, Madieh S, Darras F (2022) Synthesis, isolation, identification and characterization of a drug-excipient interaction degradation impurity in pramipexole by HPLC, LC/MS and NMR. Separations 10(1):7. https://doi.org/10.3390/separations10010007

Al-Rubaye IM (2022) A review of the literature on antimicrobial preservatives: definition, properties, classification, safety, side effects and antimicrobial effectiveness testing. Atena J Public Health 4:7

Bácskay I, Hosszú Z, Budai I, Ujhelyi Z, Fehér P, Kósa D, Haimhoffer Á, Pető Á (2023) Formulation and evaluation of transdermal patches containing BGP-15. Pharmaceutics 16(1):36. https://doi.org/10.3390/pharmaceutics16010036

Baichwal AS, Staniforth JN (1990) Sustained release excipient and tablet formulation. US5128143A

Balasubramaniam J, Bindu K, Rao VU, Ray D, Haldar R, Brzeczko AW (2008) Effect of superdisintegrants on dissolution of cationic drugs. Dissolut Technol 15(2):18. https://doi.org/10.14227/DT150208P18

Ballentine C (1981) Sulfanilamide disaster. FDA Consumer Mag 5. https://www.fda.gov/files/about%20fda/published/The-Sulfanilamide-Disaster.pdf

Banerjee S, Chattopadhyay P, Ghosh A, Datta P, Veer V (2014) Aspect of adhesives in transdermal drug delivery systems. Int J Adhes Adhes 50:70–84. https://doi.org/10.1016/j.ijadhadh.2014.01.001

Bejaoui M, Kalfat R, Galai H (2021) The effect of adding PVP to the binary solid dispersion (indomethacin: kaolin) on the formation of physically stable amorphous drug. J Pharm Innov 17:736. https://doi.org/10.1007/s12247-021-09553-6

Bhatia V, Dhingra A, Chopra B, Guarve K (2022) Co-processed excipients: recent advances and future perspective. J Drug Deliv Sci Technol 71:103316. https://doi.org/10.1016/j.jddst.2022.103316

Bjelošević M, Pobirk AZ, Planinšek O, Grabnar PA (2020) Excipients in freeze-dried biopharmaceuticals: contributions toward formulation stability and lyophilisation cycle optimisation. Int J Pharm 576:119029. https://doi.org/10.1016/j.ijpharm.2020.119029

Blundell R, Butterworth P, Charlier A, Daurio D, Degenhardt M, Harris D, Hancock B, Johnston M, Kasina R, Kaye J, Kelly R (2022) The role of titanium dioxide (E171) and the requirements for replacement materials in oral solid dosage forms: An IQ Consortium Working Group Review. J Pharm Sci 111:2943. https://doi.org/10.1016/j.xphs.2022.08.011

Bodratti AM, Alexandridis P (2018) Amphiphilic block copolymers in drug delivery: advances in formulation structure and performance. Expert Opin Drug Deliv 15(11):1085–1104. https://doi.org/10.1080/17425247.2018.1529756

Cartilier L, Chebli C (1997) Cross-linked cellulose as a tablet excipient. US5989589A

Choudhury H, Gorain B, Pandey M, Kumbhar SA, Tekade RK, Iyer AK, Kesharwani P (2017) Recent advances in TPGS-based nanoparticles of docetaxel for improved chemotherapy. Int J Pharm 529(1–2):506–522. https://doi.org/10.1016/j.ijpharm.2017.07.018

Dave VS, Saoji SD, Raut NA, Haware RV (2015) Excipient variability and its impact on dosage form functionality. J Pharm Sci 104(3):906–915. https://doi.org/10.1002/jps.24299

DeMerlis CC, Smith A, Schoneker DR (2016) Regulatory information for excipients. In: Pharmaceutical excipients: properties, functionality, and applications in research and industry, pp 241–267. https://doi.org/10.1002/9781118992432.ch6

Desai D, Kothari S, Huang M (2008) Solid-state interaction of stearic acid with povidone and its effect on dissolution stability of capsules. Int J Pharm 354(1–2):77–81. https://doi.org/10.1016/j.ijpharm.2007.11.042

Ernest Bollin J, Fletcher MG (1987) Trehalose as stabilizer and tableting excipient. US4762857A

Foss W, Shinde R (2018) Use of polymeric excipients for lyophilization or freezing of particles. US10555912B2

Gaber DA, Alnwiser MA, Alotaibi NL, Almutairi RA, Alsaeed SS, Abdoun SA, Alsubaiyel AM (2022) Design and optimization of ganciclovir solid dispersion for improving its bioavailability. Drug Deliv 29(1):1836–1847. https://doi.org/10.1080/10717544.2022.2083723

Gabrič A, Hodnik Ž, Pajk S (2022) Oxidation of drugs during drug product development: problems and solutions. Pharmaceutics 14(2):325. https://doi.org/10.3390/pharmaceutics14020325

Gonçalves RA, Holmberg K, Lindman B (2023) Cationic surfactants: a review. J Mol Liq 2023:121335. https://doi.org/10.1016/j.molliq.2023.121335

Gorain B, Choudhury H, Pandey M, Madheswaran T, Kesharwani P, Tekade RK (2018) Drug–excipient interaction and incompatibilities. In: Dosage form design parameters. Academic, pp 363–402. https://doi.org/10.1016/B978-0-12-814421-3.00011-7

Haeberlin B, Kramer A (2012) Pharmaceutical compositions comprising colloidal silicon dioxide. US20120289536A1

Indurkhya A, Patel M, Khan MA (2023) Development and characterization of trandolapril immediate release tablets using different superdisintegrating agents. Curr Res Pharm Sci 13:70–77. https://doi.org/10.24092/CRPS.2023.130106

IPEC (2023) Federation [WWW document]. https://ipec-federation.org/

Kamba EA, Itodo AU, Ogah E (2013) Utilization of different emulsifying agents in the preparation and stabilization of emulsions. Int J Mater Chem 3(4):69–74. https://doi.org/10.5923/j.ijmc.20130304.01

Kim MK, Zhao H, Lee CH, Kim DD (2001) Formulation of a reservoir-type testosterone transdermal delivery system. Int J Pharm 219(1–2):51–59. https://doi.org/10.1016/S0378-5173(01)00631-7

Kumar P, Verma M (2023) Pharmaceutical suspensions: an updated patent review on novel suspending agents and recent advancement. Recent Adv Drug Deliv Formul 17(3):193–209. https://doi.org/10.2174/0126673878246149231010085610

Kusuma SA, Abdassah MA, Maryati FI (2020) Comparison of preservatives efficacy of benzalkonium chloride, thimerosal, and benzyl alcohol in eye drop products containing chloramphenicol. Int J Appl Pharm 12:100–105. https://doi.org/10.22159/ijap.2020v12i4.37686

Lake JA, Adams SD (2008) Chemical additives to make polymeric materials biodegradable. US8513329B2

Lee JY, An J, Chua CK (2017) Fundamentals and applications of 3D printing for novel materials. Appl Mater Today 7:120–133. https://doi.org/10.1016/j.apmt.2017.02.004

Li J-X, Carlin B, Ruszkay T (2007) Co-processed microcrystalline cellulose and sugar alcohol as an excipient for tablet formulations. US8932629B2

Li J, Wang Y, Yu D (2023) Effects of additives on the physical stability and dissolution of polymeric amorphous solid dispersions: a review. AAPS PharmSciTech 24(7):175. https://doi.org/10.1208/s12249-023-02622-8

Liu N, Xiao J, Zang LH, Quan P, Liu DC (2023) Preparation of trans-crocetin with high solubility, stability, and oral bioavailability by incorporation into three types of cyclodextrins. Pharmaceutics 15(12):2790. https://doi.org/10.3390/pharmaceutics15122790

Malaquias LF, Schulte HL, Chaker JA, Karan K, Durig T, Marreto RN, Gratieri T, Gelfuso GM, Cunha-Filho M (2018) Hot melt extrudates formulated using design space: one simple process for both palatability and dissolution rate improvement. J Pharm Sci 107(1):286–296. https://doi.org/10.1016/j.xphs.2017.08.014

Manchanda S, Chandra A, Bandopadhyay S, Deb PK, Tekade RK (2018) Formulation additives used in pharmaceutical products: emphasis on regulatory perspectives and GRAS. In: Dosage form design considerations, pp 773–831. https://doi.org/10.1016/B978-0-12-814423-7.00022-8

Mandadapu G, Kolli P, Kurra VG (2022) Formulation & evaluation of cyclodextrin complexed tablets by enhancing the dissolution rate. J Innov Appl Pharm Sci 8:123–132. https://doi.org/10.37022/jiaps.v7i3.375

Mašková E, Kubová K, Raimi-Abraham BT, Vllasaliu D, Vohlídalová E, Turánek J, Mašek J (2020) Hypromellose—a traditional pharmaceutical excipient with modern applications in oral and oromucosal drug delivery. J Control Release 324:695–727. https://doi.org/10.1016/j.jconrel.2020.05.045

Mehmood Y, Farooq U (2015) Excipients use in parenteral and lyophilized formulation development. Open Sci J Pharm Pharmacol 3(3):19–27

Mishra A, Sinha VR, Sharma S, Mathew AT, Kumar R, Yadav AK (2023) A comprehensive compatibility study of ganciclovir with some common excipients. Am J Biopharm Pharm Sci 3:2. https://doi.org/10.25259/AJBPS_4_2023

Mitrevej A, Hollenbeck RG (1983) Influence of hydrophilic excipients on the interaction of aspirin and water. Int J Pharm 14(2–3):243–250. https://doi.org/10.1016/0378-5173(83)90097-2

Nakmode D, Bhavana V, Thakor P, Madan J, Singh PK, Singh SB, Rosenholm JM, Bansal KK, Mehra NK (2022) Fundamental aspects of lipid-based excipients in lipid-based product development. Pharmaceutics 14(4):831. https://doi.org/10.3390/pharmaceutics14040831

Narang AS, Mantri RV, Raghavan KS (2017) Excipient compatibility and functionality. In: Developing solid oral dosage forms. Academic, pp 151–179. https://doi.org/10.1016/B978-0-12-802447-8.00006-6

Nema S, Brendel RJ (2011) Excipients and their role in approved injectable products: current usage and future directions. PDA J Pharm Sci Technol 65(3):287–332. https://doi.org/10.5731/pdajpst.2011.00634

Pan X, Lenka S, Davis J, Nagapudi K, Mantik P, Saggu M, Dai L, Cadang L, Zhang HM, Zhang J, Izadi S (2023) Probing the protein–excipient interaction in the orally delivered protein by solid-state hydrogen–deuterium exchange mass spectrometry and molecular dynamics. Anal Chem. https://doi.org/10.1021/acs.analchem.3c04255

Panakanti R, Narang AS (2015) Impact of excipient interactions on drug bioavailability from solid dosage forms. In: Excipient applications in formulation design and drug delivery. Springer, pp 273–310. https://doi.org/10.1007/978-3-319-20206-8_10

Parekh BV, Saddik JS, Patel DB, Dave RH (2023) Evaluating the effect of glidants on tablet sticking propensity of ketoprofen using powder rheology. Int J Pharm 635:122710. https://doi.org/10.1016/j.ijpharm.2023.122710

Pérez M, Carou D, Rubio EM, Teti R (2020) Current advances in additive manufacturing. Proc CIRP 88:439–444. https://doi.org/10.1016/j.procir.2020.05.076

Pichayakorn W, Suksaeree J, Boonme P, Amnuaikit T, Taweepreda W, Ritthidej GC (2012) Nicotine transdermal patches using polymeric natural rubber as the matrix controlling system: effect of polymer and plasticizer blends. J Membr Sci 411:81–90. https://doi.org/10.1016/j.memsci.2012.04.017

Pilgaonkar PS, Rustomjee MT, Gandhi AS, Bhumbra VS (2003) Pharmaceutical excipient. US20040228932A1

Pramanick S, Singodia D, Chandel V (2013) Excipient selection in parenteral formulation development. Pharma Times 45(3):65–77

Rajabalaya R, Tor LQ, David S (2012) Formulation and in vitro evaluation of ondansetron hydrochloride matrix transdermal systems using ethyl cellulose/polyvinyl pyrrolidone polymer blends. Int J Pharmacol Pharm Sci 6(12):667–671

Rayaprolu BM, Strom JG (2013) Design and evaluation of D-α tocopheryl polyethylene glycol 1000 succinate emulsified poly-ϵ-caprolactone nanoparticles for protein/peptide drug delivery. Drug Dev Ind Pharm 39(7):1046–1052. https://doi.org/10.3109/03639045.2012.699069

Rayaprolu BM, Strawser JJ, Anyarambhatla G (2018) Excipients in parenteral formulations: selection considerations and effective utilization with small molecules and biologics. Drug Dev Ind Pharm 44(10):1565–1571. https://doi.org/10.1080/03639045.2018.1483392

Rojek B, Wesolowski M (2023) A combined differential scanning calorimetry and thermogravimetry approach for the effective assessment of drug substance-excipient compatibility. J Therm Anal Calorim 148(3):845–858. https://doi.org/10.1007/s10973-022-11849-9

Rojek B, Gazda M, Plenis A (2023) FTIR, Raman spectroscopy and HT-XRD in compatibility study between naproxen and excipients. Spectrochim Acta A Mol Biomol Spectrosc 2013:123048. https://doi.org/10.1016/j.saa.2023.123048

Rosiaux Y, Jannin V, Hughes S, Marchaud D (2014) Solid lipid excipients—matrix agents for sustained drug delivery. J Control Release 188:18–30. https://doi.org/10.1016/j.jconrel.2014.06.004

Sakdiset P, See GL, Sawatdee S, Yoon AS (2023) Preparation and characterization of lidocaine HCl-loaded proniosome gels with skin penetration enhancers. J Drug Deliv Sci Technol 86:104639. https://doi.org/10.1016/j.jddst.2023.104639

Sanchez BS, Fernandez PI, Badiola I, Marquez J (2017) Novel vehicles for the transfection of miRNAs. EP3441074A1

Saokham P, Muankaew C, Jansook P, Loftsson T (2018) Solubility of cyclodextrins and drug/cyclodextrin complexes. Molecules 23(5):1161. https://doi.org/10.3390/molecules23051161

Sharma A (2019) Effect of residual reactive impurities in excipients on the stability of pharmaceutical products. In: Innovative dosage forms: design and development at early stage, pp 91–118. https://doi.org/10.1002/9783527812172.ch3

Sherwood BE, Hunter EA, Staniforth JH (1995) Pharmaceutical excipient having improved compressibility. US5585115A

Sista HS, Espinoza YJ (2017) Excipients for use in adeno-associated virus pharmaceutical formulations and pharmaceutical formulations made therewith. US20180055939A1

Snipes WC (1988) Low-melting moldable pharmaceutical excipient and dosage forms prepared therewith. US5004601A

Soni N, Soni N, Pandey H, Maheshwari R, Kesharwani P, Tekade RK (2016) Augmented delivery of gemcitabine in lung cancer cells exploring mannose anchored solid lipid nanoparticles. J Colloid Interface Sci 481:107–116. https://doi.org/10.1016/j.jcis.2016.07.020

Vadlamudi MK, Dhanaraj S (2017) Significance of excipients to enhance the bioavailability of poorly water-soluble drugs in oral solid dosage forms: a review. In: IOP conference series: materials science and engineering, vol 263(2), p 022023. https://doi.org/10.1088/1757-899X/263/2/022023

Van Gessel AW (2010) Co-processed tablet excipient composition its preparation and use. US20130177649A1

Varshosaz J, Ghassami E, Ahmadipour S (2018) Crystal engineering for enhanced solubility and bioavailability of poorly soluble drugs. Curr Pharm Des 24(21):2473–2496. https://doi.org/10.2174/1381612824666180712104447

Velip L, Dhiman V, Kushwah BS, Samanthula G (2023) Characterization of degradation products and drug–excipient interaction products of erdafitinib by LC–Q-TOF-MS/MS and NMR. Chromatographia 86(8–9):627–638. https://doi.org/10.1007/s10337-023-04268-x

Veronica N, Heng PW, Liew CV (2022) Ensuring product stability—choosing the right excipients. J Pharm Sci 111(8):2158–2171. https://doi.org/10.1016/j.xphs.2022.05.001

Xie XY, Chang ZY, Chen CL, Zhang L, Wang M, Tang C, Xue HR, Gao XL (2023) Physicochemical characterization of three active pharmaceutical ingredients for compound glycyrrhizin solid formulation and compatibility analysis via thermal and non-thermal techniques. J Pharm Innov 18(2):485–496. https://doi.org/10.1007/s12247-022-09656-8

Zarmpi P, Flanagan T, Meehan E, Mann J, Fotaki N (2017) Biopharmaceutical aspects and implications of excipient variability in drug product performance. Eur J Pharm Biopharm 111:1–5. https://doi.org/10.1016/j.ejpb.2016.11.004

Advances in Pharmaceutical Oral Solid Dosage Forms

5

P. Saikiran, T. Pawan Kumar, Shristi Arya, Darshana Tijare, Soham Loharkar, Gopal Bajad, Deepankar Bahuguna, Pawan Devangan, Atul Mourya, Harithasree Veerabromma, Chantibabu Katta, and Jitender Madan

Abstract

The pharmaceutical and healthcare industries are constantly moving through a period of unmatched changes. Oral administration of solid dosage forms is one of the most preferred route and is highly patient compliant as well as stable when compared to other dosage forms. However, the challenge of transition from a traditional manufacturing approach to continuous automated processing approach remains unmet. Continuous modifications in materials and manufacturing technologies are gradually being implemented in the industry to avail the benefits related to process automation, improved quality and reduced costs. Therefore, it is well accepted that there is a growing demand to discover alternative processes and design formulation strategies that can significantly improve powder processing techniques, granulation methods and equipment to ensure safe, reproducible and quality products are manufactured. The intended chapter presents an overview of the state-of-the-art research equipment, material handling techniques, process analytical technique implementation and 3D printing application strategies. It covers recent advances in techniques and processes at every stage of pharmaceutical product development and methods employed to troubleshoot the processing hurdles.

Keywords

Oral route of administration · Solid dosage form · Materials and manufacturing · 3D printing

---

P. Saikiran · T. Pawan Kumar · S. Arya · D. Tijare · S. Loharkar · G. Bajad · D. Bahuguna · P. Devangan · A. Mourya · H. Veerabromma · C. Katta · J. Madan (✉)
Department of Pharmaceutics, National Institute of Pharmaceutical Education and Research, Hyderabad, Telangana, India

K. Jain, A. K. Yadav (eds.), *Advances in Pharmaceutical Product Development*, https://doi.org/10.1007/978-981-97-9230-6_5

5.1 Introduction

In the vast landscape of pharmaceutical formulations, conventional oral solid dosage forms hold a significant place as one of the most commonly utilised and accessible means of drug administration (Zhang et al. 2004). These solid preparations, designed to be swallowed and dissolved in the gastrointestinal tract, have played a crucial role in the treatment of various diseases and ailments. Furthermore, the popularity of conventional oral solid dosage forms can be attributed to their ease of administration, allowing patients to self-administer medications conveniently without the need for specialised medical procedures.

Various kinds of oral solid dosage forms such as tablets, capsules, lozenges, powders, granules and many other formulations are used for the delivery of active pharmaceutical ingredients (API) (Arshad et al. 2021). Tablets can be further categorised into immediate release (compressed tablets), extended release, chewable, effervescent, sublingual, buccal, enteric-coated, dispersible, scored and film-coated tablets, each serving specific therapeutic purposes (Lachman et al. 1976). Film-coated tablets are compressed tablets that are coated with a thin layer of film. These coatings have become the first choice of formulators due to advances in material science and polymer chemistry (Almoazen and Felton 2013). Enteric-coated/functionalised coated tablets are compressed tablets, coated with substances that resist dissolution in gastric fluid but disintegrate in the intestine (Maderuelo et al. 2019). A layer or film made out of water-soluble materials is the main idea behind capsule development which surrounds drug or active API, hence rendering it odourless, tasteless, elegant and easy to swallow. There are presently two main types of capsules, viz. hard gelatin and soft gelatin capsules (soft shells) (Hoag 2017). Soft shell capsules (soft gels) are typically used to encapsulate non aqueous liquid and semi-solid formulations, and hard shell capsules are used to encapsulate both solid and nonaqueous formulations (Gullapalli and Mazzitelli 2017). Pellets are dosage forms similar to tablets and capsule which can be prepared by different types of novel techniques. They are small solid substances composed of highly purified active ingredient with or without excipient (Kurbanoglu et al. 2017).

Despite many advantages offered by conventional dosage forms, they often suffer from some major limitations like low bioavailability, repeated dosing, peripheral adverse effects and lack of patient compliance (Laffleur and Keckeis 2020). Lack of targeted action, premature decomposition of drugs and fluctuations in plasma drug concentrations are the extended limitations of oral solid dosage forms. For very potent medicines, the precise dosing may be difficult or impossible with conventional dosage forms (Adepu and Ramakrishna 2021). Solubility, which ultimately affects the bioavailability of the medicinal product and results in less effective action, is a major problem associated with oral solid dosage forms. A complex chain of physicochemical, biological, physiological and anatomical factors acting independently and in concert with each other to limit drug bioavailability is a major obstacle to successful oral administration (Boyd et al. 2019). Another barrier to the administration of medicinal products orally is inadequate absorption and first-pass metabolism, which has an effect on the amount of drugs reaching systemic circulation necessary for their therapeutic efficacy (Devadasu et al. 2018).

To overcome the limitations associated with conventional dosage form, we need advanced technologies. The major aim of the drug delivery system is to liberate the substance at the correct time in an adequate concentration on a target site (Laffleur and Keckeis 2020). In addition, novel raw materials have been used to improve manufacturability and functionality. The development of additive manufacturing technology, 3D printing medical devices, which offer key advantages over traditional drug delivery systems, has gained increasing popularity. The ability to produce 3D structures that are custom designed and have a complex architecture is of paramount importance, as well as easy access to personalised medicines (Wang et al. 2021a). Automated medication system helps to minimise errors in drug administration process, and along with this it is cost-effective (Risør et al. 2017). Artificial intelligence (AI), which helps to increase the quality of a product and assist in its optimisation, is emerging as an important tool for pharmaceutical industry (Khanna et al. 2020). Process analytical technology (PAT) is an indispensable tool for implementing quality by design (QbD), enabling effective process parameter monitoring and the fabrication of finished pharmaceuticals of the highest standard. The development of analytical QbD techniques facilitates the development of an appropriate control plan to govern the analytical method, hence reducing variability and enhancing robustness in method performance with exceptional quality (Haneef and Beg 2021). Real-time release testing (RTRt) of the product is made possible when a thorough understanding of the process is combined with the application of quick and precise analytical sensors. To do this, identify the product's critical quality attributes (CQA), followed by the critical process parameters (CPP) and the critical material attributes (CMA), all of which have a big impact on the CQAs (Galata et al. 2021).

Progress of machine design over the years has led to the development of continuous manufacturing technologies which increases productivity by application of PAT. The intended chapter aims to summarise recent advances in the field of oral solid dosage forms highlighting key aspects of novel excipients used in its manufacturing. It throws light upon the utilisation of AI and 3D printing in solid oral dosage form.

5.2 Novel Excipients Involved in Manufacturing of Oral Solid Dosage Form

Pharmaceutical excipients are components used in the manufacturing process or included in the dosage form of a final pharmaceutical product but are not pharmacologically active drugs or prodrugs. The word excipient is derived from the Latin word excipere, meaning 'to except', which is simply described as 'other than' (van der Merwe et al. 2020). Excipients can make up to 80–90% of the formulation of a drug product and are vital to the formulation of drug products because they provide for the efficient distribution of therapeutic ingredients. A pharmaceutical excipient

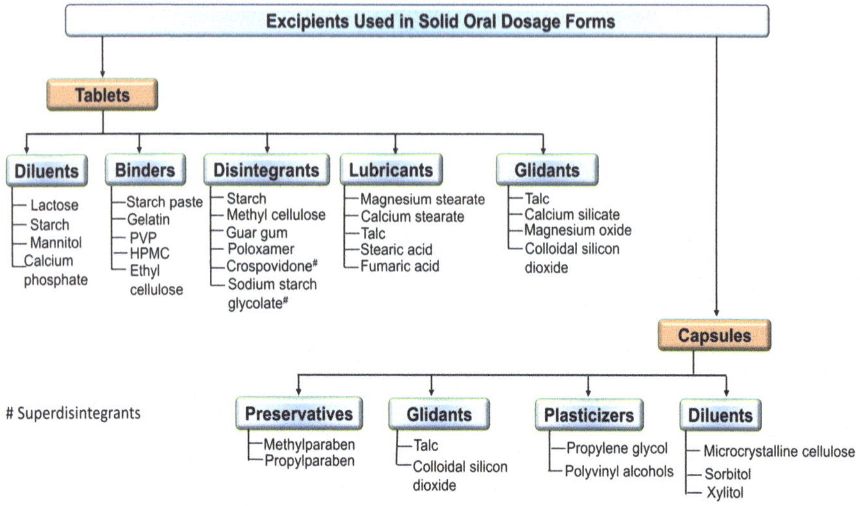

Fig. 5.1 Conventional excipients used in the manufacture of pharmaceutical solid oral dosage form

being inert offers many functions and is used to increase the bulk of medicament during its formulation, improve product's precision as well as API dose accuracy, increase bioavailability and finally produce a more palatable final pharmaceutical form, with increased patient acceptance (Fig. 5.1) (Abrantes et al. 2016).

Chemically modified excipients are categorised as novel excipients which can be listed in inactive ingredient database (IID). Single-component excipients may not always offer the necessary performance to enhance the formulation characteristics for proper manufacturing of some active pharmaceutical compounds (Chaudhari et al. 2012). Formulation scientists have an increasing number of co-processed excipients introduced into the market. New approaches for enhancing excipient functionality include new combinations of presently used excipients (van der Merwe et al. 2020), e.g. binders, disintegrants, sweeteners, colorants and co-processed excipients.

5.2.1 Binders

Pharmaceutical binders are designed to produce a product with the appropriate flow qualities. These components employed to give the powder ingredients a specific shape and size (Kar et al. 2019). Binders are applied to the powders to improve the flow qualities of the powdered raw material by reducing surface area and increasing cohesiveness (Allenspach et al. 2020). They are obtained from natural origin (starch), semisynthetic origin (hydroxyl ethyl cellulose) and synthetic origin (povidone). Now a days, different grades of binders are introduced that have more advanced properties such as better drug release and controlled pharmacokinetics than conventional forms.

5.2.1.1 Hydroxy Propyl Methyl Cellulose (HPMC)

HPMC is used in a direct compression process or for continuous production. HPMC exhibits poor flow for which new grades of direct compression HPMC have been created. Three distinct novel direct compression (DC) grades of HPMC were used to improve flow properties (K4M, K15M and K100M). Physical characteristics, bioadhesive strength, buoyancy lag time, swelling index and in vitro drug release studies were assessed for the prepared tablets (Pittu and Sharma 2013).

5.2.1.2 LYCATAB

These are pregelatinised starches, which are often analysed with around 30% cws. Conventionally, starch has limitations of low cold water solubility (Elballa and Salih 2022). Pregelatinised starch obtained from Roquette is water dispersible and acts as suitable binder.

5.2.1.3 GalenIQ (Isomalt)

GalenIQ (isomalt) is one of the most multi-functional water-soluble filler and binder that has low hygroscopicity and is available in different pharmaceutical grades based on solubility. It can be milled to desired particle size for preparing solid oral dosage forms. It can be a mixture of D-sorbitol and D-mannitol suitable for a direct compression process. It possesses excellent flowability, negligible sticking to tableting tools and a very good compatibility. It is widely used in pharmaceutical preparations for coating of oral dispersible tablets and capsules.

5.2.2 Disintegrants

Disintegrants are chemical substances or mixtures of chemicals that are added to a drug formulation to help break down or disintegrate the contents of tablets or capsules into smaller pieces which would likely dissolve more quickly than they would otherwise. Superdisintegrants are a new class of agents that are added to improve dissolution and disintegration properties of dosage form in less than 30 s, thereby enhancing drug release absorption (e.g. croscarmellose, sodium starch glycolate) (Desai et al. 2016).

Modified nanocrystalline cellulose (NCC) is a novel pharmaceutical excipient which acts as a super disintegrant. It has been derivatised from microcrystalline cellulose (MCC) via cross-linking process. Structural characteristics, such as low porosity, cause unnecessary swelling, which restricts pharmaceutical applications of MCC. However, NCC has better biocompatibility, biodegradability, low cost, nontoxic and possesses characteristics such as large specific surface area, high tensile strength, stiffness and lightweight (Sheikhy et al. 2021). Malic acid used as a cross-linking agent reacts with NCC polymer to form double bond cross-linker. In the graft copolymer nanocrystalline cellulose-poly (2-hydroxyethyl methacrylate-co-itaconic acid) as modified NCC, it was hypothesised that adding hydrophilic itaconic acid (IA) and relatively more hydrophobic poly (2-hydroxy-ethyl methacrylate) (PHEMA), HEMA, as co-monomers, may improve the HLB and increase swelling property.

5.2.3 Lubricants

Lubricants are used to attain uniform tableting force distribution and tablet density distribution. These are utilised to reduce friction between the particles or tablets and the die-hole wall (Miller and York 1988). This allows the compressed tablet to be pushed out of the die-hole smoothly while also requiring lesser force and wearing down less frequently. Sodium stearyl fumarate is a relatively new lubricant and comparatively less hydrophobic than stearic acid. It imparts lesser effect on tablet disintegration and has higher ability to lessen friction and adhesion to the punches. The particle size of sodium stearyl fumarate was more important, as it is an alternative to magnesium stearate. It is generally nontoxic and nonirritant material and can be included in IID in a concentration range of 0.2–0.5% (de Backere et al. 2022).

5.2.4 Co-processed Excipients

Co-processing is an appealing method for enhancing the material properties of APIs, which can be done at the point of API isolation. The pharmaceutical industry is becoming more and more interested in continuous manufacturing techniques (Stocker et al. 2023).

5.2.4.1 Kollitab™ DC 87 L

It is a new all-in-one tableting excipient material with four characteristics such as binder, filler, disintegrant and lubricant. It enables quick tablet disintegration, offers greater flowability and produces high tablet strength while using a wide range of compression strengths (both low and high), eases formulation procedures and decreases the complexity of production.

5.2.4.2 COMBILOSE

Lactose is the primary pharmaceutical diluent in solid oral dosage forms. Hence, lactose's inadequate compressibility and flow characteristics prevent it from being used as a direct compressible filler and binder (Somnache et al. 2023). Maltose monohydrate 10%, maize starch and lactose monohydrate 20:1 were all processed together using co-freezing and co-drying techniques which has the potential to offer improved dilution capability and compressibility with less susceptibility to lubricants.

5.2.4.3 PEARLITOL CR-H

PEARLITOL formulations are used for modern direct compression techniques. PEARLITOL® CR-H is a co-processed blend of hydroxypropyl methylcellulose (HPMCK4M) (70%) and D-mannitol (30%) that enables the controlled release of active medicinal ingredients, with excellent functional characteristics that improve tablet processibility, such as flowability (Jin et al. 2023).

5.2.4.4 PROSOLV EASYtab SP (Silicified Microcrystalline Cellulose)

It is a multifunctional homogenous lubricant-coated co-processed excipient composite. It is ready-to-use material comprising of four individual components such as silicified MCC, colloidal silicon dioxide, sodium starch glycolate and sodium stearyl fumarate which is a binder-filler, glidant, super disintegrant and a lubricant, respectively (Darzuli et al. 2019). These components maintain their chemical identities while synergistically providing increased functional performance (Buckton et al. 1999). Utilisation of PROSOLV EASYtab SP has the potential of increasing profitability through cost savings brought by increased productivity, less setup and cleaning costs, improved yield and reduced loss.

5.3 New-Age Material Handling Techniques Developed

In the pharmaceutical sector, particularly when producing solid oral dosage forms, material management is crucial (American Society of Health-System Pharmacists 2006). Throughout the production process, material handling comprises transportation, storage, management and protection of raw materials, completed items and other things (Steenweg et al. 2021). The efficacy and efficiency of material handling operations significantly affect the cost, lead time and quality of the final product (Jagtap et al. 2022). New-generation material handling solutions have emerged in recent years as a result of technological breakthroughs in material handling, offering various advantages over conventional methods (Arshad et al. 2021). Oral solid dosage forms are taken orally and then absorbed in gastrointestinal tract to provide a localised therapeutic impact in the mouth, throat, digestive tract or systemic action in the body (Goodin et al. 2011). APIs and acceptable excipients can be milled, dried, encapsulated, mixed, granulated or tableted to create oral solid dosage forms. Due to its simplicity and resulting patient compliance, a variety of solid oral dosage forms, including tablets, capsules, lozenges, powders and granules, have already been used effectively for conveying API (Ng et al. 2022). Some of the most well-liked and thoroughly researched areas of oral solid dosage form development include tablet formulations that offer a unit dose which is either immediate drug release, modulated release or taste-masked (Finke and Kwade 2021). In order to create durable tablet dosage forms, quality by design (QBD)-based formulation methodological approaches are often used to decrease variation in processes. In order to enhance the functioning and manufacturability of tablet formulations, new raw ingredients were also used (Jagtap et al. 2022). New Age Material Handling Methods 1. Automated dispensing systems, 2. Vacuum conveying systems, 3. Flexible screw conveyors.

5.3.1 Automated Dispensing System

Solid oral dosage form material handling in the present era relies heavily on automated dispensing devices. Electronic devices called automated dispensing systems are used to accurately and consistently deliver medicine (Paul et al. 2023). They have transformed the way pharmacies and pharmaceutical companies handle pharmaceuticals, making the procedure quicker, more effective and more precise (Sng et al. 2019). Automated dispensing systems (ADS) are computer-controlled machines that store, dispense and track medication inventory (Shin et al. 2023). The computer system monitors the medicine dispensing process and controls the dispensing device. The computer system may also produce reports on drug consumption, stock levels and patient information (Gharib et al. 2023). For the purpose of ensuring that prescription orders are appropriately filled and documented, ADS can also be connected with electronic health records (EHR). Medication error reduction is one of the main advantages of ADS (Brax et al. 2023). One research found that one in five drug dosages given in hospitals are delivered incorrectly, demonstrating the prevalence of pharmaceutical mistakes in the healthcare industry (Paimard et al. 2023). By ensuring that the right prescription and dosage are administered, automated dispensing systems may dramatically lower the chance of medication mistakes. The capacity of ADS to enhance drug management is another benefit. It is simpler for pharmacies to maintain their stock levels and replenish medication as necessary, thanks to ADS's ability to keep pharmaceutical inventory and track consumption (Benoit and Beney 2011). These devices automate and employ robotics to correctly distribute the right amount of material. By delivering drugs in exact amounts, ADS can assist decrease waste by lowering the quantity of medication that is thrown away (Alahmari et al. 2022). On the market, there are several varieties of automated dispensing systems (ADS). The most typical examples include the following:

5.3.1.1 Unit Dose Dispensing Systems

These systems distribute medication in premeasured amounts, which makes it simpler for patients to take their prescription as prescribed and lowers the possibility of medication mistakes (Hänninen et al. 2023). These systems deliver medication in specific dosages, making them perfect for healthcare facilities like hospitals and nursing homes (Dubourg et al. 2013).

5.3.1.2 Centralised Dispensing Systems

In hospitals, clinics and other healthcare institutions, centralised dispensing systems are frequently used to administer solid oral dose forms including tablets, capsules and powders (Balka and Nutland 2004). Compared to conventional manual dispensing techniques, these systems provide a number of benefits, including enhanced efficiency, accuracy and safety (Tsao et al. 2014).

5.3.1.3 Robotic Dispensing Systems

Robots are used in these systems to take medications out of storage and provide them to patients (Hänninen et al. 2023). These robots are set up to retrieve pharmaceuticals from predetermined storage places and distribute them into bottles or blister packs of various sizes (Batson et al. 2021).

5.3.2 Vacuum Conveying Systems

Vacuum conveying is one of the most widely used material handling techniques in solid oral dosage forms (Beaulac et al. 2022). A vacuum generator, a conveying pipeline and a receiving vessel are the three major parts of vacuum conveying systems for solid oral dosage forms. The solid dosage form is dragged through the pipeline and into the receiving vessel by the negative pressure that the vacuum generator produces in the pipeline (Verstraeten et al. 2017). Stainless steel or similar substance suited for pharmaceutical uses is often used to make the pipeline (Tu and Yan 2022).

The utilisation of advanced sensors and controls is among the major developments in vacuum conveying systems for solid dosage forms (Bhatia 2019). These systems have the ability to continuously monitor the flow of material and adapt the vacuum pressure and flow rate as necessary (Hou et al. 2023). This makes it possible to precisely manage the conveying process, which is necessary when working with delicate or sensitive materials (Chen et al. 2022). The application of specialist machinery for handling powders and other fine materials is another development in vacuum conveying systems for solid oral dosage forms (Klinzing et al. 2010). These valves are intended to stop substances from escaping into the environment or being caught in the valve. Additionally, they may be set up to reduce dust production, which is crucial for maintaining a hygienic and secure working environment (Gomes de Freitas et al. 2021). The accuracy, dependability, and adaptability of vacuum conveying systems to a variety of pharmaceutical applications are increasing due to advancements in sensor technology, specialised machinery, and system design (Kuang et al. 2020).

5.3.3 Flexible Screw Conveyors

Flexible screw conveyors are a kind of mechanical conveyor that transports goods through a tube or duct using a revolving helical screw. This approach is very adaptable and may be created to meet the particular requirements of a production facility (Chongchitpaisan and Sudsawat 2022). Since they can handle a variety of materials and offer flexibility of conveying distance, angle and routing, these conveyors are enormously beneficial in the manufacturing of pharmaceuticals. Additionally, flexible screw conveyors are kind to the goods and can reduce break or damages (Zhang et al. 2021a).

Flexible screw conveyors' capacity to handle a variety of materials is one of its main benefits. This is crucial in the production of pharmaceuticals since various materials may need to be carried at various points along the manufacturing process (Zhang et al. 2021b). Flexible screw conveyors are an adaptable option for pharmaceutical manufacturing since they can easily handle powders, granules, capsules, tablets and other products. Additionally, flexible screw conveyors are simple to clean and maintain, which is crucial in the pharmaceutical business where cleanliness and hygienic conditions are of the utmost importance (Dhaval et al. 2022). Furthermore, these conveyers are easy to clean and are made of materials that are resistant to corrosion and degradation (Dahlgren et al. 2019).

5.4 Advantages of New-Age Material Handling Techniques

5.4.1 Automation

Modern material handling techniques are characterised by automation, which provides organisations with a number of advantages (Sng et al. 2019). Automated solutions may boost productivity, cut labour expenses, increase accuracy and lower the risk of accidents and injuries from human handling (Rao and Pathak 2022).

5.4.2 Enhanced Safety

Modern techniques for material handling can also increase workplace security. Automated machinery and systems can lower the chance of disasters and injuries, which is crucial in fields that deal with big or bulky objects (Denis et al. 2020).

5.4.3 Higher Productivity

Modern approaches to material management can boost output in a number of ways (Ansari and Gangil 2022). Automated systems reduce bottlenecks and increase efficiency by handling a greater number of materials in a shorter amount of time (Björnsson et al. 2018).

5.4.4 Enhanced Accuracy

Additionally, automated devices can increase the precision of material handling operations. These systems are made to carry out repeated activities precisely, lowering the possibility of mistakes and enhancing data accuracy (Arshad et al. 2021).

5.4.5 Reduced Costs

Modern material handling techniques can also aid in cost-cutting for companies (Zaman et al. 2022). Because they require less manual effort to complete tasks, automated devices can lower labour expenses (Arvind and Gunasekaran 2014).

5.5 Limitations of New-Age Material Handling

Even though material handling advancements have many advantages, they also have significant limitations (Björnsson et al. 2018). The necessary initial investment, the possibility of technology failure and the disappearance of the need for labour are only a few of the drawbacks (Denis et al. 2020).

5.6 Utilisation of Artificial Intelligence in Solid Oral Dosage Forms

The practice of simulating human intellect with computers is known as artificial intelligence (AI). In 1956, Marvin Minsky and John McCarthy hosted a symposium where the idea was first put out (Hassanzadeh et al. 2019). Technology adoption may reduce costs and save time while also improving understanding of formulation and process characteristics (Agrawal 2018). Previously, AI was only used in the field of engineering, but more recently AI has become crucial in a number of pharmacy-related fields, along with drug discovery, the development of drug delivery formulations, marketing, management, quality management, hospital pharmacy, etc. (Grof and Štěpánek 2021). In the creation of drug delivery formulations, different artificial neural networks (ANNs), including deep neural networks (DNNs) and recurrent neural networks (RNNs), are used (Elbadawi et al. 2021a). Then again, de novo design encourages the development of much newer medicinal compounds that possess the desirable characteristics (Das et al. 2021). The pharmaceutical market is dominated by solid dosage forms (Shaikh et al. 2018). Tablets are thought to make up the majority of solid dosage forms and account for more than two thirds of the worldwide market (Xiouras et al. 2022). AI may be used to create improved formulations, which would be extremely beneficial to the pharmaceutical sector (Lou et al. 2019). Four phases make up a typical AI workflow: data gathering and preparation, AI modelling, simulation, testing and implementation (McCarthy et al. 2006). Implementing algorithms and seeing patterns in data to aid decision-making is known as machine learning, a branch of AI (Catania 2021), artificial neural networks (ANNs). Compared to traditional machine learning algorithms, ANNs (artificial neural networks) drew inspiration from the biological neuronal structure of human brains that provide superior computational and predictive power (Wang et al. 2021b). Additionally, deep learning has been extensively employed for several applications, including image classification, object recognition, image segmentation, natural language processing (NLP) and medical image analysis (Elbadawi et al. 2021b). Since it has been used so frequently in the pharmaceutical sector, AI-based drug development is viewed as a potentially effective alternative to the traditional approach (Bannigan et al. 2021). The framework known as 'Pharma 4.0' aims to address several persistent problems in the pharmaceutical manufacturing industry by using modern digital methodologies (Nagy et al. 2019). In 2021, the US FDA released the 'Artificial Intelligence/Machine Learning (AI/ML)-Based Software as a Medical Device (SaMD) Action Plan' with the goal of adjusting regulatory oversight and enabling the enhancement of patient lives (Wang et al. 2022). Solid dosage forms are the most significant form of dosage forms in pharma field, and over 50% of new molecular entities (NMEs) are being accounted repeatedly according to the Food and Drug Administration Center for Drug Evaluation and Research (FDA CDER), because of its several advantages, which include shelf stability, patient adherence, simplicity of transportation and exact dosing (Suryadinata 2017).

5.6.1 Widely Used Databases

Obtaining a database is the initial stage in executing an AI-based study. A high-quality database must first be created in order to properly create a workable formulation development model (Liu et al. 2020). There are various traditional methods for creating a modelling database with analytical techniques like the Design of Experiment (DOE) tool (Zhao et al. 2019). Scientists can identify the connection between many elements and reactions using the organised approach known as Design of Experiments (DOE) (Lee et al. 2022). They may also ascertain how various components interact and maximise the response (Palo et al. 2021).

5.6.2 Methods for Processing Data

Before creating the models, researchers must process the raw data they have obtained from public resources or internal experimental findings (Dong et al. 2021). To adapt and then evaluate the data, it is important to employ several extensively used techniques, such as data cleaning, dimension reduction, unbalanced data solutions and data splitting (Han et al. 2019). The removal of data points and their replacement with mean and median values are two ways to clean up data (Hesse et al. 2021). One more essential step in data processing is data partitioning (Mak and Pichika 2019). The entire data set will normally be separated into three subgroups using this methodology: training, validation and evaluation. Consequently, prior to performing modelling tasks, data processing and splitting procedures are required (Vamathevan et al. 2019).

5.6.3 Development of Solid Dosage Forms Using AI Algorithms

In recent times, a variety of AI-based approaches have been effectively used in the creation of pharmacological solid dosage forms. Artificial intelligence (AI) is a combination of information technology, data analysis and mathematics (Shah et al. 2019). Combining computer science, data analytics and mathematics generates artificial intelligence (Paul et al. 2021). ML is a branch of AI that is often divided into three categories: supervised learning, unsupervised learning and reinforcement learning (Aksu et al. 2012). A method known as 'deep learning' comprises of output/target variables that will be estimated from a collection of input variables (Castro et al. 2021). Throughout the training process, a relation of the input vs. intended output will be developed, resulting in the accuracy level that is wanted (Shaheen 2021). Solid dose formulations have been generated using a variety of deep learning techniques, including decision tree, logistic regression, linear regression, k-nearest neighbors (KNN), random forest, XGBoost, LightGBM and support vector (Ma et al. 2020). Unsupervised machine learning is an algorithm that controls just the input variables and uses feature-finding and grouping techniques (Patel and Shah 2022). A branch of machine learning called deep learning (DL) uses cutting-edge algorithms like convolutional neural networks to learn from a significant quantity of experimental data (van der Lee and Swen 2023). In order to make predictions, deep learning algorithms introduce a very complex model structure. In recent decades, deep learning (DL) algorithims, are successfully being used in pharmaceutical field for different purposes in development of a solid oral dosage formulation (Fig. 5.2).

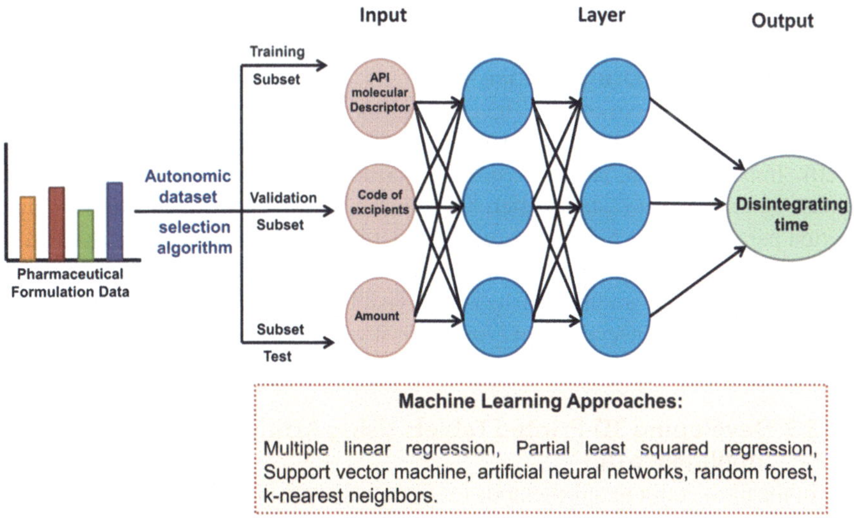

Fig. 5.2 AI algorithms used for process optimisation of oral dosage forms

5.6.4 Evaluation and Explainability of Model Predictive Performance

Assessment of the models' predicting abilities is required following the machine learning modelling procedure (Shaheen 2021). We may categorise the measures used to evaluate the predicted accuracy into regression metrics and categorisation metrics (Khanna et al. 2020). The coefficient of determination (R^2), mean squared error (MSE), root mean squared error (RMSE) and mean absolute error (MAE) are often used metrics for evaluation of regression modelling assignments (Moingeon et al. 2022). A confusion matrix will be used to construct various metrics, such as accuracy, precision and recall, for classification modelling work initially. However, in the process of unbalanced classification modelling, the outcomes of several classification metrics, such as accuracy, are deceptive (Floresta et al. 2022). As a result, new assessment measures including Cohen's kappa, receiving operating characteristic (ROC) and area under the curve (AUC) are used to evaluate models (Lowe et al. 2022).

5.6.5 Applications of Artificial Intelligence (AI) in Solid Dosage Forms

5.6.5.1 Tablets

The most important oral solid dose form is a tablet. A tablet is often created by compression or moulding and contains a combination of APIs and excipients (Lou et al. 2019). Excipients are substances added to tablets to promote tableting performance. Examples include lubricants, glidants, binders, diluents, sweeteners, food colouring and drug release modifiers which all serve to improve aesthetics and disintegrants and sustain release polymer coating (Lamberti et al. 2019).

5.6.5.2 Predicting Drug Release

During the development of a product, two of the most important preclinical trials are drug release studies, including in vitro and in vivo tests (Dropka and Holena 2020). Important material properties and critical processing variables have an impact on medication release patterns. For instance, little adjustments to the compaction parameters, such as pressure and tablet shape, or other elements, like drug loading, may have a big impact on dissolving rates (Elbadawi et al. 2021b). In addition, specialised apparatuses such as UV-visible spectrophotometers and USP-approved vessels are needed for a typical in vitro drug release investigation (Wirtz et al. 2019).

5.6.5.3 Developing 3D-Printed Tablets Using Artificial Intelligence (AI)

One of the most cutting-edge methods for customised medicine is three-dimensional (3D) printing, which has the capacity to create tablets taking patients' physiology, genetic profiles and pharmacological responses into account (Boobier et al. 2020).

Individualised 3D-printed tablets have been created using a variety of techniques, including fused filament manufacturing, binder jetting and selective laser sintering (SLS) (Duch et al. 2007). AI technologies have enormous potential to be implemented into this approach and discover the design window in order to improve the 3D printing process and decrease the test effort with many variables.

5.6.5.4 Detecting Tablet Defects

During the production process, it is normal for tablets to exhibit flaws including cracking, capping, binding and sticking. These damaged tablets often need to be filtered out manually, which calls for a sizable workforce and is difficult to scale up. X-ray computed tomography (XRCT) (Doerr and Florence 2020), for instance, can be utilised to analyse the interior structure of tablets in order to fix this problem. Researchers have effectively detected tablet faults by combining XRCT with the deep learning approach to broaden the use of this technology.

5.6.5.5 Granules

Another pharmaceutical solid dose formulation made up of aggregation of powder particles containing medicines and excipients is called granules. Granules are more shelf-stable than liquid formulations and provide variable dosing for people who have problems in swallowing pills or capsules (Zhao et al. 2021). Artificial intelligence techniques were used to assess and anticipate the medicine contents in sugar-free granules (Jiang et al. 2022). In this work, near-infrared (NIR) spectroscopy first showed that it was possible to determine the amount of medication in granules. The drug remaining was then predicted using several machine learning techniques using the near-infrared (NIR) spectrums. This proves the drug content in granules may be quantified using AI models, which are acceptable tools for doing so (Grof and Štěpánek 2021).

5.7 Continuous Manufacturing Technology

The technique of making medicinal items constantly, without breaks or stops in the assembly line, is known as continuous manufacturing technology of solid oral dosage forms (Rogers et al. 2013). This strategy is distinct from conventional batch manufacturing, which involves creating medicinal items in large batches and then placing them through additional processing procedures (Lee et al. 2015).

The continuous manufacturing (CM) approach, in contrast, has been effectively utilised for many years in the petrochemical, food, consumer goods and automotive industries to increase production productivity and save costs (O'Connor et al. 2016). In a batch mode, raw materials are loaded into a single-system operation after which the process is performed at approved parameters until the predetermined endpoint is reached (Hock et al. 2021). Until all standards of quality are satisfied and the material may be moved to the following unit process, these intermediates are momentarily kept in the warehouse (Myerson et al. 2015). For steady-state processing, a constant hold-up mass must be present in the system, and this must be done while

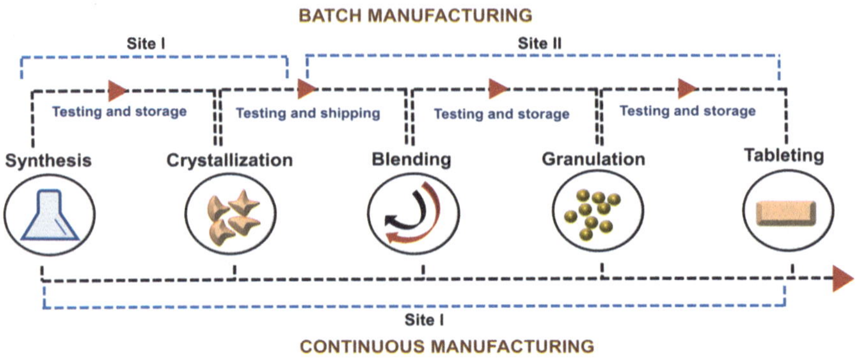

Fig. 5.3 Overview of batch and continuous manufacturing technology of solid oral dosage form

the same flow rate is constantly used to remove the final goods from the process (Yu et al. 2014). A traditional manufacturing process for pharmaceutical products is made up of several independent unit operations such as high shear granulation, tableting, drying and blending (Haimhoffer et al. 2021). The unit activities are all combined into a single production train in an integrated continuous manufacturing process, without breaks or starts between units of work (Fig. 5.3) (Rathore et al. 2015). Recalls due to poor product quality have led to medicine shortages (Nambiar et al. 2022). These quality problems were frequently brought on by out-of-date production techniques and tools as well as a shortage of efficient quality control procedures (Burcham et al. 2018). In addition to advancing technology, including high-quality medications into products throughout development will strengthen process capabilities (Poechlauer et al. 2013). The ICH (International Conference on Harmonisation) Q8 defines QBD (quality by design) as a systematic method of product development that starts with predetermined goals and places a significant emphasis on product/process understanding and control systems as a solid scientific basis for quality risk management (Welch et al. 2017). Establishing clinically applicable standards is a key goal of QBD (quality by design), i.e. relevant standards for product quality that are focused on ensuring clinical efficacy (Balogh et al. 2018). Secondly, a more capable process will result in less variation in the final output, which will lower the prevalence of flaws (Thabet et al. 2018).

5.7.1 Overcoming Obstacles to Continuous Manufacturing

When it comes to combining unit activities into a single continuous process train, the pharmaceutical sector has been somewhat cautious. A number of obstacles still need to be overcome before continuous manufacturing (CM) becomes a preferred and widely utilised technology platform throughout the industry (Medendorp et al. 2020).

5.7.1.1 Regulatory Uncertainties

For many years, the regulatory setting tended to restrict any modifications made to medicinal products after their approval (Srai et al. 2015). This contributed to a mindset whereby batch procedures were still seen as the only viable option, despite missed possibilities to accelerate the manufacture and release cycle (Srai et al. 2020). The FDA's innovative technology team is willing to support the development of this ground-breaking technology by engaging in early and participatory talks with the organisation while creating a continuous manufacturing (CM) process (Nasr et al. 2017). Process validation is the gathering and assessment of data from the method design stage to commercial operations that creates scientific proof that a process is capable of reliably producing high-quality products (Matsuda 2018).

5.7.1.2 Process Automation Technologies (PAT)

Most process automation technology (PAT) applications used nowadays in tablet secondary manufacture are based on NIR (near-infrared) or Raman spectroscopy, since these methods do not require sample preparation and provide quick, nondestructive assessments of physical and chemical characteristics (Fig. 5.4) (O'Connor et al. 2016). Both NIR (near-infrared) and Raman spectroscopies are molecular spectrometric methods, yet they complement one another since they detect functional groups in molecules differently (Allison et al. 2015). NIR (near infrared) is more sensitive to polar bonds and asymmetric vibrations, whereas Raman spectrometry is especially effective for examining nonpolar bonds and symmetric vibrations (Esmonde-White et al. 2022). Raman spectroscopy has a lower limit-of-detection than NIR spectroscopy. The most common problem for in-line monitoring for NIR (near-infrared) and Raman spectroscopy is low detection sensitivity (De Beer et al. 2011). Nevertheless, it should be considered that NIR (near-infrared) and Raman spectrometry detection sensitivities are extremely formulation dependent.

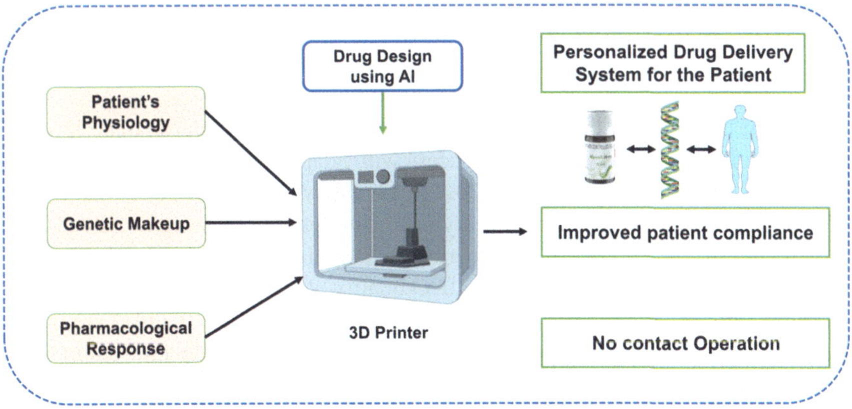

Fig. 5.4 Utilisation of PAT tools for optimisation of pharmaceutical solid dosage forms

5.7.1.3 Equipment

Although many of the unit activities used to manufacture pharmaceutical tablets are essentially continuous and have been well researched for years, integrated continuous from-powder-to-tablet systems are still relatively new (Schmidt et al. 2018).

5.7.1.4 Growth in Knowledge

There is a shortage of experienced scientists, process engineers, and operators (Verstraeten et al. 2017). Accordingly training are required to spread cutting-edge information (Ye et al. 2019). Additionally, the number of contract research (CRO) and manufacturing companies with cutting-edge continuous manufacturing (CM) facilities and knowledge is rather small, which limits the external network's capacity for production (Kallakunta et al. 2019). Pharmaceutical quality by design (QBD) seeks to create a reliable medication product that satisfies the necessary critical quality attributes (CQAs). The pharmaceutical sector is steadily embracing continuous manufacturing (CM), and more research is being done on completely integrated powder-to-tablet production lines rather than studies at the unit operation level (Van Snick et al. 2017). These studies are desperately needed because interactions between unit activities have an impact on the quality of the end output, and having this information is crucial for creating and implementing effective control loops (De Leersnyder et al. 2018).

5.7.1.5 Modern Process Control Techniques

When it came to the earliest commercial continuous manufacturing (CM) implementations, conservative control measures were used leading to a shortage of regulatory and legal standards. But several of them involved switching from batch to continuous production as a second source for pharmaceutical products, with no limitations on time and resources (Nasr et al. 2017).

5.7.1.6 Levels of Control

Three levels may commonly be distinguished among control strategy implementations. The pharmaceutical sector often uses the lowest degree of control (i.e. level 3), which is based on carefully controlled material properties and process variables. Here, rigorous finished product testing makes up for the lack of knowledge about many causes of variability (Schaber et al. 2011).

Through the creation of a multifactorial design space, an intermediate level of control (i.e. level 2) seeks to run the process with adaptable raw material properties and process variables (Chopda et al. 2022). A window of opportunity to move controls upstream and decrease the quantity of finished product testing is created here by the better product and process expertise, which also makes it easier to identify possible variability sources that might affect quality of the product. The utmost level of control (i.e. level 1) guarantees that critical quality attributes (CQAs) are monitored and controlled in real time, allowing process variables to be adjusted in reaction to disruptions and guaranteeing that quality characteristics constantly meet the predefined acceptance standards (Baxendale et al. 2015).

5.8 3D Printing Implementation in Oral Solid Dosage Form

3D printing is growing rapidly in the pharmaceutical and healthcare industries. It is an advancing technology with a great potential for both patients and healthcare industry; few examples of the fast growing applications of 3D printing in pharmaceutical and healthcare industries are rapid prototyping of orthotics, dental retainers and drug-loaded implants. The 3D printing was first applied to the development of pharmaceuticals in 1996 (Okafor-Muo et al. 2020). We can carefully manage dosage, release kinetics and a number of attractive aspects of dosage forms, including colour, shape and texture, using 3D printing (Brenan 2015). Additionally, polypills can be created with combinations of medications in one solid dosage form at completely customisable strengths that would be extremely difficult to obtain commercially (Tracy et al. 2022). At the same time that 3D printing technology and formulations are progressing, the discovery of innovative hybrid materials to make superior formulations is picking up momentum (Lopez-Vidal et al. 2022). This technique allows for the accurate creation of dosage forms, and it can help in drug product production by offering a wide variety of release modes to meet clinical needs and enable patient compliance, particularly personalised dosing, to treat particular disease conditions (Fig. 5.5) (Dos Santos et al. 2023).

A technique known as '4D printing' facilitates the creation of 3D objects through the predefined modification of intelligent materials in response to outside stimuli (Zhang et al. 2019). First, a CAD software is used to generate the product design, geometry and part sizes. The input is then converted to a machine-readable format and sliced into printable layers (Zhang et al. 2023).

When compared to conventional pharmaceutical techniques, like powder preparation, milling, blending, granulation and compression, which lack manufacturing

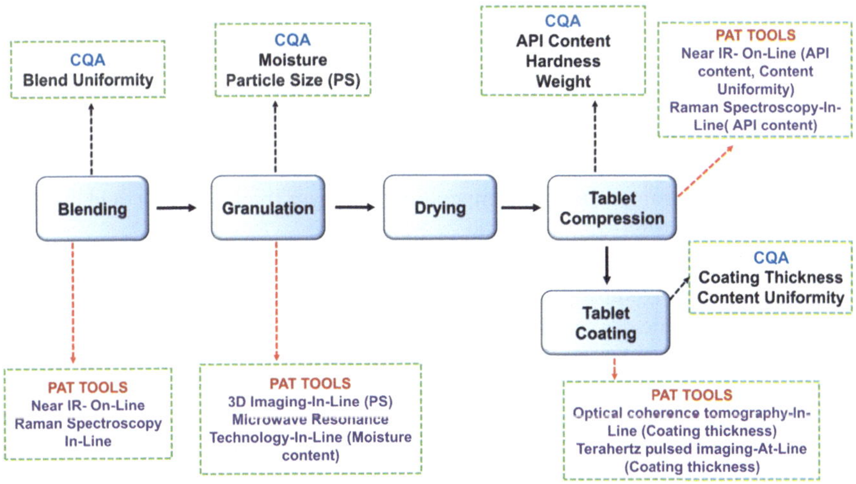

Fig. 5.5 Role of 3D printing in pharmaceutical manufacturing of solid oral dosage form

flexibility and process capabilities, 3D printing provides novel advantages (Okafor-Muo et al. 2020). Computer-aided design (CAD) software or imaging techniques can develop structures from a digital 3D file to generate personalised objects on demand (Goyanes et al. 2017). There are several 3D printing technologies available to create oral solid dosage forms.

5.8.1 Selective Laser Sintering (SLS)

Selective laser sintering (SLS) is a 3D printing method that employs the use of laser beam to fuse powder particles layer by layer. This additive manufacturing approach has numerous advantages, including excellent resolution, the ability to reuse the powder and the absence of pre-processing. The principle of selective laser sintering is consistent with all other powder bed 3D printing techniques, involving the dispersion of powder layers, typically ranging from 0.05 to 0.3 mm in thickness, followed by the selective scanning of each layer with a laser beam (Franco et al. 2010). While the support structure is created by the unsintered excess, which is effectively removed during post-printing processing, the part or final structure is generated by the sintered powder. Every plane that is processed as part of the system's 3D component represents the basic vectors used for laser scanning (Alhnan et al. 2016). The powder bed surface is decreased by a height equivalent to one layer's thickness, and a consecutive layer of powder is deposited and fused by the laser (Qiu et al. 2015).

5.8.1.1 Process Variables

In order to produce a product with the desirable qualities, the SLS process requires control over the process variables. These selective laser sintering (SLS) process variables are most widely researched in engineering fields (Akilesh et al. 2019). The SLS technique is modified to optimise the dosage forms' dimensional accuracy, surface/subsurface quality, mechanical properties and other critical quality attributes (CQAs) (Akande et al. 2016). The critical quality attributes are dependent on several factors, such as precision of the stereolithography (SLA) transcoding of files from computer-aided design (CAD) software, the division of layers, resolution by machine, layer thickness, material shrinking, length-to-width ratio and laser beam spot dimensions. The most significant processing variables affecting CQAs of dosage forms are laser intensity, bed temperature and layer thickness (Ali et al. 2019).

5.8.1.2 Characteristics of Sintered Printlets

Since there are many various processing factors involved, it is critical to determine key parameters while developing an SLS-printed dosage form in selective laser sintering (SLS) connected to the average powder diameter, size, variation, layer broadness, laser scan speed and consumed power (Kruth et al. 2007). These variables were utilised to regulate the porosity of circular SLS-printed discs for zero-order drug release. Ideally, discs featuring two concentric circular regions with varying porosities are intended to be fabricated, with a porous interior designed for drug encapsulation (Kumar 2003). The study used powder blends from two biodegradable thermoplastic polymers: poly caprolactone (PCL) and poly lactic acid (PLA), while methylene blue was used as a model drug (Mazzoli 2013).

5.8.2 Applications

Due to its versatility in manufacturing printlets with various geometries and compositions, 3D printing is anticipated to revolutionise customised medicine. For example, it can be used to make medications for orphan or uncommon illnesses, as well as for paediatric, aged or special-needs patients (Gueche et al. 2021).

5.8.2.1 Stereolithography (SLA)

The first and oldest of several prospective 3D printing techniques is SLA. It was a laser-based writing technology, which was first developed in the year 1986 (Goole and Amighi 2016). Despite this, due to the high printing resolution, it generates complex geometries and smooth-surfaced products and is widely employed (Hsiao et al. 2018). This technique uses consolidation source as UV beam rays. When focused on a container containing a photosensitive resin, an ultra violet or other light source induces cross-linking and creates a polymeric matrix. SLA is made up of two terms: stereo which means solid and photolithography which means 'writing with light'. A drug-containing photopolymerisable polymer solution is solidified using stereolithography (SLA) technology and a laser (Kaale et al. 2002).

5.8.2.2 Printing Dosage Forms

Tablets were manufactured using a commercial form SLA 3D printer from Formlabs Inc. in the United States (Chia and Wu 2015). The 3D printer software was used to import stereolithography files (.stl) created by software called Autodesk Meshmixer 2014® (Autodesk Inc., USA) as the templates for generating the tablets. The printer enables the creation of objects with a layer thickness up to 200 μm and a resolution of 300 μm (Wang et al. 2016).

5.8.3 Drawbacks and Challenges

There are some drawbacks in this new technology, due to the large and irregular sizes that must be administered orally, and a few methods that produced highly porous structures and uneven shapes have been shown to decrease patient acceptance (Pravin and Sudhir 2018).

5.8.3.1 Fused Deposition Modelling (FDM)

Fused deposition modelling (FDM) is a form of 3D printing that creates 3D objects layer by layer by depositing molten polymer on a platform. By feeding a polymer filament into the heated printhead and nozzle of the 3D printer, construction material is deposited. Heating results in the extrusion of a melted semisolid shape onto the printing surface (Khalid and Billa 2022). The stage is lowered to make space for the subsequent layer when the nozzles extrude the polymer all along x- and y-axis to form a layer. Thus, material began to accumulate along the z-axis. CAD software is used to design the printed material's shape and size (Krueger et al. 2022). There is no theoretical constraint on composition variations in all three dimensions for FDM due to the possibility of using several extrusion nozzles, each with a distinct material. Heat transfer properties and rheological properties are the most essential material selection parameters for FDM materials (Bandari et al. 2021). Because they have a low melting point, thermoplastics are often used. Wax for investment casting, PVC, nylon and ABS have all been utilised effectively. Due to its low cost and accessible equipment, FDM has become one of the most prominent 3D printing technologies in the field of pharmaceutical research (Limongi et al. 2020).

5.8.3.2 Polymer Filaments for Fused Deposition Modelling

The primary method to produce filaments that serve as the basis for FDM printing is the hot-melt extrusion technique of thermoplastic polymers. The first filaments used for FDM printers that were marketed commercially were polylactic acid (PLA), polyethylene terephthalate glycol-modified (PET-G) and high-impact polystyrene (HIPS) (Pereira et al. 2020). These polymers, which are made in the dimensions of 1.75 mm and 2.85–3 mm to suit print heads presently in use, often have high melting points and strong mechanical qualities. This guarantees that they can pass readily through the printer nozzles and resist the pressure and heat generated by the extruder and printer gears (Kempin et al. 2018).

5.8.3.3 Drawbacks

The necessity for thermoplastic polymers, which are uncommon among pharmaceutical grade polymers, is one of the downsides of FDM printing (Trenfield et al. 2018). Despite avoiding the use of solvent, FDM demonstrates relatively limited drug loading capacity, necessitating certain drying steps. The thermal FDM process limits the inclusion of thermolabile pharmaceuticals and restricts the usage of a small number of suitable excipients by melting all main medications, excipients and carriers (Winarso et al. 2022).

5.8.4 Advantages of 3D Printing Solid Dosage Forms

5.8.4.1 On-Demand Manufacturing

The use of 3D printing technology can make it simple to create high-quality products in a matter of minutes. On-demand manufacturing by 3D printing can be especially useful in situations where time and materials are limited, in drug development for quicker optimisation and in the fabrication of drug products with poor stability (Norman et al. 2017).

5.8.4.2 Improved Quality Dosage Forms

It is anticipated that using desktop printers in conjunction with computer-aided designs would improve control over the several variable parameters involved in tablet 3D printing (Giannopoulos et al. 2016). Processes such as granulation, milling, compression, coating and drying are necessary in majority of traditional medicinal product formulations, and the more steps involved in these processes, the higher the chance of batch failure (Montez et al. 2022).

5.9 Summary

In summary, advances in solid oral dosage forms bring countless opportunities in the pharmaceutical field, but however some present issues are unmet which should be resolved before a feasible integration into clinical pharmacy practices can occur. The conventional techniques widely used in the manufacturing of the solid orals often face some backlashes. Present advancements in the technology have led to the fruitful development of various techniques such as continuous manufacturing technology, artificial intelligence and 3D printing which in turn can potentially revolutionise pharmaceutics, providing a high degree of precision and control in the manufacturing steps of small-scale and/or personalised processes. To elaborate further 3D printing in combination with AI, drug design and genetics have fostered the development of personalised medicine, hence improving treatment efficacy and patient compliance. Furthermore, the development of techniques such as QbD has successfully avoided the problems associated with quality and led to the manufacturing of dosage forms with built-in quality. The agreement between AI and PAT helps in the successful management of in-process manufacturing operations in solid orals, hence avoiding errors.

Acknowledgement The authors are highly thankful to the Department of Pharmaceuticals, Ministry of Chemicals and Fertilizers, Government of India, New Delhi for providing financial assistance.

References

Abrantes CG, Duarte D, Reis CP (2016) An overview of pharmaceutical excipients: safe or not safe? J Pharm Sci 105:2019–2026

Adepu S, Ramakrishna S (2021) Controlled drug delivery systems: current status and future directions. Molecules 26:5905

Agrawal P (2018) Artificial intelligence in drug discovery and development. J Pharmacovigil 6

Akande SO, Dalgarno KW, Munguia J, Pallari J (2016) Assessment of tests for use in process and quality control systems for selective laser sintering of polyamide powders. J Mater Process Technol 229:549–561

Akilesh M, Elango PR, Devanand AA, Soundararajan R, Varthanan PA (2019) Optimization of selective laser sintering process parameters on surface quality, 3D print. Addit Manuf Technol:141–157

Aksu B, Paradkar A, de Matas M, Özer Ö, Güneri T, York P (2012) Quality by design approach: application of artificial intelligence techniques of tablets manufactured by direct compression. AAPS PharmSciTech 13:1138–1146

Alahmari AR, Alrabghi KK, Dighriri IM (2022) An overview of the current state and perspectives of pharmacy robot and medication dispensing technology. Cureus 14

Alhnan MA, Okwuosa TC, Sadia M, Wan K-W, Ahmed W, Arafat B (2016) Emergence of 3D printed dosage forms: opportunities and challenges. Pharm Res 33:1817–1832

Ali SFB, Mohamed EM, Ozkan T, Kuttolamadom MA, Khan MA, Asadi A, Rahman Z (2019) Understanding the effects of formulation and process variables on the printlets quality manufactured by selective laser sintering 3D printing. Int J Pharm 570:118651

Allenspach C, Timmins P, Sharif S, Minko T (2020) Characterization of a novel hydroxypropyl methylcellulose (HPMC) direct compression grade excipient for pharmaceutical tablets. Int J Pharm 583:119343. https://doi.org/10.1016/j.ijpharm.2020.119343

Allison G, Cain YT, Cooney C, Garcia T, Bizjak TG, Holte O, Jagota N, Komas B, Korakianiti E, Kourti D (2015) Regulatory and quality considerations for continuous manufacturing May 20–21, 2014 continuous manufacturing symposium. J Pharm Sci 104:803–812

Almoazen H, Felton L (2013) Remington: essentials of pharmaceutics. Am J Pharm Educ 77:233

American Society of Health-System Pharmacists (2006) ASHP guidelines on handling hazardous drugs. Am J Heal Pharm 63:1172–1193

Ansari A, Gangil M (2022) A critical review of the use of labour productivity in industries, material, method, application and challenges. Res J Eng Technol Med Sci 5

Arshad MS, Zafar S, Yousef B, Alyassin Y, Ali R, AlAsiri A, Chang M-W, Ahmad Z, Elkordy AA, Faheem A (2021) A review of emerging technologies enabling improved solid oral dosage form manufacturing and processing. Adv Drug Deliv Rev 178:113840

Arvind R, Gunasekaran N (2014) A literature review on cycle time reduction in material handling system by value stream mapping. Int J Res Appl Sci Eng Technol 2:70–72

Balka E, Nutland K (2004) Automated drug dispensing systems. Lit Rev

Balogh A, Domokos A, Farkas B, Farkas A, Rapi Z, Kiss D, Nyiri Z, Eke Z, Szarka G, Örkényi R (2018) Continuous end-to-end production of solid drug dosage forms: coupling flow synthesis and formulation by electrospinning. Chem Eng J 350:290–299

Bandari S, Nyavanandi D, Dumpa N, Repka MA (2021) Coupling hot melt extrusion and fused deposition modeling: critical properties for successful performance. Adv Drug Deliv Rev 172:52–63

Bannigan P, Aldeghi M, Bao Z, Häse F, Aspuru-Guzik A, Allen C (2021) Machine learning directed drug formulation development. Adv Drug Deliv Rev 175:113806

Batson S, Herranz A, Rohrbach N, Canobbio M, Mitchell SA, Bonnabry P (2021) Automation of in-hospital pharmacy dispensing: a systematic review. Eur J Hosp Pharm 28:58–64

Baxendale IR, Braatz RD, Hodnett BK, Jensen KF, Johnson MD, Sharratt P, Sherlock J, Florence AJ (2015) Achieving continuous manufacturing: technologies and approaches for synthesis, workup, and isolation of drug substance. May 20–21, 2014 continuous manufacturing symposium. J Pharm Sci 104:781–791

Beaulac P, Issa M, Ilinca A, Brousseau J (2022) Parameters affecting dust collector efficiency for pneumatic conveying: a review. Energies 15:916

Benoit E, Beney J (2011) Can new technologies reduce the rate of medications errors in adult intensive care? J Pharm Belg:82–91

Bhatia A (2019) Pneumatic conveying systems. Chem Eng (New York):1–57

Björnsson A, Jonsson M, Johansen K (2018) Automated material handling in composite manufacturing using pick-and-place systems–a review. Robot Comput Integr Manuf 51:222–229

Boobier S, Hose DRJ, Blacker AJ, Nguyen BN (2020) Machine learning with physicochemical relationships: solubility prediction in organic solvents and water. Nat Commun 11:5753

Boyd BJ, Bergström CAS, Vinarov Z, Kuentz M, Brouwers J, Augustijns P, Brandl M, Bernkop-Schnürch A, Shrestha N, Préat V (2019) Successful oral delivery of poorly water-soluble drugs both depends on the intraluminal behavior of drugs and of appropriate advanced drug delivery systems. Eur J Pharm Sci 137:104967

Brax AD, Sapko MM, Cole JW, Landgrave LC, Magers JK, Pierson SM (2023) Implementation of an electronic pharmacy scoring tool to prioritize clinical pharmacists' daily workflow at a pediatric institution. Am J Heal Pharm 80:68–74

Brenan CJH (2015) 3-D-printed pills: a new age for drug delivery [from the editor]. IEEE Pulse 6:3

Buckton G, Yonemochi E, Yoon WL, Moffat AC (1999) Water sorption and near IR spectroscopy to study the differences between microcrystalline cellulose and silicified microcrystalline cellulose before and after wet granulation. Int J Pharm 181:41–47. https://doi.org/10.1016/S0378-5173(98)00413-X

Burcham CL, Florence AJ, Johnson MD (2018) Continuous manufacturing in pharmaceutical process development and manufacturing. Annu Rev Chem Biomol Eng 9:253–281

Castro BM, Elbadawi M, Ong JJ, Pollard T, Song Z, Gaisford S, Pérez G, Basit AW, Cabalar P, Goyanes A (2021) Machine learning predicts 3D printing performance of over 900 drug delivery systems. J Control Release 337:530–545

Catania LJ (2021) AI applications in diagnostic technologies and services. Found Artif Intell Healthc Biosci:125–198

Chaudhari PD, Phatak AA, Desai U (2012) A review: co processed excipients-an alternative to novel chemical entities. Int J Pharm Chem Sci 1:1480–1498

Chen L, Sun Z, Ma H, Pan G, Li P, Gao K (2022) Flow characteristics of pneumatic conveying of stiff shotcrete based on CFD-DEM method. Powder Technol 397:117109

Chia HN, Wu BM (2015) Recent advances in 3D printing of biomaterials. J Biol Eng 9:1–14

Chongchitpaisan P, Sudsawat S (2022) A review on screw conveyors for bulk materials in various applications. Ladkrabang Eng J 39:1–12

Chopda V, Gyorgypal A, Yang O, Singh R, Ramachandran R, Zhang H, Tsilomelekis G, Chundawat SPS, Ierapetritou MG (2022) Recent advances in integrated process analytical techniques, modeling, and control strategies to enable continuous biomanufacturing of monoclonal antibodies. J Chem Technol Biotechnol 97:2317–2335

Dahlgren G, Tajarobi P, Simone E, Ricart B, Melnick J, Puri V, Stanton C, Bajwa G (2019) Continuous twin screw wet granulation and drying—control strategy for drug product manufacturing. J Pharm Sci 108:3502–3514

Darzuli N, Budniak L, Hroshovyi T (2019) Selected excipients in oral solid dosage form with dry extract of pyrola Rotundifolia L. Int J Appl Pharm 11:210–216. https://doi.org/10.22159/ijap.2019v11i6.35282

Das S, Dey R, Nayak AK (2021) Artificial intelligence in pharmacy. Indian J Pharm Educ Res 55:304–318

de Backere C, Quodbach J, De Beer T, Vervaet C, Vanhoorne V (2022) Impact of alternative lubricants on process and tablet quality for direct compression. Int J Pharm 624:122012. https://doi.org/10.1016/j.ijpharm.2022.122012

De Beer T, Burggraeve A, Fonteyne M, Saerens L, Remon JP, Vervaet C (2011) Near infrared and Raman spectroscopy for the in-process monitoring of pharmaceutical production processes. Int J Pharm 417:32–47

De Leersnyder F, Vanhoorne V, Bekaert H, Vercruysse J, Ghijs M, Bostijn N, Verstraeten M, Cappuyns P, Van Assche I, Vander Heyden Y (2018) Breakage and drying behaviour of granules in a continuous fluid bed dryer: influence of process parameters and wet granule transfer. Eur J Pharm Sci 115:223–232

Denis D, Gonella M, Comeau M, Lauzier M (2020) Questioning the value of manual material handling training: a scoping and critical literature review. Appl Ergon 89:103186

Desai PM, Liew CV, Heng PWS (2016) Review of disintegrants and the disintegration phenomena. J Pharm Sci 105:2545–2555. https://doi.org/10.1016/j.xphs.2015.12.019

Devadasu VR, Deb PK, Maheshwari R, Sharma P, Tekade RK (2018) Physicochemical, pharmaceutical, and biological considerations in GIT absorption of drugs. In: Dosage form design considerations. Elsevier, pp 149–178

Dhaval M, Sharma S, Dudhat K, Chavda J (2022) Twin-screw extruder in pharmaceutical industry: history, working principle, applications, and marketed products: an in-depth review. J Pharm Innov 17:294–318

Doerr FJS, Florence AJ (2020) A micro-XRT image analysis and machine learning methodology for the characterisation of multi-particulate capsule formulations. Int J Pharm X 2:100041

Dong J, Gao H, Ouyang D (2021) PharmSD: a novel AI-based computational platform for solid dispersion formulation design. Int J Pharm 604:120705

Dos Santos J, de Souza Balbinot G, Buchner S, Collares FM, Windbergs M, Deon M, Beck RCR (2023) 3D printed matrix solid forms: can the drug solubility and dose customisation affect their controlled release behaviour? Int J Pharm X 5:100153

Dropka N, Holena M (2020) Application of artificial neural networks in crystal growth of electronic and opto-electronic materials. Crystals 10:663

Dubourg J, Messerer M, Karakitsos D, Rajajee V, Antonsen E, Javouhey E, Cammarata A, Cotton M, Daniel RT, Denaro C (2013) Individual patient data systematic review and meta-analysis of optic nerve sheath diameter ultrasonography for detecting raised intracranial pressure: protocol of the ONSD research group. Syst Rev 2:1–6

Duch W, Swaminathan K, Meller J (2007) Artificial intelligence approaches for rational drug design and discovery. Curr Pharm Des 13:1497–1508

Elbadawi M, McCoubrey LE, Gavins FKH, Ong JJ, Goyanes A, Gaisford S, Basit AW (2021a) Harnessing artificial intelligence for the next generation of 3D printed medicines. Adv Drug Deliv Rev 175:113805

Elbadawi M, McCoubrey LE, Gavins FKH, Ong JJ, Goyanes A, Gaisford S, Basit AW (2021b) Disrupting 3D printing of medicines with machine learning. Trends Pharmacol Sci 42:745–757

Elballa W, Salih M (2022) Influence of partially and fully pregelatinized starch on the physical and sustained-release properties of Hpmc-based Ketoprofen Oral matrices. Int J Pharm Pharm Sci:29–34. https://doi.org/10.22159/ijpps.2022v14i8.45031

Esmonde-White KA, Cuellar M, Lewis IR (2022) The role of Raman spectroscopy in biopharmaceuticals from development to manufacturing. Anal Bioanal Chem 1–23

Finke JH, Kwade A (2021) Powder processing in pharmaceutical applications—in-depth understanding and modelling. Pharmaceutics. 13:128

Floresta G, Zagni C, Gentile D, Patamia V, Rescifina A (2022) Artificial intelligence technologies for COVID-19 de novo drug design. Int J Mol Sci 23:3261

Franco A, Lanzetta M, Romoli L (2010) Experimental analysis of selective laser sintering of polyamide powders: an energy perspective. J Clean Prod 18:1722–1730

Galata DL, Meszaros LA, Kallai-Szabo N, Szabo E, Pataki H, Marosi G, Nagy ZK (2021) Applications of machine vision in pharmaceutical technology: a review. Eur J Pharm Sci 159:105717

Gharib AM, Bindoff IK, Peterson GM, Salahudeen MS (2023) Computer-based simulators in pharmacy practice education: a systematic narrative review. Pharmacy 11:8

Giannopoulos AA, Mitsouras D, Yoo S-J, Liu PP, Chatzizisis YS, Rybicki FJ (2016) Applications of 3D printing in cardiovascular diseases. Nat Rev Cardiol 13:701–718

Gomes de Freitas A, Furlan de Oliveira V, Oliveira Lima Y, Borges dos Santos R, Alberto Martinez Riascos L (2021) Energy efficiency in pneumatic conveying: performance analysis of an alternative blow tank. Part Sci Technol 39:868–876

Goodin S, Griffith N, Chen B, Chuk K, Daouphars M, Doreau C, Patel RA, Schwartz R, Tamés MJ, Terkola R (2011) Safe handling of oral chemotherapeutic agents in clinical practice: recommendations from an international pharmacy panel. J Oncol Pract 7:7–12

Goole J, Amighi K (2016) 3D printing in pharmaceutics: a new tool for designing customized drug delivery systems. Int J Pharm 499:376–394

Goyanes A, Fina F, Martorana A, Sedough D, Gaisford S, Basit AW (2017) Development of modified release 3D printed tablets (printlets) with pharmaceutical excipients using additive manufacturing. Int J Pharm 527:21–30

Grof Z, Štěpánek F (2021) Artificial intelligence based design of 3D-printed tablets for personalised medicine. Comput Chem Eng 154:107492

Gueche YA, Sanchez-Ballester NM, Cailleaux S, Bataille B, Soulairol I (2021) Selective laser sintering (SLS), a new chapter in the production of solid oral forms (SOFs) by 3D printing. Pharmaceutics 13:1212

Gullapalli RP, Mazzitelli CL (2017) Gelatin and non-gelatin capsule dosage forms. J Pharm Sci 106:1453–1465

Haimhoffer Á, Vasvári G, Trencsényi G, Béresová M, Budai I, Czomba Z, Rusznyák Á, Váradi J, Bácskay I, Ujhelyi Z (2021) Process optimization for the continuous production of a gastroretentive dosage form based on melt foaming. AAPS PharmSciTech 22:187

Han R, Xiong H, Ye Z, Yang Y, Huang T, Jing Q, Lu J, Pan H, Ren F, Ouyang D (2019) Predicting physical stability of solid dispersions by machine learning techniques. J Control Release 311:16–25

Haneef J, Beg S (2021) Quality by design-based development of nondestructive analytical techniques. In: Hand book of anal disorders. Elsevier, pp 153–166

Hänninen K, Ahtiainen HK, Suvikas-Peltonen EM, Tötterman AM (2023) Automated unit dose dispensing systems producing individually packaged and labelled drugs for inpatients: a systematic review. Eur J Hosp Pharm 30:127–135

Hassanzadeh P, Atyabi F, Dinarvand R (2019) The significance of artificial intelligence in drug delivery system design. Adv Drug Deliv Rev 151:169–190

Hesse R, Krull F, Antonyuk S (2021) Prediction of random packing density and flowability for non-spherical particles by deep convolutional neural networks and discrete element method simulations. Powder Technol 393:559–581

Hoag SW (2017) Capsules dosage form: formulation and manufacturing considerations. Elsevier, Developing solid oral dosage forms, pp 723–747

Hock SC, Siang TK, Wah CL (2021) Continuous manufacturing versus batch manufacturing: benefits, opportunities and challenges for manufacturers and regulators. Generics Biosimilars Initiat J 10:1–14

Hou P, Besenhard MO, Halbert G, Naftaly M, Markl D (2023) Development and implementation of a pneumatic micro-feeder for poorly-flowing solid pharmaceutical materials. Int J Pharm 635:122691

Hsiao W-K, Lorber B, Reitsamer H, Khinast J (2018) 3D printing of oral drugs: a new reality or hype? Expert Opin Drug Deliv 15:1–4

Jagtap K, Chaudhari B, Redasani V (2022) Quality by design (QbD) concept review in pharmaceuticals. Asian J Res Chem 15:303–307

Jiang J, Ma X, Ouyang D, Williams RO III (2022) Emerging artificial intelligence (ai) technologies used in the development of solid dosage forms. Pharmaceutics. 14:2257

Jin C, Wu F, Hong Y, Shen L, Lin X, Zhao L, Feng Y (2023) Updates on applications of low-viscosity grade Hydroxypropyl methylcellulose in coprocessing for improvement of physical properties of pharmaceutical powders. Carbohydr Polym 311:120731. https://doi.org/10.1016/j.carbpol.2023.120731

Kaale E, Van Schepdael A, Roets E, Hoogmartens J (2002) Determination of capsaicinoids in topical cream by liquid-liquid extraction and liquid chromatography. J Pharm Biomed Anal 30:1331–1337. https://doi.org/10.1016/s0731-7085(02)00476-4

Kallakunta VR, Patil H, Tiwari R, Ye X, Upadhye S, Vladyka RS, Sarabu S, Kim DW, Bandari S, Repka MA (2019) Exploratory studies in heat-assisted continuous twin-screw dry granulation: a novel alternative technique to conventional dry granulation. Int J Pharm 555:380–393

Kar M, Chourasiya Y, Maheshwari R, Tekade RK (2019) Current developments in excipient science: implication of quantitative selection of each excipient in product development. In: Basic fundamentals of drug delivery. Elsevier, pp 29–83

Kempin W, Domsta V, Grathoff G, Brecht I, Semmling B, Tillmann S, Weitschies W, Seidlitz A (2018) Immediate release 3D-printed tablets produced via fused deposition modeling of a thermo-sensitive drug. Pharm Res 35:1–12

Khalid GM, Billa N (2022) Solid dispersion formulations by FDM 3D printing—a review. Pharmaceutics. 14:690

Khanna V, Ahuja R, Popli H (2020) Role of artificial intelligence in pharmaceutical marketing: a comprehensive review. J Adv Sci Res 11:54–61

Klinzing GE, Rizk F, Marcus R, Leung LS, Klinzing GE, Rizk F, Marcus R, Leung LS (2010) An overview of pneumatic conveying systems and performance. Pneum Conveying Solids A Theor Pract Approach:1–33

Krueger L, Miles JA, Popat A (2022) 3D printing hybrid materials using fused deposition modelling for solid oral dosage forms. J Control Release 351:444–455
Kruth J-P, Levy G, Klocke F, Childs THC (2007) Consolidation phenomena in laser and powder-bed based layered manufacturing. CIRP Ann 56:730–759
Kuang S, Zhou M, Yu A (2020) CFD-DEM modelling and simulation of pneumatic conveying: a review. Powder Technol 365:186–207
Kumar S (2003) Selective laser sintering: a qualitative and objective approach. JOM 55:43–47
Kurbanoglu S, Uslu B, Ozkan SA (2017) Carbon-based nanostructures for electrochemical analysis of oral medicines. In: Nanostructures for oral medicine. Elsevier, pp 885–938
Lachman L, Lieberman HA, Kanig JL (1976) The theory and practice of industrial pharmacy. Lea & Febiger Philadelphia
Laffleur F, Keckeis V (2020) Advances in drug delivery systems: work in progress still needed? Int J Pharm 590:119912
Lamberti MJ, Wilkinson M, Donzanti BA, Wohlhieter GE, Parikh S, Wilkins RG, Getz K (2019) A study on the application and use of artificial intelligence to support drug development. Clin Ther 41:1414–1426
Lee SL, O'Connor TF, Yang X, Cruz CN, Chatterjee S, Madurawe RD, Moore CMV, Yu LX, Woodcock J (2015) Modernizing pharmaceutical manufacturing: from batch to continuous production. J Pharm Innov 10:191–199
Lee H, Kim J, Kim S, Yoo J, Choi GJ, Jeong Y-S (2022) Deep learning-based prediction of physical stability considering class imbalance for amorphous solid dispersions. J Chem 2022:1
Limongi T, Susa F, Allione M, Di Fabrizio E (2020) Drug delivery applications of three-dimensional printed (3DP) mesoporous scaffolds. Pharmaceutics. 12:851
Liu L, Ouyang W, Wang X, Fieguth P, Chen J, Liu X, Pietikäinen M (2020) Deep learning for generic object detection: a survey. Int J Comput Vis 128:261–318
Lopez-Vidal L, Real JP, Real DA, Camacho N, Kogan MJ, Paredes AJ, Palma SD (2022) Nanocrystal-based 3D-printed tablets: semi-solid extrusion using melting solidification printing process (MESO-PP) for oral administration of poorly soluble drugs. Int J Pharm 611:121311
Lou H, Chung JI, Kiang Y-H, Xiao L-Y, Hageman MJ (2019) The application of machine learning algorithms in understanding the effect of core/shell technique on improving powder compactability. Int J Pharm 555:368–379
Lowe M, Qin R, Mao X (2022) A review on machine learning, artificial intelligence, and smart technology in water treatment and monitoring. Water 14:1384
Ma X, Kittikunakorn N, Sorman B, Xi H, Chen A, Marsh M, Mongeau A, Piché N, Williams RO III, Skomski D (2020) Application of deep learning convolutional neural networks for internal tablet defect detection: high accuracy, throughput, and adaptability. J Pharm Sci 109:1547–1557
Maderuelo C, Lanao JM, Zarzuelo A (2019) Enteric coating of oral solid dosage forms as a tool to improve drug bioavailability. Eur J Pharm Sci 138:105019
Mak K-K, Pichika MR (2019) Artificial intelligence in drug development: present status and future prospects. Drug Discov Today 24:773–780
Matsuda Y (2018) PMDA activities for implementation of continuous manufacturing. In: Proceedings of the ISPE continuous manufacturing work, Arlington, VA. pp 6–7
Mazzoli A (2013) Selective laser sintering in biomedical engineering. Med Biol Eng Comput 51:245–256
McCarthy J, Minsky ML, Rochester N, Shannon CE (2006) A proposal for the Dartmouth summer research project on artificial intelligence, august 31, 1955. AI Mag 27:12
Medendorp J, Shapally S, Vrieze D, Tolton K (2020) Process control of drug product continuous manufacturing operations—a study in operational simplification and continuous improvement. J Pharm Innov 1–12:85
Miller TA, York P (1988) Pharmaceutical tablet lubrication. Int J Pharm 41:1–19. https://doi.org/10.1016/0378-5173(88)90130-5
Moingeon P, Kuenemann M, Guedj M (2022) Artificial intelligence-enhanced drug design and development: toward a computational precision medicine. Drug Discov Today 27:215–222

Montez M, Willis K, Rendler H, Marshall C, Rubio E, Rajak DK, Rahman MH, Menezes PL (2022) Fused deposition modeling (FDM): processes, material properties, and applications. In: Tribology of additively manufactured materials. Elsevier, pp 137–163

Myerson AS, Krumme M, Nasr M, Thomas H, Braatz RD (2015) Control systems engineering in continuous pharmaceutical manufacturing May 20–21, 2014 Continuous manufacturing symposium. J Pharm Sci 104:832–839

Nagy B, Farkas A, Borbás E, Vass P, Nagy ZK, Marosi G (2019) Raman spectroscopy for process analytical technologies of pharmaceutical secondary manufacturing. AAPS PharmSciTech 20:1–16

Nambiar AG, Singh M, Mali AR, Serrano DR, Kumar R, Healy AM, Agrawal AK, Kumar D (2022) Continuous manufacturing and molecular modeling of pharmaceutical amorphous solid dispersions. AAPS PharmSciTech 23:249

Nasr MM, Krumme M, Matsuda Y, Trout BL, Badman C, Mascia S, Cooney CL, Jensen KD, Florence A, Johnston C (2017) Regulatory perspectives on continuous pharmaceutical manufacturing: moving from theory to practice: September 26-27, 2016, international symposium on the continuous manufacturing of pharmaceuticals. J Pharm Sci 106:3199–3206

Ng LH, Ling JKU, Hadinoto K (2022) Formulation strategies to improve the stability and handling of Oral solid dosage forms of highly hygroscopic pharmaceuticals and nutraceuticals. Pharmaceutics. 14:2015

Norman J, Madurawe RD, Moore CMV, Khan MA, Khairuzzaman A (2017) A new chapter in pharmaceutical manufacturing: 3D-printed drug products. Adv Drug Deliv Rev 108:39–50

O'Connor TF, Lawrence XY, Lee SL (2016) Emerging technology: a key enabler for modernizing pharmaceutical manufacturing and advancing product quality. Int J Pharm 509:492–498

Okafor-Muo OL, Hassanin H, Kayyali R, ElShaer A (2020) 3D printing of solid oral dosage forms: numerous challenges with unique opportunities. J Pharm Sci 109:3535–3550

Paimard G, Ghasali E, Baeza M (2023) Screen-printed electrodes: fabrication, modification, and biosensing applications. Chemosensors 11:113

Palo HK, Sahoo S, Subudhi AK (2021) Dimensionality reduction techniques: principles, benefits, and limitations, data anal. Bioinforma A Mach Learn Perspect:77–107

Patel V, Shah M (2022) Artificial intelligence and machine learning in drug discovery and development. Intell Med 2:134–140

Paul S, Baranwal Y, Tseng Y-C (2021) An insight into predictive parameters of tablet capping by machine learning and multivariate tools. Int J Pharm 599:120439

Paul P, Sobhan Gupta DADR, Mian S (2023) EVALUATING THE EFFECTIVENESS OF INTERVENTIONS TO REDUCE MEDICATION ERRORS: A SYSTEMIC REVIEW. J Pharm Negat Results:3064–3074

Pereira GG, Figueiredo S, Fernandes AI, Pinto JF (2020) Polymer selection for hot-melt extrusion coupled to fused deposition modelling in pharmaceutics. Pharmaceutics. 12:795

Pittu V, Sharma J (2013) FORMULATION AND EVALUATION OF GASTRORETENTIVE FLOATING TABLETS OF VALSARTAN

Poechlauer P, Colberg J, Fisher E, Jansen M, Johnson MD, Koenig SG, Lawler M, Laporte T, Manley J, Martin B (2013) Pharmaceutical roundtable study demonstrates the value of continuous manufacturing in the design of greener processes. Org Process Res Dev 17:1472–1478

Pravin S, Sudhir A (2018) Integration of 3D printing with dosage forms: a new perspective for modern healthcare. Biomed Pharmacother 107:146–154

Qiu C, Adkins NJE, Hassanin H, Attallah MM, Essa K (2015) In-situ shelling via selective laser melting: modelling and microstructural characterisation. Mater Des 87:845–853

Rao D, Pathak P (2022) Evolving robotic process automation (RPA) & artificial intelligence (AI) in response to Covid-19 and its future. In: AIP conference proceedings. AIP Publishing

Rathore AS, Agarwal H, Sharma AK, Pathak M, Muthukumar S (2015) Continuous processing for production of biopharmaceuticals. Prep Biochem Biotechnol 45:836–849

Risør BW, Lisby M, Sørensen J (2017) Cost-effectiveness analysis of an automated medication system implemented in a Danish hospital setting. Value Heal 20:886–893

Rogers AJ, Hashemi A, Ierapetritou MG (2013) Modeling of particulate processes for the continuous manufacture of solid-based pharmaceutical dosage forms. PRO 1:67–127

Schaber SD, Gerogiorgis DI, Ramachandran R, Evans JMB, Barton PI, Trout BL (2011) Economic analysis of integrated continuous and batch pharmaceutical manufacturing: a case study. Ind Eng Chem Res 50:10083–10092

Schmidt A, de Waard H, Kleinebudde P, Krumme M (2018) Continuous single-step wet granulation with integrated in-barrel-drying. Pharm Res 35:1–16

Shah P, Kendall F, Khozin S, Goosen R, Hu J, Laramie J, Ringel M, Schork N (2019) Artificial intelligence and machine learning in clinical development: a translational perspective. NPJ Digit Med 2:69

Shaheen MY (2021) Applications of artificial intelligence (AI) in healthcare: a review, Sci. Prepr

Shaikh R, O'Brien DP, Croker DM, Walker GM (2018) The development of a pharmaceutical oral solid dosage forms. In: Computer aided chemical engineering. Elsevier, pp 27–65

Sheikhy S, Safekordi AA, Ghorbani M, Adibkia K, Hamishehkar H (2021) Synthesis of novel superdisintegrants for pharmaceutical tableting based on functionalized nanocellulose hydrogels. Int J Biol Macromol 167:667–675. https://doi.org/10.1016/j.ijbiomac.2020.11.173

Shin S, Koo J, Kim SW, Kim S, Hong SY, Lee E (2023) Evaluation of robotic systems on cytotoxic drug preparation: a systematic review and meta-analysis. Medicina (B. Aires) 59:431

Sng Y, Ong CK, Lai YF (2019) Approaches to outpatient pharmacy automation: a systematic review. Eur J Hosp Pharm 26:157–162

Somnache SN, Vasantakumar KP, Godbole AM, Gajare PS, Pednekar AS (2023) COMBILOSE: a novel lactose-based co-processed excipient for direct compression. J Appl Pharm Sci 0:1–8. https://doi.org/10.7324/japs.2023.37613

Srai JS, Badman C, Krumme M, Futran M, Johnston C (2015) Future supply chains enabled by continuous processing—opportunities and challenges. May 20–21, 2014 continuous manufacturing symposium. J Pharm Sci 104:840–849

Srai JS, Settanni E, Aulakh PK (2020) Evaluating the business case for continuous manufacturing of pharmaceuticals: a supply network perspective. In: Continuous pharmaceutical processing. Springer, pp 477–512

Steenweg C, Seifert AI, Böttger N, Wohlgemuth K (2021) Process intensification enabling continuous manufacturing processes using modular continuous vacuum screw filter. Org Process Res Dev 25:2525–2536

Stocker MW, Tsolaki E, Harding MJ, Healy AM, Ferguson S (2023) Combining isolation-free and co-processing manufacturing approaches to access room temperature ionic liquid forms of APIs. J Pharm Sci 112:2079. https://doi.org/10.1016/j.xphs.2023.01.030

Suryadinata HU (2017) The benefits of automated dispensing machine as solutions for hospital pharmacy in Indonesia: a systematic review. Proc Int Conf Appl Sci Heal 1:151–159

Thabet Y, Lunter D, Breitkreutz J (2018) Continuous manufacturing and analytical characterization of fixed-dose, multilayer orodispersible films. Eur J Pharm Sci 117:236–244

Tracy T, Wu L, Liu X, Cheng S, Li X (2022) 3D printing: innovative solutions for patients and pharmaceutical industry. Int J Pharm 631:122480

Trenfield SJ, Awad A, Goyanes A, Gaisford S, Basit AW (2018) 3D printing pharmaceuticals: drug development to frontline care. Trends Pharmacol Sci 39:440–451

Tsao NW, Lo C, Babich M, Shah K, Bansback NJ (2014) Decentralized automated dispensing devices: systematic review of clinical and economic impacts in hospitals. Can J Hosp Pharm 67:138

Tu P, Yan F (2022) Experimental and numerical study on the horizontal-vertical pneumatic conveying system with pulse excitation flow based on POD and recursive analysis. Adv Powder Technol 33:103524

Vamathevan J, Clark D, Czodrowski P, Dunham I, Ferran E, Lee G, Li B, Madabhushi A, Shah P, Spitzer M (2019) Applications of machine learning in drug discovery and development. Nat Rev Drug Discov 18:463–477

van der Lee M, Swen JJ (2023) Artificial intelligence in pharmacology research and practice. Clin Transl Sci 16:31–36
van der Merwe J, Steenekamp J, Steyn D, Hamman J (2020) The role of functional excipients in solid oral dosage forms to overcome poor drug dissolution and bioavailability. Pharmaceutics. 12:393
Van Snick B, Holman J, Cunningham C, Kumar A, Vercruysse J, De Beer T, Remon JP, Vervaet C (2017) Continuous direct compression as manufacturing platform for sustained release tablets. Int J Pharm 519:390–407
Verstraeten M, Van Hauwermeiren D, Lee K, Turnbull N, Wilsdon D, Am Ende M, Doshi P, Vervaet C, Brouckaert D, Mortier STFC (2017) In-depth experimental analysis of pharmaceutical twin-screw wet granulation in view of detailed process understanding. Int J Pharm 529:678–693
Wang J, Goyanes A, Gaisford S, Basit AW (2016) Stereolithographic (SLA) 3D printing of oral modified-release dosage forms. Int J Pharm 503:207–212
Wang J, Zhang Y, Aghda NH, Pillai AR, Thakkar R, Nokhodchi A, Maniruzzaman M (2021a) Emerging 3D printing technologies for drug delivery devices: current status and future perspective. Adv Drug Deliv Rev 174:294–316
Wang W, Ye Z, Gao H, Ouyang D (2021b) Computational pharmaceutics-A new paradigm of drug delivery. J Control Release 338:119–136
Wang G, Su W, Hu B, Al-Huqail AA, Majdi HS, Algethami JS, Jiang Y, Ali HE (2022) Assessment in carbon-based layered double hydroxides for water and wastewater: application of artificial intelligence and recent progress. Chemosphere 308:136303
Welch CJ, Faul MM, Tummala S, Papageorgiou CD, Hicks F, Hawkins JM, Thomson N, Cote A, Bordawekar S, Wittenberger SJ (2017) The enabling technologies consortium (ETC): fostering precompetitive collaborations on new enabling technologies for pharmaceutical research and development. Org Process Res Dev 21:414–419
Winarso R, Anggoro PW, Ismail R, Jamari J, Bayuseno AP (2022) Application of fused deposition modeling (FDM) on bone scaffold manufacturing process: a review. Heliyon 8:e11701
Wirtz BW, Weyerer JC, Geyer C (2019) Artificial intelligence and the public sector—applications and challenges. Int J Public Adm 42:596–615
Xiouras C, Cameli F, Quilló GL, Kavousanakis ME, Vlachos DG, Stefanidis GD (2022) Applications of artificial intelligence and machine learning algorithms to crystallization. Chem Rev 122:13006–13042
Ye X, Kallakunta V, Kim DW, Patil H, Tiwari RV, Upadhye SB, Vladyka RS, Repka MA (2019) Effects of processing on a sustained release formulation prepared by twin-screw dry granulation. J Pharm Sci 108:2895–2904
Yu LX, Amidon G, Khan MA, Hoag SW, Polli J, Raju GK, Woodcock J (2014) Understanding pharmaceutical quality by design. AAPS J 16:771–783
Zaman R, Arefeen A, Quarnstrom J, Barman S, Yang J, Xiang Y (2022) Optimization-based biomechanical lifting models for manual material handling: a comprehensive review. Proc Inst Mech Eng Part H J Eng Med 236:1273–1287
Zhang GGZ, Law D, Schmitt EA, Qiu Y (2004) Phase transformation considerations during process development and manufacture of solid oral dosage forms. Adv Drug Deliv Rev 56:371–390
Zhang B, Gao L, Ma L, Luo Y, Yang H, Cui Z (2019) 3D bioprinting: a novel avenue for manufacturing tissues and organs. Engineering 5:777–794
Zhang Y, Abatzoglou N, Hudon S, Lapointe-Garant P-P, Simard J-S (2021a) Dynamics of heat-sensitive pharmaceutical granules dried in a horizontal fluidized bed combined with a screw conveyor. Chem Eng Process Intensif 167:108516
Zhang Y, Liu T, Kashani-Rahimi S, Zhang F (2021b) A review of twin screw wet granulation mechanisms in relation to granule attributes. Drug Dev Ind Pharm 47:349–360
Zhang Z, Feng S, Almotairy A, Bandari S, Repka MA (2023) Development of multifunctional drug delivery system via hot-melt extrusion paired with fused deposition modeling 3D printing techniques. Eur J Pharm Biopharm 183:102–111

Zhao L, Kim M, Zhang L, Lionberger R (2019) Generating model integrated evidence for generic drug development and assessment. Clin Pharmacol Ther 105:338–349

Zhao J, Tian G, Qiu Y, Qu H (2021) Rapid quantification of active pharmaceutical ingredient for sugar-free Yangwei granules in commercial production using FT-NIR spectroscopy based on machine learning techniques. Spectrochim Acta Part A Mol Biomol Spectrosc 245:118878

Advances in Tablet Production and Tablet Coating

6

Nagphase Nakshatra Jitendra, Rohit Garg, Md Imtiyaz Alam, and Awesh K. Yadav

Abstract

Tablets have long been a preferred pharmaceutical dosage form due to their convenience, stability, and ease of administration. Significant advancements in tablet production and coating technologies have revolutionized the pharmaceutical and nutraceutical industries in recent years. This book chapter provides an overview of key innovations in tablet manufacturing and coating processes, highlighting their impact on drug delivery, product quality, and patient compliance. Advancements in granulation methods such as spray drying, fluid bed granulation, and hot-melt extrusion have improved drug content uniformity, dissolution rates, and overall product stability. The adoption of these technologies promises continued progress in the field of tablet manufacturing, offering new opportunities for personalized medicine and enhanced therapeutic outcomes. Notably, coating technologies have also evolved significantly to enhance tablet appearance, taste-masking, and drug release profiles. Film coating using aqueous and organic solutions has become more efficient, environmentally friendly, and cost-effective. Furthermore, automation involves the utilization of machinery and tools to execute both physical and cognitive tasks within a production process. The growing focus on automated technology in the pharmaceutical industry is driven by the promising trend of fully automating tablet production. This book chapter also emphasized issues concerning the process of tablet manufacturing and recent advancements in the tablet coating process.

Nagphase Nakshatra Jitendra, Rohit Garg, and Md Imtiyaz Alam contributed equally to this work.

N. N. Jitendra · R. Garg · M. I. Alam · A. K. Yadav (✉)
Department of Pharmaceutics, National Institute of Pharmaceutical Education and Research (NIPER) Raebareli, Lucknow, Uttar Pradesh, India
e-mail: awesh.yadav@niperraebareli.edu.in

K. Jain, A. K. Yadav (eds.), *Advances in Pharmaceutical Product Development*, https://doi.org/10.1007/978-981-97-9230-6_6

Keywords

Tablet · Coating technologies · Granulation methods · Film coating · Tablet automation

6.1 Introduction

For convenience and effectiveness, a variety of oral solid dosage forms, including tablets, capsules, powders, and granules, are typically utilized. Tablets stand out among them as the most popular solid oral dosage form due to their advantages, which encompass excellent stability, therapeutic effectiveness, and ease of storage and transport. So, it's crucial to regulate the tablet preparation procedure. The process of preparing tablets is a complex procedure that can be divided into multiple operational stages. These production processes involve crushing, screening, drying, granulating, lubricant/disintegrating agent mixing, tableting, coating, and mixing with auxiliary ingredients. When making tablets, a variety of process issues or material traits are present that may have an impact on the end product's quality. For example, factors like the material's size, its initial water content, drying temperature, humidity, air velocity, and various other variables can influence the transfer of mass and heat between the liquid and gas phases during the drying process. The final blended product's quality is influenced by how the particles interact while being mixed, which is controlled by the particle size, shape, and speed of the mixer. Traditional tools for analyzing processes frequently encounter challenges when it comes to measuring the transfer of mass and heat, particle interactions, and fluid multiphase flow within a system, which are expensive as well as time-consuming. As a result, the analysis of the manufacturing process for tablets may be done better with numerical simulation technologies. It can forecast the viability of drug manufacturing procedures, optimize them, and choose processing conditions that are safer and more efficient. The spray-fluidized bed granulation process was simulated by using the CFD (computational fluid dynamics) model. Investigations were done into how starting particle size and nozzle positioning affected spray-fluidized bed granulation. In order to forecast the mixing quality between particles, studied the flow of particles in the Ross-type static mixer and the vertical feed mixer using the DEM model. Numerical simulation can be thought of as a computer-based experiment that can faithfully depict different flow situations inside a system, the allocation of force and the transmission of energy, and other challenging issues. Despite the widespread use of numerical technology, the fundamental steps in solving these issues remain the same. They are as follows: (1) create model equations for response problems and establish the conditions under which they can be solved; (2) select an effective and precise approach to solving the issue; and (3) program and compute (Li et al. 2023).

6.2 Excipients

Traditionally, excipients such as binders, disintegrants, lubricants, fillers, and glidants have been utilized as formulation aids. However, in recent years, functional excipient applications have been thoroughly investigated for modulating drug release, prolonging the tablet's presence through mucoadhesion, concealing taste, and enhancing solubility. Important technical factors play a role in determining the selection of excipients for oral solid dosage forms, integrating the critical pharmacokinetic efficacy of the formula with the physical or chemical properties of the active pharmaceutical ingredient (API), as well as the dosage levels. The excipients chosen are those that have ability to achieve desired effects, such as improved bioavailability, modified release, or drug stability (Sohail Arshad et al. 2021). Moreover, co-processed excipients that are prepared for immediate use have been created to enhance tablet manufacturing.

6.2.1 Superdisintegrants

It is important for a tablet to disintegrate because this causes the drug component to dissolve. Because natural polymers like starch are only partially soluble, their inclusion in a formulation as a disintegrant may increase the viscosity of the ambient medium, preventing disintegration and dissolution. Incorporating cross-links within the polymer chains can mitigate the influence of the disintegrant on moderate viscosity; introducing carboxyl groups into the polymer's structure will improve its ability to attract and interact with water molecules, thereby increasing its hydrophilicity, aiding in the resolution of this issue (Quodbach and Kleinebudde 2016). The characteristics of the disintegrant pertaining to its ability to swell and regain its shape after deformation and its washing effect are among the mechanisms engaged in tablet disintegration. In swelling, tablet disintegration is caused by the particles expanding in all directions, which increases pressure within the system and puts stress on it. Water enters the system by capillary action during wicking, which causes van der Waals force, hydrogen bonds, and electrostatic interactions to break down (Markl and Zeitler 2017). Van Kamp examined the disintegration capabilities of potato starch (20%), SSG (4%), and CP (4%). The disintegration times of CP (26 s) and SSG (49 s) were much less than the time measured with potato starch (149 s), demonstrating the improved performance of cross-linked superdisintegrants.

6.2.2 Fillers and Binders

Fillers significantly increase the volume of the formulation, allowing it to be processed into dosage forms and conveniently administered in unit dosages. The pharmaceutical industry uses two types of fillers: water-soluble (such as α-lactose monohydrate, sucrose, and PEG (polyethylene glycol) 6000) and water-insoluble (such as calcium hydrogen phosphate in either an anhydrous or dihydrate form) (van der Merwe et al. 2020). Through the utilization of cohesive and adhesive forces, such as van der Waals and electrostatic forces, hydrogen bonding, solid bridges, and mechanical interlocking, binders enhance flexibility and strengthen the cohesion of the ingredients within the formulation at the interparticulate level (Adolfsson et al. 1998).

In tablet production, a range of binders is employed, including wet binders like gelatin, pregelatinized starch, starch, PEG, gum acacia, okra, and xanthan, as well as dry binders like cellulose, MCC (microcrystalline cellulose), methyl cellulose, PVP (polyvinyl pyrrolidone), and PEG. Different materials that have undergone co-processing, such as silicified MCC, α-lactose monohydrate-MCC, hydroxypropyl methylcellulose (HPMC)-α-lactose monohydrate, vinyl pyrrolidone-vinyl acetate, and corn starch-MCC-α-lactose monohydrate, have been utilized as binding agents in tablet compositions that include tramadol HCl, hydrochlorothiazide, and acetyl salicylic acid (Komersová et al. 2016; Mužíková et al. 2014, 2017).

6.2.3 Lubricants/Anti-adherents

To prevent sticking, picking, and capping concerns, lubricants are added in minor quantities to the tablet formulation (0.25–0.5% w/w). They function by enclosing particles or surfaces with a stable layer. The tablet compression process is influenced by a number of variables, including the type and quantity of lubricant, the way the lubricating agent is incorporated, as well as the lubrication technique, including methods like either within the formulation (internal) or applied externally by spraying onto punches and dies (external). A variety of lubricants consist of metallic fatty acid salts like magnesium stearate, zinc stearate, and aluminum stearate. Additionally, they encompass fatty compounds such as fatty acids, fatty alcohols, and hydrocarbons, like stearic acid. Fatty acid esters like glyceryl behenate, sodium stearyl fumarate, and sucrose monopalmitate are also part of these lubricants. Alkyl sulfates such as magnesium lauryl sulfate and sodium lauryl sulfate are included, alongside inorganic substances like magnesium silicate. Polymers, including polyoxyethylene-polyoxypropylene copolymer, polytetrafluoroethylene, and PEG 4000, are further examples of components used in these lubricants (Wang et al. 2010; Li and Wu 2014). Research indicated the promise of L-leucine and hexagonal boron nitride as innovative lubricating substances (Sohail Arshad et al. 2021).

6.2.4 Solubility/Dissolution Enhancers

A significant number of novel pharmaceutical entities fall under BCS class II, which is distinguished by limited dissolution due to low solubility and high permeability, with the former being the primary factor restricting overall bioavailability. To improve the solubility of these drugs, a variety of approaches have been used, including physical and chemical changes of the active component. Modifications to a pharmaceutical compound at a physical level involve decreasing particle size (micronization, nanosuspension), changing crystal structure (co-crystallization, polymorphism, amorphization), and dispersing the drug within carriers (solid solutions/dispersions, eutectic mixtures, cryogenic methods). The production of salts or prodrugs, the addition of buffer, derivatization, and complexation are examples of chemical modifications to the drug molecule. Additional strategies utilized to enhance drug dissolution encompass supercritical fluid techniques and integrating additives such as solubilizers, hydrotropic agents, surfactants, and cosolvents for the formulation of solid dispersions (Khadka et al. 2014; Savjani et al. 2012). The literature discusses the enhancement of solubility for diverse drugs (such as paclitaxel, Adriamycin, doxorubicin, lonidamine, famotidine, ondansetron, furosemide, itraconazole, ibuprofen) through innovative additives like sulfobutylether-b-cyclodextrin, HP-b-cyclodextrin, HPMC acetate succinate, graft copolymers of polyethylene glycol/polyvinyl acetate/polyvinylcaprolactam, xyloglucan, dextran, chondroitin sulfate/pluronic copolymer, and silica (Cho and Jung 2015; Basha et al. 2020; Havel 2018).

6.2.5 Drug Release Rate Modifiers

In many situations, controlled drug release is preferred because it enables appropriate plasma concentrations, a longer duration of the therapeutic effect, reduced dosage administrations, and, ultimately, increased adherence from the patient. Monolithic, reservoir, osmotic, ion-exchange, and membrane diffusion systems are a few examples of different formulation strategies that are used to produce modified releases of drugs (Wen and Park 2010).

Numerous factors regulate the discharge of the drug from the potential formulations. These factors encompass the characteristics of both the excipients and the active component, especially the loading of drugs and their solubility. Additionally, the dimensions of the tablet (such as shape, size, and surface area) as well as the properties of the coating membrane (including material and thickness) play a significant role (Mužíková et al. 2014; Kaur et al. 2018).

A range of polymers, including polyvinyl alcohol, polymethacrylate, ethyl cellulose, chitosan, polyethylene oxide, polyacrylic acid, polydextrose, gelatin, pectin, sodium alginate, xanthan gum, tragacanth gum, and poloxamers, have been studied for their ability to create structures or barrier membranes in oral controlled-release systems such as multiparticulate, matrix, osmotic, and enteric-coated tablets (Sohail Arshad et al. 2021).

Hydrophilic molecules called pyrogens, known for their substantial ability to swell within a matrix, are integrated into a formulation based on hydrophobic polymers. This integration aims to enhance the formulation's porosity, as these porogens later exit through dissolution. Hydrophilic polymers, polyethylene glycols (PEGs), sodium chloride, sugars, and sugar alcohols, as well as L-menthol and sodium chloride, are some of the substances that result in pore formation (Vasvári et al. 2018; Raza et al. 2020). Diverse lipid substances are also employed in the formulation of controlled-release matrices. Examples of such lipids comprise caprylocaproyl polyoxyl-8 glycerides, linoleyl polyoxyl-6 glycerides, lauroyl polyoxyl-32 glycerides, and stearoyl polyoxyl-32 glycerides. Certain commercially accessible medications formulated with these polymers and lipids consist of felodipine, gliclazide, ritonavir, verapamil, fenofibrate, and enzalutamide (Siepmann et al. 2019). Inorganic substances having adsorbent qualities, such as montmorillonite and mesoporous silica, are the carriers that recently have been recognized for their role in controlled drug release (Sohail Arshad et al. 2021).

6.2.6 Co-processed Excipients

Co-processed excipients are utilized to enhance drug loading capacity, tensile strength, and operational efficiency, along with improving flow properties, moisture resistance, sensitivity to lubricants, disintegration performance, and responsiveness to strain rates. A co-processed product often consists of a mix of brittle and plastic components that interact at the subparticle level. Melt extrusion, granulation, spheronization, co-crystallization, co-drying, co-milling, spray drying, and coprecipitation are just a few examples of co-processing techniques that may be used (Saha and Shahiwala 2009). Innovative combined excipients comprise calcium carbonate-MCC, cellulose-lactose (Cellactose), silicified MCC, magnesium silicate-maize starch, MCC-dicalcium phosphate dihydrate, and alpha-lactose monohydrate cross-linked with PVP-HPMC E3 (Saha and Shahiwala 2009; Viscasillas Clerch et al. 2013; Wang et al. 2015; Rashid et al. 2011).

These multipurpose co-processed excipients might perhaps be used in continuous manufacturing. During continuous manufacture, excipients are fed at the right pace and in the right proportion; it might be expensive to dedicate a separate feeder to each component.

Therefore, using multifunction composite excipients can significantly contribute to lowering both the quantity and variability of input materials in a continuous process (Challener 2020).

6.3 Advances in Tablet Manufacturing Processes

Over a period of time, routines for tablet manufacturing have been developed that incorporate advanced granulation techniques, hot-melt extrusion, extrusion/spheronization, injection molding, spray drying, spray congealing, coprecipitation, and nanotechnology-based methods.

6.3.1 Advanced Granulation Approaches

Granulation is a multistep process used to produce tablets. Recent advancements in wet and dry granulation methods have successfully reduced process variability and enhanced productivity.

1. **Dry Granulation**

   A significant progression in dry granulation technology involves the utilization of pneumatic dry granulation (PDG). This approach merges the conventional roller compaction method with a specialized pneumatic system to create granules that exhibit enhanced flow characteristics and compressibility. In the context of powder dispersion granulation (PDG), a mild compressive pressure is exerted using a roller compactor on the powder particles, leading to the formation of a consolidated mixture comprising both fine particles and granules. Through the utilization of a pneumatic system, the granules of the intended size range are sorted within a fractioning chamber. PDG permits the incorporation of substantial drug quantities (ranging from 70 to 100%). Various additional benefits encompass rapid processing pace; minimal to no material loss; compatibility with drugs sensitive to moisture, solvents, or heat; enhanced flow; and compressibility. However, notable limitations of this production method involve the impact of recycling on the quality of granules and concerns related to the fragility of granules (Shanmugam 2015; Sandler and Lammens 2011).

   (a) **Hot-Melt Extrusion**

      Hot-melt extrusion (HME) engages in the mixing of formulation constituents within a barrel using rotating screws at elevated temperatures. The melted mixture is then forced through an extruder, and the resulting extrusions can be formed into pellets or compressed to create tablets (Bruce et al. 2005). The quality of the prepared formulation is affected by various factors like the rate of feeding, temperature at different points, duration of residence, the design or velocity of the extruder screw, and the rate of cooling. The pharmaceutical sector extensively investigates the application of the HME method for crafting solid dispersions. The outcome of the hot-melt extrusion process is an amorphous solid solution characterized by elevated innate energy that is naturally present, resulting in enhanced drug solubility and bioavailability. Nonetheless, this procedure requires substantial energy and is unsuitable for drugs sensitive to high temperatures (Bruce et al. 2005; Simões et al. 2019). The hot-melt extrusion (HME) approach was employed to produce orally disintegrating tablets containing sildenafil citrate, with a focus on taste-masking. Ethyl cellulose was chosen as the matrix material. It was noted that the taste-masking effectiveness of the formulation is influenced by the screw configuration, which modifies the physical condition of the active substance (Morott et al. 2015). Tablet formulations employing hot-melt extrusion (HME) were developed to create sustained-release tablets of acetohydroxamic acid and chlorpheniramine maleate. These tablets are designed to float in the stomach for over 24 h. Eudragit RS PO, Eudragit E PO, and sodium bicarbonate were used in the formulation process (Fukuda et al. 2006). Tablets containing itraconazole and nimodipine, formulated through the solid dispersion method using HPMC acetate succinate, HPMC, and PVP, were prepared via the HME technique. The FDA (Food and Drug Administration) granted approval to two tablet dosage forms relying on solid dispersion technology: Viekirax/Technivie (containing ombitasvir, paritaprevir, and ritonavir) in 2015 and

Maviret/Mavyret (comprising glecaprevir and pibrentasvir) in 2017 (Simões et al. 2019).

(b) **Pneumatic Dry Granulation (PDG)**

Pneumatic dry granulation (PDG) is an innovative dry granulation technique that integrates roller compaction with a unique air sorting method to produce granules. These granules possess an exceptional balance of flow properties and the ability to be compressed (Shanmugam 2015). This approach generates granules from powdered particles by first applying a gentle compression force using a roller compactor. As a consequence, there is a consolidated mixture comprising a combination of fine particles and granules. The pneumatic system in a fractioning chamber segregates the smaller granules and fine particles by carrying them away within a stream of gas. At the same time, the granules of the desired size advance into the fractionation chamber, where they undergo compression to form tablets. The fine particles and smaller granules carried in the gas stream are subsequently moved to a tool like a cyclone. The granules can either be promptly recycled or reused in the roller compactor or stored in a container for later processing, to achieve the desired granule size (Heilakka et al. 2010). The PDG method has shown its effectiveness in creating granules that exhibit consistent flow characteristics across all formulations, resulting in compacts with an estimated tensile strength of 0.5 MPa. Furthermore, this technique allows for the incorporation of elevated drug quantities ranging from 70 to 100%. This is possible as satisfactory flowability can be attained even with reduced roll compaction forces (lower solid fractions) in contrast to standard roller compaction methods (Sandler and Lammens 2011). Apart from these advantages, this technology also offers several other benefits like quicker processing, cost-effectiveness, and minimal to zero material wastage, and the enclosed system design leads to decreased exposure to dust. However, the potential effects of recycling on granule quality, its compatibility with low-dose formulations, and its impact on friability and related aspects continue to be notable concerns associated with the pneumatic dry granulation (PDG) technology. Figure 6.1 illustrated the pneumatic dry granulation mechanism.

2. **Recent Advancements in Wet Granulation**

The extensively employed method is wet granulation, where granules are created by moistening the excipients and API using a granulating liquid, optionally with or without using a binder. Wet granulation has experienced a range of technical and technological advancements, including methods like steam granulation, moisture-activated dry granulation (or moist granulation), thermal adhesion granulation, melt granulation, freeze granulation, foamed binder (or foam granulation), and reverse wet granulation.

(a) **Reverse Wet Granulation**

Reverse wet granulation, alternatively referred to as reverse-phase wet granulation, is an emerging innovation within the wet granulation process. It involves immersing the dry powder formulation in the binder liquid and then

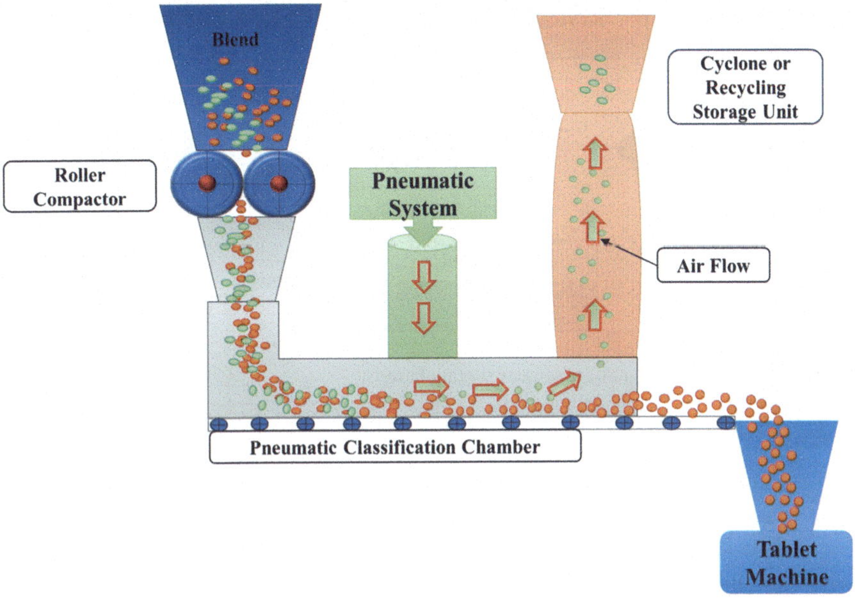

Fig. 6.1 Schematic diagram of pneumatic dry granulation

gently breaking it to form granules (Li and Reynolds 2010). As outlined by this innovation, initially, the binder solution was prepared, followed by the introduction of the dry powder excipients into the binder solution while being blended within a granulator. As a different approach, the drug was combined with a hydrophilic polymer and/or a solution of binder, creating a slurry of drug-polymer/binder utilized as the granulating fluid. This slurry was then used to create granules by submerging a blend of additional dry excipients into it. After drying, the wet granules that were obtained underwent a milling process. The granules yielded by this procedure demonstrated favorable flow and handling attributes similar to those achieved through the wet granulation process. Moreover, tablets originating from these granules exhibited a more consistent erosion pattern during dissolution testing in comparison to the usual wet granulation method. The main process of granule formation in the reverse wet granulation method is believed to involve controlled fracture (Wade et al. 2014, 2015). This method is claimed to enhance the dissolution properties of drugs with low water solubility. It achieves this by ensuring even dispersion of the binder, which serves as a wetting agent. This, in turn, facilitates the effective moistening of the drug substance throughout the granulation procedure. Figure 6.2 shows schematic presentation of reverse wet granulation process. Furthermore, it raises the probability of sufficient and consistent interaction between the pharmaceutical compound and the hydrophilic polymer, leading to enhanced dissolution. These enhanced granule attributes lead to uniform tablet

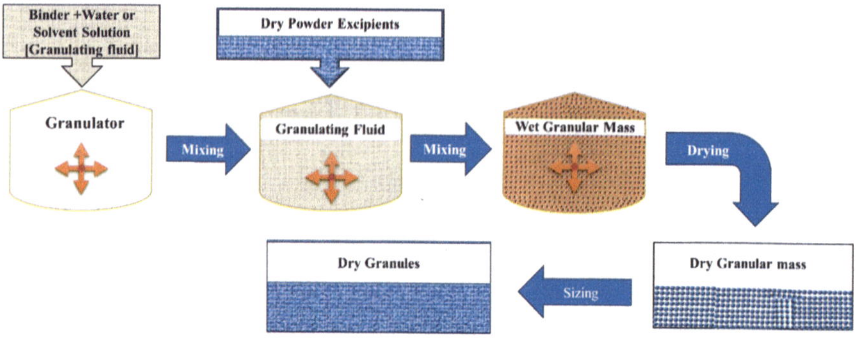

Fig. 6.2 Schematic diagram of reverse wet granulation

disintegration during the dissolution process. The benefits of this approach compared to traditional wet granulation encompass the generation of compact, round granules that exhibit enhanced flow characteristics. Additionally, there is uniform saturation and disintegration of these granules. This method might be appropriate for drugs with limited water solubility due to the close connection between the drug and the polymer. The practicality of existing equipment, like the high-speed mixer, is an added advantage of this method. Nevertheless, at lower binder concentrations, this approach yields granules with larger average mass diameter and reduced internal porosity compared to conventional wet granulation (Shanmugam 2015).

(b) **Steam Granulation**

Steam granulation is an innovative method of wet granulation; steam is used as a binding agent instead of the typical granulation liquid, which is typically water. Pure steam is a clear gaseous substance that offers enhanced permeability into the powder and a more advantageous heat equilibrium throughout the drying phase. Following the steam condensation, water generates a heated, delicate layer on the powder particles. This layer necessitates only a minor additional energy input for removal and undergoes evaporation more smoothly (Rodriguez et al. 2002). This method provides various advantages, including the effective distribution and penetration of powder particles by steam, the creation of round granules with enhanced surface area, and a faster, eco-friendly process without the use of organic solvents. A high-shear mixer connected to a steam generator would be adequate for executing this method. Nonetheless, significant energy inputs are essential for generating steam using this method. Furthermore, this procedure is not appropriate for all binding agents and is responsive to heat-sensitive drugs. The granules yielded by this technique exhibit an enhanced dissolution rate due to their larger granule surface area in contrast to the traditional wet granulation process (Rodriguez et al. 2002; Cavallari et al. 2002; Albertini et al. 2003; Vialpando et al. 2013a, b).

(c) **Moisture-Activated Dry Granulation (MADG)**

This method is a modification of the traditional wet granulation technique. It utilizes a small amount of water to trigger binding agents and initiate the agglomeration process (Ullah et al. 1987). This method consists of a dual step: (1) moist agglomeration of the powder particles and (2) moisture absorption or distribution. The combination of the medication, the binder, and extra additives is made by blending with a small quantity of water, frequently below 5% (with a preference for 1–4%). Agglomeration occurs when the binder is activated by the granulating fluid (water). Following the agglomeration, moisture-absorbing materials like silicon dioxide, microcrystalline cellulose, etc. are added to make it easier for more moisture to absorb. The moisture-absorbent materials remove the moisture from agglomerates, causing the powder mixture's moisture content to be redistributed, resulting in a comparatively dry granules mixture. While some of the bigger agglomerates may break down resulting in a more evenly distributed particle size, some smaller agglomerates may remain unchanged in size during this moisture redistribution process (Railkar and Schwartz 2000, 2001a, b). Using MADG for manufacturing controlled-release and immediate-release dosage forms demonstrated the benefits of wet granulation, such as enhanced flow properties, greater compressibility, and larger particle sizes. Moreover, this technique provides broad versatility, time and energy savings, and fewer process variables while ensuring consistent process effectiveness (Railkar and Schwartz 2001a, b).

(d) **Melt Granulation**

Melt granulation, also known as thermoplastic granulation, is a technique that enables the compaction of powder particles utilizing binders with the ability to melt or become soft at relatively low temperatures, typically ranging from 50 to 90 °C (Haramiishi et al. 1991). The last phase of the granulation process involves cooling the gathered powder and subsequently solidifying the melted or softened binding agent (Maejima et al. 1998a, b). The inclusion of binders with lower melting points into the granulation procedure can occur in two ways: as solid particles that undergo melting as part of the process (referred to as the melt-in technique or in situ melt granulation) and as liquid molten binders, potentially with the drug dispersed within them (known as the spray-on or pump-on technique). These approaches offer a range of possibilities for tailoring the characteristics of the eventual granules. To be more precise, the melt-in method within the melt granulation process entails heating a mixture of drugs, binding agents, and extra additives to a temperature that matches or exceeds the melting point of the binder. In contrast, the spray-on method involves the application of a melted binding agent, potentially carrying the drug, onto the powder that is being heated (Abberger 2001; Passerini et al. 2010; Aleksić et al. 2014). Melt granulation serves as a suitable substitute for various wet granulation methods utilized with materials sensitive to moisture (Watanabe et al. 2016).

Furthermore, when contrasted with the standard wet granulation procedure, it presents numerous benefits (Shanmugam 2015).

Typically, the melt granulation process doesn't necessitate the use of organic or aqueous solvents. This eliminates the need for managing and recycling organic solvents, meeting environmental criteria. Additionally, the absence of water removes the wetting and drying stages, resulting in a process that consumes less time and energy. The melt granulation process can effectively enhance the inadequate physical characteristics of the drug substance and improve the stability of drugs sensitive to moisture (Kowalski et al. 2009; Shah et al. 2013). The main drawback of this method is the need for elevated temperatures throughout the procedure that may result in oxidative instability and/or deterioration of the ingredients, particularly of the drugs that are thermolabile. In this technique, either hydrophilic or hydrophobic binders can be utilized. Choosing a meltable binder, regardless of whether it has hydrophilic or hydrophobic properties, significantly impacts the dissolution behavior of drugs. Appropriate equipment for melt granulation comprises a high-shear mixer and a fluidized bed granulator (Abberger and Henck 2000; Aoki et al. 2015; Van Melkebeke et al. 2006). Figure 6.3 shows graphical presentation of melt granulation process.

(e) **Foam Granulation**

Comparable to spray agglomeration, the foam granulation or foamed binder granulation process includes introducing a liquid or aqueous binder in a foamy state, as opposed to spraying or pouring it onto the powder particles. The Dow Chemical Company (Midland, MI) initially launched this foam binder technology in 2003 for the delivery of aqueous binder systems in high-shear and fluid bed wet granulation processes (Keary and Sheskey 2004). The binder can be introduced as foam rather than being sprayed or poured over the moving powder particles by installing a foam generator inside the tank containing the binder solution, where there is either a high-shear granulator or a fluid bed granulator. Utilizing the binder solution in the form of foam rather than a spray prevents irregular and unforeseen distribu-

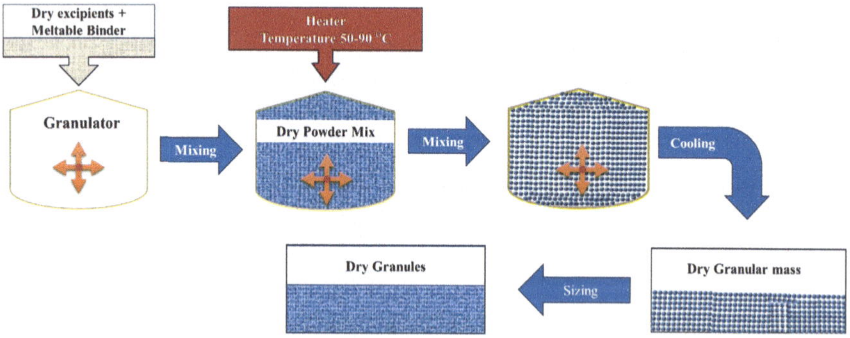

Fig. 6.3 Schematic diagram of melt granulation

tion of the binder, which could impact tablet firmness and drug release characteristics. Compared to sprayed water, the foamed binder/water has significantly greater surface area and overall volume. This method utilizes the properties of the foamed binder to promote uniform allocation of the binder across the powder particles. This is achieved even when using less binder compared to what's typically needed in the traditional spray granulation technique. Additionally, the liquid droplets that are sprayed have a poor spread-to-soak ratio, which means that instead of spreading on the surface of the particles, they prefer to soak into powders and produce over-wetting, which requires a lot of water and binder followed by drying to remove extra water. Instead, because of the high spread-to-soak ratio of foamed binders, which are deposited over the particles instead of being soaked, less binder is used, and the dispersion of the binder is more evenly distributed. These elements decrease processing time while enhancing repeatability. The most significant benefit of this technique is the removal of spray nozzles, along with the processing variations and clogging issues they cause (Tan et al. 2013). Incorporating this technology, apart from the advantages mentioned earlier, would prove beneficial for drug formulations with high potency and low dosage requirements because it ensures uniform drug distribution. In addition to formulations for instant release and controlled release, water-sensitive formulations may also be made using this technique because of the minimal water need and quick processing time. For this technique, a foam generator might be combined with common tools like a fluid bed granulator, high/low shear mixers, etc. Although this technique has several benefits, more research has to be done on foam quality, process variables, equipment, flow patterns, mixing behavior, etc. Additionally, getting regulatory permission would be a major challenge (Keary and Sheskey 2004; Tan et al. 2013; Koo et al. 2012; Rocca et al. 2015; Thompson et al. 2012).

6.4 Process Automation

Automation involves the utilization of machinery and tools to execute both physical and cognitive tasks within a production process, substituting the need for human involvement (Gandu et al. 2023). The growing focus on automated technology in the pharmaceutical industry is driven by the promising trend of fully automating tablet production. This trend is fueled by the desire for enhanced tablet quality, streamlined manufacturing processes, and effective validation of the entire production process. The paramount characteristic of an automated compression system lies in its capacity for flexible design, primarily concerning both the tablet press and, more crucially, the control system. No automated compression system is suitable for all production scenarios. In each production area, there is an existing system already in operation, and it is the responsibility of the vendor to offer an automated design that replicates this system through automation (Bardin et al. 2004).

The direct compaction continuous tablet manufacturing process encompasses unit operations such as feeder, comill, blender, and tablet press. These operations are managed on-site through their respective user interfaces, indicating that pharmaceutical manufacturing plants still rely on manual or semiautomatic operation. The existing operating system doesn't facilitate easy data extraction, presenting a significant obstacle to implementing a control system in pharmaceutical manufacturing plants due to the lack of automation. Integrating control hardware, software, and sensors with process equipment is a difficult endeavor because of the absence of standardized controls in pharmaceutical equipment. Hence, there is a strong need for a structured framework that can effectively automate a pharmaceutical plant (Singh 2018).

6.4.1 Automation in Tablet Manufacturing

The advantages of incorporating automation in tablet manufacturing are discussed below.

1. Improved material handling

   Examples of material handling improvement:

   The computerized systems oversee the granulation and tableting stages in the tablet manufacturing process at Merck Sharp & Dohme (MSD) and Eli Lilly and Company.

   Advantages:

   - Materials are handled without human intervention.
   - The structure, situated within a three-story building, is designed to maximize the use of gravity through vertical drops and employs pumps, vacuum systems, and bucket conveyors to facilitate the upward movement of materials as needed.

2. Improvement in a specific step of tablet manufacturing known as "unit operations" like sieving, particle size reduction, mixing, drying, compression, coating, and packaging

   Examples include automatic weigher and recording system for weighing and measuring, high-speed rotary press for compression, and H I-COATER for aqueous coating.

3. Coating process improvement

   The application of a coating on tablets and pills using a coating pan relies heavily on the operator's skills and control.

   The standard Accela Cota is designed for coating tablets or pills with aqueous or solvent-based films. It features a coating pan with side vents or perforations.

   Advantages:

   - Air flows in a singular direction through the tablet bed, exiting through the perforations in the pans. This significantly minimizes or eradicates the rebound of atomized spray and the drying of spray droplets, a common issue with solvent-based coating in traditional pans.

- This aids in the coating process by promoting efficient drying due to the increased airflow through the bed.
- Utilized for the uninterrupted coating of film and sugar systems.

4. **Tableting Improvement**

   The Thomas Tablet Sentinel (TTS) is an automated tablet production unit that incorporates real-time control and monitoring of tablet weight. This unit makes use of readily accessible strain gauge technology, with strain gauges positioned on the tablet machine to observe the strain generated during the compression process. The pressure or compression force at each station is determined by the quantity of powder in the respective die, establishing a link between compression force and tablet weight. Consequently, measuring this compression force provides an indirect means of monitoring the weight of the tablets.

6.4.2 Fundamental Process Control Instruments

- Sensors: These are instruments designed to gauge the process variables essential for making control determinations.
- Input-output (I-O) devices: These convert sensor signals into formats that can be effectively processed by the computer. This category includes computer devices such as programmable controllers, microcomputers, and minicomputers.
- Activators: They receive directives from the computer through the I-O devices and translate them into actions executed within the process.

6.4.2.1 Automation in Direct Compaction Continuous Pharmaceutical Manufacturing Process

If adequately automated, a centralized control platform can operate the continuous pharmaceutical manufacturing pilot plant. Automation additionally enables the collection of process data measured by sensors integrated into each unit operation. In order to achieve full automation, it is essential to integrate the operating platform with each unit operation of the pilot plant. Pharmaceutical unit operations can be categorized into three groups depending on their required automation approach: (1) fieldbus connectivity; (2) serial port connectivity; and (3) OLE process control (OPC) connectivity. Three types of communication protocols were employed to link the unit operations with the operating platform. The integration of the feeder was accomplished using Profibus communication, while the blender and comill were integrated using serial ports communication. OPC communication was used to integrate the tablet press. Both platforms are linked to the plant through a switch, allowing easy selection of either platform to operate the plant. It's important to emphasize that at any given time, the plant only necessitates a single control platform.

6.4.2.2 Rotary Tablet Press

Utilizing computer technology, the pharmaceutical control units, equipped with a microprocessor, automatically rectify weight variations, monitor tablet values, and manage the rejection of faulty tablets. Additionally, they can trigger quick-stop controls if specific tolerances are exceeded.

This system also oversees and optimizes speed, efficiently utilizes tools, assesses tableting data, documents errors, and performs statistical evaluations throughout the manufacturing process.

The Fette Perfecta 2000 Cooltex, a tablet press system, enables compression at lower temperatures by connecting a well-designed press to a cooling unit, allowing the press to reach temperatures as low as −6 °C (Sambhamurti 2005).

6.5 Issues Concerning the Process of Tablet Manufacturing

6.5.1 Capping

Capping occurs when either the top or bottom of a tablet separates horizontally from its central section, either partially or entirely, and it separates in a manner akin to a cap, while being discharged either from the tablet press or through subsequent handling.

Reason: Capping commonly occurs because air gets trapped within the tablet during compression, leading to the tablet expanding when it's ejected from the die (Remington 2006; Shah 2011).

6.5.1.1 The Causes and Remedies of Capping Related to Formulation (Granulation)

Causes

1. Excessive presence of fine particles within the granulated material.
2. Inadequate moisture content, resulting in ineffective binding.
3. Granules are not adequately dried.
4. Insufficient binder quantity or unsuitable binder type.
5. Insufficiency or incorrect use of lubricants.
6. Granule mass at an excessively low temperature.

Remedies

1. Eliminate a portion or all of the fine particles using a 100–200 mesh screen.
2. Adequately add moisture to the granules by introducing a hygroscopic substance like sorbitol, methylcellulose, or PEG 4000.
3. Ensure thorough drying of the granules.
4. Incorporate a dry binding agent like pregelatinized starch, gum acacia, powdered sorbitol, PVP, hydrophilic silica, or powdered sugar to improve granule cohesion.

6.5.1.2 The Causes and Remedies of Capping Related to Machine (Dies, Punches, and Tablet Press)

Causes

1. Inadequately polished dies.
2. Punches with significant concavity or punches featuring beveled edges.

3. The lower punch doesn't rise above the die surface during ejection.
4. Incorrectly set sweep-off blade.
5. Elevated rotation speed of the turret.

Remedies

1. Ensure thorough polishing of dies. Explore alternative steel or material options.
2. Opt for punches with flat surfaces.
3. Set the lower punch appropriately for ejection.
4. Ensure effective ejection by finely tuning the sweep-off blade.
5. Extend the dwell period by decreasing the rotational speed of the turret.

6.5.2 Lamination

The term "lamination" refers to the splitting of a tablet into multiple separate horizontal layers.

Reason: The phenomenon of air getting trapped during compression and subsequently released upon ejection becomes more noticeable as the turret speed increases (Picta 2011).

6.5.2.1 The Causes and Remedies of Lamination Related to Formulation (Granulation)

Causes

1. Presence of greasy or oily substances within the granules
2. Excessive use of water-repellent lubricants
3. Magnesium stearate

Remedies

1. Make modifications to the mixing procedure by adding an adsorbent or absorbent material.
2. Decrease the quantity of lubricating substance used or transition to an alternative lubricant variant.

6.5.2.2 The Causes and Remedies of Lamination Related to Machine (Dies, Punches, and Tablet Press)

Causes

1. Quick relaxation of the tablet's outer edges as it is ejected from a mold
2. Quick depressurization

Remedies

1. Use dies that have a tapered configuration, where the upper section of the die bore slopes outward, incline ranging from 3 to 5°.
2. Integrate an initial compression stage, lower the rotational speed of the turret, and decrease the final compression pressure.

6.5.3 Chipping

Chipping is the term used to describe the breaking of tablet edges, which can occur either when the tablet is ejected from the press or during later handling and coating stages.

Reason: Inadequate machine settings, specifically the improper calibration of the ejection take-off mechanism (Remington 2006).

6.5.3.1 The Causes and Remedies of Chipping Related to Formulation (Granulation)

Causes

1. Adherence to the punch surfaces.
2. Granules with insufficient moisture.
3. Excessive binding leads to chipping at the base.

Remedies

1. Ensure thorough drying of the granules or enhance lubrication.
2. Add moisture to the granules for plasticization.
3. Incorporate substances with hygroscopic properties.
4. Fine-tune the binding process or employ dry binders.

6.5.3.2 The Causes and Remedies of Chipping Related to Machine (Dies, Punches, and Tablet Press)

Causes

1. Wear on the die's groove where compression occurs.
2. Conical die (the center of the die is broader than the ends).
3. The punch face's edge has been reversed inward.
4. Excessive depth of concavity hinders proper compression.

Remedies

1. Polish the die by exposing its open end, turning it around, or obtaining a new one.
2. Refine the die's surface to achieve a cylindrical shape.
3. Buff the edges of the punch.
4. Minimize the indentation on the punch faces and employ flat punches.

6.5.4 The Defect Related to the Machine

6.5.4.1 Double Impression

Double impression occurs exclusively with punches that bear a monogram or some form of engraving.

Reason: Throughout the compression process, the tablet takes on the impression of the punch. In specific machines, the lower punch descends without restraint, briefly moves freely, and then ascends along the ejection cam to eject the tablet from the die. While undergoing this unrestricted movement, the punch may rotate, potentially leading to a fresh imprint on the tablet's underside, leading to a dual imprint (Lachman et al. 1986).

Cause: During tablet ejection, the lower punch can spin freely without restriction.

Remedies

1. Include keying within the tooling, which involves inserting a key beside the punch to align with and hinder punch rotation.
2. More recent presses feature anti-rotation mechanisms that inhibit punch rotation.

6.6 Tablet Coating

Coating plays a significant role in the manufacturing process of various solid oral dosage forms like tablets and granules. The main goal of film coating is to apply a fine layer of polymer onto the surface of a tablet or granule that holds active pharmaceutical ingredients (APIs). Recent years have seen over 50% of all pharmaceutical tablets being subjected to coating. In the present day, manufacturing sugar-coated tablets is infrequent because the process is intricate and requires a high level of skill and expertise from operators. Instead, tablets are commonly coated using contemporary machinery like drum and pan coaters, employing polymer films of diverse compositions (Hemchand et al. 2017). Figure 6.4 shows the different types of coatings for tablet.

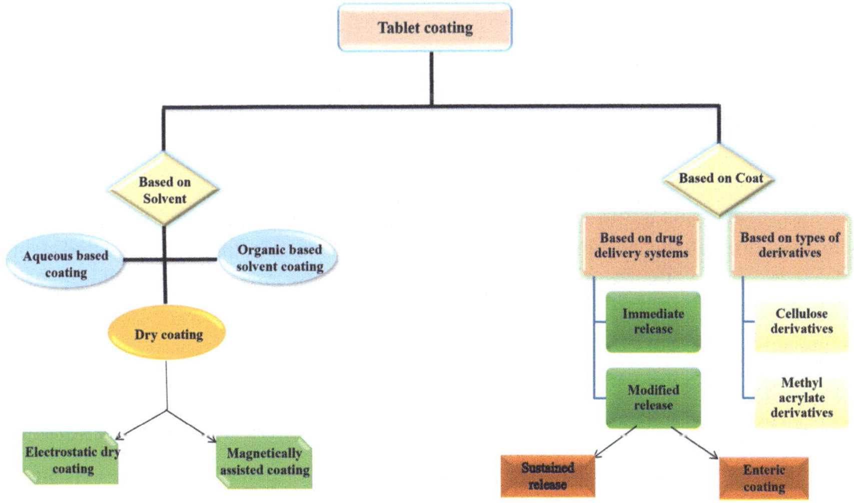

Fig. 6.4 Types of tablet coating

Abbott Laboratories introduced the initial pharmaceutical film-coated tablet for commercial use in 1954. These tablets were manufactured within a fluidized bed coating column using the Wurster principle as the foundation (Ahmed et al. 2021).

6.6.1 Sugar Coating

Within the pharmaceutical sector, sugar-coating tablets are utilized to enhance the aesthetics, taste, and longevity of oral medications. However, sugar coating has become less prevalent and has been substituted with polymer film coating. Typically, the sugar coating on tablets dissolves easily in liquid environments like gastrointestinal fluid due to its water-soluble nature. An objective of sugar coating is to safeguard the drug within the tablet and serve as a protective barrier against potential environmental contamination.

Advantages

- It is not a critical process as compared to film coating.
- It provides the attractiveness to tablet coating.
- It is easily swallowed due to its sugar-coating mask.
- It is inexpensive compared to other methods.

Disadvantages

- It requires specialized personnel.
- It consumes a long processing time.
- It increases the weight and size of the tablet (Qiao et al. 2010).

6.6.2 Film Coating

There are several methods for applying film coatings, including spraying a liquid, immersion, application of a powder via an electrostatic method, or condensing from supercritical fluids. This method has been substituted with film coating technology because the sugar-coating approach requires a lot of time. The procedure includes applying a solution containing polymer, pigments, and plasticizer onto a tablet bed in rotation, creating a thin and even film on the tablet's surface (Bagade et al. 2014). The selection of the polymer predominantly relies on whether the intended drug release location is the stomach or the intestine, or on the desired rate of release. The most commonly used technique for film coating is liquid spraying, which usually includes three stages:

- This process involves the application of atomized liquid onto a moving target surface through spraying.
- It maintains precise control by introducing heated airflow to the target surface.
- It achieves the desired coating amount through this procedure (Basu et al. 2013).

6.7 Recent Advancements in Tablet Coating Technology

6.7.1 Electrostatic Coating

Historically, a pan-coater system was employed to implement an electrostatic dry powder coating method for tablets. This approach enables achieving a polished tablet surface, excellent uniformity in the coating, and controlled release in a particular solvent grade. The electrostatic coating technique plays a crucial role in various production fields, such as paint technology, the food industry, metal coating, and the pharmaceutical sector, particularly in coating solid dosage forms. Following the principle of electrostatic powder coating, a blend of small granules and polymers is applied to the substrate surface without requiring any solvent. The substrate is then heated in an oven for drying, causing the powder mixture to melt into a film (Pawar Avinash et al. 2010). Electrostatic dry coating system is shown in Fig. 6.5.

There are two categories of spraying units, determined by the charging mechanism:

- Corona charging
- Tribo charging

6.7.1.1 Mechanism of Corona Charging

In this technique, high voltage is applied to a sharp, needle-like electrode (also known as a charging tip) at the gun's outlet, which causes electrical disruption and ionizes the air. The negative ions are picked up by the granular particles as they travel from the cannon to the substrate. A mixture of electrical and mechanical energies transports the particles between the substrate and the charging cannon. The electrical field can be modified to regulate the design's size, form, and volume as the powder is discharged from the gun (Ahmed et al. 2021).

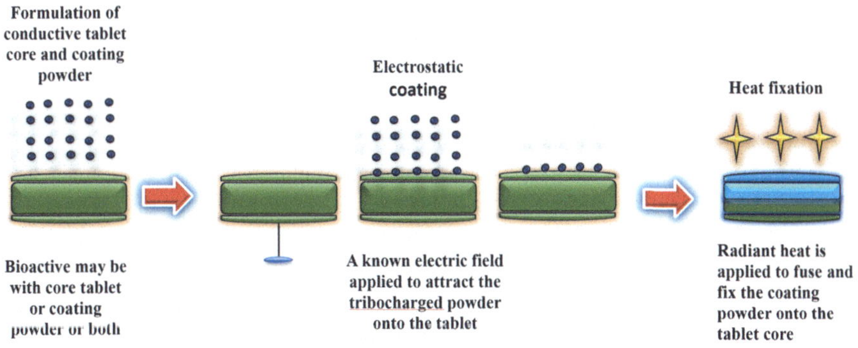

Fig. 6.5 Diagrammatic representation of electrostatic dry coating system

6.7.1.2 Mechanism of Tribo Charging

Tribo charging works based on friction charging coupled with dielectric attributes. Therefore, neither free ions nor electrical fields exist among the grounded element and the spray cannon. The repulsion force present between the particles in the tribo charging cannon is caused by electrical forces. When forces of attraction occur among the ground substrate and the surrounding space, charged substances can enter the region, which causes the particles to settle on the substrate, and charged atom homogeneity on the earthy substrate is caused by mechanical forces. Forces ultimately become equal because the forces of attraction and repulsion are equal, which prevents particles from sticking to the substrate and inhibits the thickness of the coating substance from increasing. Pharmaceutical tablet cores and electrically nonconducting materials are significantly more challenging to electrostatically coating than conventional coating methods (Remington 2006).

The efficacy of powder coating, including transfer efficiency, film thickness, association, and physical form, is significantly influenced by the chemical makeup, tribo and corona charging properties, electrical resistance, hygroscopicity, fluidity, and shape uniformity. The electrostatic coating method is greatly influenced by distance, nozzle shape, and the prior solution composition (Porter 2001).

6.7.2 Aqueous Film Coating Technology

The film coating method has substituted the time-consuming sugar-coating procedure, which is dependent on the coating operator's expertise. Due to environmental and regulatory concerns, aqueous film coating has substituted the organic solvents previously employed in this process, such as methylene salt. In addition, the price for any particular organic solvent is significantly higher compared to the price of pure water. Consequently, transitioning from coating with organic solvent to using aqueous solvent improves the effectiveness of the coating process. The disadvantages of film coating using organic solvents and the advantages of utilizing aqueous-based systems have been extensively recognized. As a result of recent advancements in film coating technology, aqueous coating is now more preferred instead of the exception (Ganguly et al. 2022).

Development of Film Coating Formulation

The development of a film coating may need to be optimized to raise coating durability, reduce bridging of integrations, enhance coating adherence to the core material, or enhance any additional characteristic that the formulator considers to be lacking. Flexibility, tensile force, and interaction between the film and the tablet interface are the three main variables that the formulation scientist must consider while determining the performance of the film. Because of these crucial elements, it is essential to utilize the optimal coating preparations to obtain the greatest results (Aulton 2002).

6.7.3 Supercell Coating Technology (SCT)

A **disruptive** tablet with Supercell coating technology can endure excessively hygroscopic materials and have hard coating elements put on top of it. In some situations, this method produces nonhomogeneous results because of deficiencies and unpredictable results. The edges of the tablets have uneven coating thickness compared to their surfaces, and their ends are trimmed using the Supercell coating technique to achieve a uniform finish. This prevents the stacking of tablets in the rotating pan and prevents airflow through the pan. As a result of the coating components being deposited, the modified release of coating is reduced. Niro Pharma Systems developed the Supercell coating technique, which uses a compact modular architecture to aid with a variety of issues (Kumar et al. 2011).

Moreover, the current technology is incapable of notably coating hygroscopic tablets, as well as tablets with flat shapes or unconventional configurations. For the prevention of "twinning," in which two or more tablets attach to one another, this procedure must be carried out carefully. The purpose of this study is to look into Supercell coating, a virtual tablet coater system that uses a distinctive airflow arrangement. To determine the impact of various moisture conditions on the level of quality of the coats generated, tablets coated at various spray velocities (4, 6, 8, 10, and 12 mL/min) are studied. The degree of roughness is observed to be lower at 6 mL/min compared to different spray rates, and the coat looks smoothest, with the droplets appearing to be joined together. When sprayed at faster rates, the droplets resemble scale-like patterns with branching arms.

The continuous small-batch coating technique used by SCT is reliable and effective. In SCT, the number of units of tablets that are coated ranges from 30 to 120 g, and they scale up uniformly to reach production levels. To achieve a more effective process, the coating spray is applied to the tablets in the same way as the drying gas. The tablets move rapidly and uniformly across the spray zone due to the special air circulation plate design used by SCT. This allows for enhanced coating accuracy because the tablets receive only a tiny quantity of coating per pass. Since this procedure takes less time, in minutes or seconds rather than hours, it is easier to the tablets.

Additionally, Supercell coating technology can be used to cover brittle tablets in addition to flat or extremely oval tablet shapes. Particularly hygroscopic tablets can be coated using this method since drying occurs very quickly. The Niro Company asserts that the traditional approaches to tablet coating yield irregular and suboptimal outcomes, causing inconsistencies that could potentially impact the tablet's performance. This outcome could introduce a degree of diversity, which becomes more important when a limited batch of tablets is manufactured for clinical testing. In the usual coating machines, tablets are placed inside big rotating pans and then exposed to hot air for drying. However, this process can result in the abrasion of tablet edges, filling in any engraved markings with the coating material, and uneven coating thickness on edges and corners compared to the tablet faces. According to Niro, these kinds of errors restrict the adoption of specific release coatings (Ahmed et al. 2021).

Distinctive characteristics of the Supercell coating technology:

1. Coating without interruption
2. Reduced processing duration
3. Adaptable design with modular flexibility
4. Technology enhancement
5. Coating with multiple layers
6. Challenging shapes for coating
7. Easily breakable tablets
8. "Low humidity process" suitable for materials sensitive to moisture
9. Empowering technology

6.7.4 Magnetically Assisted Impaction Coating (MAIC)

A method is created to predict the duration needed for coating within a magnetically assisted impaction coating (MAIC) apparatus. The combination of guest, host, and magnetic particles is believed to remain in a fluidized condition, characterized by a velocity distribution following the Maxwell-Boltzmann model. The assumption is made that particle collisions play a crucial role in directing the guest particles onto the host particle surfaces, thereby creating a partially lasting coating on those surfaces. The duration of coating relies on various factors, such as the host particle's number density, the ratio of host particle diameter to guest particle diameter, the height of the fluidized particle bed, and the material characteristics of both host and guest particles. There is a specific height for a bed where the coating process takes the least amount of time. When the bed height strays from the ideal value, whether it's smaller or larger, and when the size of the host particles increases, the coating time experiences a noticeable and substantial increase. Several dry coating techniques have emerged, including compression coating, plasticizer dry coating, heat dry coating, and electrostatic dry coating. Typically, these approaches allow for the application of increased shearing stresses, powerful impact forces, or higher temperatures during the coating process. These intense mechanical forces, coupled with the resulting heat, can lead to the guest particles being layered or even incorporated into the surface of the host particles. Magnetically assisted impaction coating (MAIC) devices offer the capability to coat delicate organic host and guest particles without altering their physical dimensions. Despite the occurrence of slight particle collisions within the MAIC process, resulting in minimal heat production, this effect is insignificant. This attribute becomes especially beneficial when working with temperature-sensitive powders like pharmaceuticals. The ongoing advancement of magnetically assisted impaction coating (MAIC) aims to enhance the efficiency of blending powders containing nano-sized particles, all without requiring solvents or heat. Generally, achieving consistent blending of nanoscale materials poses greater challenges compared to larger particles. As this technology continues to be refined, it holds the potential to assist manufacturing processes in generating superior-quality products (Pawar Avinash et al. 2010).

6.7.4.1 Mechanism of Coating in the MAIC Process

The process of MAIC coating involves various stages:

Stage I: Activation of magnetic particles
Stage II: Dispersal of guest particles (coating material)
Stage III: Alignment and dispersion of guest particles on the host particle surface (material to be coated)
Stage IV: Interaction between magnetic host particles
Stage V: Interaction between magnetic host particles and the chamber wall
Stage VI: Generation of coated products

6.7.5 Dip Coating

The application of coating involves immersing the tablets in a liquid coating, after which the damp tablets are dried in traditional coating pans. The procedure might involve multiple rounds of dipping and drying to attain the aspired coating thickness. However, this method limits the speed, adaptability, and dependability of spray coating methods (Ahmed et al. 2021).

6.7.6 Vacuum Film Coating

This innovative coating method utilizes a custom-designed pan with baffles, featuring a hot water jacket and the ability to be vacuum-sealed. Tablets are positioned within the pan, and nitrogen is introduced to displace the air until the desired vacuum level is reached. An airless spray system is employed to apply the coating solution, while a vacuum system removes the vapors of evaporated solvents. This technique enables the efficient utilization of organic solvents, accompanied by environmental solid safety measures (Ahmed et al. 2021).

6.7.7 One-Step Dry-Coating (OSDrC®)

To address challenges associated with traditional dry coating methods and the formation of Tablet-in-Tablet, a groundbreaking development known as the one-step dry-coating (OSDrC®) equipment was introduced, sparking a significant transformation in tablet manufacturing. OSDrC® is a registered trademark of Sanwa Kagaku Kenkyusho Co. Ltd. in Japan. This innovative technique allows scientists to regulate drug release by altering the thickness of the outer coating layer formulation (Maiti 2014).

At first, the space between the lower-center punch and lower-outer punch is filled with powder for the outer coating layer. Subsequently, precompression is carried out using the upper-center punch. Subsequently, as the upper-center punch advances the pre-compressed first-outer layer, the lower-center punch moves downward. The upper-center punch is then retracted to create space, into which the core powder is introduced and pre-compressed by the upper-center punch. In the end, the lower-outer punch descends, permitting the second-outer layer powder to fill and encase the pre-compressed core and first-outer layer. The combination of the core and first-outer layer, along with the second-outer layer, is compressed utilizing the upper and lower punches, with the center punches and outer punches functioning cooperatively (Ozeki et al. 2004).

Advantages of OSDrC® Technology

- Improves the safety and efficacy of drugs
- Enhanced pharmaceutical treatment
- Improved branding and product distinction
- Enhanced controlled-release formulations

6.8 Defects of Tablet Coating

The tablet coating has several defects, including blistering, chipping, picking, twinning, pitting, orange peel, and mottling. These defects and their remedies are summarized in Table 6.1.

Table 6.1 Overview of defects of tablets, their definitions, reasons, and remedies

Defects	Definition	Reason	Remedies	Reference
Blistering	It is a localized separation of the film from the substrate's surface that results in a blister	Gas retention in the film as a result of excessive heat throughout spraying as well as the final step of the coating process	In this case, slower drying conditions are necessary	Ganguly et al. (2022)
Chipping	It is a problem in which the film appears chipped and dented, typically along the tablet's edges	Reduced fluidizing air or a decrease in drum rotation speed in pan coating	Exercise caution to avoid excessive drying of the tablets during the preheating phase, as this could render the tablets fragile and encourage capping	Ganguly et al. (2022)
Picking	This problem occurs when certain regions of the film detach from the surface due to tablets adhering together and subsequently separating	Similar to cratering, an excessively moist tablet bed can cause subsequent tablets to fuse together before breaking	The issue is typically resolved by slowing down the liquid application speed or enhancing the drying air heat, air volume, and temperature of the air	Ganguly et al. (2022)

Table 6.1 (continued)

Defects	Definition	Reason	Remedies	Reference
Twinning	Two tablets that adhere to one another are referred to as twinning	This issue generally arises in capsule-shaped tablets	If preserving the tablet shape is a priority, the issue can be resolved by adjusting the interplay between the pan speed and the spray rate. Experiment with decreasing the spray rate while elevating the pan speed. Occasionally, it might be essential to slightly alter the tooling design by making a subtle adjustment to the radius. This modification, nearly imperceptible, effectively mitigates the occurrence of tablet adhesion	Ganguly et al. (2022)
Pitting	This flaw entails the formation of indentations on the tablet core surface, even when there is no apparent damage to the film coating	The temperature of the tablet core is higher than the materials' melting points when the tablets are made	During the entire production process, regulate the temperature of the tablet core	Ganguly et al. (2022)
Orange peel	This surface irregularity leads to a film texture that is uneven and lacks glossiness, resembling the appearance of an orange peel	Improperly applying the coating solution before it has dried	This issue might be resolved by adding more solvent to thin the solution	Ganguly et al. (2022)
Mottling	The term "mottling" describes an uneven distribution of color over the tablet's surface, leaving spots that are both deeper and brighter	It is primarily caused by excipients with varied colorations or tablets with colored degradation products	An adequate amount of coating solution is prepared	Ganguly et al. (2022)

6.9 Conclusion

Tablets have long been a preferred pharmaceutical dosage form because of their convenience, stability, and ease of administration. Remarkable advancements in tablet production and coating technologies have revolutionized the pharmaceutical and nutraceutical industries in recent years. Advancements in granulation methods such as spray drying, fluid bed granulation, and hot-melt extrusion have improved drug content uniformity, dissolution rates, and overall product stability. The adoption of these technologies promises continued progress in the field of tablet manufacturing, offering new opportunities for personalized medicine and enhanced therapeutic outcomes. Coating technologies have also evolved significantly to enhance tablet appearance, taste-masking, and drug release profiles. Automation involves the utilization of machinery and tools to execute both physical and cognitive tasks within a production process, substituting the need for human involvement.

Acknowledgments The authors thank the National Institute of Pharmaceutical Education and Research-Raebareli (NIPER-R), Department of Pharmaceutics, Ministry of Chemicals and Fertilizers, Govt. of India for the entire support. Also, the NIPER-Raebareli has provided the communication number for this manuscript and the communication number is **NIPER-R/Communication/567**.

References

Abberger T (2001) Influence of binder properties, method of addition, powder type and operating conditions on fluid-bed melt granulation and resulting tablet properties. Pharmazie 56(12):949–952

Abberger T, Henck JO (2000) Kornbildungsmechanismen bei der Schmelzgranulierung in der Wirbelschicht und ihre Auswirkungen auf die Tablettierung [Granule formation mechanisms in fluid-bed melt granulation and their effects on tablet properties]. Pharmazie 55(7):521–526

Adolfsson Å, Caramella C, Nyström C (1998) The effect of milling and addition of dry binder on the interparticulate bonding mechanisms in sodium chloride tablets. Int J Pharm 160(2):187–195

Ahmed SAN, Patil SR, Khan MKS, Khan MS (2021) Tablet coating techniques: concept and recent trends. Int J Pharm Sci Rev Res 66(1):43–53

Albertini B, Cavallari C, Passerini N, González-Rodríguez ML, Rodriguez L (2003) Evaluation of beta-lactose, PVP K12 and PVP K90 as excipients to prepare piroxicam granules using two wet granulation techniques. Eur J Pharm Biopharm 56(3):479–487. https://doi.org/10.1016/s0939-6411(03)00114-0

Aleksić I, Duriš J, Ilić I, Ibrić S, Parojčić J, Srčič S (2014) In silico modeling of in situ fluidized bed melt granulation. Int J Pharm 466(1–2):21–30. https://doi.org/10.1016/j.ijpharm.2014.02.045

Aoki H, Iwao Y, Uchimoto T, Noguchi S, Kajihara R, Takahashi K et al (2015) Fine granules showing sustained drug release prepared by high-shear melt granulation using triglycerin full behenate and milled microcrystalline cellulose. Int J Pharm 478(2):530–539

Aulton ME (2002) Pharmaceutics: the science of dosage form design

Bagade OM, Pujari RR, Nemlekar NA et al (2014) Tablet coating and its outcome with complementary sprouting technology. Res J Pharm Biol Chem Sci Apprais 5(5):298–315

Bardin M, Knight PC, Seville JPK (2004) On control of particle size distribution in granulation using high-shear mixers. Powder Technol 140(3):169–175

Basha M, Salama A, Noshi SH (2020) Soluplus® based solid dispersion as fast disintegrating tablets: a combined experimental approach for enhancing the dissolution and antiulcer efficacy of famotidine. Drug Dev Ind Pharm 46(2):253–263. https://doi.org/10.1080/03639045.2020.1716376

Basu A, De A, Dey S (2013) Techniques of tablet coating: concepts and advancements. A comprehensive review. RRJPPS 2(4):1–6

Bruce LD, Shah NH, Malick AW, Infeld MH, McGinity JW (2005) Properties of hot-melt extruded tablet formulations for the colonic delivery of 5-aminosalicylic acid. Eur J Pharm Biopharm 59(1):85–97. https://doi.org/10.1016/j.ejpb.2004.06.007

Cavallari C, Albertini B, González-Rodríguez ML, Rodriguez L (2002) Improved dissolution behaviour of steam-granulated piroxicam. Eur J Pharm Biopharm 54(1):65–73. https://doi.org/10.1016/s0939-6411(02)00021-8

Challener CA (2020) Key ingredients needed to drive the success of continuous manufacturing. PharmTech. https://www.pharmtech.com/view/keyingredients-needed-to-drive-the-success-of-continuous-manufacturing. Accessed 23 Nov 2020

Cho E, Jung S (2015) Supramolecular complexation of carbohydrates for the bioavailability enhancement of poorly soluble drugs. Molecules (Basel, Switzerland) 20(10):19620–19646. https://doi.org/10.3390/molecules201019620

Fukuda M, Peppas NA, McGinity JW (2006) Floating hot-melt extruded tablets for gastroretentive controlled drug release system. J Control Release 115(2):121–129. https://doi.org/10.1016/j.jconrel.2006.07.018

Gandu S, Gandla K, Vaishnavi M, Repudi L (2023) A review on implementation of automation in the pharmaceutical industries. World J Adv Res Rev 18(2):606–615

Ganguly D, Ghosh S, Chakraborty P, Mitra S, Chatterjee S, Panja S, Choudhury A (2022) A brief review on recent advancement of tablet coating technology. J Appl Pharm Res 10(1):07–14

Haramiishi Y, Kitazawa Y, Sakai M, Kataoka K (1991) Study on fluidized melt-granulation. I. Examination of the factors on the granulation. Yakugaku Zasshi J Pharm Soc Jpn 111(9):515–523

Havel H (2018) Aqueous solubility-enhancing excipient technologies: a review of recent developments. Eur Pharm Rev

Heilakka E, Rahja P, Lammens R, Sandler N (2010) Pneumatic dry granulation (PDG) in solid dosage form manufacture. In: AAPS annual meeting and exposition, pp 14–18

Hemchand P, Doye Avinash R, Tukaram MS, Mahavir B (2017) Recent advances in different aspects of tablet coating. Asian J Pharm Res Develop 16:89–92

Kaur G, Grewal J, Jyoti K, Jain UK, Chandra R, Madan J (2018) Oral controlled and sustained drug delivery systems: concepts, advances, preclinical, and clinical status. In: Drug targeting and stimuli sensitive drug delivery systems. William Andrew Publishing, pp 567–626

Keary CM, Sheskey PJ (2004) Preliminary report of the discovery of a new pharmaceutical granulation process using foamed aqueous binders. Drug Dev Ind Pharm 30(8):831–845. https://doi.org/10.1081/ddc-200030504

Khadka P, Ro J, Kim H, Kim I, Kim JT, Kim H et al (2014) Pharmaceutical particle technologies: an approach to improve drug solubility, dissolution and bioavailability. Asian J Pharm Sci 9(6):304–316

Komersová A, Lochař V, Myslíková K, Mužíková J, Bartoš M (2016) Formulation and dissolution kinetics study of hydrophilic matrix tablets with tramadol hydrochloride and different co-processed dry binders. Eur J Pharm Sci 95:36–45. https://doi.org/10.1016/j.ejps.2016.08.002

Koo OM, Ji J, Li J (2012) Effect of powder substrate on foaml drainage and collapse: implications to foam granulation. J Pharm Sci 101(4):1385–1390. https://doi.org/10.1002/jps.23053

Kowalski J, Kalb O, Joshi YM, Serajuddin AT (2009) Application of melt granulation technology to enhance stability of a moisture sensitive immediate-release drug product. Int J Pharm 381(1):56–61. https://doi.org/10.1016/j.ijpharm.2009.05.043

Kumar KV, Sivakumar T, Mani TT (2011) Colon targeting drug delivery system: a review on recent approaches. Int J Pharm Biomed Sci 2:11–19

Lachman L, Liberman H, Kanig J (1986) The theory and practice of industrial pharmacy, 3rd edn. Lippincott Williams & Wilkins, pp 293–345, 346–373

Li B, Reynolds TD (2010) U.S. Patent Application No. 12/560,668

Li J, Wu Y (2014) Lubricants in pharmaceutical solid dosage forms. Lubricants 2(1):21–43

Li Z, Peng WH, Liu WJ, Yang LY, Naeem A, Feng Y, Ming LS, Zhu WF (2023) Advances in numerical simulation of unit operations for tablet preparation. Int J Pharm 634:122638. https://doi.org/10.1016/j.ijpharm.2023.122638

Maejima T, Kubo M, Osawa T, Nakajima K, Kobayashi M (1998a) Application of tumbling melt granulation (TMG) method to prepare controlled-release fine granules. Chem Pharm Bull 46(3):534–536. https://doi.org/10.1248/cpb.46.534

Maejima T, Osawa T, Nakajima K, Kobayashi M (1998b) Application of tumbling melt granulation method to prepare controlled-release beads by coating with mixture of functional non-meltable and meltable materials. Chem Pharm Bull 46(3):531–533. https://doi.org/10.1248/cpb.46.531

Maiti S (2014) OSDrC®: a revolution in drug formulation technology. J PharmaSciTech 4(1):12–13

Markl D, Zeitler JA (2017) A review of disintegration mechanisms and measurement techniques. Pharm Res 34(5):890–917. https://doi.org/10.1007/s11095-017-2129-z

Morott JT, Pimparade M, Park JB, Worley CP, Majumdar S, Lian Z, Pinto E, Bi Y, Durig T, Repka MA (2015) The effects of screw configuration and polymeric carriers on hot-melt extruded taste-masked formulations incorporated into orally disintegrating tablets. J Pharm Sci 104(1):124–134. https://doi.org/10.1002/jps.24262

Mužíková J, Hávová Š, Ondrejček P, Komersová A, Lochař V (2014) A study of tablets with a co-processed dry binder containing hypromellose and α-lactose monohydrate. J Drug Deliv Sci Technol 24(1):100–104

Mužíková J, Srbová A, Svačinová P (2017) A study of a novel coprocessed dry binder composed of α-lactose monohydrate, microcrystalline cellulose and corn starch. Pharm Dev Technol 22(8):964–971. https://doi.org/10.3109/10837450.2015.1131717

Ozeki Y, Ando M, Watanabe Y, Danjo K (2004) Evaluation of novel one-step dry-coated tablets as a platform for delayed-release tablets. J Control Release 95(1):51–60. https://doi.org/10.1016/j.jconrel.2003.10.028

Passerini N, Calogerà G, Albertini B, Rodriguez L (2010) Melt granulation of pharmaceutical powders: a comparison of high-shear mixer and fluidised bed processes. Int J Pharm 391(1–2):177–186. https://doi.org/10.1016/j.ijpharm.2010.03.013

Pawar Avinash S, Bageshwar Deepak V, Khanvilkar Vineeta V, Kadam Vilasrao J (2010) Advances in pharmaceutical coatings. Int J ChemTech Res 2:16–22

Picta R (2011) Problems associated with tablet manufacturing

Porter SC (2001) Novel drug delivery: review of recent trends with oral solid dosage forms. Am Pharm Rev 4:28–36

Qiao M, Zhang L, Ma Y, Zhu J, Chow K (2010) A novel electrostatic dry powder coating process for pharmaceutical dosage forms: immediate release coatings for tablets. Eur J Pharm Biopharm 76(2):304–310. https://doi.org/10.1016/j.ejpb.2010.06.009

Quodbach J, Kleinebudde P (2016) A critical review on tablet disintegration. Pharm Dev Technol 21(6):763–774. https://doi.org/10.3109/10837450.2015.1045618

Railkar AM, Schwartz JB (2000) Evaluation and comparison of a moist granulation technique to conventional methods. Drug Dev Ind Pharm 26(8):885–889. https://doi.org/10.1081/ddc-100101313

Railkar AM, Schwartz JB (2001a) The effects of formulation factors on the moist granulation technique for controlled-release tablets. Drug Dev Ind Pharm 27(9):893–898. https://doi.org/10.1081/ddc-100107669

Railkar AM, Schwartz JB (2001b) Use of a moist granulation technique (MGT) to develop controlled-release dosage forms of acetaminophen. Drug Dev Ind Pharm 27(4):337–343. https://doi.org/10.1081/ddc-100103733

Rashid I, Al-Remawi M, Leharne SA, Chowdhry BZ, Badwan A (2011) A novel multifunctional pharmaceutical excipient: modification of the permeability of starch by processing with magnesium silicate. Int J Pharm 411(1–2):18–26. https://doi.org/10.1016/j.ijpharm.2011.03.011

Raza A, Hayat U, Wang HJ, Wang JY (2020) Preparation and evaluation of captopril loaded gastro-retentive zein based porous floating tablets. Int J Pharm 579:119185. https://doi.org/10.1016/j.ijpharm.2020.119185

Remington J (2006) Remington: the science and practice of pharmacy, vol II, 19th edn. Lippincott Williams & Wilkins, pp 1615–1641

Rocca KE, Weatherley S, Sheskey PJ, Thompson MR (2015) Influence of filler selection on twin screw foam granulation. Drug Dev Ind Pharm 41(1):35–42. https://doi.org/10.3109/03639045.2013.845839

Rodriguez L, Cavallari C, Passerini N, Albertini B, González-Rodríguez M, Fini A (2002) Preparation and characterization by morphological analysis of diclofenac/PEG 4000 gran-

ules obtained using three different techniques. Int J Pharm 242(1–2):285–289. https://doi.org/10.1016/s0378-5173(02)00189-8
Saha S, Shahiwala AF (2009) Multifunctional coprocessed excipients for improved tabletting performance. Expert Opin Drug Deliv 6(2):197–208. https://doi.org/10.1517/17425240802708978
Sambhamurti BS (2005) Pharmaceutical engineering. New Age International
Sandler N, Lammens RF (2011) Pneumatic dry granulation: potential to improve roller compaction technology in drug manufacture. Expert Opin Drug Deliv 8(2):225–236. https://doi.org/10.1517/17425247.2011.548382
Savjani KT, Gajjar AK, Savjani JK (2012) Drug solubility: importance and enhancement techniques. ISRN Pharm 2012:195727. https://doi.org/10.5402/2012/195727
Shah A (2011) Coating tablet defects: the cause and the remedies. Coating Polymers
Shah S, Maddineni S, Lu J, Repka MA (2013) Melt extrusion with poorly soluble drugs. Int J Pharm 453(1):233–252. https://doi.org/10.1016/j.ijpharm.2012.11.001
Shanmugam S (2015) Granulation techniques and technologies: recent progresses. Bioimpacts 5(1):55–63. https://doi.org/10.15171/bi.2015.04
Siepmann J, Faham A, Clas SD, Boyd BJ, Jannin V, Bernkop-Schnürch A, Zhao H, Lecommandoux S, Evans JC, Allen C, Merkel OM, Costabile G, Alexander MR, Wildman RD, Roberts CJ, Leroux JC (2019) Lipids and polymers in pharmaceutical technology: lifelong companions. Int J Pharm 558:128–142. https://doi.org/10.1016/j.ijpharm.2018.12.080
Simões MF, Pinto RMA, Simões S (2019) Hot-melt extrusion in the pharmaceutical industry: toward filing a new drug application. Drug Discov Today 24(9):1749–1768. https://doi.org/10.1016/j.drudis.2019.05.013
Singh R (2018) Automation of continuous pharmaceutical manufacturing process. In: Computer aided chemical engineering, vol 41. Elsevier, pp 431–446
Sohail Arshad M, Zafar S, Yousef B, Alyassin Y, Ali R, AlAsiri A, Chang MW, Ahmad Z, Ali Elkordy A, Faheem A, Pitt K (2021) A review of emerging technologies enabling improved solid oral dosage form manufacturing and processing. Adv Drug Deliv Rev 178:113840. https://doi.org/10.1016/j.addr.2021.113840
Tan MX, Nguyen TH, Hapgood KP (2013) Drug distribution in wet granulation: foam versus spray. Drug Dev Ind Pharm 39(9):1389–1400. https://doi.org/10.3109/03639045.2012.719233
Thompson MR, Weatherley S, Pukadyil RN, Sheskey PJ (2012) Foam granulation: new developments in pharmaceutical solid oral dosage forms using twin screw extrusion machinery. Drug Dev Ind Pharm 38(7):771–784. https://doi.org/10.3109/03639045.2011.633265
Ullah I, Corrao R, Wiley G, Lipper R (1987) Moisture activated dry granulation: a general process. Pharm Technol 11(9):48–54
van der Merwe J, Steenekamp J, Steyn D, Hamman J (2020) The role of functional excipients in solid Oral dosage forms to overcome poor drug dissolution and bioavailability. Pharmaceutics 12(5):393. https://doi.org/10.3390/pharmaceutics12050393
Van Melkebeke B, Vermeulen B, Vervaet C, Remon JP (2006) Melt granulation using a twin-screw extruder: a case study. Int J Pharm 326(1–2):89–93. https://doi.org/10.1016/j.ijpharm.2006.07.005
Vasvári G, Kalmár J, Veres P, Vecsernyés M, Bácskay I, Fehér P, Ujhelyi Z, Haimhoffer Á, Rusznyák Á, Fenyvesi F, Váradi J (2018) Matrix systems for oral drug delivery: formulations and drug release. Drug discovery today. Technologies 27:71–80. https://doi.org/10.1016/j.ddtec.2018.06.009
Vialpando M, Albertini B, Passerini N, Bergers D, Rombaut P, Martens JA, Van Den Mooter G (2013a) Agglomeration of mesoporous silica by melt and steam granulation. Part I: a comparison between disordered and ordered mesoporous silica. J Pharm Sci 102(11):3966–3977. https://doi.org/10.1002/jps.23700
Vialpando M, Albertini B, Passerini N, Vander Heyden Y, Rombaut P, Martens JA, Van Den Mooter G (2013b) Agglomeration of mesoporous silica by melt and steam granulation. Part II: screening of steam granulation process variables using a factorial design. J Pharm Sci 102(11):3978–3986. https://doi.org/10.1002/jps.23699

Viscasillas Clerch A, Fernandez Campos F, Del Pozo A, Calpena Campmany AC (2013) Pharmaceutical design of a new lactose-free coprocessed excipient: application of hydrochlorothiazide as a low solubility drug model. Drug Dev Ind Pharm 39(7):961–969. https://doi.org/10.3109/03639045.2012.686507

Wade JB, Martin GP, Long DF (2014) Feasibility assessment for a novel reverse-phase wet granulation process: the effect of liquid saturation and binder liquid viscosity. Int J Pharm 475(1–2):450–461. https://doi.org/10.1016/j.ijpharm.2014.09.012

Wade JB, Martin GP, Long DF (2015) Controlling granule size through breakage in a novel reverse-phase wet granulation process: the effect of impeller speed and binder liquid viscosity. Int J Pharm 478(2):439–446. https://doi.org/10.1016/j.ijpharm.2014.11.067

Wang J, Wen H, Desai D (2010) Lubrication in tablet formulations. Eur J Pharm Biopharm 75(1):1–15. https://doi.org/10.1016/j.ejpb.2010.01.007

Wang S, Li J, Lin X, Feng Y, Kou X, Babu S, Panicucci R (2015) Novel coprocessed excipients composed of lactose, HPMC, and PVPP for tableting and its application. Int J Pharm 486(1–2):370–379. https://doi.org/10.1016/j.ijpharm.2015.03.069

Watanabe A, Tanaka H, Sakurai Y, Tange K, Nakai Y, Ohkawara T, Takeda H, Harashima H, Akita H (2016) Effect of particle size on their accumulation in an inflammatory lesion in a dextran sulfate sodium (DSS)-induced colitis model. Int J Pharm 509(1–2):118–122. https://doi.org/10.1016/j.ijpharm.2016.05.043

Wen H, Park K (2010) Introduction and overview of oral controlled release formulation design. In: Oral controlled release formulation design and drug delivery: theory to practice. Wiley, pp 1–19

7 Formulation Evaluation and Development of Specialized Tablets

Sofiya Tarannum and Keerti Jain

Abstract

Oral solid formulations are the most popular and convenient among all kinds of dosage forms available in the market. Tablets are the solid dosage forms that ensure fixed dose, tamper resistance, and easy handling and administration. The merits of tablets make them survive in the pharmaceutical market with huge profit. Tablets have been modified from time to time into several types to be accepted by all groups and kinds of patients. Based on release profile, tablets are basically divided into immediate, sustained, delayed release, and many more. Chewable, effervescent, and orodispersible tablets hold a specific position in the market. In this chapter we have discussed formulation, evaluation, and development of these specialized tablets to cater needs of particular groups of patients in brief. The future of the tablets starts with the exploration of the world of 3D printing which makes the customized tablets easily available for the patients as per their needs, hence improving the patient compliance and acceptability.

Keywords

Chewable tablet · Effervescent tablet · Orodispersible tablet · 3D printing

S. Tarannum · K. Jain (✉)
Drug Delivery and Nanomedicine Research Laboratory, Department of Pharmaceutics, National Institute of Pharmaceutical Education and Research (NIPER)—Raebareli, Lucknow, Uttar Pradesh, India
e-mail: keertijain.02@niperraebareli.edu.in

K. Jain, A. K. Yadav (eds.), *Advances in Pharmaceutical Product Development*,
https://doi.org/10.1007/978-981-97-9230-6_7

Abbreviations

3D	Three-dimensional
API	Active pharmaceutical ingredient
BCS	Biopharmaceutics Classification System
DSC	Differential scanning calorimetry
ET	Effervescent tablet
FLX	Fluoxetine
HPMC	Hydroxypropyl methylcellulose
ODT	Orodispersible tablet
PVP	Polyvinylpyrrolidone
PXRD	Powder X-ray diffraction
RH	Relative humidity
SARS-CoV-2	Severe acute respiratory syndrome coronavirus 2
STP	Stiripentol
SUPAC IR	Scale-up and Post-Approval Changes Immediate Release
US FDA	United States Food and Drug Administration
USP	United States Pharmacopeia

7.1 Tablet Dosage Form

Orally administered solid dosage forms are highly preferred forms of medication in recent times. Solid dosage forms that are administered orally have accurate dosages as well as stable components. The most appreciable advantages of solid dosage forms are patient compliance, tamper resistance, and ease of handling and use (Hemamanjushree and Tippavajhala 2020). Tablets are well-designed in appearance, and they are produced with some identification using distinctive embossing and debossing on their surface. They are economical formulations as their production is simple and highly robust and appropriate for packaging, shipping, as well as transportation. The tablets are the utmost stable dosage form compared to the liquid capsules, solutions, or suspensions in terms of physicochemical and microbial characteristics. Tablets are swallowed conveniently and processing steps in their production are quite few as well as simple (Arshad et al. 2021).

7.2 History of Tablet

In 1500 BCE (Before the Common Era), the most basic oral solid dosage form, pills, came into the picture. In 4000 BCE, medical preparations like liquid, herbs, spices, plant powders and metals were mostly prescribed as compared to solids. In primitive Egypt, the pills were mentioned on papyruses. Pills at that time were composed of honey, grease, and bread dough. Pills were handmade with the help of medicated constituents like plant powders or spices. Medications were referred to as katapotia by the ancient Greece people (Advankar et al. 2019). Pills were termed as pilula (little ball) by the Roman scholar Pliny. Slimy constituents of plants were utilized for coating the pills in the olden days. Sugar coating and gelatin coating were invented in 1800 as gelatin capsules.

William Brockedon shaped the pills, lozenges, and other oral dosage forms using a device that applies pressure in dies and patented the same (Bueno and Nozal 2010). This machine can compact the powder in the form of a tablet with no use of adhesive. In 1844 the first compacted tablets were formulated by Professor Brockedon in England. The produced tablets were exceptionally hard with no mention of their solubility and disintegration time. The inspiration for compressing the medications and additives in the form of tablets came from compressing the lead for making pencils. Thereby, he carried out the compaction of potassium hydrogen carbonate and sodium hydrogen carbonate which became the first illustration of modern tablet compression. The tablets then came into the market in England and managed to have decent sales (Skaftason and Jóhannesson 2013). Messrs. Newbery procured William Brockedon's business in 1871. At the same time, Jacob Dunton from Philadelphia started the compaction of several formulations, like quinine tablets, using Brockedon's die-compression methods (Salawi 2022).

In the mid-nineteenth century, compressed powders were becoming very popular as they could be easily administered. The negligible use of liquid in these formulations makes them very stable as compared to conventional pills; the ease of storage, as well as transportation, makes them highly desirable in military pharmacies. Conventionally the production of pills and lozenges at the industrial scale is much more complicated than the manufacturing of tablets by powder compression. From the 1870s to the present, several kinds of tablets were manufactured. The manufacturing was significantly accelerated by the Burroughs Wellcome Company, established in the year 1879 by S. M. Burroughs and H. S. Wellcome (Helmstaedter 2020). In 1872, John Wyeth & Brother and Mr. Henry Bower developed an advanced nonautomatic instrument that could produce tablets at a cheaper rate (Kebler 1914). Dr. Robert R. Fuller from New York first suggested the filling of molds with medicated milk sugar. In 1883, Mr. Fraser manufactured molded tablets entirely in the exact approach it is manufactured today (Wood 1906).

Between the 1940s and 1990s, polymers of synthetic and semisynthetic nature became popular for enteric coating. This comprises the oral sustained-release formulations by Smith, Kline & French utilizing the Spansule mechanism: Dexedrine (dextroamphetamine sulfate) (1952) (Maurya et al. 2014; Park 2016). Within the time span of the 1970s and 1980s, the creation and marketing of osmotic drug delivery systems via the oral route were carried out by the Alza Corporation (Santus and Baker 1995). Tablets are leading pharmaceutical dosage form globally as compared to other kinds of pharmaceutical formulations. Its market analysis is described in brief in below section.

7.3 Global Market Analysis

The market size of oral solid dosage forms administered orally was evaluated to be US$ 679.53 billion in 2022, and the gross income is expected to surge at cross annual growth rate of 7.10% from 2023 to 2029, reaching nearly US$ 1098.34 billion. The vital market companies producing oral solid dosage forms are Takeda Pharmaceutical Company Limited, Merck & Co., Corealis Pharma Inc., Astellas Pharma Inc., Corden Pharma International, Bayer AG, AbbVie Inc., Thermo Fisher Scientific Inc., Lonza, Catalent Inc., Alcami Corporation, Halo Pharmaceutical Inc., Arnet Pharmaceutical, Jubilant Life Sciences, Pfizer Inc., Piramal Enterprises Ltd., and many more. The tablet division is governing the world market with a stake of 55.6% in 2022.

The oral solid dosage forms market analysis report provides a complete outlook throughout the globe with unique importance in North America, Europe, Asia Pacific, Latin America, the Middle East, and Africa. North America region had the leading market share in the oral solid dosage form market in 2021. Asia Pacific held the second position which was followed by the other provinces (Oral Solid Dosage Pharmaceutical Formulation Market—Global 2023). Tablets are modified from time to time to improve its applicability, popularity, patient's compliance, and profitability. Several kinds of tablets that are available in the market are listed below.

7.4 Types of Tablets

Tablets have been modified several times to cater needs of patients, which led to the existence of a huge collection of varieties of tablets available for commercialization. Traditional tablets are basically mono-layered tablets with basic constituents like active pharmaceutical ingredients (APIs), diluents, binders, disintegrants, and lubricants (Waterman et al. 2009). Bilayer tablets are intended to avoid chemical incompatibility by physically separating the two active substances. Further, two-layered tablets prompt two diverse release kinetics, with one layer for immediate release and the second for controlled release (Patel et al. 2007). As two or more incompatible active agents need to be managed concurrently, multilayered tablets can be prepared. Therefore numerous granulated preparations are formulated which undergo compression and produced a single tablet with either double or additional layers of variable colors generating a unique appearance (Abebe et al. 2014).

With time, the art of formulation of tablets headed in the direction of specialized tablets. Specialized tablets can overcome the limitations of normal tablets as the consistent drug levels in the blood can be achieved, adverse effects can be minimized, and inappropriate physicochemical properties of the API, like low absorption and solubility, presystemic metabolism, and narrow therapeutic index, can be tackled. Specialized tablets can be classified broadly in two ways: organ-targeted tablets and drug-release profile-specific tablets (Advankar et al. 2019). Apart from this, several types of tablets come under the category of miscellaneous like matrix system; reservoir system; chewable, orodispersible, and effervescent tablets; lozenges; osmotic tablets; and many more.

7.4.1 Organ-Targeted Tablets

Organ-targeted tablets are specifically designed for a particular organ or for localized action and make sure that other body parts remain unexposed to the drug. Common organ-specific tablets include buccal tablets, sublingual tablets, floating tablets, enteric-coated tablets, vaginal tablets, rectal tablets, dental cones, implantable tablets, and ocular tablets (Orlu et al. 2006). Some of these are shown in Fig. 7.1.

7.4.2 Modified Release Tablets

The pharmaceutical companies have explored several approaches like modified drug delivery that involves timed, sustained, delayed, as well as programmed-release systems, to satisfy particular market requirements. The principle of modified release indicates certain modifications in the release profile like faster, delayed, sustained, controlled, or targeted. As per the regulatory bodies, modified release preparations need to be efficacious, safe, and with minimum side effects. The high crests and low troughs in the plasma drug concentration curve should be negligible with the main focus on reducing the dosing administration frequency, which will eventually improve the patient's understanding of medication (Şengel et al. 2006).

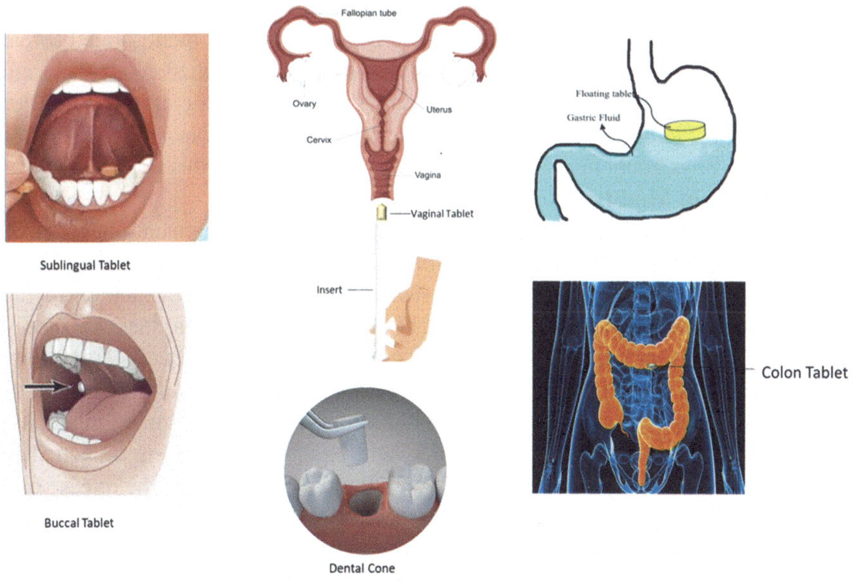

Fig. 7.1 Organ-specific tablets

Modified release formulations are fabricated in such a manner that their rate or site of release of therapeutics is controlled, unlike the immediate release formulations. This thoughtful alteration is attained by implementing changes in the formulation and production approaches. Modified release formulations are administered by different routes like oral, intramuscular, subcutaneous, etc. The modified release version of any medicament is made if the need for a more precise formulation is well analyzed (like patient compliance or safety). The report submitted for marketing approval needs to give a thorough explanation of the physical structure of the modified release substitute and the release principle; the reason for the selection of the specific dosage form in terms of indication and posology; and the reason for the selection of the quantity of API per unit of the dosage form (European Medicines Agency 2014).

1. **Formulation of Modified Release Tablets**

   Modified release dosage forms are of diverse types. These are (a) prolonged release; (b) delayed release; (c) multiphasic release; (d) biphasic release; (e) pulsatile release; (f) multiple-unit, (g) single-unit, e.g., osmotic tablet, and (h) intramuscular/subcutaneous depot formulations; (i) transdermal drug delivery systems; and so on.

   Hydrogels like cross-linked polyvinyl alcohol (PVA), cross-linked polyvinyl pyrrolidone (PVP), and polyethylene oxide (PEO); hydrophilic polymers like polyvinylpyrrolidone (PVP) and hydroxypropyl methylcellulose (HPMC); biopolymers like polylactic acid (PLA) and polyglycolic acid (PGA); mucoadhesive polymers like polyacrylic acid, tragacanth, methylcellulose, and pectin; nonbiodegradable polymers like cellulose acetate and ethyl cellulose; and natural polysaccharides like guar gum, karaya gum, and locust bean gum are the common excipients which can modify the release profile of incorporated active agents leading to the fabrication of various kinds of modified release dosage forms available in the market (Murugesan et al. 2019; Zarate et al. 2010). For observing the tablet production quality, quantitative evaluations and assessments of tablet's physicochemical as well as bioavailability properties must be made.
2. **Evaluation of Modified Release Tablets**

   The evaluation of modified release tablets is similar to conventional tablets. Thickness, hardness, friability, content uniformity, weight variation, swelling index, and dissolution studies are carried out for the proper evaluation of the prepared modified release tablets (Wen et al. 2021). Tablet's hardness is usually studied as the force needed for carrying out the disruption of the tablet in a particular axis. Hardness index is used which establishes a relation between hardness and weight to break the tablet so that a numerical value can be generated which can be utilized for relating tablets with their ease of getting broken. The dissolution process is conducted using apparatus 2 to investigate the release of a modified release tablet at various pH levels that mimic the pH conditions throughout the gastrointestinal tract. The medium is maintained at a constant temperature of 37 °C ± 0.5 °C and stirred at a speed of 50 rpm. The release is carried out at pH 1.2, 4.5, 7, and more alkaline pH, respectively, using 2 M HCl and 2 M NaOH. The time points for sampling are 0.5, 1, 2, 3, 4, 5, 6, 7, 8, 9, 10, 11, and 12 h. At every time point, 2 mL of the sample is extracted, resulting in a further dilution to 20 mL, and subsequently evaluated using a spectrophotometer. After each sample is extracted, it is substituted with a new dissolving media. The dissolution of tablet is performed in triplicate, and the average of these values is calculated. The evaluation of drug-release pattern involves the investigation of kinetics and mechanisms using several model-dependent approaches (such as zero and first models, Higuchi, Hixson-Crowell, and Korsmeyer-Peppas models) and model-independent approaches (such as mean dissolution time determination), as previously indicated.

Many researchers formulated modified release tablets of different drugs and evaluated them to analyze their performance. Some instances of the formulation development of modified release tablets are given. Mahapatra et al. (2015) developed delayed release tablets of zolpidem tartrate via the biphasic delivery system process, using sodium starch glycolate as a super disintegrant in instant-release fragment and HPMC as a delayed release matrix in the central part. Tablets were directly compressed with active agents present in both the central part and in the coat. Optimized formulations follow Fick's law of diffusion and exhibited similar release profiles as mentioned in the United States Pharmacopeia guidelines for extended-release dosage forms of zolpidem tartrate. Accelerated stability tests were carried out on the optimized formulation which ensures that it can tolerate the typical stressful circumstances of humidity and temperature (Mahapatra et al. 2015). Gaikwad et al. (2020) formulated modified release affordable tablets of capecitabine with no use of coating practices. The tablets were developed by the wet granulation process. Hydroxypropyl cellulose (HPC) was employed which acts as an extended-release matrix agent, and sodium alginate (SA) gave the sustained-release effect and exhibited mild gelling properties. Characterization using differential scanning calorimetry (DSC) and powder X-ray diffraction (PXRD) highlighted the retention of crystallinity of capecitabine within the optimized tablet. The selected formulation delayed the release up till 24 h (Gaikwad et al. 2020). Nagasamy et al. (2021) developed sustained-release tablets of isradipine, comprised of HPMC using the wet granulation method. Its in vitro as well as in vivo characterization was matched with the normal immediate release tablet as a primary evaluation. Tablets having 15% (w/w) HPMC showed a sustained release for 24 h. The release pattern of isradipine followed the Fickian diffusion abiding the first-order reaction. The absorption was significantly greater than the typical tablets, which exhibited immediate drug release. In vivo studies performed on rabbits indicated that the isradipine absorption rate from the formulated tablets was better than the immediate release tablets (Nagasamy Venkatesh et al. 2021). Some types of tablets do not fit in either the category organ-specific or modified release tablets. These tablets (chewable, effervescent, and orodispersible) are discussed in the miscellaneous part in the upcoming section.

7.4.3 Miscellaneous

7.4.3.1 Chewable Tablets

Chewable tablets are subjected to be masticated in between the teeth and the rate of release of API depends on the number of chews. These tablets are prescribed mostly for kids, elders, and patients with dysphagia. These are basically desirable for local action as compared to systemic action. Wet granulation or direct compression techniques are well established for the manufacturing of chewable tablets. Additionally, therapeutics in micronized and submicronized forms are used to improve the absorptivity of less soluble therapeutics (El-Gazayerly et al. 2004).

As per the USP, chewable tablets are divided into two types: chewable tablets which are (1) masticated for easy administration, for example, Nicorette gum (nicotine) and Aspergum (aspirin) and (2) those that need to be masticated before gulping down in order to circumvent choking or to make sure the release of the API, for example, niclosamide tablets. Chewable tablets are preferred for the APIs whose therapeutic dose is adequately large, and thereby, they cannot be administered as an intact tablet because of the large size. Chewable tablets are also proven to have better bioavailability as chewing tablets prior to swallowing improves the dissolution of the medication. Similarly, because of the pleasant taste and mouthfeel, the compliance of pediatric patients improves (Suzuki et al. 2004).

However chewable tablets have a few limitations like their hygroscopic nature and compromising mechanical strength which necessitates their vigilant handling. Moreover, continued mastication sometimes causes pain in the facial muscles. Bowel congestion occurs when patients ingest intact or partially chewed tablets, whereas dental injury or denture fracture can arise from tablets that are excessively hard. Further, a few constituents, like sorbitol, are responsible for flatulence and diarrhea, while certain flavoring agents can cause mouth ulcers (Dahiya et al. 2015). But these side effects are very rare and can be avoided easily.

1. **Formulation of Chewable Tablets**

   The main components of chewable tablets are therapeutic agents, diluents, sweetening agents, flavors, and colorants. The major difficulty that arises in the formulation of these tablets is taste masking as numerous medicines identified up to now are bitter in taste. Accordingly, taste-masking substances, certain flavors, and sweetening agents are used. Flavor masking comprises the inclusion of flavors, acids, and sweetening agents (Elder et al. 2016). Taste masking using granulation, solid dispersion, molecular complexes, inclusion complexes, ion exchange, microencapsulation, preparation of distinct salts/derivatives, spray cooling, and spray coating are also employed (Pardhi and Jain 2021). During the chewable tablet formulation, the API's taste is examined foremostly followed by the usage of respective sweeteners as the optimum quantity of sweetness is essential for sapidity. Moreover, the bitter taste is concealed with the use of suitable flavoring agents (Sohi et al. 2004) as shown in Fig. 7.2.

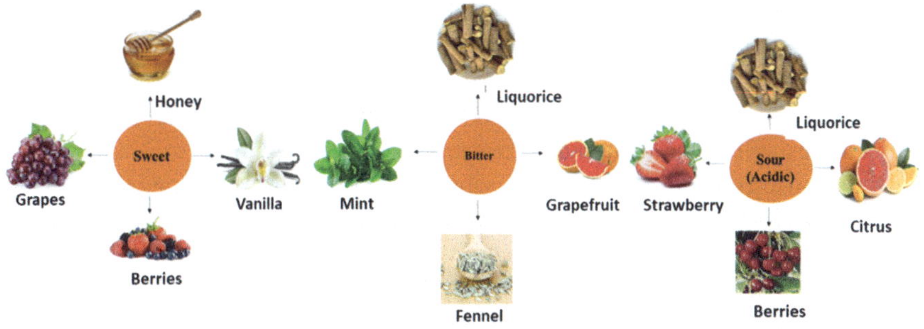

Fig. 7.2 Flavor groups for different types of taste

2. **Evaluation of Chewable Tablets**

Essential quality parameters for chewable tablets consist of hardness, disintegration, and dissolution, along with all factors that may affect bioavailability as well as bioequivalence of active agents. The size, thickness, friability, and taste of the tablet mainly decide the acceptance of the tablet by the patients. The efficiency of a chewable tablet does not depend on one quality characteristic. Rather, the prime focus must be to incorporate the appropriate blending of the mentioned parameters so that the chewable tablets show optimum efficiency for its expected use (SahabUddin et al. 2016).

Hardness: Chewable tablets need to have a hardness in a range so that they can resist the harshness of production, packaging, transportation, and distribution and at the same time can be masticated in no time by the subjected patients (Gupta et al. 2015).

Disintegration: For chewable tablets, disintegration time has to be short to avoid intestinal blockage in case a tablet is masticated in an improper manner by the patient. Generally, the incorporation of a suitable disintegrant based on its type, quality, and quantity enables quick disintegration of the tablet (FDA 2018).

Dissolution: Absorption of active agents from chewable tablets is based on the release of the same from the intact tablets; thus, in vitro dissolution study of chewable tablets must adhere to the dissolution testing fundamentals of typical immediate release tablets. The APIs of the chewable tablets must be effectively released into solution from the dosage form. From the characterization point of view of chewable tablet in the course of development, in vitro dissolution studies need to be performed on entire tablets in minimum four media, that is, water and aqueous buffer solutions at pH 4.5, 6.8, and 1.2, respectively, with well-known dissolution approaches by means of USP apparatus type 1 (basket), type 2 (paddle), or type 3 (reciprocating cylinder).

(a) Performance in Simulated Physiological Media

The in vitro approaches using simulated physiological media like saliva as well as gastric fluids are likely to mimic the in vivo performance of tablets, specifically the ones prepared with low-soluble APIs (Dressman et al. 1998; Markopoulos et al. 2015).

Certainly, chewable tablets need to be assessed with dissolution media like simulated fasted as well as fed state gastric and intestinal fluids with enzymes. Hardness testing is carried out after a transitory exposure (30 s) to approximately 1 mL of simulated salivary fluid, and the result can be useful for establishing the hardness requirement.

(b) In Vitro Bioequivalence Data Usability and Post-approval Concerns

The solubility and permeability of the APIs basically decide the position of the drug in the Biopharmaceutics Classification System (BCS). The BCS class of the active agents mainly determines whether the in vitro figures will suffice the bioequivalence or not for chewable tablets (U.S. Department of Health and Human Services et al. 2023). Changes in the chemistry, manufacturing, and controls after approval of the chewable tablets must be made in accordance with the principles instructed in the Scale-up and Post-Approval Changes Immediate Release (SUPAC IR) guidelines (U.S. Department of Health and Human Services et al. 2023). Numerous drugs in the dosage form of chewable tablets are available in the market like carbamazepine, Motrin, Alka-Seltzer, montelukast, etc. with a profitable business. Some of the recent developments in the field of chewable tablets are discussed in the below subsections (Gaikwad 2023).

3. **Formulation Development of Chewable Tablets**

Many researchers formulated chewable tablets of different drugs and evaluated them to analyze their performance. Some instances of the formulation development of chewable tablets are given. Dille et al. (2018) produced a gelatin-based, soft chewable formulation of APIs (ibuprofen, acetaminophen, meloxicam) as a substitute for conventional tablets for dysphagia patients. Soft-chewy tablets displayed better stability for all three therapeutics (till 24 months); the chewable tablet of ibuprofen showed similar dissolution as that of normal tablet along with better resistance to microbial growth. The rheology of the ibuprofen/gelatin tablet was highly comparable to the true gelatin gel, even though the formulation has greater storage and young moduli as well as melting point (Dille et al. 2018). Kimaro et al. (2019) formulated a novel generic albendazole chewable tablet with a better dissolution profile. They carried out the optimization with the help of the wet granulation process. The formed product was similar to the innovator tablet available in the market on the basis of quality considerations and thereby can be marketed as a generic product in the local market with the due support of the pharmaceutical industry in manufacturing (Kimaro et al. 2019).

Yoo et al. (2022) developed a new chewable chocolate-based tablet of tramadol for children. Tramadol is a bitter opioid analgesic, used for the treatment of postsurgical pain in children. Still, in several countries, no approved tramadol formulation is accessible for children. The relative bioavailability of this chewable tablet was 1.25-fold greater as compared to tramadol hydrochloride oral liquid, and its acceptability was better in children due to its taste when analyzed by the children, parents, and nurses as well (Yoo et al. 2022). Karaoglu et al. (2023) designed a standard, effective, and safe masticable tablet (using propolis and three plant extracts) with promising inhibition of variants (Wuhan B.1.36 and Omicron BA.1.1) of severe acute respiratory syndrome coronavirus 2 (SARS-CoV-2) as well as additional viral diseases. Green tea, bilberry, and dried pomegranate peel, along with propolis extracts, were used. Cytotoxic and antiviral effects of every single constituent, and the formulated chewable tablet, were studied against the SARS-CoV-2 by means of Vero E6 cells with the xCELLigence real-time cell analysis multiple plates system. The antiphlogistic and analgesic effects and mutagenic and antimutagenic properties of the chewable tablet were also assessed. Chewable formulations of 110 and 55 μg/mL concentrations had antiviral action rates of 101% and 81%, respectively, for the Wuhan variant and 112% and 35%, consecutively, for the Omicron variant. The blend (plant and propolis concentrate) was highly efficacious compared to the single extract (Karaoglu et al. 2023). Al-Madhagi et al. (2023) developed a novel chewable tablet by combining meclizine and pyridoxine which ensures instant absorption of therapeutics as well as reduces travel sickness. The chewable tablet was highly acceptable as it acted in a relatively faster manner, and it was favorable for patients of any age category (Al-Madhagi et al. 2023).

7.4.3.2 Effervescent Tablets

Effervescent tablets (ETs) produce solutions with the release of carbon dioxide instantaneously. Basically, these tablets are formulated by compressing the APIs with a blend of sodium bicarbonate, sodium carbonate, and organic acids like citric acid, adipic acid, malic acid, ascorbic acid, fumaric acid, maleic acid, succinic acid, or tartaric acid, thereby fastening the release as well as enhancing the bioavailability of the incorporated therapeutics (Zheng et al. 2019; Jaipal et al. 2016). Furthermore, sodium bicarbonate in a dose-dependent manner improves gastric emptying, which along with fast dissolution accelerates absorption. Mostly, those drugs whose immediate action is required are incorporated within these tablets, so that the dissolution happens in seconds whenever they come in contact with water which resulted in a clear and palatable solution (Dubray et al. 2021). Patients who face difficulty in swallowing capsules or tablets, or the ones who want instant relief, generally prefer ETs. The incorporation of a greater amount of dose than the conventional tablets is possible with the ETs which is quite acceptable in elderly patients and children as there is no need for swallowing during the administration.

Nevertheless, some limitations (recurrent sticking, high absorptivity, and tender nature) of the ETs halt their manufacturing and applicability. Specifically, sticking is very common in tablet manufacturing because the acidic constituent sticks on the punch tips of the tablets throughout compression (Sendall and Staniforth 2011). Hence, the surface of the formulated tablet becomes rough and may even undergo cracking gradually because of continuous compaction runs. All these can lead to the production of bad-quality tablets with low yield and productivity (Chattoraj et al. 2018; Uchimoto et al. 2013).

1. **Formulation of Effervescent Tablets**

   API can be incorporated within the effervescent granule, or it can be transformed into the salt form if it has low solubility at the time of dissolution. ETs are manufactured by blending the active agents or its salt forms with binders, diluents, as well as lubricants, followed by compaction in the form of tablets. Highly soluble lubricants are utilized like sodium benzoate, polyethylene glycol, and adipic acid. Disintegrants are not required in these tablets as the release of in situ CO_2 accelerates the disintegration process (Clark and Witzel 1978). Plasticizers are added to improve the rate of effervescence. With the increase in the plasticizer concentration, the rate of effervescence increases. Additionally, the effervescence can be changed by adjusting the hydrophilicity of binding agents used in the hot melt extrusion method. As the concentration of nonpolar binder increases, the effervescent rate decreases. The rate of effervescence increases with a minor increase in the concentration of either acid or alkali used than the condition in which both the components are added in equal amounts. Additionally, the ETs are coated with suitable substances for manipulating the release profile of active agents at the specific site in gastrointestinal tract (Rotthäuser et al. 1998). Effervescent granules are formulated at sites where the relative humidity (RH) is below 40%. These granules are blended in a V-type/ribbon mixer. Wet granulation is carried out which is followed by the shifting of granules to the oven. All the needed equipment should be free of traces of water or moisture so that the effervescent mixture remains stable. Drying is followed by the sizing of granules, blending them with hydrophilic lubricants, and then at last compression is performed to collect the tablets. Humidity should be monitored stringently in the production units (18–25 °C, 10% RH). Packaging of ETs should be in such a way that it must protect the tablet from environmental stress and air entrapment should be minimal as moisture that gets entrapped results in physicochemical degradation of the tablet (Altomare et al. 1997).
2. **Evaluation of Effervescent Tablets**

   The formulated ETs are evaluated thoroughly so that the standard integrity can be maintained. Their quality can be controlled and assured to the regulatory bodies. Most importantly the patient compliance can be managed. Several evaluation tests are performed which are mentioned in brief.

   (a) **Pre-compression evaluation**: These evaluations are done before the compression of tablets takes place. Particle size is measured using the sieve method. These parameters help in understanding the interparticulate interactions and in turn the flow properties. The flow of powder is crucial in the process of tableting, as it needs to flow smoothly and evenly into the tablet dies in order to achieve uniform tablet weight and produce tablets with consistent and replicable characteristics. Bulk density, tapped density, compressibility index, and Hausner's ratio are determined. The angle of repose is found using the fixed funnel method (Aslani and Sharifian 2014).

   (b) **Post-compressional evaluation**: Weight variation, friability, thickness, hardness, assay, and content uniformity are measured similar to that of conventional tablets. Apart from these, a few evaluation tests specific to ETs are given in brief.

pH evaluation: A single tablet is dissolved in water. As the dissolution ended, the pH of the obtained solution is determined using a pH meter. The test is performed in triplicate.

Carbon dioxide content: The individual tablet is weighed and solvated in 1 N H_2SO_4 (100 mL) followed by the measurement of weight after dissolution. CO_2 content is calculated in milligram. This test is performed on three tablets.

Effervescent time: A single tablet was taken in a beaker filled with 200 mL water and effervescent time is calculated using a timer. The test is performed in triplicate.

Water content: Ten tablets are weighed before and after insertion in a desiccator having activated silica gel for 4 h. The water content is determined using the equation:

$$\text{Water content}\left(\%\right)=\frac{\text{Weight before drying}-\text{Weight after drying}}{\text{Weight before drying}}*100$$

Equilibrium moisture content: Three tablets are kept in different desiccators having saturated saline solutions, potassium nitrate (RH, 90%), sodium chloride (RH, 71%), and sodium nitrite (RH, 60%) at 18 °C. After 1 and 7 days, the equilibrium moisture content in terms of percentage is calculated using the Karl Fischer titration by means of an automatic titrator.

3. **Formulation Development of Effervescent Tablets**

Many researchers formulated effervescent tablets of different drugs and evaluated them to analyze their performance. Some instances of the formulation development of effervescent tablets are given. Mohammed et al. (2016) formulated risperidone ETs for simplifying its administration and covering its bitter taste. The solubility of the active agent was improved by the solid dispersion method. Optimization was done with water-soluble excipients trehalose, inulin, pregelatinized starch, carboxymethylcellulose sodium, and Eudragit E100 at different ratios. Rotary evaporation, freeze-drying, and the kneading oven were employed for evaporating the solvent. The relative bioavailability of risperidone in the optimized batch was 161.41% with a considerable enhancement in absorption than the existing marketed normal tablets (Abu Bakr Mohammed et al. 2016).

Zheng et al. (2019) worked on overcoming the shortcomings of ETs (sticking, high hygroscopicity) so that their production as well as applicability can be improved. Basically, most of the time recurrent sticking critically damages the ETs. Hence, the work highlighted the use of polyvinylpyrrolidone (PVP) at variable usage levels, grades, or spray solution concentrations as a coater for the acidic and alkaline granules of effervescent formulations for combating the stickiness with the aid of fluidized-bed coating method. The conclusions drawn from the work were that a uniform layer of PVP over the granules solved the sticking issue and efficiently reduced the hygroscopicity and improved the compactness. Further the coating did not affect the disintegration time of the effervescent formulations (Zheng et al. 2019). Dubray et al. (2021) developed the effervescent formulation of paracetamol which was found to have faster absorption as compared to conventional tablets, which in turn was able to relieve the pain in less time. Additionally, ETs were highly acceptable as they were safe; further, they reduce the overdose issue, and in consequence, the chances of hepatotoxicity were lower than the normal tablets. ETs possibly will be helpful in encouraging the proper use of paracetamol, thus lowering the considerable public health burden related to the paracetamol overdoses (Dubray et al. 2021).

Rosch et al. (2021) studied the effect of various saccharides on the appearance as well as mechanical strength and the fragmentation of hydrogen-producing effervescent formulations produced by dry granulation. Hydrogen is a medical gas, used for the therapy of diseases associated with inflammation and oxidative damage. Dihydrogen is produced when hydrogen ion is reduced using a metal. Elemental magnesium was chosen as it can reduce hydrogen ions into hydrogen gas under acidic conditions. Mannitol was selected among saccharides as it can produce hydrogen ions very rapidly along with satisfactory mechanical properties (Rosch et al. 2021). Wang et al. (2022a) produced the ETs by compressing stiripentol (STP)-loaded enteric solid dispersions enhancing its solubility as well as stability. STP-loaded solid dispersions (STP-SDs) ETs were produced with the help of solvent evaporation and dry granulation methods, simultaneously, followed by the optimization process. STP-SDs remarkably enhanced the solubility as well as the stability of STP. The bioavailability of STP-SD-ETs was 138.71% greater than the STP suspensions. Additionally, STP-SD-ETs considerably improved the intestinal absorption rate of STP (Wang et al. 2022a).

7.4.3.3 Orodispersible Tablets

An orodispersible tablet (ODT) is supposed to be positioned in the mouth, where they quickly release the incorporated active agents, thereby producing a suspension or solution of it in the saliva. Hence, ODT enhances medication compliance by the patients as they are effortlessly swallowed without drinking or chewing and ensures precise dosing than the liquid formulation. ODT can transform the pharmacokinetics of the active agents as per their physicochemical characteristics (Cilurzo et al. 2018).

Conventional tablets and capsules are frequently related with dysphagia and pseudodysphagia, and, thereby, the numbers of conditions, where these dosage forms are not accepted by the patients, are increasing. This is mostly related to the children, geriatrics, and patients having swallowing problems (Patel et al. 2010). Patients having mental disorders or the ones who are nauseated or not supportive do not prefer the conventional tablets and capsules (Seager 1998). Hence, for those patients, ODTs are better option. But ODTs have limitations too. ODTs absorb moisture rapidly, so they must be placed in a cool and dry place. They must be packed in specialized packaging to avoid instability and damage. Dose uniformity is also another practical challenge.

1. **Formulation of Orodispersible Tablets**

   Taste is the critical attribute for the acceptability of ODTs by the patients. Taste masking in ODTs is carried out using sweetening agents, flavors, microcapsules, or complex formation (Douroumis 2011). A proper equilibrium between disintegration time and hardness is a must for ODTs which get variated by the process as well as formulation parameters. Accordingly, the packaging is optimized for protecting the orodispersive final formulation from environmental stresses. ODTs look analogous to normal tablets; however, the quick entry of water via the capillary action within the porous structure results in their disintegration within 30 s–3 min. The manufacturing approaches comprise freeze-drying/lyophilization, molding, and direct compression with particular excipients.

Lyophilization is the preferred technique for the production of ODT. Solvent sublimation from a frozen solution or suspension of an active agent with matrix-generating excipients basically leads to a porous and fluffy mass, which liquefies immediately and releases the therapeutics as soon as positioned in the mouth. For low-soluble therapeutics, lyophilization can be helpful in achieving a finished dosage form with the appropriate physicochemical properties by reducing the crystal dimension or by amorphization of crystals (Pardhi et al. 2024). Furthermore, lower temperature diminish the thermal degradation of active agents; thereby, such a process in current times was projected for use in the manufacturing of ODT loaded with vaccines. Usually, matrix-forming constituents are gelatin, dextran, or alginates. Mannitol is frequently utilized for improving the porosity of the freeze-dried sample, and glycine is utilized for preventing the shrinkage in the course of lyophilization (Patel et al. 2010).

Compression comprises the dampening of the powder mixture with a hydroalcoholic solution, accompanied by compaction within the mold plates, resulting in the formation of a moist mass, which is air-dried at the end.

In **heat molding** method, dispersed or dissolved therapeutic agents are incorporated within the molded mass and directly transferred to blister packets which is followed by the solidification at room temperature.

2. **Evaluation of Orodispersible Tablets**

The ODTs are evaluated by performing tests like hardness strength, friability, wetting time, moisture absorption, disintegration, and dissolution test.

Hardness strength: The hardness of the ODT is examined utilizing a Monsanto hardness tester (Dey and Maiti 2010). The limit is at the lower side as rapid disintegration in the mouth is required.

Friability: ODTs prepared by any methods have high percentage of friability ranging between 0.1% and 0.9%; hence maintaining the friability within this limit is a challenge. Roche friabilator is generally used.

Wetting time: It specifies the interior structure of the tablets and the hydrophilicity of the excipients. Hence, wetting time of a formulation is associated with the wetting angle. The lesser the wetting time, the faster is the disintegration of the tablets (Bandari et al. 2008). The wetting time is determined by using five circular tissue papers of 5 cm radius positioned in a petri dish of a 5 cm radius (Gohel et al. 2004). An eosin solution of 10 mL is added to the petri dish. An ODT is cautiously placed onto the tissue paper. The time interval at which water from the petri dish touches the upper surface of the tablet is called the wetting time. To determine the water-absorption ratio (*R*), the initial weight of the tablet is recorded before placing it in the petri dish (Wi). The tablet that has been moistened in the petri dish is removed and weighed again (Wf). Now, the *R* is calculated using the formula:

$$R = 100(\mathrm{Wf} - \mathrm{Wi}) / \mathrm{Wi}$$

Moisture uptake study: It is an essential study for ODTs to assess their stability. Briefly, ten tablets are placed in the desiccators over $CaCl_2$ at 37 °C for 24 h. The tablets are weighed and subjected to 75% RH (using saturated NaCl solution in a desiccator for 72 h), at normal temperature for 14 days (Bandari et al. 2008). One tablet (with no super disintegrant) is placed as a control for the analysis of the moisture uptake. Tablets are weighed again and the increment in weight in terms of percentage is noted.

Disintegration study: The time taken to disintegrate the ODTs is commonly <1 min and the real disintegration time experienced by the patients is 5–30 s. Six ODTs are inserted in individual six tubes of the disintegration apparatus, respectively, and one disc is placed in each tube. The disintegration test for ODTs is anticipated to replicate the process of disintegration that occurs in the oral cavity, namely, within the salivary secretions.

Dissolution test: It is necessary as it helps in understanding the drug-release profile. Dissolution of ODTs is very rapid. So, USP 2 (paddle apparatus) is used for dissolution testing as it ensures reproducible dissolution profile (Swamy et al. 2009).

3. **Formulation Development of Orodispersible Tablets**

A multitude of researchers developed ODTs containing various medications and assessed their performance for analysis. Several examples of the formulation development of ODTs are provided. ODTs dissolve rapidly and are promptly get absorbed because of the inclusion of super disintegrants such as croscarmellose, crospovidone, sodium starch glycolate, and magnesium aluminum silicate (VEEGUM HV). This leads to a rapid onset of action for the medicine. The drug's bioavailability rises because it is absorbed straight from the mouth. ODTs include drugs that are not subject to first pass effect. Binders are generally added for the appropriate mechanical strength that will prevent the breaking (Badgujar and Mundada 2011). The methods used to create ODTs that meet mechanical standards involve the incorporation of excipients capable of inducing rapid disintegration, such as effervescent agents or super disintegrants (Muñoz et al. 2014). Alternatively, sublimation agents like menthol, camphor, thymol, and ammonium bicarbonate (Kumar et al. 2009) or melting binders that have a melting point at body temperature like PEG-6 stearate and a-tocopheryl polyethylene glycol 1000 succinate (TPGS) are used (Al-Khattawi and Mohammed 2013). Durasolv® and Orasolv® technologies based on direct compression for ODTs are available in the market.

Oliveira et al. (2020) designed lactose-free ODTs of ketoprofen. While manufacturing the ODTs of ketoprofen, the researchers explored super disintegrants: croscarmellose, crospovidone, or starch glycolate. Croscarmellose as the super disintegrants was selected as it lessens the disintegration time and the ketoprofen was released in 20 minutes (Oliveira et al. 2020). Ortega et al. (2020) proposed and prepared mini ODTs of enalapril maleate for pediatric use. The effects of diverse excipients and APIs at varied doses were examined. Excipients like Tablettose® 80, MicroceLac® 100, and StarLac® were selected followed by their direct compression to form mini-tablets. The formulations complied with the US Pharmacopeia. The ODTs produced by incorporating StarLac® exhibited better hardness, flow properties, as well as fast disintegration. The ODT with 0.1 mg of enalapril maleate was selected after optimizing attributes like hardness, friability, disintegration time, and drug content uniformity, along with wetting time (Ortega et al. 2020). Marzouk et al. (2021) formulated the ODT of fluoxetine (FLX), first by complexing it with β-cyclodextrin for disguising the taste and then using different super disintegrants like crospovidone, croscarmellose sodium, sodium starch glycolate, and indion. The FLX-super disintegrant mixtures have good flowability, and the ODTs prepared out of them follow the pharmacopeia for friability, hardness, etc. with acceptable palatability. The release order of the FLX was as follows: crospovidone > croscarmellose sodium > sodium starch glycolate > indion, respectively. Moreover, the in vivo data highlighted the efficiency of ODTs of FLX as an antidepressant (Marzouk et al. 2021). Few popular specialized tablets are shown in Fig. 7.3.

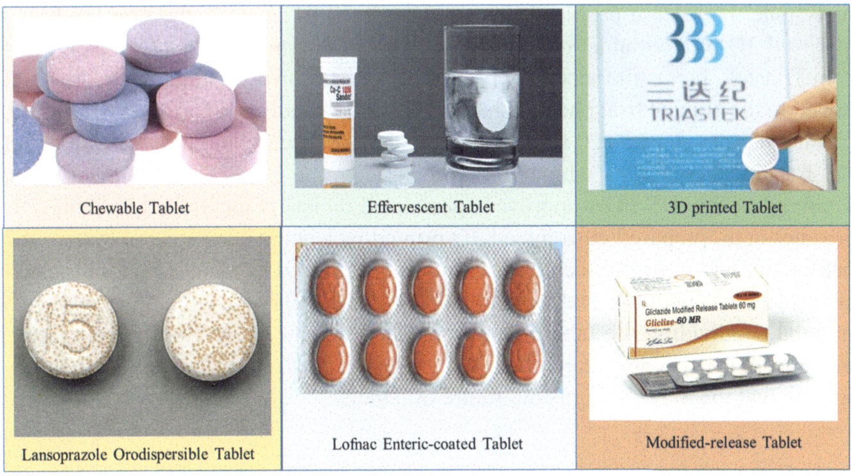

Fig. 7.3 Few popular specialized tablets

7.5 Future Prospective and Conclusion

Solid oral dosage forms are very convenient and very well ensure the dose uniformity. Tablets are the dosage form which are highly acceptable by the patients. Tablets are modified from time to time in order to fulfil the needs of every kind of patients. Its transformation from a simple pill into specialized tablets is discussed in this chapter in detail. Tablet types are divided basically into three groups which are organ-specific, modified-release, and miscellaneous (all other kinds of tablets). Chewable, effervescent, and orodispersible tablets along with their formulation development and evaluation are described in a detailed manner.

The conventional tablets have some disadvantages. The dosage and dosage combination of the tablets manufactured by these methods are same for everyone and can't be personalized (Wang et al. 2022b). In the era of pharmacogenomics, where the genetic makeup of every individual decides how the medication will affect his/her body, the concept of conventional tablets being fit for all is too old. The advent of the concept of personalized medicine has transformed this scenario and directed the revolution of medical therapy toward customized treatment (Liang et al. 2019; Jain et al. 2021). The United States Food and Drug Administration (US FDA) approved some personalized medicines, for example, SPRITAM® (levetiracetam) that got approval in 2015 and some others that are in the pipeline. The applicability of numerous three-dimensional (3D) printing techniques in the therapeutic arena has fascinated interest and support in favor of personalized medicines. This technique fabricates 3D tablets by layer-by-layer deposition, polymerization, or binding of substances, controlled by the computer software. Primarily, quick fabrication in easy steps can be attained, using the 3D printing (Afsana et al. 2019; FDA n.d.).

Moreover, the 3D printing can enhance the solubility of active agents and accurately tune the release profile (Lim et al. 2018). SPRITAM® is the first US FDA-approved 3D printed tablet developed by Aprecia Pharmaceuticals Company. T19 from Triastek, Inc. is the second approved 3D printed investigational new drug and moved in the clinical trial phase in January 2021 for rheumatoid arthritis. Currently, Triastek has developed T20 for the therapy of heart disease and thrombophilia and T21 for ulcerative colitis which got approval from the US FDA for clinical trials. Summary of few US FDA-approved specialized tablets is mentioned in Table 7.1. Goyanes et al. prepared customized 3D printed isoleucine chewable tablets for the therapy of maple syrup urine disease (Goyanes et al. 2019). Zheng et al. fabricated spironolactone subdivided tablets for children and circumvented the practice of cutting and subdividing the tablet manually (Zheng et al. 2020). So, the conclusion drawn from the recent research of tablet as the dosage form is unequivocally highlighting that the 3D printing is the future of tablets.

Table 7.1 Summary of few US FDA-approved specialized tablets

S No.	Types of tablet	Brand name	Active ingredient	Application
1.	Modified release tablet	AcipHex	Rabeprazole sodium	Gastroesophageal reflux disease (GERD)
2.		APLENZIN	Bupropion hydrobromide	Depression and seasonal affective disorder (SAD)
3.	Chewable tablet	Vyvanse	Lisdexamfetamine dimesylate	Attention deficit hyperactivity disorder (ADHD)
4.		Suprax	Cefixime	Bacterial infections
5.		Isentress	Raltegravir	Antiretroviral medication for HIV/AIDS
6.	Effervescent tablet	Binosto	Alendronate sodium	Osteoporosis
7.		CETYLEV	Acetylcysteine	Antidote for acetaminophen overdose
8.	Orodispersible tablet	FazaClo	Clozapine	Severe schizophrenia symptoms
9.		STAXYN	Vardenafil hydro chloride	Erectile dysfunction
10.	3D printed tablet	SPRITAM	Levetiracetam	Seizures (partial-onset, myoclonic, tonic-clonic seizures)
11		T19	–	Rheumatoid arthritis

Acknowledgments The authors are thankful to the Department of Pharmaceuticals (DoP), Ministry of Chemicals and Fertilizers, Government of India, for providing facilities to write this chapter. NIPER-Raebareli communication number for this chapter is NIPER-R/Communication/620.

Conflict of Interest The authors report no conflict of interest related to this chapter.

References

Abebe A, Akseli I, Sprockel O, Kottala N, Cuitiño AM, Cuitiño C (2014) Review of bilayer tablet technology. Int J Pharm 461:549–558

Abu Bakr Mohammed K, Kamal Ibrahim H, Mohammed Ghorab M (2016) Effervescent tablet formulation for enhanced patient compliance and the therapeutic effect of risperidone. Drug Deliv 23(1):297–306

Advankar A, Maheshwari R, Tambe V, Todke P, Raval N, Kapoor D et al (2019) Specialized tablets: ancient history to modern developments. Drug Deliv Syst 2019:615–664

Afsana JV, Haider N, Jain K (2019) 3D printing in personalized drug delivery. Curr Pharm Des 24(42):5062–5071

Al-Khattawi A, Mohammed AR (2013) Compressed orally disintegrating tablets: excipients evolution and formulation strategies. Expert Opin Drug Deliv 10(5):651–663

Al-Madhagi WM, Alshargabi A, Alzomor AKY, Sharhan O (2023) Formulation of new chewable oral dosage forms of meclizine and pyridoxine hydrochloride. Adv Pharmacol Pharm Sci 2023:5512379

Altomare E, Vendemiale G, Benvenuti C, Andreatta P (1997) Bioavailability of a new effervescent tablet of ibuprofen in healthy volunteers. Eur J Clin Pharmacol 52(6):505–506

Arshad MS, Zafar S, Yousef B, Alyassin Y, Ali R, Alasiri A et al (2021) A review of emerging technologies enabling improved solid oral dosage form manufacturing and processing 178:113840
Aslani A, Sharifian T (2014) Formulation, characterization and physicochemical evaluation of amoxicillin effervescent tablets. Adv Biomed Res 3:209
Badgujar BP, Mundada AS (2011) The technologies used for developing orally disintegrating tablets: a review. Acta Pharm 61(2):117–139
Bandari S, Mittapalli RK, Gannu R, Rao YM (2008) Orodispersible tablets: an overview. Asian J Pharm 2(1):159
Bueno AG, Nozal RR (2010) Innovation vs. tradition: the election of an European way toward pharmaceutical industrialisation, 19th–20th centuries. An R Acad Nac Farm 76(4):459–478
Chattoraj S, Daugherity P, McDermott T, Olsofsky A, Roth WJ, Tobyn M (2018) Sticking and picking in pharmaceutical tablet compression: an IQ Consortium review. J Pharm Sci 107(9):2267–2282
Cilurzo F, Musazzi UM, Franzé S, Selmin F, Minghetti P (2018) Orodispersible dosage forms: biopharmaceutical improvements and regulatory requirements. Drug Discov Today 23(2):251–259
Clark KW, Witzel F (1978) Effervescent tablet and method. United States Patent US4127645A, Spring Valley
Dahiya J, Jalwal P, Singh B (2015) Chewable tablets: a comprehensive review. Pharma Innov J 4(5):100–105
Dey P, Maiti S (2010) Orodispersible tablets: a new trend in drug delivery. J Nat Sci Biol Med 1(1):2
Dille MJ, Hattrem MN, Draget KI (2018) Soft, chewable gelatin-based pharmaceutical oral formulations: a technical approach. Pharm Dev Technol 23(5):504–511
Douroumis D (2011) Orally disintegrating dosage forms and taste-masking technologies. Expert Opin Drug Deliv 8(5):665–675
Dressman JB, Amidon GL, Reppas C, Shah VP (1998) Dissolution testing as a prognostic tool for oral drug absorption: immediate release dosage forms. Pharm Res 15(1):11–22
Dubray C, Maincent P, Milon JY (2021) From the pharmaceutical to the clinical: the case for effervescent paracetamol in pain management. A narrative review. Curr Med Res Opin 37(6):1039–1048
Elder DP, Kuentz M, Holm R (2016) Pharmaceutical excipients—quality, regulatory and biopharmaceutical considerations 87:88–99
El-Gazayerly ON, Rakkanka V, Ayres JW (2004) Novel chewable sustained-release tablet containing verapamil hydrochloride. Pharm Dev Technol 9(2):181–188
European Medicines Agency (2014) Guideline on the pharmacokinetic and clinical evaluation of modified release dosage forms. pp 1–46. www.ema.europa.eu/contact
FDA (n.d.) CDER researchers explore the promise and potential of 3D printed pharmaceuticals. FDA
FDA, CDER, Purdie FP (2018) Quality attribute considerations for chewable tablets guidance for industry
Gaikwad A (2023) A review on chewable tablet. Int J Pharm Res Appl 8:599
Gaikwad SS, Avhad RD, Kalkotwar RS (2020) Formulation, development and in vitro characterization of modified release tablets of capecitabine. Drug Dev Ind Pharm 46(1):20–30
Gohel M, Patel M, Amin A, Agrawal R, Dave R, Bariya N (2004) Formulation design and optimization of mouth dissolve tablets of nimesulide using vacuum drying technique. AAPS PharmSciTech 5(3):10
Goyanes A, Madla CM, Umerji A, Duran Piñeiro G, Giraldez Montero JM, Lamas Diaz MJ, Gonzalez Barcia M, Taherali F, Sánchez-Pintos P, Couce ML, Gaisford S, Basit AW (2019) Automated therapy preparation of isoleucine formulations using 3D printing for the treatment of MSUD: first single-centre, prospective, crossover study in patients. Int J Pharm 567:118497
Gupta A, Chidambaram N, Khan MA (2015) An index for evaluating difficulty of Chewing Index for chewable tablets. Drug Dev Ind Pharm 41(2):239–243
Helmstaedter A (2020) Controlling the quality of tablets: from their invention to the dissolution test. Pharm Hist 50(20):53–58

Hemamanjushree S, Tippavajhala VK (2020) Simulation of unit operations in formulation development of tablets using computational fluid dynamics. AAPS PharmSciTech 21(3):1–13
Jain K, Shukla R, Yadav A, Ujjwal RR, Flora SJS (2021) 3D printing in development of nanomedicines. Nanomaterials (Basel) 11(2):420. https://doi.org/10.3390/nano11020420
Jaipal A, Pandey MM, Charde SY, Sadhu N, Srinivas A, Prasad RG (2016) Controlled release effervescent buccal discs of buspirone hydrochloride: in vitro and in vivo evaluation studies. Drug Deliv 23(2):452–458
Karaoglu O, Serhatlı M, Pelvan E, Karadeniz B, Demirtas I, Çakırca G et al (2023) Chewable tablet with herbal extracts and propolis arrests Wuhan and Omicron variants of SARS-CoV-2 virus. J Funct Foods 105:105544
Kebler LF (1914) The tablet industry—its evolution and present status—the composition of tablets and methods of analysis. J Am Pharm Assoc 3(6):820–848
Kimaro E, Shedafa R, Temu M, Kaale E (2019) Formulation development of chewable albendazole tablets with improved dissolution rate. Heliyon 5(12):e02911
Kumar R, Patil S, Patil MB, Patil SR, Paschapur MS (2009) Formulation evaluation of mouth dissolving tablets of fenofibrate using sublimation technique. Int J ChemTech Res CODEN 1(4):840–850
Liang K, Brambilla D, Leroux JC (2019) Is 3D printing of pharmaceuticals a disruptor or enabler? Adv Mater 31(5):1805680
Lim SH, Kathuria H, Tan JJY, Kang L (2018) 3D printed drug delivery and testing systems—a passing fad or the future? Adv Drug Deliv Rev 132:139–168
Mahapatra AK, Sameeraja NH, Murthy PN (2015) Development of modified-release tablets of zolpidem tartrate by biphasic quick/slow delivery system. AAPS PharmSciTech 16(3):579
Markopoulos C, Andreas CJ, Vertzoni M, Dressman J, Reppas C (2015) In-vitro simulation of luminal conditions for evaluation of performance of oral drug products: choosing the appropriate test media. Eur J Pharm Biopharm 93:173–182
Marzouk MA, Osman DA, Mohamed OS (2021) In vitro and in vivo evaluation of taste-masked orodispersible tablets of fluoxetine hydrochloride for the treatment of depression. Drug Dev Ind Pharm 47(4):645–653
Maurya R, Kumar Sharma P, Malviya R (2014) A review on controlled drug release formulation: spansules. Int J Pharma Sci Res 5(3):78–81
Muñoz H, Castan H, Clares B, Ruiz MA (2014) Obtaining fast dissolving disintegrating tablets with different doses of melatonin. Int J Pharm 467(1–2):84–89
Murugesan S, Gowramma B, Lakshmanan K, Reddy Karri VVS, Radhakrishnan A (2019) Oral modified drug release solid dosage form with special reference to design; an overview. Curr Drug Res Rev 12(1):16–25
Nagasamy Venkatesh D, Meyyanathan SN, Shanmugam R, Kamatham SS, Campos JR, Dias-Ferreira J et al (2021) Physicochemical, pharmacokinetic, and pharmacodynamic characterization of isradipine tablets for controlled release. Pharm Dev Technol 26(1):92–100
Oliveira LJ, Veiga A, Stofella NCF, Cunha AC, da Graça T, Toledo M, Andreazza IF et al (2020) Development and evaluation of orodispersible tablets containing ketoprofen. Curr Drug Deliv 17(4):348–360
Oral Solid Dosage Pharmaceutical Formulation Market—Global (2023). https://www.maximizemarketresearch.com/market-report/oral-solid-dosage-pharmaceutical-formulation-market/121388/
Orlu M, Cevher E, Araman A (2006) Design and evaluation of colon specific drug delivery system containing flurbiprofen microsponges. Int J Pharm 318(1–2):103–117
Ortega CA, Favier LS, Cianchino VA, Cifuente DA (2020) New orodispersible mini tablets of enalapril maleate by direct compression for pediatric patients. Curr Drug Deliv 17(6):505–510
Pardhi VP, Jain K (2021) Impact of binary/ternary solid dispersion utilizing poloxamer 188 and TPGS to improve pharmaceutical attributes of bedaquiline fumarate. J Drug Deliv Sci Technol 62:102349

Pardhi VP, Patel P, Vaisya A, Jain K (2024) Inclusion complexes of bedaquiline fumarate with β-cyclodextrin and its derivatives: in silico, in vitro and in vivo evaluation. J Drug Deliv Sci Technol 91(105253):1773–2247

Park K (2016) Drug delivery of the future: chasing the invisible gorilla 28(240):2–8

Patel VM, Prajapati BG, Patel MM (2007) Formulation, evaluation, and comparison of bilayered and multilayered mucoadhesive buccal devices of propranolol hydrochloride. AAPS PharmSciTech 8(1):E147–E154

Patel AR, Prajapati DS, Raval JA (2010) Fast dissolving films (FDFs) as a newer venture in fast dissolving dosage forms. Int J Drug Dev Res 2(2):232–246

Rosch M, Lucas K, Al-Gousous J, Pöschl U, Langguth P (2021) Formulation and characterization of an effervescent hydrogen-generating tablet. Pharmaceuticals 14(12):1327

Rotthäuser B, Kraus G, Schmidt PC (1998) Optimization of an effervescent tablet formulation containing spray dried l-leucine and polyethylene glycol 6000 as lubricants using a central composite design. Eur J Pharm Biopharm 46(1):85–94

SahabUddin M, Al Mamun A, Tasnu T, Asaduzzaman M (2016) In-process and finished products quality control tests for pharmaceutical capsules according to pharmacopoeias. J Pharm Res Int 7(9):1–9

Salawi A (2022) Pharmaceutical coating and its different approaches, a review. Polymers (Basel) 14(16):3318

Santus G, Baker RW (1995) Osmotic drug delivery: a review of the patent literature. J Control Release 35:1–21

Seager H (1998) Drug-delivery products and the Zydis fast-dissolving dosage form. J Pharm Pharmacol 50(4):375–382

Sendall FEJ, Staniforth JN (2011) A study of powder adhesion to metal surfaces during compression of effervescent pharmaceutical tablets. J Pharm Pharmacol 38(7):489–493

Şengel CT, Hasçiçek C, Gönül N (2006) Development and in-vitro evaluation of modified release tablets including ethylcellulose microspheres loaded with diltiazem hydrochloride. J Microencapsul 23(2):135–152

Skaftason JF, Jóhannesson T (2013) Tablets and tablet production—with special reference to Icelandic conditions. Laeknabladid 99(4):197–202

Sohi H, Sultana Y, Khar RK (2004) Taste masking technologies in oral pharmaceuticals: recent developments and approaches. Drug Dev Ind Pharm 30(5):429–448

Suzuki H, Onishi H, Hisamatsu S, Masuda K, Takahashi Y, Iwata M et al (2004) Acetaminophen-containing chewable tablets with suppressed bitterness and improved oral feeling. Int J Pharm 278(1):51–61

Swamy PV, Divate SP, Shirsand SB, Rajendra P (2009) Preparation and evaluation of orodispersible tablets of pheniramine maleate by effervescent method. Indian J Pharm Sci 71(2):151

U.S. Department of Health and Human Services, Food and Drug Administration, Center for Drug Evaluation and Research (CDER) (2023) Waiver of in vivo bioavailability and bioequivalence studies for immediate-release solid oral dosage forms based on a biopharmaceutics classification system. Digital collections. Guidance for industry immediate release solid oral. National Library of Medicine

Uchimoto T, Iwao Y, Yamamoto T, Sawaguchi K, Moriuchi T, Noguchi S et al (2013) Newly developed surface modification punches treated with alloying techniques reduce sticking during the manufacture of ibuprofen tablets. Int J Pharm 441:128–134

Wang Y, Xu S, Xiao Z, Jiang Y, Jiang Q, Li J et al (2022a) Stiripentol enteric solid dispersion-loaded effervescent tablets: enhanced dissolution, stability, and absorption. AAPS PharmSciTech 23(5):1–14

Wang N, Shi H, Yang S (2022b) 3D printed oral solid dosage form: modified release and improved solubility. J Control Release 351:407–431

Waterman KC, Macdonald BC, Roy MC (2009) Extrudable core system: development of a single-layer osmotic controlled-release tablet 134(3):201–206

Wen X, Deng Z, Xu Y, Yan G, Deng X, Wu L et al (2021) Preparation and in vitro/in vivo evaluation of orally disintegrating/modified-release praziquantel tablets. Pharmaceutics 13(10):1567

Wood JR (1906) Tablet manufacture; its history, pharmacy and practice. Lippincott

Yoo O, Tang EKY, Salman S, Nguyen MN, Sommerfield D, Sommerfield A et al (2022) A randomised controlled trial of a novel tramadol chewable tablet: pharmacokinetics and tolerability in children. Anaesthesia 77(4):438–448

Zarate J, Igartua M, Hernandez R, Pedraz J (2010) Polymeric materials and formulation technologies for modified-release tablet development. Mini Rev Med Chem 9(13):1504–1517

Zheng X, Wu F, Hong Y, Shen L, Lin X, Feng Y (2019) Improvements in sticking, hygroscopicity, and compactibility of effervescent systems by fluid-bed coating 9(54):31594–31608

Zheng Z, Lv J, Yang W, Pi X, Lin W, Lin Z et al (2020) Preparation and application of subdivided tablets using 3D printing for precise hospital dispensing. Eur J Pharm Sci 149:105293

Suspensions: Theory, Formulation Considerations, Flocculated and Deflocculated Suspensions, and Evaluation of Suspension Stability

8

Anjali Rajora and Kalpana Nagpal

Abstract

Pharmaceutical formulations like suspensions are highly beneficial for delivery of drugs. Several challenges arise during development due to their inherent structural instability. These formulations typically comprise small solid particles (sized between 0.5 and 5 μm) suspended in a liquid or semiliquid that acts as the continuous phase and desired carrier. Theoretical concerns linked to pharmaceutical suspensions (such as interfacial properties, surface free energy, and surface potential along with electric double layer, zeta, and Nernst potential) are conceptualized. Certain other crucial aspects like wetting, electrokinetic phenomena, and DLVO (Derjaguin, Landau, Verwey, and Overbeek) theory with theory of sedimentation are addressed. From a stability point of view, a comprehensive understanding of these factors is crucial since both the characteristics of the carrying liquid and the properties of dispersed solids impact the solidity of the suspension. This chapter delivers an encompassing survey of pharmaceutical suspensions, their introduction, classification, theoretical exploration, assessments, and stability considerations. It is an attempt to equip researchers, practitioners, and stakeholders with a comprehensive understanding of leveraging suspensions for efficient drug delivery.

Keywords

Suspensions · Flocculated suspensions · Deflocculated suspensions · DLVO theory

A. Rajora · K. Nagpal (✉)
Amity Institute of Pharmacy, Amity University Uttar Pradesh, Noida, Uttar Pradesh, India
e-mail: knchaswal@amity.edu

K. Jain, A. K. Yadav (eds.), *Advances in Pharmaceutical Product Development*, https://doi.org/10.1007/978-981-97-9230-6_8

Abbreviations

DLVO	Derjaguin, Landau, Verwey, and Overbeek
EDL	Electric double layer
L	Contact angle
PCS	Photon correlation spectroscopy
SDS	Sodium dodecyl sulfate
SPOS	Single particle optical sensing
V	Potential energy
VA	van der Waals attraction
V_{defloc}	Final volume of sedimentation in deflocculated suspension
V_{floc}	Final volume of sedimentation in flocculated suspension
V_{i}	Actual suspension volume
VR	van der Waals repulsion
V_{s}	Final sediment volume
z	Zeta

8.1 Introduction

A pharmaceutical suspension is formed when a solid particle can't dissolve in the solvent liquid. Generally, these suspensions have particles larger than 0.5 μm in diameter (Jones 2016). A clear distinction between suspensions and dispersions with smaller particles cannot be made because it is difficult and impractical (Brunaugh et al. 2019). They come with several advantages like effectively delivering hydrophobic drugs; giving intramuscular depot therapy; covering up unpleasant tastes of certain ingredients; avoiding the use of cosolvents; protecting against drug degradation from things like oxidation, hydrolysis, or microbes; and making it easier for young and elderly patients to take the medicine (Iyer et al. 2006). Furthermore, compared to solutions, suspensions can take in much higher drug concentrations (Wong et al. 2008). This study gives a summary of various suspension-related aspects including suspension theories, their categories, and their stability aspect. It also touches on some formulation-related subjects (Arora et al. 2022).

The two phases make up heterogeneous systems known as suspensions. Solid particles that are mostly insoluble in the external phase make up the internal or dispersed phase (Jermain et al. 2018). It also contains the continuous phase also called dispersion medium which is usually in the form of a semisolid (e.g., gels) or liquid (such as liquid suspensions). Although organic or greasy liquids are occasionally utilized, an aqueous dispersion medium makes up most pharmaceutical suspensions (Leleux and Williams III 2014).

8.2 Theoretical Considerations

8.2.1 Interfacial Properties

Pharmaceutical suspensions contain a solid phase in the form of tiny fragments dispersed in the dispersion medium. Because of this, there are many interlinks within the formulation that influences the firmness of the suspension. As a result, change in the physical characteristics of the dispersion largely depends on its surface qualities. Surface free energy plus surface potential is among the important characteristics of interface (Nutan and Reddy 2010).

8.2.1.1 Surface Free Energy

Solids that have been finely divided frequently have a large surface area and a lot of free energy there. The relation between surface area and free energy could be stated as:

$$\Delta G = \gamma \Delta A$$

where γ is the interface tension linking the continuous phase and solid phase, ΔG is the rise in free energy, and ΔA is surface area change. The suspension is more thermodynamically stable reflected by the lower value of ΔG (Kulshreshtha et al. 2010). Thus, a system with incredibly small particle due to the huge total surface area is thermodynamically unstable. The system tends to aggregate to lower the surface area with the surplus free energy. This free energy can be minimized to prevent particle grouping, which is achieved by decreasing the interface tension. Addition of a wetting agent causes its absorption at the interface. As the interfacial tension decreases, the system will become more stable (Aulton and Taylor 2013).

8.2.1.2 Surface Potential

The existence or nonexistence of a surface potential can generally be used to explain the strength of lyophobic colloidal systems. Suspension systems can also benefit from this concept (Garad et al. 2010). Surface potential survives when the dispersed solid particles within suspension have a charge relative to the liquid medium in which they are suspended. Several processes can lead to charged particulate matter. The possibility of selective adsorption by solid particles of a certain ionic type exists if the suspension contains electrolytes. Several processes can lead to charged particulate matter. The possibility of selective adsorption by solid particles of a certain ionic type exists if the suspension contains electrolytes. As a result, charged particles will eventually form. Because of the disproportionate nature of the OH^- ion, it is possible, for example, that rubber particles dispersed in water will adsorb more OH^- ions than H_3O^+ ions (Brunaugh et al. 2019). Surfactants are adsorbed on the interface of liquid and solid. In rare cases, they can ionize and give the particles a negative or positive charge. For example, sodium dodecyl sulfate (SDS) is an anion in water medium. Solid objects can also be charged by ionizing their functional groups. In this case, the total charge is affected by the pH of the nearby vehicle.

Ionizable groups such as -COOH and -NH_2 are found in peptide and protein molecules. Dispersed particles of these compounds will ionize. The pH of the surrounding vehicle has a significant effect on the direction and intensity of ionization (Bhattacharjee 2016).

8.2.2 Electric Double Layer (EDL)

Once dispersed particles encounter an aqueous electrolyte mixture, they can choose only one type of charge to adsorb. The particles are often negatively charged when an anion is absorbed. The ions, in this case the anions that determine the charge on the particle, are known as co-ions or potential-determining ions as shown in Fig. 8.1 (Ohshima 2014). The residual ionic species in the liquid consist of the remaining anions and sum of the added cations. This means that the dispersion medium will include more cations than anions. These cations are called counterions or gegenions because their charge is different from that of the ions that determine the potential. Electric forces attract them to a negatively charged surface. Once the initial adsorption is complete, the gegenions also repel any additional approaching anions from the particle surface. All the ions in the solution are evenly distributed due to these electrical forces and thermal motion. This creates an equilibrium situation where

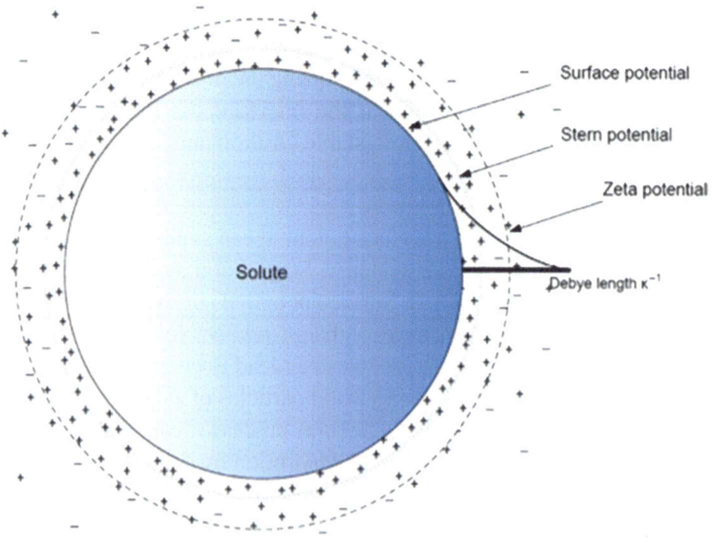

Fig. 8.1 Schematic representation showing the electric double layer in a liquid in contact with a negatively charged particle. The zeta potential is the potential at the slipping plane that separates the mobile fluid from the fluid that remains attached to the solute (Barhoum et al. 2018). (Reprinted from Emerging Applications of Nanoparticles and Architecture Nanostructures, Ahmed Barhoum, M. Luisa García-Betancourt, Hubert Rahier, Guy Van Assche, Physicochemical characterization of nanomaterials: polymorph, composition, wettability, and thermal stability, page no 255–278. Copyright (2018), with permission from Elsevier)

certain additional cations move toward the surface and the remaining cations are transferred away from charged surface as dispersed in reduced amounts (Eyley et al. 2015).

In the region of the liquid closest to the particles, almost all counterions will be present. The solvent component that is firmly attached to the particle surface along with these counterions is called the Stern layer. This layer moves as the particles move through the continuous phase, so Stern layer is near to the shear plane. The co-ions, which get adsorbed on the surface of a solid body, are more abundant than counterions in a compactly bound layer (Ferrar et al. 2020). Thus, the potential in the shear plane is however negatively charged. Stern layer is surrounded by a diffuse layer containing additional counterions than co-ions. Relatively mobile ions in this layer continuously move into and out of the main body of the dispersion medium due to thermal energy. Electrical neutrality appears wherever the moving diffuse layer stops. Outside the diffuse layer, the concentrations of counter as well as covalent ions are identical, indicating an electrically neutral state for the rest of the dispersion medium. At the liquid-solid interface, the electrical distribution can be thought of as a charged EDL. Stern layer consists mainly of counterions and is firmly attached to the surface of the solid. More counterions and greater mobility are present in the second layer. The EDL thickness rests on the form and number of ions present in the solution. It's vital to keep in mind that the suspension is electrically neutral, even though the charges are unevenly distributed in the EDL. Two additional options are possible (Vo et al. 2020). If the counterion concentration in the rigidly bound layer is identical to the concentration of cations at the surface of solid, the shear plane will be electrically neutral, and there will be only one layer of medium and ions, as opposed to two layers. If the entire charge of counterions in the Stern layer is greater than the charge derived from the co-ions, the total charge in the shear plane will be positive rather than negative. Electrical neutrality is attained where the electrical EDL terminates, and the diffuse layer contains additional co-ions than counterions (Kulshreshtha et al. 2009). The difference between the concentration of positive and negative ions at any given distance from the surface is used to calculate the charge density at this point (Frigaard 2019).

8.2.3 Zeta and Nernst Potential

Electro-thermodynamic potential, surface potential, or Nernst potential refers to the difference in electrical potential between the real or true surface of the particle and the electroneutral zone. The electric potential at the particle's surface thereby protects the Nernst potential because of the potential-regulating ions. The electrokinetic or zeta (z) potential is the potential difference between the electroneutral zone and the shear plane. The z potential, in contrast to the Nernst potential, has a significant effect on the establishment of a stable suspension (Hebishy et al. 2015). To neutralize the charged particles in suspension, an EDL is created. The electrical potential at any point in the suspension system depends on its precise location, as was explained in the paragraphs before. The potential in the diffuse layer gradually varies with increasing distance from the solid object. The strength of the attraction between neighboring, equally charged particles in a solid dispersion is controlled by the zeta potential. The van der Waals force, the attractive force between particles, outweighs the repulsive forces and leads to the formation of floccules when the zeta potential falls to a specific level, which depends on the particular system being examined. We refer to this process as flocculation. Both surface charge and EDL width have an impact on zeta potentials and surface area. Microelectrophoresis, which involves watching the movement of particles in a voltage field using a microscope, can be used to measure zeta potential (Loftsson 2014; Haines and Martin 1961a).

8.2.4 Wetting

Due to the elevated interface tension among the particles and dispersion medium, hydrophobic powders exhibit weak adhesion. The contact angle between the liquid and solid phases is maintained at an unusually high level because of this deformation. Surfactants can be used to reduce the interfacial tension, thereby reducing the contact angle and achieving excellent wetting (Haines and Martin 1961b). The hydrocarbon chain from the surfactants is adsorbed on the lipophilic surface of particles, while the polar end is retained in the liquid medium. Therefore, the necessary wetting of solid materials can be accomplished by reducing the interfacial tension. Young's equation is used to determine whether a solid has become wet (Haines and Martin 1961c).

$$\cos\Theta = \frac{\gamma_{\mathrm{sv}} - \gamma_{\mathrm{sl}}}{\gamma_{\mathrm{lv}}}$$

where γ is the interface tension between vapor (v), solid (s), and liquid phases and Θ is the contact angle and the contact angle (*L*). Wetting agents can reduce the contact angle by reducing *sl* in addition to the lv ratio. Wetting agents can be made using surfactants having an HLB (Hydrophilic Lipophilic Balance) value of 7–9 (Kayes 1977). Most surface-active agents are employed as wetting agents in concentrations up to 0.1%. Polysorbates and sorbitan esters are two commonly used oral humectants. One of the surfactants that is applied externally is SDS. Lecithin, in addition to other ingredients, can be part of parenteral drugs along with polysorbates and some poloxamers. Excess surfactant can lead to deflocculation or foamed systems, neither of which is considered ideal (Matthews and Rhodes 1968a).

8.2.5 Electrokinetic Phenomena

Four electrokinetic processes, including electroosmosis, sedimentation potential, electrophoresis, and streaming potential, can exist due to the existence of interfacial potentials. In fact, all these features are the result of the motion of the charged surface of the adjoining liquid phase (Khan et al. 2022). When a potential difference is applied, charged particles move through the liquid, and this movement is measured by electrophoresis (Roura and Fort 2004). The z potential of a suspension can be determined using:

$$Z = \frac{v}{e} \times \left(9 \times 10^4\right) \times \frac{4\pi\eta}{\varepsilon}$$

where e is the dielectric constant of the medium, η is the viscosity of the medium in poises, Z is the zeta potential of the medium in volts, v is the migration speed of particles in the electrophoresis tube in cm/s, and E is the potential gradient in V/cm (Matthews and Rhodes 1968b).

8.2.6 DLVO Theory

The renowned DLVO (Derjaguin, Landau, Verwey, and Overbeek) theory, formulated by Derjaguin, Landau, Verwey, and Overbeek during the 1940s (Derjaguin and Landau 1941; Verwey and Overbeek 1948), elucidates scenarios in which van der Waals forces coexist with electrostatic forces (Adair et al. 2001; Kellaway and Najib 1981). This theory operates on the premise that electrostatic double-layer forces and van der Waals forces act independently. Hence, these forces can be combined or summed at each interacting distance between two particles (Lieberman et al. 1987). Several researchers have considered the method of so-called controlled flocculation for the preparation of drug suspensions and this is well-known (Trefalt and Borkovec 2014; Yotsumoto and Yoon 1993). The idea of "controlled flocculation" is based on the knowledge of the so-called DLVO theory of colloidal stability (Pal et al. 2010; Strum et al. 1978). This implies that the van der Waals attraction (VA) and the repulsion (VR) combine to create potential energy (V) because of contact among the two particles (Venkateshwarlu et al. 1990).

$$V = \text{VR} + \text{VA}$$

The phrases "coagulation" and "flocculation" are often used interchangeably in the literature to refer to the accumulation of particles in suspensions. The term "coagulation" should be used only to describe primary minimal aggregation to provide the most satisfactory definition of these concepts (Short and Rhodes 1973). The term "flocculation" refers to the grouping of particles in a secondary minimum. The latter term is also used to describe the occlusion of most particles and polymer bridges caused by interactions between metal ions and polyelectrolytes. It has long been believed that chemicals such as nonionic surfactants can stabilize a dispersion if they are adsorbed on the particle surface in the absence of a detectable zeta potential (Liu et al. 2019). The interaction potential energy should also consider the additional steric stabilization term Vs (Jangde et al. 2011). Thus

$$V = \text{VR} + \text{VA} + \text{Vs}$$

However, the term "controlled flocculation" is possibly a misnomer as:

1. It occurs when a secondary minimum is reached and the primary maximum is reduced by the addition of electrolyte, deepening the secondary minimum
2. It is caused by polymer or metal ion polyelectrolyte bridges
3. Primary minimal coagulation is induced by the addition of electrolyte or other charged species, the depth of which is limited by the steric effect of the additional surfactant or polymer, which is more precisely called controlled coagulation (Khan et al. 2022)

The essence of the DLVO theory can be easily understood in Fig. 8.2. The figure depicts the opposite forces: electrical repulsion originating from the electric double layer and attraction originating from the van der Waals force. These forces are drawn in opposite directions because of their different natures. If the net energy curve is consistently above the baseline, this indicates a repulsive state. Conversely, curves below the baseline show attraction. The highest point of repulsion is called the energy barrier itself (Kulshreshtha et al. 2010). To facilitate agglomeration, the two particles on a collision course must possess sufficient momentum to overcome this barrier. When particles successfully move through repulsion, they are brought together by the force of attraction. In the future, they can be considered captured by van der Waals-London forces (Martin 2001).

8.2.7 Theory of Sedimentation

Velocity of sedimentation is represented by Stoke's equation:

$$V_{\text{Sed}} = \frac{2r2(\rho_s - \rho_o)g}{9\eta} = \frac{d2(\rho_s - \rho_o)g}{18\eta}$$

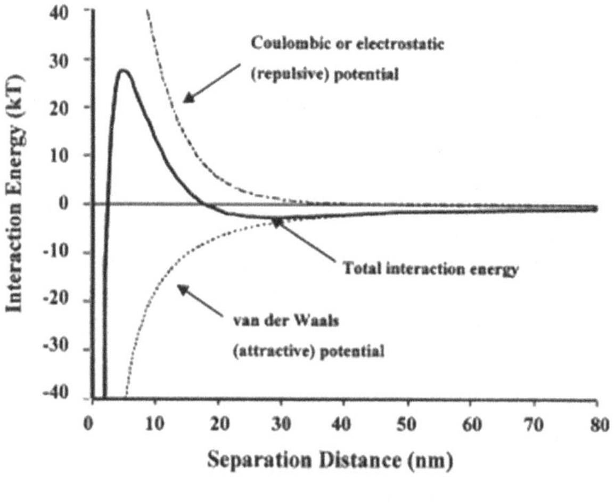

Fig. 8.2 Schematic interaction energy versus distance profiles of DLVO interaction. The attractive van der Waals and the repulsive electrostatic potentials form the total interaction energy (Adair et al. 2001). (Reprinted from Encyclopedia of Materials: Science and Technology, Second edition, J.H. Adair, E. Suvaci, J. Sindel, Surface and Colloid Chemistry, page no. 1–10. Copyright (2001) with permission from Elsevier)

where d = particle diameter, η = viscosity of the dispersed medium in the poise, v = sedimentation velocity in cm/s, g = acceleration due to gravity, r = radius of the particle, ρ_0 = density of the dispersed medium, and ρ_s = density of the dispersed phase. Particle size is the most important variable because it is squared in equation. A modest settling velocity is produced by smaller particles. To reduce settling, one can increase the viscosity of the medium. Cellulose compounds such as methyl cellulose and hydroxypropyl methyl cellulose are often used as viscosity enhancers (Cooper and Gun 2002). Natural gums are additional examples (e.g., acacia and tragacanth). The disparity in densities between the dispersed phase and the continuous phase may have an impact on the rate of settling. Zero denotes the absence of a delay. Since the density of the dispersed phase cannot be altered, the medium would need to be made denser (Banker and Rhodes 1979).

8.2.7.1 Limitation of Stoke's Equation

- Only spherical particles in a highly dilute solution obey Stoke's equation (0.5–2 g per 100 mL).
- The particles should fall freely and independently of each other (without collision).
- The particles should not have any physical or chemical attraction or affinity to the dispersion medium, but since most pharmaceutical suspension formulations are 5%, 10%, or higher in consistency, there is a possibility that particle settling will be difficult (Ansel et al. 2005).

8.2.8 Important Considerations for Suspension

The creation of a pharmaceutical suspension requires knowledge of the parameters of the dispersion component and the dispersion medium. When choosing ingredients for the creation of a suspension, it's crucial to take the route of administration, the intended purpose, and any potential side effects into account (Martin and Swarbrick 1966). The following considerations should be made when making pharmaceutical suspensions:

1. Suspended material's nature: In the formulation of a slurry, the boundary qualities of the suspended material are crucial. Low surface tension particles can be easily wetted by water and suspended as a result. High interfacial tension material particles do not wet readily, nevertheless. Surfactants are typically used to suspend such materials. Surfactants make particles more wettable (Soci and Parrot 1980).
2. Size of suspended particles: According to Stoke's rule, a smaller particle size causes a slower rate of deposition for suspended particles. Processes including grinding, screening, and crushing can reduce the particle size. The speed and volume of medication absorption, dissolution, and biodistribution are also influenced by particle size. However, above a certain point, particle size reduction might cause a compact cake to form during deposition (Ludwing and Van Ooteghen 1988).

3. Viscosity of the continuous phase: Higher viscosity of the continuous phase ensures lower sedimentation. Nevertheless, this may negotiate additional necessary properties like syringeability of parenteral suspensions, dispersibility of topical suspensions, and ease of pourability of oral suspensions. The shear thinning property is extremely necessary as it will offer comparatively higher viscosity of suspension during storage (low shear); and thus, the settling is slow. After mixing (high shear), the viscocity is lower which permit better pourability of the suspension from the bottle (Pennington et al. 1988).

8.3 Classification of Suspensions

Particles suspended within a liquid medium have the potential to experience both van der Waals forces that attract them and electrostatic forces that repel them. Consequently, these forces can lead to the dispersion of particles, their clumping into aggregates, or their formation into clusters. Some differences between flocculated and deflocculated suspension are shown in Fig. 8.3. The interplay of attractive and electrostatic repulsive forces at the particle's surface dictates the extent of both aggregation and clustering processes.

8.3.1 Flocculated Suspension

The structure of flocculation is due to a combination of a strong attractive force and a reduction in electrical repulsive forces between dispersed particles in suspension. When this happens, the particles come closer to each other with less attraction, resulting in a loosely aggregated structure known as a floccules. The rate of

FLOCCULATED SUSPENSIONS	DEFLOCCULATED SUSPENSIONS
• The loose collection of particles creates a network-like structure.	• Particles do not form floccules; they are separate entities.
• Rate of sedimentation is high.	• Sedimentation rate is slow
• It is simple to re-disperse sediment.	• Sediment is difficult to re-disperse
• Sediment is loosely packed and does not form a hard cake.	• Sediment is tightly packed and forms a hard cake.
• Supernatant is clear.	• Supernatant is cloudy.
• Floccules stick to the sides of the bottle.	• Particles do not stick to the sides of the bottle.
• The suspension is not pleasing in appearance.	• The suspension is pleasing in appearance.

Fig. 8.3 A comparison of flocculated and deflocculated suspension

sedimentation is always high because the floccules are composed of an extensive network of small individual particles. The dispersion medium can pass through the flocs during sedimentation due to their open porous structure (Dicolo et al. 1980). In addition, flocs capture a significant part of the liquid phase. As a result, the finished precipitate will still have a significant volume and will be easily dispersed by gentle shaking. Although flocs settle faster than individual particles, they do so in the form of a lattice that prevents complete settling, making them less prone to densification and cake formation (Doye et al. 2017).

8.3.2 Deflocculated Suspension

The individual particles in the deflocculated suspension remain as separate individual units and gradually settle. The sluggish rate of particle settling prevents individual suspension particles from capturing the liquid medium and causes them to become compact, leading to the formation of a cake (Jani 2004). With gentle stirring, it can be very difficult to re-disperse this cake. Clumping is a very serious physical stability problem that occurs in this resulted slurry. The constant turbidity of the supernatant after stirring is another distinguishing quality of this suspension. This is mainly due to the extremely low rate of sedimentation of the smallest suspension particles (Lachman 1996).

8.4 Pharmaceutical Suspension Stability Study

Formulating suspensions requires a comprehensive assessment of both chemical and physical stability considerations. As with other pharmaceutical formulations, it is very important to investigate the potential degradation of drugs in suspension. To ensure chemical stability, some suspensions are stored in powder form and reconstituted immediately before use. Physical stability involves a variety of factors, including particle settling, a particle growth phenomenon known as Ostwald ripening, and particle aggregation. These aspects are evaluated by measuring the settling velocity and volume of the particles, as well as the homogeneity of the suspension after stirring. Changes in particle size distribution can be analyzed using techniques such as microscopy or light scattering techniques.

8.4.1 Particle Settling

The sedimentation process can be mitigated by using a structured agent, which refers to a thick aqueous solution containing natural and/or synthetic polymers (such as gums). These polymers are specifically designed to slow particle settling and minimize deposition by increasing the viscosity of the continuous phase of the liquid carrier.

8.4.2 Particle Aggregation

Suspensions are inherently unstable due to the relatively large surface area of the particles resulting from processes such as particle size reduction or milling. This greater surface area leads to an increase in surface free energy, which can be thought of as the energy required to create new surfaces and break intermolecular bonds. Dispersion of particles in a liquid medium, when their solubility is low, requires a significant amount of energy due to the increased liquid-solid interface and interfacial surface tension.

8.4.3 Particle Growth (Ostwald Ripening)

Ostwald ripening is a phenomenon where particles in a suspension tend to enlarge over time and during storage as shown in Fig. 8.4. Temperature fluctuations during suspension storage can cause slight shifts in drug solubility. Minor temperature increases can lead to the dissolution of the smallest particles in the suspension. When the temperature returns to lower levels, the dissolved drug preferentially recrystallizes on the surfaces of larger particles that are still suspended. This process alters the overall particle size distribution toward larger particles. Suspensions with a narrow particle size distribution experience reduced Ostwald ripening because all particles exhibit similar solubility (Brunaugh et al. 2019).

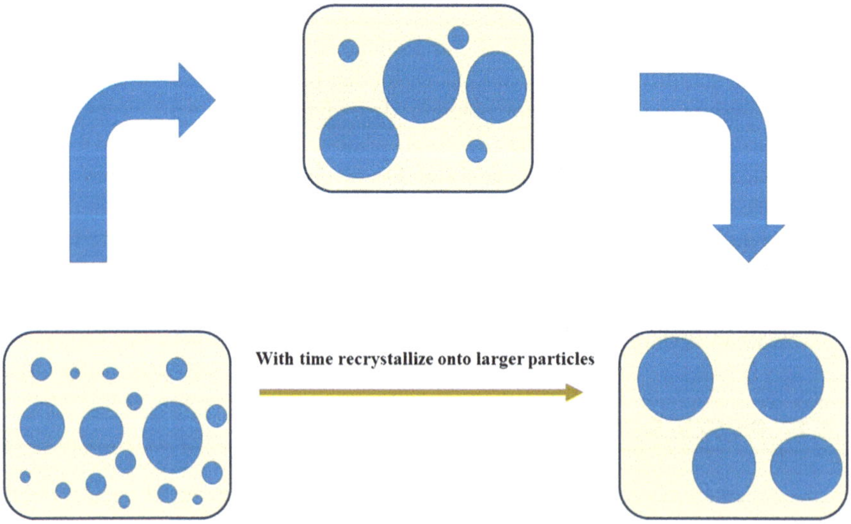

Fig. 8.4 Ostwald ripening transpires when minute particles within a suspension dissolve and subsequently crystallize onto larger particles

8.5 Evaluation parameters of Suspension

The stability of a pharmaceutical suspension is evaluated using the following techniques that are found to be extremely useful in ascertaining the stability aspect of suspension.

8.5.1 Determination of the pH of the Suspension

The pH is the negative logarithm (base 10) of the activity (the product of the molar concentration and the activity coefficient) of the hydrogen ions (H^+) in the solution. Suspensions must be in desired pH range to avoid precipitation. Using a digital pH meter, we can estimate the pH of each composition (Aulton 2002).

8.5.2 Amount of Sedimentation

Settling of solid particles or floccules under gravitation force in liquid at the bottom of the container is referred as sedimentation. Suspending agents decrease sedimentation by keeping solid particles suspended in a liquid medium and impart viscosity to the liquid medium.

$$F = H_u / H_o \text{ is used to calculate sediment volume,}$$

where H_o is the initial suspension height and H_u is the final or final height of the sludge as it settles (Liebermann 1989).

8.5.3 Redispersibility

Since dilute suspensions tend to settle, redispersibility is an important aspect of their pharmaceutical quality. The content of active ingredient of single doses of a suspension depends to a large extent upon the redispersibility of the product (Deicke and Süverkrüp 2000). A fixed volume of each suspension (50 mL) should be stored in calibrated tubes that have been kept at room temperature for varying periods of time (1, 5, 10, 15, 20, 30, 45 days). To do this, one test tube is removed and shaken vigorously to redistribute the precipitate. If any deposits were present, they should be noted (Remington 2000).

8.5.4 Flow Rate (*F*)

Flow rate of suspensions decreases with increase in concentration of suspending agent. The following equation was used to compute the flow rate and determine how long it took a 10 mL sample of suspension to pass through a 10 mL pipette (Chaudhari et al. 2014):

$$F = \text{Volume of pipette}(\text{mL}) / \text{Flow time}(\text{s})$$

8.5.5 Viscosity Determination

Viscosity of suspensions is of great importance for stability and pourability of suspensions. As we know suspensions have least physical stability among all dosage forms due to sedimentation and cake formation. So as the viscosity of the dispersion medium increases, the terminal settling velocity decreases, thus the dispersed phase settles at a slower rate, and they remain dispersed for longer time yielding higher stability to the suspension (Kumar and Yagnesh 2016). On the other hand, as the viscosity of the suspension increases, its pourability decreases and inconvenience to the patients for dosing increases. The Brookfield viscometer was set to 100 revolutions per minute to measure the viscosity of liquid samples. At least three duplicates of each determination were made, and the findings were expressed as mean values (Balakrishnan et al. 2009).

8.5.6 Degree of Flocculation (β)

If the parenteral suspensions are flocculated, their syringeability will be less. For the flocculation to occur, repulsive forces must be diminished until the same attractive forces prevail, where V_{floc} and V_{defloc} are the final volume of sedimentation in flocculated and deflocculated suspensions, respectively. The following equation was used to measure the degree of flocculation (Kumar and Yagnesh 2016).

$$\beta = \frac{V_{\text{floc}}}{V_{\text{defloc}}}$$

8.5.7 Sedimentation Volume and Rate

The suspended particles should not settle rapidly and sediment produced must be easily resuspended by the use of moderate amount of shaking. Sedimentation volume is defined as the ratio of the final sediment volume (V_s) to the actual suspension volume (V_i). For a flocculated suspension, the F value is small (i.e., less than 1), while the precipitation volume is close to 1, and in certain cases greater than 1, when the floccules produced are loose or fluffy such that the volume they occupy exceeds their actual volume (Manosroi et al. 2008).

$$F = \frac{V_s}{V_i}$$

8.5.8 Temperature Effect

Fluctuations in temperature might affect stability of suspensions. Thus, it is important to note the changes occurred in suspensions at different temperatures. In addition, the effect of temperature (30–60 °C) on the viscosity of the suspension of all formulations can be investigated to determine stability (Ahmed et al. 2005).

8.5.9 Drug Content

The determination of drug content in a pharmaceutical formulation is a critical aspect of quality control and assurance in the pharmaceutical industry. It involves measuring the quantity of active pharmaceutical ingredient (API) present in a given dosage form. This process is essential to ensure that each unit of the product contains the intended amount of the active ingredient and meets regulatory standards (Sriamornsak et al. 2010). Weighed 10 mL of the suspension should be transferred to a measuring cup with a volume of 100 mL and add 0.1 N HCl. Another 1 mL of the above suspension was removed and added to a 10 mL beaker with 0.1 N HCl. Absorbance was recorded at a maximum wavelength of 280 nm using a dual-beam UV-Vis spectrophotometer. By comparing the absorbance with the reference curve, the drug content was determined (Iyer et al. 2006).

8.5.10 In Vitro Dissolution Studies

This involves simulating the process of drug release from suspensions. By mimicking the conditions of the gastrointestinal tract, in vitro dissolution studies provide valuable insights into a drug's release profile (Azarmi et al. 2007). Carefully add 10 mL of the suspension to the base of the apparatus. At 5-min intervals, 5 mL aliquots were taken for analysis and replaced with an equal volume of blank sample. Aliquots were filtered using Whatman filter paper before being subjected to further analysis using a dual-beam UV-visible spectrophotometer at the appropriate frequency (Okafo et al. 2022).

8.5.11 Zeta Potential

Zeta potential is often used as an indicator of the droplet stability, where values more positive than +30 mV and more negative than −30 mV indicate good stability against coalescence (Kadu et al. 2011; Krstić et al. 2018). To determine the zeta potential, 1 mL of the prepared suspension was taken. Measurements were performed after 1 mL of the suspension was further diluted to 50 mL with distilled water and placed in a cleaned capillary tube (Sarafraz and Safaei 2019).

8.5.12 Particle Size and Shape

The flocculating and settling behavior of a suspension is a function of the size of its suspended particles, the forces of attraction/repulsion between them, and the viscosity of the continuous liquid phase. Thus, determination of particle size and shape is highly important. A microscope allows you to observe changes in physical characteristics, crystal shape, and particle size distribution. A change in the color of the suspension indicates poor distribution and a change in particle size. Instead of using water when diluting for microscopic examination, use a diluted dispersion medium because water can change the way the drug crystallizes. Recently, techniques for measuring the size of suspension particles, such as photon correlation spectroscopy (PCS), have gained favor. This method also provides data on the polydispersity of the suspension. Ultrasonic attenuation, single particle optical sensing (SPOS), and laser diffraction are additional tools for particle size measurement (Tscharnuter 2006).

8.5.13 Odor and Taste

Odor and taste are significant characteristics of pharmaceutical suspensions. A variation in any of them may show stability problems, a modification in crystal habit, and subsequent change in particle solubility (Jangde et al. 2011).

8.5.14 Density

Degradation in particle size sometimes results from increased suspension density. For larger particles, the effect of gravity becomes significant, especially if there is a sizeable difference in density between the dispersed and continuous phases (Larson et al. 1968). A thoroughly mixed and homogenized suspension should be used to measure the density of the suspension. A precision hydrometer is a valuable tool for determining density. The presence of air in the liquid mass is usually indicated by a decrease in density (Jangde et al. 2011).

8.5.15 Freezing and Thawing

The physical stability of suspensions can be understood by subjecting them to freeze-thaw cycles. However, since pharmaceutical suspensions often do not withstand freezing during their shelf life, a widely used similar product as sold should be included in the test for comparison (Ali et al. 2010).

8.6 Conclusion

To transport the active therapeutic agent to the site of action, dosage forms and delivery systems, together with several excipients, are used to deliver the drug in therapeutic usage. Unexpectedly, many novel medication candidates have poor bioavailability in solution dose form because they are mainly insoluble in water. For solvent-insoluble drug moieties, medicinal dosage forms known as suspensions can be taken orally, topically, parenterally, and intraocularly for therapeutic use. There are a few requirements that a well-made suspension must meet, even though suspensions are a viable formulation choice for many medications, particularly water-insoluble hydrophobic medicinal compounds. The theoretical issues of pharmaceutical coarse dispersions are covered in this chapter (e.g., interfacial properties, EDL, sedimentation, etc.). As the field of pharmaceutical sciences continues to evolve, the insights provided in this chapter are poised to navigate the challenges such as stability issues and drive innovation in pharmaceutical suspension formulations.

References

Adair JH, Suvaci E, Sindel J (2001) Surface and colloid chemistry. In: Jürgen Buschow KH, Cahn RW, Flemings MC, Ilschner B, Kramer EJ, Mahajan S, Veyssière P (eds) Encyclopedia of materials: science and technology. Elsevier, pp 1–10

Ahmed SG, Guinea A, Mortada ND, Mansour S et al (2005) Preparation and evaluation of reverse phase evaporation and multilamellar niosomes as ophthalmic carriers of acetazolamide. Int J Pharm 306:71–82

Ali Y, Kimura A, Coffey MJ, Tyle P (2010) Pharmaceutical development of suspension dosage form. In: Kulshreshtha A, Singh O, Wall G (eds) Pharmaceutical suspensions. Springer

Ansel C, Allen LV, Popovich NG (2005) Disperse systems. In: Ansel C, Allen LV, Popovich NG (eds) Pharmaceutical dosage forms & drug delivery systems, 8th edn. Lippincott Williams and Wilkins, pp 387–389, 398

Arora K, Vats V, Verma PK (2022) A review on pharmaceutical suspension and its advancement. Ann Clin Case Rep 7:2321

Aulton EM (2002) Suspension. In: Michael E (ed) Pharmaceutics, the science of dosage form design, 2nd edn. Churchill Livingstone, pp 271–278

Aulton ME, Taylor KMG (2013) Aulton's pharmaceutics. In: E-book: The design and manufacture of medicines. Elsevier, pp 417–425

Azarmi S, Roa W, Löbenberg R (2007) Current perspectives in dissolution testing of conventional and novel dosage forms. Int J Pharm 328(1):12–21

Balakrishnan P, Shanmugam S, Lee WS et al (2009) Formulation and in vitro assessment of Minoxidil niosomes for enhanced skin delivery. Int J Pharm 377:1–8

Banker GS, Rhodes CT (1979) Dispersed systems. In: Modern pharmaceutics, vol 72. Marcel Dekker, Inc., New York, pp 345–346

Barhoum A, García-Betancourt ML, Rahier H, Van Assche G (2018) Physicochemical characterization of nanomaterials: polymorph, composition, wettability, and thermal stability. In: Emerging applications of nanoparticles and architecture nanostructures. Elsevier, pp 255–278

Bhattacharjee S (2016) DLS and zeta potential–what they are and what they are not? J Control Release 235:337–351

Brunaugh AD, Smyth HD, Williams RO III (2019) Disperse systems: suspensions. In: Essential pharmaceutics. Springer, pp 91–110

Chaudhari SP, Akuskar G, Salvankar SS, Bangar J (2014) Evaluation of suspending and emulsifying properties of Citrullus lanatus seeds gum. Evaluation 7(5):181

Cooper, Gun (2002) Dispersed system. In: Tutorial pharmacy, 6th edn. CBS Publishers, pp 75–78

Deicke A, Süverkrüp R (2000) Dose uniformity and redispersibility of pharmaceutical suspensions 2: Assessment of three commercial erythromycin ethyl succinate oral liquids. J Pharm Sci 49(1):73–78

Derjaguin B, Landau LD (1941) Theory of the stability of strongly charged lyophobic sols and of the adhesion of strongly charged particles in solutions of electrolytes. Acta Physicochim URSS 14:633–662

Dicolo G, Carelli V, Giannacciini B, Serafini MF, Bottari F (1980) J Pharm Sci 69:387

Doye PA, Mena TA, Das NI (2017) Formulation and bio-availability parameters of pharmaceutical suspension. Int J Curr Pharm Res 9(3):8–14

Eyley S, Vandamme D, Lama S, Van den Mooter G, Muylaert K et al (2015) CO_2 controlled flocculation of microalgae using pH responsive cellulose nanocrystals. Nanoscale 7(34):14413–14421

Ferrar JA, Sellers BD, Chan C, Leung DH (2020) Towards an improved understanding of drug excipient interactions to enable rapid optimization of nanosuspension formulations. Int J Pharm 578:119094

Frigaard I (2019) Background lectures on ideal visco-plastic fluid flows. In: Lectures on visco-plastic fluid mechanics. Springer, pp 1–40

Garad S, Wang J, Joshi Y, Panicucci R (2010) Preclinical development for suspensions. In: Pharmaceutical suspensions. Springer, pp 127–176

Haines BA, Martin AN (1961a) J Pharm Sci 50:228–232

Haines BA, Martin AN (1961b) J Pharm Sci 50:753–756

Haines BA, Martin AN (1961c) J Pharm Sci 50:756–759

Hebishy E, Buffa M, Guamis B, Blasco-Moreno A, Trujillo A-J et al (2015) Physical and oxidative stability of whey protein oil-in-water emulsions produced by conventional and ultra high-pressure homogenization: effects of pressure and protein concentration on emulsion characteristics. Innovative Food Sci Emerg Technol 32:79–90

Iyer AK et al (2006) Exploiting the enhanced permeability and retention effect for tumor targeting. Drug Discov Today 11(17):812–818

Jangde R, Daharwal SJ, Sahu RK, Singh J (2011) Formulation development and evaluation of suspension of Gatifloxacin using suspending agent. Pharmacology Online 2:1161–1170

Jani GK (2004) Liquid dosage forms. In: Pharmaceutics-II (dispensing pharmacy), 4th edn. B. S. Shah Prakashan, Ahmedabad, p 202.44

Jermain SV, Brough C, Williams RO III (2018) Amorphous solid dispersions and nanocrystal technologies for poorly water-soluble drug delivery—an update. Int J Pharm 535(1–2):379–392

Jones DS (2016) FASTtrack Pharmaceutics dosage form and design. Pharmaceutical Press, London, pp 25–35

Kadu PJ, Kushare SS, Thacker DD, Gattani SG (2011) Enhancement of oral bioavailability of atorvastatin calcium by self-emulsifying drug delivery systems (SEDDS). Pharm Dev Technol 16:65–74

Kayes JB (1977) Pharmaceutical suspensions: relation between zeta potential, sedimentation volume and suspension stability. J Pharm Pharmacol 29(1):199–204

Kellaway IW, Najib NM (1981) The effect of hydrophilic polymers on the electrophoretic mobility of suspended particles. Int J Pharm 7(4):285–292

Khan SA, Baseer A, Khan S, Hussain M (2022) Oral suspensions. In: Khan SA (ed) Essentials of industrial pharmacy. Springer International Publishing, Cham, pp 81–93

Krstić M, Medarević Đ, Đuriš J, Ibrić S (2018) Self-nanoemulsifying drug delivery systems (SNEDDS) and self-microemulsifying drug delivery systems (SMEDDS) as lipid nanocarriers for improving dissolution rate and bioavailability of poorly soluble drugs. In: Lipid nanocarriers for drug targeting. William Andrew Publishing, pp 473–508

Kulshreshtha AK, Singh ON, Wall GM (2009) Pharmaceutical suspensions: from formulation development to manufacturing. Springer, pp 20–50

Kulshreshtha AK, Singh ON, Wall GM (2010) General principles of suspensions, Chapter 2. In: Kulshreshtha AK, Singh ON, Wall GM (eds) Pharmaceutical suspensions, pp 39–65. https://doi.org/10.1007/978-1-4419-1087-5

Kumar RS, Yagnesh TN (2016) Pharmaceutical suspensions: patient compliance oral dosage forms. World J Pharm Pharm Sci 7(12):1471–1537

Lachman L (1996) Pharmaceutical suspension. In: The theory and practice of industrial pharmacy, 3rd edn. Varghese Publishing House, pp 488–489

Larson MA, Timm DC, Wolff PR (1968) Effect of suspension density on crystal size distribution. AICHE J 14(3):448–452. https://doi.org/10.1002/aic.690140320

Leleux J, Williams RO III (2014) Recent advances in mechanical reduction methods: particulate systems. Drug Dev Ind Pharm 40(3):289–300

Lieberman HA, Lachman L, Joseph BS, Kanig JL (1987) Preformulation. In: The theory and practice of industrial pharmacy, 3rd edn. Varghese Publishing House, pp 183–184

Liebermann AH (1989) Oral aqueous suspension. In: Pharmaceutical dosage forms, dispersed systems, vol 2. Marcel Dekker, Inc, New York, pp 246–250

Liu L, Gonzalez-Olivares MA, Bai H, Yi H, Song S (2019) Colloidal stability of silica and graphite in aqueous suspensions. Chem Phys 525:110405

Loftsson T (2014) Drug stability for pharmaceutical scientists. Academic, Cambridge, MA, p 50

Ludwing A, Van Ooteghen M (1988) Drug Dev Ind Pharm:142–267

Manosroi A, Chutoprapat R, Abec M, Manosroi J et al (2008) Characteristics of niosomes prepared by supercritical carbon dioxide ($scCO_2$) fluid. Int J Pharm 3(52):248–255

Martin A (2001) Coarse dispersion. In: Physical pharmacy, 4th edn. Lippincott Williams and Wilkins, Philadelphia, pp 479–481

Martin A, Swarbrick J (1966) In: Sprowls A (ed) American pharmacy, 6th edn. Lippincott, p 205

Matthews BA, Rhodes CT (1968a) Some studies of flocculation phenomena in pharmaceutical suspensions. J Pharm Sci 57:569–573

Matthews BA, Rhodes CT (1968b) Coagulation and flocculation in suspensions of griseofulvin and polystyrene latex. J Pharm Pharmacol 20:204s–212s

Nutan MTH, Reddy IK (2010) General principles of suspensions. In: Kulshreshtha AK, Singh ON, Wall GM (eds) Pharmaceutical suspensions: from formulation development to manufacturing. Springer, New York, pp 39–65

Ohshima H (2014) Approximate analytic expression for the stability ratio of colloidal dispersions. Colloid Polym Sci 292:2269–2274

Okafo SE, Anie CO, Nwankwo LU, Agada EI (2022) Physicochemical and microbiological evaluation of ibuprofen suspension produced using Detarium microcarpum as suspending agent. J Drug Deliv Therap 12(5):80–86

Pal D, Nayak AK, Kalia S (2010) Studies on Basella alba L. leaves mucilage: evaluation of suspending properties. Int J Drug Discov Tech 1(1):15–20

Pennington AK, Ractlife JH, Wilson CG, Hardy JG (1988) Int J pharm, vol 43, p 221

Remington (2000) Pharmaceutical necessities. In: The science and practice of pharmacy, 20th edn. Lippincott Williams and Wilkins, Philadelphia, pp 1017–1021

Roura P, Fort J (2004) Local thermodynamic derivation of Young's equation. J Colloid Interface Sci 272(2):420–429

Sarafraz MM, Safaei MR (2019) Diurnal thermal evaluation of an evacuated tube solar collector (ETSC) charged with graphene nanoplatelets-methanol nano-suspension. Renew Energy 142:364–372

Short MP, Rhodes CT (1973) Can J Pharm Sci 8:46–48

Soci MM, Parrot EL (1980) J Pharm Sci 69:403

Sriamornsak P, Nunthanid J, Cheewatanakornkool K, Manchun S (2010) Effect of drug loading method on drug content and drug release from calcium pectinate gel beads. AAPS PharmSciTech 11:1315–1319

Strum JD, Colaizzi JL, Goehl TJ, Jaffe JM, Pitlick WH, Shah VP, Poust RI (1978) Bioavailability of sulfonamide suspensions I: Dissolution profiles of sulfamethizole using paddle method. J Pharm Sci 67(12):1399–1402

Trefalt G, Borkovec M (2014) Overview of DLVO theory. Laboratory of Colloid and Surface Chemistry, University of Geneva, Switzerland, p 304
Tscharnuter W (2006) Photon correlation spectroscopy in particle sizing. https://doi.org/10.1002/9780470027318.a1512
Venkateshwarlu G, Rambu D, Ramesh BR, Rao VVS (1990) Dissolution kinetics of rifampicin aqueous suspension. Indian Drugs 28:8–12
Vo A, Feng X, Patel D, Mohammad A, Kozak D, Choi S et al (2020) Factors affecting the particle size distribution and rheology of brinzolamide ophthalmic suspensions. Int J Pharm 586:119495
Wong J et al (2008) Suspensions for intravenous (IV) injection: a review of development, preclinical and clinical aspects. Adv Drug Deliv Rev 60:939–954
Yotsumoto H, Yoon RH (1993) Application of extended DLVO theory: I. Stability of rutile suspensions. J Colloid Interface Sci 157(2):426–433

Liquid and Polydisperse Systems: Emulsions

9

K. Shubham, B. Behera, and Meenakshi K. Chauhan

Abstract

Many cosmetic, food, and pharmaceutical applications rely on emulsions. Emulsions are thermodynamically unstable systems containing two immiscible liquid phases. In emulsion, one phase is dispersed as globules in another phase. Creams, ointments, gels, and pastes are emulsions that are commonly used for the delivery of drugs. Simple water-in-oil (W/O) emulsions contain hydrophilic drugs, while hydrophobic drugs are encapsulated within the oil droplets of oil-in-water (O/W) emulsions. The primary objective is to protect the drug and bioactive substances, thus increasing the bioavailability. Emulsions intended for oral use can mask the drug's bitter taste, while the topical delivery increases the permeability of bioactive agents. For pharmaceutical applications including taste masking and vaccine adjuvants, multiple emulsions or double emulsions such as water-in-oil-in-water (W/O/W) and oil-in-water-in-oil (O/W/O) are attractive systems. These emulsions have smaller droplets in the dispersed phase that are composed of the same components as the external phase. The variations in droplet size of the dispersed phase leads to the formation of microemulsion and nanoemulsions. Microemulsions are a dispersion of oil, water, and an amphiphile. They are optically isotropic and thermodynamically stable dispersions of either W/O or O/W types. Microemulsions act as multifunctional carriers in addition to tailoring nanoparticles. Miniemulsions, nanoemulsions, ultrafine emulsions, and submicron emulsions are defined as emulsions with droplet sizes between 20 and 200 nm. These emulsions are clear because of the small droplets of the dispersed phase, unlike coarse emulsions that appear milky. Emulsification refers to the process of formation of emulsions. The aim of emulsification is to make a stable system of two immiscible phases. The droplet size in a conventional emulsion

K. Shubham · B. Behera (✉) · M. K. Chauhan (✉)
Delhi Institute of Pharmaceutical Sciences and Research, DPSRU, New Delhi, Delhi, India
e-mail: beauty.behera@dpsru.edu.in; mkchauhan@dpsru.edu.in

K. Jain, A. K. Yadav (eds.), *Advances in Pharmaceutical Product Development*,
https://doi.org/10.1007/978-981-97-9230-6_9

grows with time leading to a separation of phases. Therefore, a third substance such as stabilizers or emulsifiers is added during the emulsification for long-term stability. Emulsifiers, which include phospholipids, polysaccharides, proteins, small molecule surfactants, and other surface-active polymers, are often amphiphilic compounds with both hydrophobic and hydrophilic groups on the same molecule. The stability of an emulsion is defined by the capability of the phases to remain mixed together and maintain their properties. The stability of such systems is dependent on the dispersed phase's droplet size, droplet size distribution, and differences in density, volume, and viscosities of the dispersed as well as continuous phase. An emulsion may lose its stability through coalescence, creaming, cracking, phase inversion, flocculation, and Ostwald ripening. Several theories have been proposed for the formation of stable emulsions considering the various factors and emulsifying agents used. These theories explain the behavior and action of the emulsifiers at the interface of two immiscible liquids. Some of these theories may be applicable to specific emulsifying agents under some specific circumstances such as pH and proportion of both continuous and dispersed phases. Molecular adsorption of small and large molecules to the droplet surface reduces the aggregation of the dispersed phase. The emulsion stability is thus improved by the formation of a protective film around the drop. This chapter discusses recent advances in emulsion types, theories of emulsification, formulations, and stability aspects of these biphasic systems.

Keywords

Emulsion · Flocculation · Ostwald ripening · Rheology · Spectrometry

9.1 Introduction

Many technological processes are based on the formation of a polydisperse group of particles interacting with their surroundings. A collection of particles with varying sizes that can evolve and form in a metastable environment make up this group. For instance, such events and processes include the nucleation and development of crystals in the metastable gas or liquid phase (Alexandrov et al. 2022). Dispersion refers to systems that include at least two immiscible phases. A disperse system consists of a dispersed phase in a dispersion medium also known as continuous phase (Goodarzi and Zendehboudi 2019). The term "dispersions" includes generation of reversible agglomerates containing two or more substances by hydrogen bond, van der Waals forces, hydrophobic interaction, and physical attachments instead of covalent bonds (Zhang et al. 2018). According to the state and type of the dispersed phase, there are three basic types of dispersions (Fig. 9.1) such as coarse dispersions, molecular dispersions, and colloidal dispersions (Zhang et al. 2018). Emulsions are included in colloidal dispersions which consist of two immiscible liquids stabilized by surfactant or emulsifying agents (Nikmaram et al. 2017). The size of dispersed phase in an emulsion generally varies from 0.5 to 100 μm. Emulsions are typically developed by a variety of chemical processes and use

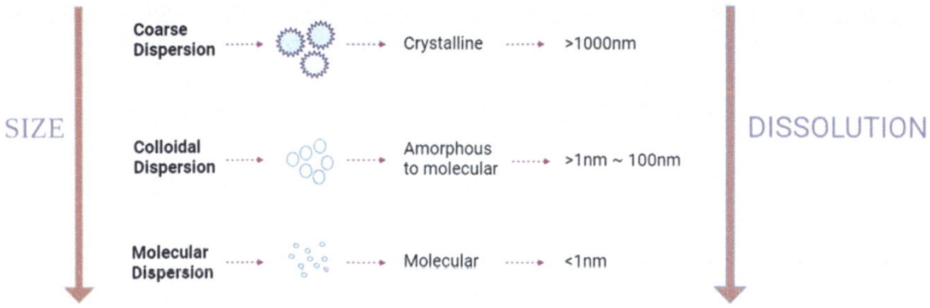

Fig. 9.1 Classification of dispersions based on the size of dispersed phase

small- and large-scale equipment, including homogenizers, ultrasonic processor, magnetic stirrer, and rotor-stator mixer (Goodarzi and Zendehboudi 2019; Chen et al. 2020a).

Majority of emulsions used in real-world applications are complex. The International Union of Pure and Applied Chemistry (IUPAC) has developed the definition of an emulsion. The IUPAC system states that in an emulsion liquid droplet and/or liquid crystals are dispersed in a "liquid" (Friberg et al. 1996). The unfavourable interaction between the oil and water phases makes emulsion an unstable system. In the absence of emulsifiers, the emulsions begin to split/crack in distinct layers due to differences in densities of the medium (Nour 2018). It is thus necessary that stabilizers such as emulsifiers are added to these formulations in order to increase their long-term stability. Emulsifiers, which include small molecule surfactants, phospholipids, proteins, polysaccharides, and other surface-active polymers, are generally amphiphilic compounds that include both hydrophilic and hydrophobic groups on the same molecule (McClements and Jafari 2018).

9.2 Classification of Emulsions

The colloidal dispersions, also known as colloids, are combinations in which one substance is very finely mixed with another. They have particle size between 1 and 100 nm which is smaller than the particles present in a suspension. The emulsions are colloidal dispersions which consist of two immiscible liquids stabilized by third agent (Nikmaram et al. 2017). Commonly used examples of emulsions are milk, cream, desserts, sauces, and mayonnaise. Based on the fluid that becomes a continuous phase, emulsions are categorized as water-in-oil (W/O) and oil-in-water (O/W). The O/W emulsions are produced when oil is dispersed in the water or aqueous phase. These emulsions are commonly used to protect, encapsulate, or control the release of bioactive agents (Lu et al. 2021). Emulsions of W/O are formed when water or aqueous phase of droplets is dispersed in the oil phase. In W/O emulsions, the arrangement of emulsifier molecules is opposite to that observed in O/W type. The non-polar tails are pointed towards the oil phase and polar heads are oriented towards the water droplets. The use of emulsifiers depends on the type of emulsions to be prepared (Table 9.1). Emulsions can be also classified on the basis of the size of dispersed phase into macroemulsions, microemulsions, nanoemulsions, and pickering emulsions (Aslam et al. 2023; Juneja et al. 2022; Lipei et al. 2021; Suthar et al. 2023). The following section discusses in brief the various emulsion systems.

9.2.1 Macroemulsion

Macroemulsions are classified as W/O, O/W, and multiple or complex or double emulsions. If the dispersed droplets in the emulsion are between 0.1 and 100 μm, it is referred to as a macroemulsion (Nikmaram et al. 2017). This group includes most of the emulsions that are manufactured in the market. Macroemulsions with diameters more than 100 μm are thermodynamically unstable and rarely maintain their initial drop state (Kim et al. 2022). Generally, a phase is labelled as the dispersed phase if its volume percentage is lower than that of the other phase, which is labelled as the continuous phase. In W/O type macroemulsion, oil is the continuous oil phase. When it comes to oil-based emulsions, the most widely manufactured emulsions fall under the W/O category. A number of variables, including temperature and the W/O ratio, affect these types of emulsions. Sometimes, the reverse emulsions are also known as O/W emulsions (Glatter and Salentinig 2020). Reverse emulsions are used as a vehicle for parenteral drug administration as well as in food and cosmetic industries (Prankerd and Stella 1990).

The complex and heterogeneous dispersion systems known as multiple or double emulsions are made of small droplets in bigger droplets and are suspended in a continuous phase. Oil-in-water-in-oil (O/W/O) and water-in-oil-in-water (W/O/W) multiple emulsions are the two main forms of these systems. In an O/W/O emulsion, the oil droplets are suspended in bigger water droplets that are suspended in a

Table 9.1 Classification of emulsions

Emulsions	Types	Emulsifying agent	Size range
Macroemulsion	O/W	Gelatin, tween 80 (polyoxyethylene sorbitan monooleate)	0.1–5 μm
	W/O	Beeswax, span 80 (sorbitan monooleate)	0.1–5 μm
Multiple/double emulsion	O/W/O	Polyoxyethylene di-isostearate, polyoxyethylene di-oleate, microcrystalline wax	0.1–100 μm
	W/O/W	Guar gum, sodium caseinate, maltodextrin	0.1–100 μm
Microemulsion	W/O	Span 20 and/or absolute butanol as co-surfactant	1–100 nm
	O/W	Tween 20/80 (polysorbate 20) and/or Cremophor EL (polyoxyl 35 castor oil)	1–100 nm
Nanoemulsion	O/W	β-Lactoglobulin	10–500 nm
	W/O	Span 80 (sorbitan monooleate)	10–500 nm
Pickering emulsion	O/W	Sodium stearoyl lactylate	0.1–4200 nm
	W/O	Whey protein-polyphenol complex particles	0.1–5 μm

continuous oil phase. Similarly, W/O/W is made of water droplets suspended in bigger oil droplets which is further emulsified in a continuous aqueous phase. Both the O/W/O and W/O/W multiple emulsion systems have been used for drug delivery applications by the pharmaceutical industries (Jena et al. 2018). These multiple systems require two or more emulsifiers during their formation. One of the emulsifiers stabilizes the W/O emulsion primarily through hydrophobic interactions, whereas the other stabilizes the O/W emulsion primarily through hydrophilic interactions. In addition to interfering with each other's ability to stabilize, both surfactants may interact with both surfaces. Furthermore, the composition of the binary surfactant mixture controls the lifetime of the films at the interfaces as well as their permeation characteristics (Cunha et al. 2014).

9.2.2 Microemulsion

A different class of emulsions known as microemulsions exists, which is distinct from macroemulsions. When two immiscible liquids come in contact, it results in the formation of this type of emulsion, due to their low interfacial energy (Fig. 9.2). The stability and formulation aspects of microemulsions are often different from those of macroemulsions (Goodarzi and Zendehboudi 2019; Pardhi et al. 2022). Microemulsions are isotropic, transparent solutions and are thermodynamically stable. It consists of at least a hydrophobic, a hydrophilic, and an amphiphilic component. Their thermodynamic stability and significantly lower structural size

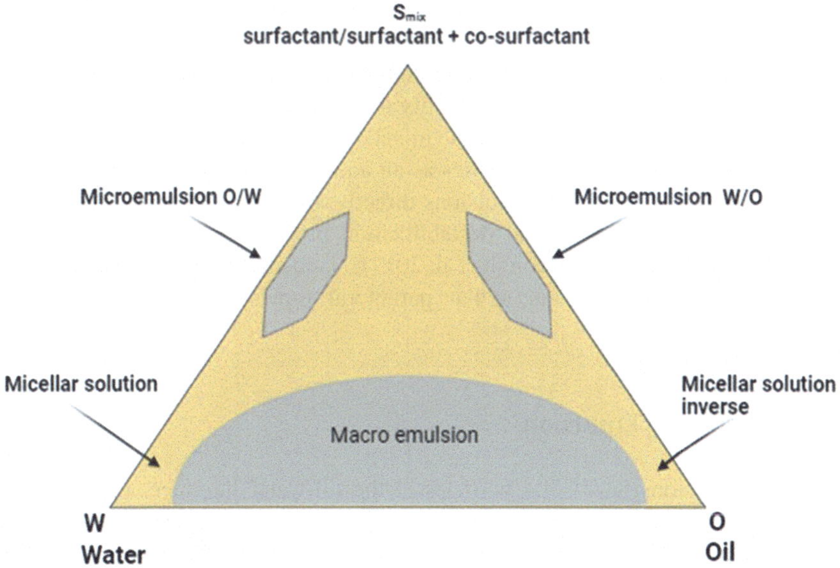

Fig. 9.2 Ternary phase diagram for the formation of microemulsion

(3–30 nm) provide them a unique edge over conventional emulsions. These two characteristics translate them into the long-term stabilization of mixed apolar/polar systems that would be impossible to achieve in a conventional emulsion (Hejazifar et al. 2020). Microemulsions are a flexible carrier with a number of prominent characteristics, including very low surface tension, effective absorption, improved bioavailability of poorly soluble drugs, penetration due to its small droplet size, and inexpensive approach of formulation. Typically, the formulation of a microemulsion is characterized by using the ternary phase diagram. This ternary phase diagram is represented by a triangle with concentration of oil, water, and surfactants at three sides. Co-surfactants are considered as pseudo-components and used in microemulsions at a predetermined ratio with surfactants (Shukla et al. 2018). Apart from pharmaceutical and nutraceutical, microemulsion is one of the most significant chemical flooding agents and an effective tool for oil recovery because of its high degree of oil extraction efficiency (Pal et al. 2019).

9.2.3 Nanoemulsion

Nanoemulsions are also categorized into O/W and W/O dispersion systems of two immiscible liquids stabilized using a suitable emulsifier. The mean droplet diameter of dispersed phase in a nanoemulsion is usually less than 500 nm. Compared to the milky white colour of a coarse emulsion (whose tiny droplets participate in multiple light scattering processes), their appearance is clear or hazy due to the small droplet size. Miniemulsion or submicron emulsion is occasionally used interchangeably with the term nanoemulsion; however, it should not be confounded with microemulsion. Although nanoemulsions and microemulsions have the same range of droplet size, they differ greatly in terms of structural features and prolonged thermodynamic stability (Kumar et al. 2021; Singh et al. 2017; Sudhar et al. 2022). Nanoemulsions have higher kinetic stability than coarse emulsions. They possess a large solubilization capacity than simple micellar dispersions. They have been used in the pesticide and cosmetic industries as an aqueous base for organic deliveries. The small droplet size of nanoemulsions directly affects their long-term physical stability by hindering traditional destabilization processes such as coalescence, creaming, and sedimentation (Singh et al. 2017). Nanoemulsions are used to deliver a variety of drugs through oral, ocular, parenteral, topical, and transdermal route (Vv et al. 2018).

9.2.4 Pickering Emulsion

Pickering emulsions have solid particles at the oil-water interface. Unlike traditional emulsions that use surfactants, solid particles are adsorbed at the interface in a pickering emulsions. The solid particles such as clay, protein particles, or colloidal silica irreversibly adsorb to the interface and form a barrier that prevents the droplets from aggregating or coalescing. The three-phase contact angle of the particles with the interface determines their wettability, which in turn influences the type of pickering emulsion that forms such as W/O or O/W. The W/O emulsions are stabilized by hydrophobic particles (contact angle is greater than 90°), whereas O/W emulsions are stabilized by more hydrophilic particles (contact angle is less than 90°). The benefits of these systems are that they offer greater stability against coalescence and good biocompatibility when employing food-grade particles and have the capacity to introduce new functionality to the emulsion. They are used in a number of fields, including cosmetics, medicines, food, and increased oil recovery (Yang et al. 2017).

9.3 Theories of Emulsions

The formulation and stability of the emulsions depend critically on the adsorption of nanoparticles, proteins, polymers, or surfactants to the interface. During the emulsification process, a greater interfacial area is created more favourably by the surfactant adsorption, which reduces the interfacial tension and, consequently, the surface free energy of the liquid-liquid system (Llamas et al. 2018). There are various indispensable factors that result in the production of a stable emulsion. Numerous theories have been postulated earlier to explain the stabilization behaviour of an emulsion. The following section discusses these theories.

9.3.1 Fischer's Theory of Hydrates and Solvates

Fischer's hydration theory states that the emulsifying agents are used in particular proportion to continuous phase, resulting in the formation of colloidal hydrate. The other phase is dispersed in the colloidal hydrate. This theory is unable to explain certain observable facts and contradicts others. Therefore, it cannot be considered an acceptable emulsion theory (Roberts 2002).

9.3.2 Surface Tension Theory

After much research, it was found that the surface tension theory, or, more accurately, the interfacial tension theory, was sufficient to explain the development and stability of numerous emulsions. This theory states that the emulsifying agents reduce the interfacial tension between the two immiscible liquids, which reduce the repelling force between them and eliminate the attraction that the liquids have for their own molecules. Interfacial tension is the force that prevents two immiscible liquids from breaking when they come into contact. When two liquids have a high interfacial tension, emulsification is challenging; nevertheless, emulsification is made easier when the tension is decreased. The theory states that with the dispersion of oil in water, the interfacial tension is so high that as two globules of the dispersed phase approach one another, the liquid is drawn out from between them, causing them to coalesce. The globules remain separated when the addition of an emulsifier significantly lowers the interfacial tension. This hypothesis has been incorporated into the more recent and generally recognized adsorption film theory, which explains emulsion development and stability on the creation of an adsorbed film at the interface. According to the plastic or interfacial film theory, the emulsifying agent surrounds the internal phase droplets as a thin layer of film adsorbed on their surface, located at the interface between the oil and water. This film keeps the dispersed phase from coming into contact and merging. The formulation of a W/O or O/W emulsion is dependent on the degree of solubility of the agent in the two phases. The water-soluble emulsifiers promote O/W emulsions, while oil-soluble emulsifiers promote W/O emulsions (Roberts 2002).

9.3.3 Molecular Adsorption Theory

Amphiphiles, or surface-acting agents, form a monomolecular film by their adsorption at the oil-water interface which lowers the interfacial tension between two liquids. As per the theory, surfactants are examples of emulsifying agents that have both non-polar (hydrophobic) and polar (hydrophilic) components in their molecules. When these are added to an oil-water system, the molecules of surfactant align themselves at the interface, with their non-polar portions in the oil phase and their polar portions in the water phase. This orientation lowers the interfacial tension between the two liquids by producing a monomolecular layer of surfactant molecules at the interface. The decrease in the interfacial tension during homogenisation causes large droplets to break into smaller ones, forming a stable emulsion. Nowadays, emulsions are usually prepared using a blend of emulsifiers rather than a single agent. A complicated film forms at the interface between a hydrophobic agent in the oil phase and a hydrophilic emulsifier in the aqueous phase. Surfactant adsorption at the liquid interfaces thus influences the stability of emulsions by reducing interfacial tension, enhancing electric double-layer repulsion (ionic surfactants), and improving surface elasticity. Additionally, the nature of the surfactant can regulate how the phases in an emulsion are arranged, i.e. which phase will form the continuous phase and which will form the dispersed phase (Umar et al. 2018).

9.3.4 Oriented Wedge Theory

This theory describes how an emulsifying agent's curved monomolecular layer forms around a droplet of the emulsion's internal phase (Pratishtha et al. 2021). Emulsifying agents align themselves based on solubility since they contain both hydrophilic and hydrophobic portions. Emulsifying agents having a higher proportion of hydrophobic portions than hydrophilic ones promote the W/O emulsion and vice versa. Surfactants, which have a small group within and a bulky group outside, are always found on the surface of a droplet, such as a wedge, due to packing restrictions at the water/oil interface of emulsion drops (Wang et al. 2023).

9.4 Formulation

The emulsification process involves the breakdown of one liquid phase within the other, resulting in the production of an immense number of tiny droplets. Emulsions can be produced in research or industrial fields, using high or low energy procedures. They are prepared on a large scale using micro-fluidizers, shear mixers, sonicators, and high-pressure valve or membrane homogenizers. The technique selected relies on the emulsifier type and the field of application (Ravera et al. 2021). Emulsification involves two competing processes that occur at the same time. The first process involves the breaking of the dispersed phase into droplets, while the second is coalescence of the produced droplets. This can cause a condition in mixing devices called "morphological hysteresis". The emulsifiers serve the purpose of preventing the droplets from merging with one another, which is necessary to guarantee that the droplet size distribution stays steady. The interfacial tension between the dispersed and continuous phases is lowered in the presence of appropriate emulsifiers, which lowers the energy needed to build a new interface and promotes the production of droplets. Furthermore, as previously indicated, the emulsifiers can give the final emulsified products an appropriate lifetime and an energy barrier against coalescence (Ho et al. 2024).

9.4.1 Method of Preparation

9.4.1.1 Dry Gum Method

Dry gum method is also known as gum acacia method. This method includes four parts of oil, two parts of water, and one part of gum as an emulsifying agent. The gum is initially mixed with the right amount of oil so that the powder is thoroughly wetted. The mixture is then vigorously triturated with water until the emulsion becomes creamy white and has a "crackling" sound. The preservatives, colorant, and flavouring agents can be added to improve the palatability (Kale and Deore 2017).

9.4.1.2 Wet Gum Method

This method uses a 4:2:1 ratio of oil, water, and emulsifier, respectively. However, the sequence of mixing is different. The gum is first triturated with water to form mucilage. The oil is then slowly added while triturating. Then, the mixture is triturated for few minutes to formulate the emulsion. This method is more difficult to execute well, particularly with more viscous oils, but it may yield a more stable emulsion (Kale and Deore 2017).

9.4.1.3 Bottle Method

This approach can be used to create emulsions of volatile oils and oleaginous compounds having very low viscosities. This method is a modification of the dry gum method. In this method, gum is added to a dry bottle, followed by oil. The bottle is capped and shaken vigorously. Desired volume of water is then added all at once, and the mixture is vigorously agitated to form emulsion (Gowda et al. 2013).

9.4.1.4 In Situ Soap Method

Emulsions can be prepared using in situ soap method, in which soap molecules are formed while the emulsion is being prepared. In this method the aqueous phase contains water and an alkaline material like potassium or sodium hydroxide. The alkaline material creates soap molecules in situ by reacting with the fatty acids in the oil phase. These soap molecules act as emulsifiers and stabilize the resulting emulsion (Saygı-Arslan et al. 2009).

9.4.1.5 Phase Titration Method

The required quantity of medication is dispersed in the accurate amount of oil to create a microemulsion, which is necessary for the drug to become soluble. The mixture is then thoroughly mixed and the appropriate amount of surfactant added, along with a small amount of co-surfactant blend which is added with stirring. Drop-by-drop double-distilled water is added while stirring around 10 min to form emulsion. The stirring rate can be adjusted to meet the requirements of the particle size (Kale and Deore 2017).

9.4.1.6 Phase Inversion Temperature Method

Phase inversion temperature (PIT) method is a low-energy emulsification technique that modifies temperature to alter the optimal curvature of surfactants at constant composition to form nanoemulsions. Nanoemulsions are formed by the preparation of microemulsions at its PIT followed by quickly cooling them to room temperature. This can be divided into three main steps. First, the oil, water, and non-ionic surfactant are stirred at room temperature to produce a coarse emulsion. Second, the mixture is slowly heated to a temperature near or above the PIT. Lastly, the solution is rapidly cooled to room temperature while continuously stirred, resulting in the creation of nanoemulsions (Jintapattanakit 2018).

9.4.1.7 Spontaneous Emulsification

The transfer of water-miscible components (surfactant, co-surfactant, and solvent) from an organic phase into the aqueous phase is caused by diffusion, which occurs when the system is diluted. In spontaneous emulsification, the aqueous phases that contain water and co-surfactant are typically combined with an organic phase that contains oil and hydrophilic surfactant. The rapid migration of water-miscible components into the aqueous phase generates enormous turbulence at the interface of the two phases, resulting in a significant increase in the oil-water interface area. It causes the spontaneous production of oil droplets surrounded by an aqueous phase via the budding process (Jintapattanakit 2018).

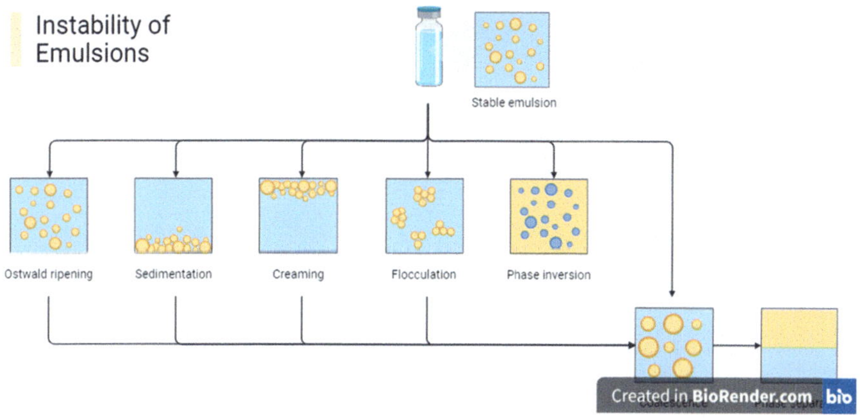

Fig. 9.3 Various destabilization process in an emulsion

9.5 Stability

The term "emulsion stability" describes an emulsion's capacity to withstand long-term alterations to its physicochemical characteristics. Differences in the formulation circumstances affect the properties even if the concentrations of the compositions are identical. Additionally, an emulsion has various collapse processes, including creaming, aggregation, Ostwald ripening, and coalescence. These processes are dependent on the ingredients and preparation circumstances as well as their concentrations (Fig. 9.3). Creaming and sedimentation are mostly due to the differences in density between the two liquids, but the other phenomena are related to the characteristics of the liquid-liquid interface. Flocculation is the development of droplet aggregates. It is caused by the interaction of adsorbed layers on the surface of the emulsion droplets. The coalescence is the process by which two droplets merge to become one. Ostwald ripening is a mass transfer process from small to large droplets which occurs due to partial breakdown of the dispersed liquid phase brought on by capillary pressure. Though primarily determined by the solubility of the dispersed liquid and mostly dependent on droplet size, this phenomenon is also impacted by the presence of surfactants and the rheological properties of the interfacial layers. The aforementioned destabilization mechanisms exhibit interdependence and have the potential to impact one another as the emulsion ages. For instance, flocculation and creaming tend to compact the droplet population, which may speed up the flocculation and coalescence processes. But they have no effect on the droplet size distribution, which is determined by Ostwald ripening and coalescence. However, flocculation and the rise in average droplet sizes brought on by Ostwald ripening and coalescence serve to accelerate the process of creaming and, ultimately, the emulsion destabilization (Huck-Iriart et al. 2016). The various aspects of emulsion stability are discussed in the following sections.

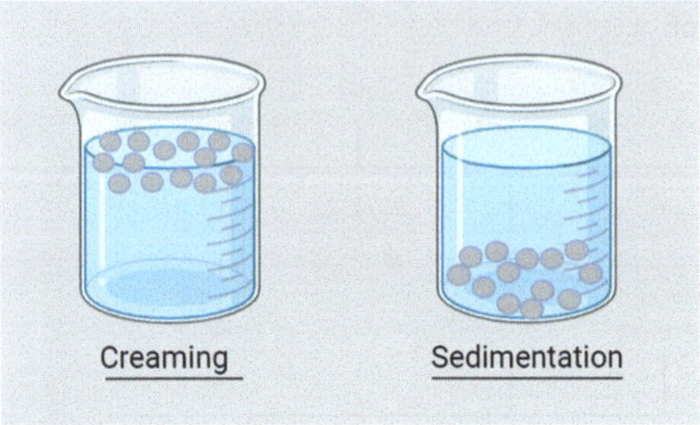

Fig. 9.4 Schematic representation of gravitational separation

9.5.1 Gravitational Separation

9.5.1.1 Creaming

Gravitational separation can be categorized into two mechanisms such as creaming and sedimentation (Hu et al. 2017). The phenomenon known as "creaming" is one of the collapse processes, in which the droplets of an emulsion are transferred to the upper or lower regions due to the variations in density between the continuous and dispersed phase. The creaming rate of an emulsion is affected by the viscosity and droplet size of the dispersed phase and so on (Fig. 9.4).

9.5.1.2 Sedimentation

Sedimentation is a process in which the dispersed phase's droplets move downwards due to the higher density as compared to the continuous phase. Creaming and sedimentation are reversible in nature as droplets can be redispersed with simple shaking, but coalescence results into permanent damage of emulsion. The sedimentation rate of an emulsion is regulated by Stoke's law (Eq. 9.1).

$$V_s = \frac{2(\rho_c - \rho_d)r^2 g}{9\eta_c} \tag{9.1}$$

where V_s represents the terminal velocity of settling drop, g is the acceleration due to gravity, ρ_d and ρ_c are individual densities of the dispersed phase and continuous phase, respectively, η_c is viscosity of continuous phase, and r is the droplet's radius, which is used to monitor its settling velocity. According to Stoke's law, the settling velocity of drops is directly proportional to the square of the radius of droplets. The lower radius of emulsion droplets is resistant to destabilizing phenomena such as sedimentation (Singh et al. 2017).

9.5.1.3 Flocculation

The stability of an emulsion against coalescence and flocculation is especially important for emulsions that are stabilized by macromolecules such as biomolecules, polymers, and, most notably, proteins (Huck-Iriart et al. 2016). Proteins are often good O/W emulsion stabilizers because they may form dense, impermeable adsorption layers at the oil drop surface. These interactions prevent the oil droplets from coalescing by generating repulsive steric and electrostatic interactions. Nonetheless, flocculation is a relatively typical event for these systems, particularly when emulsions are stored for a long period of time with potential environmental disturbances, as in food applications (Ravera et al. 2021). The flocculation behaviour of emulsion stabilized by protein is often explained by using theoretical descriptions of colloidal interactions (Delahaije et al. 2013). Therefore, flocculation might be considered as a form of aggregation. The occurrence of flocculation frequently affects the structure, stability, and rheology of O/W emulsions that contain food proteins (Dickinson 2019). Flocculation is influenced by pH and concentration and is often seen above a critical concentration of the stabilizing polymer (Huck-Iriart et al. 2016). This has been detected in the emulsion gel stabilized by chitosan polymer. It has been observed that the degree of droplet flocculation in the emulsions with pH 6.8 and 7.0 increases. The majority of droplets were significantly flocculated but did not coalesce. The aforementioned observation implies that the emulsion structuring and gel network strengthening at pH values higher than 6.5 are mostly due to the flocculation of aggregated chitosan-coated droplets and their sustained connection over extended storage periods (Wang et al. 2020).

9.5.2 Non-gravitational Separation

9.5.2.1 Coalescence

The process of droplet coalescence, or fusion, in an emulsion is a crucial destabilization mechanism that eventually causes increase in droplet size with time (Langevin 2019). This phenomenon is caused by the thinning and rupture of the liquid film between them and is greatly influenced by the behaviour of adsorbed layers. Emulsions may undergo coalescence as a result of flocculating droplets or when the droplets come in contact and merge with each other. In an emulsion, droplets' movement is relative to each other due to Brownian motion, gravity, and external mechanical forces such as centrifugation. Coalescence of droplets occurs because of their collisions with each other during the movement. However, only a fraction of all collisions may result into a coalescence event (Ho et al. 2024). There are three stages of flow-induced coalescence of emulsified liquid droplets. In the first stage, the droplets approach under a continuous flow. The second stage includes the collision of droplets resulting in the creation of a thin liquid film, which further thins due to fluid drainage between the two droplets. The last stage is the rupturing of thin film that results in coalescence (Narayan et al. 2020). Within a protein-stabilized emulsion, coalescence is a crucial stability parameter. This phenomenon is influenced by both pH and protein concentration in an emulsion. It is generally accepted that stability against coalescence during emulsion formation is largely dependent on the protein concentration. At low protein concentration, the coalescence takes place because of insufficient amounts of protein to completely cover the interface. At high protein concentration, coalescence does not take place because they completely cover the interface during the emulsification timescale (Delahaije et al. 2017).

9.5.2.2 Droplet Aggregation

The droplet aggregation is typically triggered by several interactions between droplets such as electrostatic interaction, hydrophobic interaction, etc. This results in an increase in droplet size (Shiraki et al. 2020). This phenomenon affects the stability of emulsions specially in nanoemulsions where the loss of all unique properties is at a nanoscale level. This increased aggregation may eventually lead to phase separation, which damages the system irreversibly. An empirical method for understanding the energetics underlying the aggregation process has been developed, utilizing classical Newtonian mechanics (Eq. 9.2). Overall interaction (I_E) between droplets can be characterized as the sum of electrostatic (I_{el}), hydrophobic (I_h), van der Waals (I_{VDW}), and steric (I_s) interactions occurring in between droplets. However, there are various other factors which affect the level of interactions.

$$I_E I_{el} + I_h + I_{VDW} + I_S \tag{9.2}$$

Meanwhile, developing a nanoemulsion, the interaction between droplets should be minimized by calibrating all the parameters listed above and plotting a potential energy curve against inter-droplet distance. The distances at which any two droplets might approach one another without interacting are represented by the minimum in this potential energy curve, which gives an approximate idea of the nanoemulsion's overall stability (Singh et al. 2017).

9.5.2.3 Ostwald Ripening

Ostwald ripening is also a major stability problem in emulsions. This is typically due to the coarsening of small droplets into large droplets, i.e. disappearing of small droplets which is energetically favourable (Zwicker et al. 2015). This process results in a net mass transfer of small to large droplets. This is caused by the difference in the Laplace pressure of the droplets, which is inversely proportional to their radius (Llamas et al. 2018). Ostwald ripening is more challenging to manage and can be seen in both macro- and nanoemulsions, but it typically appears in those emulsions where water is continuous phase (Koroleva and Yurtov 2021). The Ostwald ripening rates of emulsions are determined by the characteristics of the dispersed oils (solubility, composition, etc.) and their diffusion rates throughout the aqueous phase. The diffusion rate of oil molecules across oil droplet surfaces is a significant component in influencing the Ostwald ripening rates of vital oil emulsions (Park et al. 2020). The Kelvin equation illustrates how varying droplet sizes can affect the solubility of dispersed fluids (Eq. 9.3).

$$c(r) = c(\infty) e^{\frac{2\gamma V_m}{rRT}} \tag{9.3}$$

where, for O/W emulsion, $c(r)$ denotes the water solubility of the oil (dispersed phase) contained in a droplet of radius r, whereas $c(\infty)$ represents the corresponding solubility at a planar interface. V_m denotes molar volume of the dispersed phase, γ is the interfacial tension between the two liquids, R is the gas constant, and T is the absolute temperature (Khedr and Striolo 2019). Based on the above-mentioned equation, as the droplet radius falls, solubility increases; the fluid in the smaller droplets dissolves predominantly into the continuous phase and diffuses into the bigger droplets. The overall size of the emulsion droplets grows as the fluid condenses on the larger droplet surfaces, ultimately resulting in a reduction of total interfacial area (Khedr and Striolo 2019).

9.5.2.4 Phase Inversion

The process of inter-converting two forms of simple emulsions, W/O and O/W emulsions, is known as emulsion phase inversion (Kumar et al. 2015). The industry has utilized this method extensively to create a range of emulsions, and it has significant interest due to its underlying mechanism (Chen et al. 2020b). Traditional phase inversion, which occurs when the emulsifier affinity shifts from one phase to the other, can cause phase inversion. Traditional phase inversion is started by changing the way the emulsifier and the oil-water combination system interact. For example, temperature can alter the hydrophilic-lipophilic balance (HLB) of non-ionic surfactants, which can result in emulsion inversion (Chen et al. 2020b). In contrast to W/O emulsions, which form at higher temperatures, non-ionic surfactants are more hydrophilic at low temperatures and generate O/W emulsions. In addition, HLB varies in response to light irradiation, pH, and brine salinity. Catastrophic phase inversion is caused by a shift in the emulsion's water-to-oil ratio (Kumar et al. 2015). The coalescence of internal phases brought on by the presence of excess internal phases leads to catastrophic phase inversion. Phase inversion of pickering emulsions has also drawn a lot of interest. The primary component influencing these types of emulsions is particle wettability, which functions similarly to surfactant HLB. In simple terms, particles that are hydrophilic tend to stabilize O/W emulsions, while those that are hydrophobic help W/O emulsions form. Therefore, by altering the proportion of hydrophilic to hydrophobic particles, phase inversion can be accomplished in a pickering emulsion (Chen et al. 2020b).

9.6 Evaluation

9.6.1 Macroscopic Evaluation

The macroscopic evaluation of an emulsion is performed through naked eyes or under a microscope. One of the macroscopic evaluations is the determination of creaming index. Creaming changes the qualities of an emulsion product over time and temperature (Hong et al. 2018). The creaming index of an emulsion is determined by placing the system in a test tube. After a particular time, there is a separation of the system into an opaque cream layer and a transparent serum layer at the bottom (Hosseini et al. 2015). The creaming index is then determined by the following formula (Eq. 9.4):

$$\text{Creaming index }\left[\%\right] = \times 100 \quad (9.4)$$

where H_s is the height of a serum layer and H_t represents the total height of an emulsion sample.

It has been observed that creaming index is directly proportional to the gum concentration. Recently the creaming behaviour has also been evaluated using various methods like ultraviolet-visible spectroscopy and multiphoton ionization time-of-flight mass spectrometry (MPI-TOFMS). The MPI-TOFMS offers information about the oil component concentration in an emulsion. The change in the oil component caused by creaming can be evaluated by determining the mass signals. Evaluation of creaming by spectrophotometer involves the measurement of degree of turbidity. The UV-visible transmittance is measured across a wavelength (λ_{max}) from 200 to 800 nm (Shinoda and Uchimura 2018).

9.6.2 Microscopic Evaluation

The most common method for the structural analysis of the emulsions involves visualization under a microscope. As emulsions are multiphase systems, imaging under a normal light microscope provides a clear understanding about the microstructure. The other techniques that have been used in the imaging of emulsions are fluorescence and the confocal microscopy. Advanced microscopic imaging using electron beams such as transmission electron microscope (TEM) and scanning electron microscope (SEM) are also used to investigate the structure of the emulsion gels. Fluorescence imaging is a simple and non-destructive method. The light sources that have been used in fluorescence microscope are the xenon arc lamp or a mercury vapour lamp. The major principle underlying the imaging of the substances in this technique is based on the processes occurring during the electronic transitions. In this technique, the fluorophores present in the samples absorb photons and are excited to a higher energy state. When the electrons return to the ground state, they emit another light in the visible region. This phenomenon can be well understood by Jablonski diagram (Fig. 9.5).

In situ fluorescence imaging of miniemulsion revealed that the droplet size of emulsion influences the particle size and macroscopic reaction rate during miniemulsion polymerization (Liu et al. 2020). Confocal microscopy works on the same principle as that of fluorescence microscope. The difference in the two techniques lies in the use of light source. The source of light in a confocal microscope is laser. Laser of a particular wavelength is used for the imaging of the specimen. In this microscopy, the light which is not from the microscope's focal plane is excluded during imaging. This technique overcomes the difficulty of fluorescence and conventional microscopy by creating more detailed sharp 3D image. These two methods help identify the location of adsorbed molecule and encapsulated drug/microorganism (Zhang et al. 2022). SEM and TEM are not quantitative methods but these

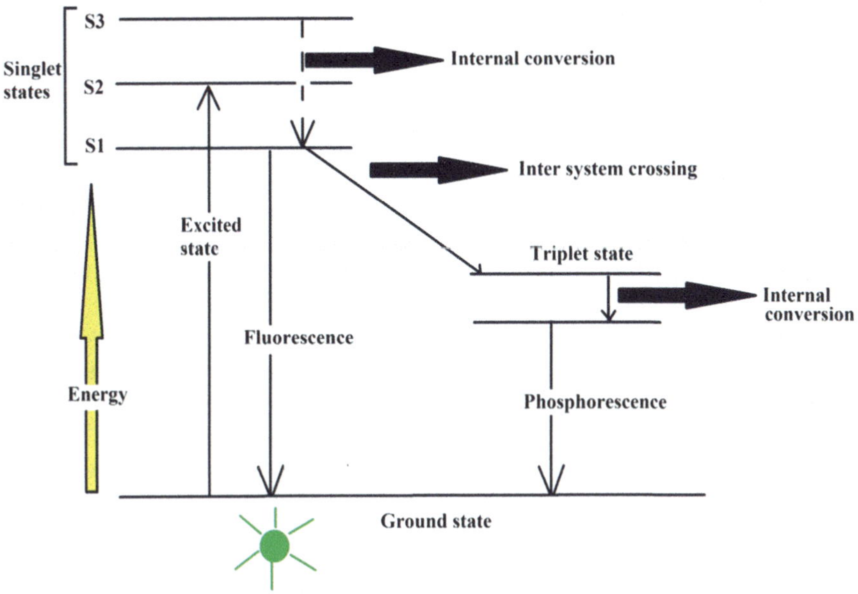

Fig. 9.5 Jabonski diagram showing the phenomenon of fluorescence and phosphorescence

Table 9.2 Some basic techniques for the visualization of gel microstructure

Technique	Advantages
Optical microscopy	Direct method requires no sample preparation
Scanning electron microscopy (SEM)	Sample preparation is easy
Transmission electron microscopy (TEM)	Easier visualization of small crystal structure
Cryo-TEM	Samples resistant to temperature

techniques are helpful in studying the microstructure morphology of the emulsion droplets. These methods provide an insight about the superstructures of nano- or microdimension formed in these systems (Tian et al. 2022).

Cryogenic-temperature transmission electron microscopy (cryo-TEM) is a method that can be used for the micrometre to nanometre characterization of emulsion droplets. The high-resolution images obtained from this technique enable in-depth analysis of structures. The basic building blocks of structures can be elucidated by the direct images. However, a major limitation of cryo-TEM is that the sample thickness should be approximately 300 nm. Another disadvantage of this technique is that it avoids the imaging of the strands in the gel network placed on the same focal plane. Table 9.2 shows some basic techniques and their advantages for visualization of emulsions.

9.6.3 Droplet Size Analysis

One of the most important factors affecting an emulsion's stability and physicochemical qualities is droplet size. Typically, small-angle X-ray scattering (SAXS), dynamic light scattering (DLS), electrical pulse counting, or ultrasonic spectrometry techniques are used to examine the size and dispersion of emulsion droplets (Kale and Deore 2017; Hong et al. 2018). The smaller particles move more quickly than larger ones and produce a greater rate of intensity fluctuation. The DLS devices measure the intensity fluctuations of light scattered by the droplets of emulsions. Droplet size from 3 nm to 5 μm, concentration, size distribution, and polydispersity index (PDI) can all be computed with it. The uniformity of droplet size is described by the PDI, which has values ranging from 0 to 1. A value of 0 indicates a system that is entirely monodisperse (Gurpreet and Singh 2018). Higher emulsion stability and a more uniform distribution of droplet sizes are indicated by a lower PDI value, and vice versa. The droplet size changes greatly in emulsions where flocculation and coalescence take place, which raises the PDI. Similar to the DLS approach, small-angle X-ray scattering (SAXS) techniques use a monochromatic X-ray beam that is size-dependently scattered by droplets of emulsion in place of a light beam. The detection of changes in the sample's electrical conductivity brought on by the droplets of emulsion moving between two electrodes is the foundation of electrical pulse counting techniques. Electrical pulses are produced when droplets moving between the electrodes alter the electrical current going through the emulsion because oils have a far lower electrical conductivity than water. The assumption used to calculate droplet size is that larger particles move more slowly and produce larger electrical pulses. This method works well for measuring droplets with diameters between 0.4 and 1200 μm. Prior to any analysis, all emulsion samples must be diluted (Hu et al. 2017). Dilution can disturb the collected droplets and can provide erroneous interpretations of emulsion stability data. Although this approach works well for droplet size analysis, it is not optimal for research in the flocculation process. The use of ultrasonic spectrometry provides a dilution-free way to determine droplet size in emulsions with greater droplet concentrations up to 50% of the total volume (Silva et al. 2022). This method calculates the size and concentration of emulsion droplets by analysing the scattering of ultrasonic waves by the emulsion. Droplet sizes from 10 nm to 1000 μm can be measured using ultrasonic spectrometry.

9.6.4 Rheology and Flow Behaviour

Emulsions are a class of complex fluid that behaves like both solids and liquids (Dekker et al. 2022). Rheology is the study of science of deformation. These studies give an idea about the flow behaviour and rigidity of the emulsion. It is crucial to understand the flow behaviour of these systems from both an industrial and a basic standpoint. An O/W emulsion exhibits Newtonian behavior for low volume fractions ($\phi < 0.1$) of oil droplets. When the droplet interacts, the oil percentage rises

($\phi > 0.1$) and the emulsions are non-Newtonian (Fuhrmann et al. 2022). Newtonian flow is exhibited by low molecular weight substances such as organic and inorganic liquids, salts, molten metal, and gases. In these types of systems, the shear stress (σ) is proportional to the rate of shear ($\gamma^{\cdot}$) at constant temperature and pressure. The constant of proportionality is the dynamic viscosity (η). Multiphase systems such as foams, emulsions, dispersions, suspensions, slurries, solutions (both natural and synthetic), and polymeric melts do not follow the Newtonian postulate of the linear relationship between ($\gamma^{\cdot}$) and (σ); such materials are thus known as non-Newtonian or rheologically complex systems. The non-Newtonian behaviour of the emulsions is classified as plastic, pseudoplastic, and dilatant flow. Plastic systems are also known as Bingham bodies. These systems exhibit a flow after a particular stress value called the yield value. Emulsions exhibiting pseudoplastic flow are known as shear thinning systems, that is, the apparent viscosity decreases as the shear rate increases (Santos et al. 2017). Another type of flow behaviour is known as the dilatancy. This behaviour is characterized by an increase in the viscosity with the shear rate; some examples include sand/water mixtures, candy compounds, clay slurries, corn starch in water, etc. Dilatancy is also referred to as shear thickening systems. The major flow behaviour exhibited by the multiphase systems is the pseudoplastic or plastic flow (Tao et al. 2020). The "power law" model can be used to calculate the viscosities of emulsions (Eq. 9.5):

$$\sigma = K\dot{\gamma}^{n} \tag{9.5}$$

where σ is shear stress, $\dot{\gamma}$ is shear rate, K is consistency index, and n is power law index (Chalapud et al. 2018). A straight line obtained from the log-log shear rate/shear-stress plot signifies the shearing behaviour of the system. The above equation gives an indication about the flow of the system. For Newtonian fluids "n" and "K" values are 1, for a pseudoplastic (shear thinning) system, $n < 1$, and for dilatant or shear thickening systems, $n > 1$. The rheological analysis of an emulsion is carried out using a rheometer (Anton Paar) equipped with a cone plate or a parallel plate geometry.

9.6.5 Determination of Emulsion Optical Properties

Transparency, opacity, turbidity, and colour are examples of optical properties of emulsions that are dependent on the amount of light that is absorbed and scattered by the sample (Rahn-Chique et al. 2012). The refractive index of an emulsion is determined using refractive indexometers (Moghaddam et al. 2023). Refractive index is defined as the product of the phase velocity of light in the evaluated medium and the speed of light in a vacuum. An emulsion is deemed transparent if its refractive index is either the same as or nearly the same as that of water, i.e. 1.333 (Kupikowska-Stobba et al. 2024). Since droplet size, distribution, and concentration have a significant impact on how an emulsion appears, tracking the optical characteristics of an emulsion over time might yield important insights into potential destabilizing processes in the system (Mirhosseini and Tan 2010). Any variation in the droplet size distribution during the emulsion's coalescence changes the overall appearance of an emulsions. The coalescence process causes larger droplets to develop, which makes the emulsion less clear, opaque, and turbid. Furthermore, the merging of the droplets may result in an increase in colour intensity. Floc formation or droplet aggregates, during flocculation, result in increased light scattering and decreased emulsion transparency. The emulsion appears hazy or unclear due to its higher turbidity, which is a result of the increase in light scattering. The colour intensity can also vary because the optical density or hue of flocs differs from that of the individual droplets. Similar changes are also observed in Ostwald ripening. Ostwald ripening causes changes in light scattering and absorption as well as a reduction in transparency because the average droplet size in the emulsion rises. All these optical changes can be detected quantitatively or visually by UV-vis spectrophotometers, colorimeters, and refractometers (Kupikowska-Stobba et al. 2024). Spectrophotometers detect the absorption and reflection of light by emulsions over a wide range of wavelengths. Colorimeters quantify parameters such as brightness and saturation to determine the appearance of emulsions.

9.6.6 Determination of Zeta Potential

When ionizable or charged emulsifier molecules are adsorbed to the surface of emulsion droplets, the droplets become electrically charged. An electric double layer is formed by two layers that encircle the droplets. The outer diffuse layer is made up of loosely connected ions, and the inner layer is composed of strongly bound ions. Ions inside the boundary follow the moving emulsion droplet, whereas those outside the boundary stay in the bulk dispersion. The shear plane is the interface between diffuse and stationary layers of counterions, and the zeta potential represents the potential of this interface (Kamble et al. 2022). The factors that contribute to the droplet charge include the type and quantity of molecules that have been adsorbed. The composition of the continuous phase and its pH, particularly the existence of other charged species like ions or macromolecules, also affect the charge distribution. The zeta potential is the most frequently examined parameter that describes the charge on the droplet surface. It is defined as the potential difference between the continuous phase and charged droplet surface. The zeta potential is an important indicator of emulsion stability. A stronger repulsive electrostatic interaction between the droplets hinders the droplet flocculation and coalescence. The zeta potential eventually improves the stability of the emulsion (Kasprzak et al. 2023). A stable emulsion is generally indicated by a high negative or positive zeta potential of around ±30 mV (Gurpreet and Singh 2018). The zeta potential of an unstable emulsions is around zero (Pinto and Buss 2020). It is important to remember that molecular mass distributions in nanoemulsions are also affected by the zeta potential; higher zeta potential nanoemulsions resist flocculation and coagulation and are electrically stabilized, exhibiting improved physical stability (Ravindran et al. 2018). Zeta potential of an emulsion is calculated by a Zetasizer (*Nano-ZSP,*

*Malvern Instruments, Worcestershire, UK). There is no direct method for the measurement of zeta potential. Theoretical models or experimental calculation of zeta potential is performed by measuring the electrophoretic mobility generated through electroacoustic or electrokinetic phenomena. The electrokinetic effects are observed due to electro-osmosis, electrophoresis, colloidal vibrations, and changes in sedimentation potential (*Kamble et al. 2022*).*

9.6.7 Determination of Electrical Conductivity

When describing emulsions, electrical conductivity is vital because it gives details about their composition, stability, and flow characteristics. The stability of emulsions can be altered over time by measuring electrical conductivity of an emulsion. Enhanced surfactant concentrations and charge densities generally result in a higher electrical conductivity in the emulsion (Chiesa et al. 2008). A conductivity metre or probe (Mettler Toledo SevenCompact™ Conductivity Meters) can be used to measure electrical conductivity by inserting it into the emulsion sample and measuring its capacity to conduct an electric current. Conductivity measurements are useful for understanding the mechanism of emulsion separation, predicting the long-term stability, and examining the interactions between surfactants (Kasprzak et al. 2023). Emulsions with higher electrical conductivity have been found to be more stable. This is because a barrier against droplet aggregation and flocculation is formed by the electric double layers that form around the droplets. Coalescence is essentially hindered by these multiple layers, which create repulsive forces between the droplets. Moreover, conductive substances like carbon nanotubes can be added to emulsions to increase their stability and electrical conductivity (Briggs et al. 2018).

9.7 Conclusion

In this chapter emulsions are described in detail. The formation of O/W and W/O emulsions is governed by adsorption, interfacial film, and oriented wedge theories. According to these proposed theories, interfacial interactions and wettability of the emulsifiers at the oil-water interfaces favour the formation of a stable emulsion. This chapter has also focused on the major types of advanced emulsions currently available, including macroemulsions, microemulsions, nanoemulsions, multiple emulsions, and pickering emulsions. Emulsions are unstable systems and are often difficult to prepare. Commercially feasible large-scale preparation techniques and materials are still needed to develop a stable emulsion. Micro- and nanoemulsions are superior to macroemulsions when it comes to delivering bioactive substances and drugs. They have higher retention efficiency, stability, better bioavailability, and the capacity to regulate or trigger the release of drugs. One of the frontier topics in structured fats and oils is double emulsions with a distinct structure, typically formed through multiple-interface stabilization. It can trap significant volumes of liquid oils inside a single internal space. However, the biological fate of these emulsions is currently not well understood, and further study is needed in this field. The phenomena of instability, effects of interfacial films on the long-term stability of emulsions, and evaluation using various techniques have been summarized. Various instruments used in the evaluation of emulsions have also been discussed in detail. In general, the chapter has the potential to shape forthcoming developments in the creation of innovative methods for producing sophisticated emulsions appropriate for use in pharmaceutical applications.

Acknowledgements The authors thank their respective universities for providing support in constructing the manuscript.

Conflicts of Interest The authors declare no conflict of interest.

Funding: Not applicable.

Author Contributions: B.B. contributed to the conceptualization, investigation, methodology, and writing and editing of the review; M.K.C. took part in the conceptualization, investigation, methodology, and writing and editing of the review, as well as supervision.

References

Nour AH (2018) Emulsion types, stability mechanisms and rheology: a review. Int J Innov Res Sci Stud 1(1):14–21

Alexandrov DV, Ivanov AA, Nizovtseva IG, Lippmann S, Alexandrova IV, Makoveeva EV (2022) Evolution of a polydisperse ensemble of spherical particles in a metastable medium with allowance for heat and mass exchange with the environment. Crystals 12(7):949

Aslam M, Deeb G, Ahmad MZ, Jain K, Ahmad J (2023) Functionalized nanoemulsions: could be a promising approach for theranostic applications. In: Jain K, Jain NK (eds) Multifunctional and targeted theranostic nanomedicines. Springer, Singapore. https://doi.org/10.1007/978-981-99-0538-6_7

Briggs N, Raman AKY, Barrett L, Brown C, Li B, Leavitt D et al (2018) Stable Pickering emulsions using multi-walled carbon nanotubes of varying wettability. Colloids Surf Physicochem Eng Asp 537:227–235

Chalapud MC, Baümler ER, Carelli AA (2018) Emulsions of sunflower wax in pectin aqueous solutions: physical characterization and stability. Food Res Int 108:216–225

Chen L, Ao F, Ge X, Shen W (2020a) Food-grade Pickering emulsions: preparation, stabilization and applications. Molecules 25(14):3202

Chen H, Wang D, Wang X et al (2020b) Triple phase inversion of emulsions stabilized by amphiphilic graphene oxide and cationic surfactants. ACS Omega 5(37):23524–23532. https://doi.org/10.1021/acsomega.0c01398

Chiesa M, Garg J, Kang YT, Chen G (2008) Thermal conductivity and viscosity of water-in-oil nanoemulsions. Colloids Surf Physicochem Eng Asp. 326(1):67–72

Cunha AG, Mougel J-B, Cathala B, Berglund LA, Capron I (2014) Preparation of double Pickering emulsions stabilized by chemically tailored nanocelluloses. Langmuir 30(31):9327–9335. https://doi.org/10.1021/la5017577

Dekker RI, Kibbelaar HV, Deblais A, Bonn D (2022) Rheology of emulsions with polymer solutions as the continuous phase. J Non-Newton Fluid Mech 310:104938

Delahaije RJBM, Gruppen H, van Nieuwenhuijzen NH et al (2013) Effect of glycation on the flocculation behavior of protein-stabilized oil-in-water emulsions. Langmuir 29(49):15201–15208. https://doi.org/10.1021/la403504f

Delahaije RJBM, Hilgers RJ, Wierenga PA, Gruppen H (2017) Relative contributions of charge and surface coverage on pH-induced flocculation of protein-stabilized emulsions. Colloids Surf Physicochem Eng Asp 521:153–160
Dickinson E (2019) Strategies to control and inhibit the flocculation of protein-stabilized oil-in-water emulsions. Food Hydrocoll 96:209–223
Friberg SE, Quencer LG, Hilton ML (1996) Theory of emulsions. In: Pharmaceutical dosage forms, 2nd edn. CRC Press
Fuhrmann PL, Breunig S, Sala G, Sagis L, Stieger M, Scholten E (2022) Rheological behaviour of attractive emulsions differing in droplet-droplet interaction strength. J Colloid Interface Sci 607:389–400
Glatter O, Salentinig S (2020) Inverting structures: from micelles via emulsions to internally self-assembled water and oil continuous nanocarriers. Curr Opin Colloid Interface Sci 49:82–93
Goodarzi F, Zendehboudi S (2019) A comprehensive review on emulsions and emulsion stability in chemical and energy industries. Can J Chem Eng 97(1):281–309
Gowda D, Sarathchandraprakash NK, Mahendra C et al (2013) Emulsions and emulsifiers—a review. Asian J Exp Chem 8:30–45
Gurpreet K, Singh SK (2018) Review of nanoemulsion formulation and characterization techniques. Indian J Pharm Sci 80(5):781
Hejazifar M, Lanaridi O, Bica-Schröder K (2020) Ionic liquid based microemulsions: a review. J Mol Liq 303:112264
Ho TM, Razzaghi A, Ramachandran A et al (2024) Emulsion characterization via microfluidic devices: a review on interfacial tension and stability to coalescence. Adv Colloid Interf Sci 299:102541
Hong IK, Kim SI, Lee SB (2018) Effects of HLB value on oil-in-water emulsions: droplet size, rheological behavior, zeta-potential, and creaming index. J Ind Eng Chem 67:123–131
Hosseini A, Jafari SM, Mirzaei H, Asghari A, Akhavan S (2015) Application of image processing to assess emulsion stability and emulsification properties of Arabic gum. Carbohydr Polym 126:1–8
Hu YT, Ting Y, Hu JY, Hsieh SC (2017) Techniques and methods to study functional characteristics of emulsion systems. J Food Drug Anal 25(1):16–26
Huck-Iriart C, Montes-de-Oca-Ávalos J, Herrera ML, Candal RJ, Pinto-de-Oliveira CL, Linares-Torriani I (2016) New insights about flocculation process in sodium caseinate-stabilized emulsions. Food Res Int 89:338–346
Jena AK, Nayak AK, De A, Mitra D, Samanta A (2018) Development of lamivudine containing multiple emulsions stabilized by gum odina. Future J Pharm Sci 4(1):71–79
Jintapattanakit A (2018) Preparation of nanoemulsions by phase inversion temperature (PIT) method. Pharm Sci Asia 45(1):1–12
Juneja M, Suthar T, Pardhi VP, Ahmad J, Jain K (2022) Emerging trends and promises of nanoemulsions in therapeutics of infectious diseases. Nanomedicine (Lond) 17(11):793–812. https://doi.org/10.2217/nnm-2022-0006. Epub 2022 May 19. PMID: 35587031
Kale SN, Deore SL (2017) Emulsion micro emulsion and nano emulsion: a review. Syst Rev Pharm 8:39
Kamble S, Agrawal S, Cherumukkil S, Sharma V, Jasra RV, Munshi P (2022) Revisiting zeta potential, the key feature of interfacial phenomena, with applications and recent advancements. ChemistrySelect 7(1):e202103084
Kasprzak MM, Berski W, Krystyjan M et al (2023) Effects of fibre addition and processing on the stability, rheology and in vitro gastric digestion of whey protein-xanthan gum stabilised emulsions with high oil phase. LWT 178:114465
Khedr A, Striolo A (2019) Quantification of ostwald ripening in emulsions via coarse-grained simulations. J Chem Theory Comput 15(9):5058–5068. https://doi.org/10.1021/acs.jctc.9b00296
Kim H, Park D, Jiang Z, Wei Y, Woong KJ (2022) Microfluidic macroemulsion stabilization through *in situ* interfacial coacervation of associative nanoplatelets and polyelectrolytes. J Colloid Interface Sci 614:574–582

Koroleva MY, Yurtov EV (2021) Ostwald ripening in macro- and nanoemulsions. Russ Chem Rev 90:293. https://doi.org/10.1070/RCR4962
Kumar A, Li S, Cheng C-M, Lee D (2015) Recent developments in phase inversion emulsification. Ind Eng Chem Res 54(34):8375–8396. https://doi.org/10.1021/acs.iecr.5b01122
Kumar D, Yadav S, Ujjwal RR, Jain K (2021) "Chapter 6: Interfacial characterization of nanoemulsion" in International Book "Nanoemulsions in food technology: development, characterization and applications" CRC Press, USA. ISBN: 9780367614928
Kupikowska-Stobba B, Domagała J, Kasprzak MM (2024) Critical review of techniques for food emulsion characterization. Appl Sci 14(3):1069. https://doi.org/10.3390/app14031069
Langevin D (2019) Coalescence in foams and emulsions: similarities and differences. Curr Opin Colloid Interface Sci 44:23–31
Lipei F, Ma Q, Liao K, An J, Bai J, He Y (2021) Application of Pickering emulsion in oil drilling and production. Nanotechnol Rev 11(1). https://doi.org/10.1515/ntrev-2022-0003/html
Liu H, Shi J, Pu Y, Wang JX, Wang D, Chen JF (2020) In situ visualization and real-time tracking of emulsion and miniemulsion polymerization at the microscale via fluorescence imaging. Chem Eng Sci 211:115288
Llamas S, Santini E, Liggieri L, Salerni F, Orsi D, Cristofolini L et al (2018) Adsorption of sodium dodecyl sulfate at water–dodecane interface in relation to the oil in water emulsion properties. Langmuir 34(21):5978–5989
Lu T, Yu F, Xue C, Han B (2021) Identification, classification, and quantification of three physical mechanisms in oil-in-water emulsions using AlexNet with transfer learning. J Food Eng 288:110220
McClements DJ, Jafari SM (2018) Improving emulsion formation, stability and performance using mixed emulsifiers: a review. Adv Colloid Interf Sci 251:55–79
Mirhosseini H, Tan CP (2010) Discrimination of orange beverage emulsions with different formulations using multivariate analysis. J Sci Food Agric 90(8):1308–1316. https://doi.org/10.1002/jsfa.3928
Moghaddam MK, Moezzi M, Dannandeh F (2023) Oil–water separation in kitchen wastewater filtration using milkweed floss fiber. Biomass Conv Bioref. https://doi.org/10.1007/s13399-023-05190-w
Narayan S, Metaxas AE, Bachnak R, Neumiller T, Dutcher CS (2020) Zooming in on the role of surfactants in droplet coalescence at the macroscale and microscale. Curr Opin Colloid Interface Sci 50:101385
Nikmaram N, Roohinejad S, Hashemi S, Koubaa M et al (2017) Emulsion-based systems for fabrication of electrospun nanofibers: food, pharmaceutical and biomedical applications. RSC Adv 7(46):28951–28964
Pal N, Kumar S, Bera A, Mandal A (2019) Phase behaviour and characterization of microemulsion stabilized by a novel synthesized surfactant: implications for enhanced oil recovery. Fuel 235:995–1009
Pardhi VP, Suthar T, Sharma A, Jain K (2022) Bedaquiline fumarate microemulsion: formulation optimization, rheological characterization and in vitro studies. Nanomedicine (Lond) 17(21):1529–1546. https://doi.org/10.2217/nnm-2022-0132. Epub 2022 Nov 23. PMID: 36416115
Park SH, Hong CR, Choi SJ (2020) Prevention of Ostwald ripening in orange oil emulsions: impact of surfactant type and Ostwald ripening inhibitor type. LWT 134:110180
Pinto I, Buss A (2020) ζ potential as a measure of asphalt emulsion stability. Energy Fuel 34(2):2143–2151
Prankerd RJ, Stella VJ (1990) The use of oil-in-water emulsions as a vehicle for parenteral drug administration. PDA J Pharm Sci Technol 44(3):139–149
Pratishtha, Gupta MK, Gupta S (2021) Multiple emulsions: emphasizing on industrial applications. In: Patel JK, Pathak YV (eds) Emerging technologies for nanoparticle manufacturing. Springer International Publishing, Cham, pp 333–347. https://doi.org/10.1007/978-3-030-50703-9_14
Rahn-Chique K, Puertas AM, Romero-Cano MS, Rojas C, Urbina-Villalba G (2012) Nanoemulsion stability: experimental evaluation of the flocculation rate from turbidity measurements. Adv Colloid Interf Sci 178:1–20

Ravera F, Dziza K, Santini E, Cristofolini L, Liggieri L (2021) Emulsification and emulsion stability: the role of the interfacial properties. Adv Colloid Interf Sci 288:102344
Ravindran S, Williams MAK, Ward RL, Gillies G (2018) Understanding how the properties of whey protein stabilized emulsions depend on pH, ionic strength and calcium concentration, by mapping environmental conditions to zeta potential. Food Hydrocoll 79:572–578
Roberts C (2002) A new theory of emulsions. J Phys Chem 36(12):3087–3107
Santos ICVM, Oliveira PF, Mansur CRE (2017) FACTORS THAT AFFECT CRUDE OIL VISCOSITY AND TECHNIQUES TO REDUCE IT: A REVIEW. Braz J Pet Gas 11(2):115. https://doi.org/10.5419/bjpg2017-0010
Saygı Arslan Ö, Sudol ED, Daniels ES, El-Aasser MS, Klein A (2009) In situ surfactant generation as a means of miniemulsification? J Appl Polym Sci 111(2):735–745
Shinoda R, Uchimura T (2018) Evaluating the creaming of an emulsion via mass spectrometry and UV–Vis spectrophotometry. ACS Omega 3(10):13752–13756. https://doi.org/10.1021/acsomega.8b02283
Shiraki K, Mimura M, Nishinami S, Ura T (2020) Effect of additives on liquid droplets and aggregates of proteins. Biophys Rev 12(2):587–592
Shukla T, Upmanyu N, Agrawal M, Saraf S, Saraf S, Alexander A (2018) Biomedical applications of microemulsion through dermal and transdermal route. Biomed Pharmacother 108:1477–1494
Silva CAM, Saraiva SV, Bonetti D, Higuti RT, Cunha RL, Pereira LOV et al (2022) Measurements of bimodal droplet size distribution of emulsions using ultrasonic spectroscopy in the long and intermediate wavelength regimes. Chem Eng Sci 252:117274
Singh Y, Meher JG, Raval K, Khan FA, Chaurasia M, Jain NK et al (2017) Nanoemulsion: concepts, development and applications in drug delivery. J Control Release 252:28–49
Suthar T, Jain VK, Popli H, Jain K (2022) "Chapter 12: Nanoemulsions as effective carriers for targeting brain tumors" in International Book "Nanocarriers for drug-targeting to brain tumors" Elsevier 1–17. ISBN: 9780323907736; https://doi.org/10.1016/B978-0-323-90773-6.00008-7
Suthar T, Patel P, Singh P, Datusalia AK, Yadav AK, Jain K et al (2023) Hesperidin microemulsion: formulation optimization, characterization, and in vitro evaluation. J Drug Deliv Technol 80(104166):1–12. https://doi.org/10.1016/j.jddst.2023.104166
Tao C, Kutchko BG, Rosenbaum E, Massoudi M (2020) A review of rheological modeling of cement slurry in oil well applications. Energies 13(3):570
Tian Y, Yuan C, Cui B, Lu L, Zhao M, Liu P et al (2022) Pickering emulsions stabilized by β-cyclodextrin and cinnamaldehyde essential oil/β-cyclodextrin composite: a comparison study. Food Chem 377:131995
Umar AA, Saaid IBM, Sulaimon AA, Pilus RBM (2018) A review of petroleum emulsions and recent progress on water-in-crude oil emulsions stabilized by natural surfactants and solids. J Pet Sci Eng 165:673–690
Vv H, Vishal P, Borawake D, Hs N (2018) Nanoemulsion: a novel platform for drug delivery system. J Mater Sci Nanotechnol 6:1–11
Wang XY, Wang J, Rousseau D, Tang CH (2020) Chitosan-stabilized emulsion gels *via* pH-induced droplet flocculation. Food Hydrocoll 105:105811
Wang M, Zhang J, Fan L, Li J (2023) Fabrication of fat-reduced water-in-oil emulsion and the application in 3D printing. Food Res Int 172:113118
Yang Y, Fang Z, Chen X, Zhang W, Xie Y, Chen Y et al (2017) An overview of Pickering emulsions: solid-particle materials, classification, morphology, and applications. Front Pharmacol 8:287. https://doi.org/10.3389/fphar.2017.00287
Zhang X, Xing H, Zhao Y, Ma Z (2018) Pharmaceutical dispersion techniques for dissolution and bioavailability enhancement of poorly water-soluble drugs. Pharmaceutics 10(3):74. https://doi.org/10.3390/pharmaceutics10030074
Zhang Y, Xie Y, Liu H, McClements DJ, Cheng C, Zou L et al (2022) Probiotic encapsulation in water-in-oil high internal phase emulsions: enhancement of viability under food and gastrointestinal conditions. LWT 163:113499
Zwicker D, Hyman AA, Jülicher F (2015) Suppression of Ostwald ripening in active emulsions. Phys Rev E 92:012317. https://doi.org/10.1103/PhysRevE.92.012317

Sterile Products and Admixtures

10

Satish Shilpi, Amit K. Dubey, Shalini Bajaj, Shashank Shekar Mishra, Chetan Ram, Debasis Gantayat, Vibhooti Pippola, and Khyati Saini

Abstract

The aseptically prepared large volume parenteral containing important necessary nutritional components for beneficial effect such as vitamins, glucose, amino acids, fats, etc. with pharmaceutical active drug is called parenteral admixture. It is manufactured directly by pharmaceutical companies after exhaust analysis of potential chemical or pharmaceutical interactions within the admixture and drugs. Before that nurses and physicians manufacture the admixtures in the hospital for the particular patient during their hospitalization using individually packed sterile components to maintain aseptic condition. The purpose of parenteral admixture is to enhance the therapeutic effectiveness as well as the stability of the product, reduce the side effect of the drug, and maintain the nutritional and electrolyte balance in the body. Parenteral admixture-utilizing countries regulate the manufacturing practice to assure the quality, stability, and microbiological integrity of the parenteral admixture. In the chapter, brief discussion of additives, their application, manufacturing, and regulatory aspects has been explored.

Keywords

Admixtures · Parenterals · Therapeutic benefits · Sterile · Nutritional value

S. Shilpi (✉) · A. K. Dubey · S. S. Mishra · C. Ram · D. Gantayat · V. Pippola · K. Saini
School of Pharmaceuticals and Population Health Informatics, Faculty of Pharmacy, DIT University, Dehradun, India

S. Bajaj
School of Pharmaceuticals Sciences, Jaipur National University, Jaipur, Rajasthan, India

K. Jain, A. K. Yadav (eds.), *Advances in Pharmaceutical Product Development*,
https://doi.org/10.1007/978-981-97-9230-6_10

10.1 Introduction

10.1.1 Additives in Parenteral Products

Parenteral admixtures refer to mixtures of medications or solutions that are prepared for administration through routes other than the digestive tract, typically by injection. This can include intravenous (IV), intramuscular (IM), subcutaneous (SC), or other routes. These admixtures are carefully prepared to ensure the proper dosing, compatibility of ingredients, sterility, and safety for the patient. Drug delivery methods designed for injection or infusion, known as parenteral formulations, have advanced significantly in the last several years (Ezike et al. 2023). When precision doses, fast onset, and high bioavailability are needed, parenteral formulations are essential for delivering drugs. Parenteral products are a class of pharmaceuticals designed to be administered directly into the body, bypassing the digestive system (Manchanda et al. 2020; Nikam Nikita et al. 2023). These products, including injections, infusions, and implants, are formulated to deliver medications, vaccines, or nutrients directly into the bloodstream, tissues, or organs (Nema et al. 2007). By circumventing the digestive process, parenteral products can provide rapid and targeted relief, making them ideal for treating emergencies, chronic conditions, and diseases that require immediate intervention. The drug then moves from the place of administration to the site of action. To deliver drugs that are ineffective when taken orally, that are destroyed in the gastrointestinal tract, or that are utilized to treat patients who are uncooperative, nauseous, or unconscious, parenteral administration provides an instant physiological reaction (Kapp 2018). Parenteral administration can have localized effects, offer a longer-lasting pharmacological effect, and ultimately restore significant imbalances in fluid and electrolyte levels (Ezike et al. 2023). However, parenteral dose forms involve aseptic procedures, require more time than other routes, must be administered by trained personnel, and typically cause pain at the injection site.

Parenteral products are an essential part of contemporary healthcare since they can also provide better bioavailability, fewer side effects, and increased patient compliance. Pharmaceutical excipients, often known as additives, are substances that are added to completed pharmaceutical products to fulfill specific purposes (Pramanick et al. 2013). In the large volume parenterals, some substances with nutritional value or other beneficial effect are added which is called parenteral admixtures. The substances may be vitamins, glucose, amino acids, and fats which are added as an admixture in the parenteral formulation. They are included to increase bulk, facilitate production, enhance stability, improve drug targeting and administration, and alter the pharmacokinetic or safety profile of the drug (Nema et al. 2007; Rayaprolu et al. 2018). To distinguish them from active pharmaceutical compounds, excipients are generally referred to as inactive or inert ingredients. Some countries have imposed limitations on the kind or quantity of excipient that can be used in the formulation of parenteral medicinal products because of safety concerns (Pramanick et al. 2013). For instance, even though these additives are present in products in other countries, the use of amino mercuric chloride or thimerosal is restricted in Japan, the United States, and the European Union (Nema and Brendel 2016).

The aim of the addition of additives to parenteral products is:

- To enhance or uphold the stability of the product
- To increase or maintain the solubility of active ingredients
- To ensure safety and sterility
- To reduce pain and irritation during the injection
- To control or extend the release of drugs
- To stabilize the therapeutic molecules
- To synergize the therapeutic effects of drugs

The composition of any parenteral dosage form includes drugs, additives, vehicles, and containers (Ahmed et al. 2018). The ingredient in the product that gives it its pharmacological effect is the active drug. Parenteral products are designed to be reconstituted as solutions. These formulations include liposomes, suspensions, emulsions, powders, and nanosystems. These are frequently utilized in freeze-dried products and parenteral formulations that concentrate a solution. Developing a reliable and safe parenteral dosage form requires extensive investigation of the active drug's or drugs' physical and chemical characteristics. Parenteral additives are compounds that are added to parenteral solutions to stabilize parenteral products, control their properties, maintain sterility, or facilitate delivery by lessening tissue irritation or injection discomfort (Ludwig 2019). To prevent interfering with the therapeutic efficacy of the drugs and assays or tests, as well as to ensure that the concentrations of these additives are safe and nontoxic, they should be used in the smallest number and quantities possible, especially in LVP (large volume parenteral) (Dongare et al. 2015). Furthermore, various additives used in parenteral products are summarized here under the scope of additives in parenteral products. Parenteral products have a wide range of additives that are specifically intended for specific formulation criteria. This ensures the safety, efficacy, stability, and patient acceptance of these vital treatments. Throughout every phase of their administration and shelf life, each additive is essential to maintaining the effectiveness and quality of parenteral formulations (Table 10.1).

Table 10.1 Role of admixture in parenterals

Benefit	Description
Nutritional support	Provides essential nutrients to patients unable to consume sufficient nutrients orally
Remediation of deficiencies	Addresses specific deficiencies in vitamins, minerals, and other nutrients
Therapeutic benefits	Antioxidants reduce oxidative stress; amino acids aid in tissue repair and immune function
Stabilization of solutions	Electrolytes maintain osmolarity and pH balance, ensuring solution stability

10.1.2 Admixture in Parenteral: A Stability Monitor

Regardless of the nation, intravenous admixture compounding is standard procedure in the most hospitals worldwide. Working in an aseptic setting presents challenges for the compounder, and compounding intravenous drugs carries a risk due to the significant potential for error resulting from their complexity (Allen 2021).

The effects of six distinct independent parameters on the stability of intravenous feeding were examined. Prentral admixtures were prepared with the following components so they could be given intravenously: amino acids, should be 2–7%; hydrated glucose, up to 20%; lipid emulsion, 5%; monovalent cations, starting from 0 to 150 meq/L; divalent cations, in the range of 4–20 meq/L; and the last, but not least, trivalent cations, ranging from 0 to 10 mg of elemental iron/L. Stability assessments included eye inspection, pH monitoring, and particle-size analysis (Driscoll et al. 1995; Boullata et al. 2022).

Parenteral nutrition (PN) is a high-alert medication that can be utilized to treat patients during a challenging clinical procedure. Once the specific clinical guidelines were produced by consensus, the Board of Directors of the American Society for Parenteral and Enteral Nutrition (ASPEN) evaluated and approved them. The following questions were addressed: Does better PN ordering result from education for prescribers? When administering PN admixtures via the peripheral vein, what is the maximum safe osmolarity? What are the appropriate calcium intake and calcium-phosphate ratios in PN for optimal infant bone mineralization? What are the therapeutic advantages and disadvantages of using commercially available prefabricated ("premixed") multichambered PN formulations over traditional or customized PN formulations (Otero-Millán et al. 2024a, b; Bakrey et al. 2024)? Other questions to think about are the following: Which recommendations work best to maximize the compatibility of phosphate (Na^+ or K^+) and calcium (gluconate) in PN admixtures? What kind of parenteral stock solutions that are used to make PN admixtures now have micronutrient contamination? Is it okay to administer non-nutrient drugs using the PN admixture as a delivery system? Is heparin added to the PN admixture required to reduce the incidence of Cerebral venous thrombosis (CVT)? Which methods of repackaging intravenous fat emulsion (IVFE) into customized, smaller volumes are safe? Which expiration date is applicable to IVFE that has been repackaged and (a) is administered in its original container for a separate infusion (Boullata et al. 2014)?

10.1.3 Rates and Intensity of Severity of Intravenous Admixture in Healthcare

When it comes to the installation of laryngeal mask airway (LMA), a drug propofol (an anesthetic and a sedative) was contrasted with a solution of sodium pentothal thiopental. A randomization process was used to assign patients, ages 18–65, to receive 7.5 mg of oral midazolam as a premedication. After then, they received either 1% propofol or a 0.5% or 1.2% admixture. The doses given were at a rate of 0.25 mL/kg of body weight. The frequency and severity of gagging (the body's natural response to keep someone from choking or eating anything disagreeable), coughing, and insufficient jaw relaxation were satisfactory, and no significant difference was found between pre- and post-therapy. Laryngospasm, a transient spasm of the vocal cords that impairs speech or breathing, was one of the other conditions that did not exhibit significant differences (Yeo et al. 2001). In another case related to admixtures in parenterals, according to recent statistics, total nutritional admixtures (TNAs) might be deemed pharmaceutically inappropriate for human administration. It was estimated that the percentage of fat (PFAT) globules with a diameter greater than 5 μm accounts for more than 0.4% of the total fat content. TNAs are unstable in this situation (Driscoll et al. 2000).

There were few proofs that the intravenous admixture preparation method and location affect the number of errors that occur. Specifically, it seems that intravenous preparation in central pharmacy settings compared to nursing wards seems to pose fewer risks. Indeed, automated versus manual preparation results in lower error rates. An observational study was done by Dehmel and colleagues in which errors rate was observed in the parentral admixture preparation. In case of automated preparation which was prepare at central phamacy, the erors rate was 16% while 53% errors rate was observed in manual preparation which was prepaed in nursing unit. They found a noticeably higher proportion of incorrect concentration mistakes (53% vs. 16%, respectively) (Dehmel et al. 2011). In another study, Khalili and colleagues found no cases of contamination in admixtures made in central pharmacies. They found a low rate of bacterial contamination (1.1%) in admixtures manufactured on nursing wards, despite the use of manual preparation processes in each setting (Khalili et al. 2013).

Intravenous admixture preparation errors (IAPEs) might cause harm to the sufferer. To inspire preventive actions, further research is required to determine the prevalence and contributing factors of various IAPEs. Any mistake made during the IV admixture preparation process, such as a departure from the medication order, the local electronic admixture preparation instructions, or the manufacturer's drug information page, was classified as an IAPE. A study conducted by Jessurun and co-workers led to the conclusion that IAPE occurs rarely. The factors associated with IAPEs illustrate the components linked to working conditions and preparation difficulties. The solutions aimed at reducing the incidence of IAPEs and, in turn, patient harm should center on the identified causes (Jessurun et al. 2022).

The information that is now available points to the necessity for standardization, preventive strategic formulation, and implementation. The other factors to be considered may be staff training on ideal admixture protocols and ongoing optimization of the intravenous preparation process. Future studies should concentrate on creating uniform definitions for error subtypes and standardizing reporting procedures. Repeatable techniques and surveillance are needed to monitor risk factors and connect them to the damage these errors cause (Hedlund et al. 2017).

This research aimed to establish a reliable and efficient method for measuring ampicillin concentrations in total parenteral nutrition (TPN) admixtures using reverse-phase high-performance liquid chromatography with diode-array detection (RP-HPLC-DAD). They developed and validated an RP-HPLC-DAD method, evaluating its performance through various parameters such as accuracy, precision, specificity, linearity, and robustness. Additionally, we assessed the stability of ampicillin in TPN admixtures under different storage conditions. The results demonstrated the method's reliability and accuracy in quantifying ampicillin, ensuring the safe and effective administration of this antibiotic in clinical settings (Stawny et al. 2019).

The investigators evaluated the physical and chemical stability of ibuprofen when mixed with various lipid emulsions and PN solutions, assessing factors like particle size, pH, and concentration. Their findings indicated that ibuprofen was compatible with certain lipid emulsions and PN solutions, but incompatibilities were observed with others, leading to precipitation or phase separation. These results provide valuable guidance for healthcare professionals to ensure the safe and effective administration of intravenous ibuprofen in combination with lipids and PN solutions, avoiding potential complications and improving patient care (Garcia et al. 2018).

Physical and chemical stability of ciprofloxacin was evaluated in combination with various parentral solutions, assessing factors like precipitation, crystallization, and concentration. Findings revealed that ciprofloxacin compatibility with PN admixtures was dependent on factors like pH, temperature, and nutrient composition. Specifically, they identified potential incompatibilities and instability when ciprofloxacin was mixed with certain PN solutions, which could lead to reduced efficacy or adverse effects. These results provide essential guidance for healthcare professionals to ensure the safe and effective co-administration of ciprofloxacin and PN admixtures, optimizing patient care and outcomes (Gostyńska et al. 2019).

They focused on ensuring the safe co-administration of loop diuretics with parenteral nutrition (PN) solutions and conducted in vitro compatibility tests to investigate potential interactions between these medications and PN admixtures. They evaluated the physical and chemical stability of loop diuretics (such as furosemide and bumetanide) when mixed with various PN solutions, assessing factors like precipitation, crystallization, and concentration changes. The results revealed potential incompatibilities and instability when certain loop diuretics were combined with specific PN solutions, which could impact drug efficacy and patient safety. This provides valuable insights for healthcare professionals, informing the development of evidence-based guidelines for the safe co-administration of loop diuretics and PN solutions in clinical practice (Tomczak et al. 2020).

This investigation highlighted a critical safety concern regarding the co-administration of sodium valproate, an anticonvulsant medication, with parenteral nutrition (PN) admixtures. Researchers conducted in vitro tests to evaluate the compatibility of sodium valproate with PN solutions, revealing significant incompatibilities that could compromise patient safety. Specifically, they found that sodium valproate precipitated out of solution when mixed with certain PN admixtures, leading to reduced drug concentration and potential treatment failure. Moreover, the precipitate formed could potentially cause infusion-related reactions or other adverse effects. This study's findings underscore the importance of careful evaluation of drug-PN compatibility to ensure safe and effective treatment and highlight the need for alternative administration strategies to avoid potential harm to patients (Piwowarczyk et al. 2022; Tomczak et al. 2023).

The antiemetic medications, such as ondansetron and metoclopramide were mixed with parentral solutions for pediatric use and further studied for its physio-chemical stability. Their findings revealed potential incompatibilities and instability, including precipitation and crystallization, which could impact drug efficacy and patient safety. The study identified specific compatibility issues and provided recommendations for alternative administration strategies, such as separate infusion lines or adjusted drug concentrations. These results contribute to the development of evidence-based guidelines for the safe co-administration of antiemetic drugs and PN solutions in pediatric care, minimizing potential risks and improving treatment outcomes (Tomczak et al. 2023; Riera et al. 2018).

This study examined the physical and chemical stability, as well as sterility, of standard parenteral nutrition (PN) solutions and simulated Y-site admixtures specifically designed for neonatal care. The effects of mixing PN solutions with other medications and nutrients on the stability and sterility of the admixtures showed that certain combinations and concentrations led to precipitation, crystallization, and microbial growth, compromising the safety and efficacy of the PN solutions. The study identified optimal conditions and guidelines for preparing and administering PN solutions and Y-site admixtures in neonatal care, ensuring the delivery of stable and sterile nutrition to vulnerable newborns (Riera et al. 2018).

The compatibility of various calcium and phosphate concentrations in different solution types, considering factors like pH, temperature, and nutrient composition, established safe limits for calcium and phosphate content in neonatal parenteral solutions, ensuring that these critical nutrients can be delivered effectively and safely to newborns without compromising solution stability or patient well-being and providing valuable guidance for healthcare professionals along with nutritionists to optimize parenteral nutrition regimens for neonates (Watrobska-Swietlikowska 2019).

10.2 Additives in Large Volume Parenteral

10.2.1 Antimicrobial Preservatives

To prevent microbial growth in multiple-dose as well as single-dose parenteral products, an appropriate antimicrobial agent is added (Table 10.2). The selection of antimicrobial agents is very important; it basically depends on the preservatives, function, and effect of the active pharmaceutical ingredient of the parenteral product. Phenol and benzyl alcohol are the two most common antimicrobial preservatives used in peptide and protein products, while phenoxyethanol is the most frequently used preservative in vaccines. Benzyl alcohol or a combination of methylparaben and propylparaben is generally found in small-molecule parenteral formulations (Meyer et al. 2007; Pramanick et al. 2013).

Table 10.2 Concentration range and antimicrobial activity of preservatives commonly used in pharmaceutical products

S. no.	Additives	Concentration range (%)	Antimicrobial activity (minimum inhibitory concentration (MIC), mg/mL) Gram-positive bacteria *Staphylococcus aureus* (*S. aureus*)	Gram-negative bacteria *Escherichia coli* (*E. coli*)/ *Pseudomonas aeruginosa* *P. aeruginosa*)
1.	Benzyl alcohol	0.75–5%	25	2000/2000
2.	Chlorobutanol	0.25–0.5%	650[a]	1000[a]
3.	*m*-Cresol	0.1–0.315%	Not specified	Not specified
4.	Methylparaben	0.05–0.18%	2000	1000/4000
5.	Phenol	0.15–0.5%	Not specified	Not specified
6.	2-Phenoxyethanol	0.50%	8500	3600/3200
7.	Propylparaben	0.005–0.1%	500	(100–500)/(>1000)
8.	Thimerosal	0.003–0.012%	0.2	4/8
9.	Benzalkonium chloride	0.02% w/v	Not specified	Not specified
10.	Benzethonium chloride	0.01%	Not specified	Not specified
11.	Myristyl-gamma-picolinium chloride	0.0195–0.169% w/v	Not specified	Not specified
12.	Phenylmercuric nitrate	0.001%	Not specified	Not specified

[a] MIC values are not specific for *S. aureus*, *E. coli*, and *P. aeruginosa*

10.2.2 Antioxidants

The antioxidants are added in parenteral product to prevent the oxidation of API and other additives which enhance the shelf life of parenteral product. Antioxidants are classified into four classes such reducing agent, blocking agent, synergistic agent, and chelating agent (Table 10.3).

The most commonly used antioxidants in the sterile formulations are ascorbic acid, acetylcysteine, sulfurous acid salts (bisulfite, metabisulfite), monothioglycerol, acetone sodium bisulfite (0.2–0.4% w/v), argon (100%), ascorbyl palmitate, and sodium ascorbate.

10.2.3 Buffers

The solubility and stability of the substance of many drugs depend on the pH of the solution. The pH of parenteral preparation may be changed during storage due to leaching the content of rubber or plastic closure in the prepration, entrapment of gases and vapors, chemical reaction with glass container, and reaction within formulation. Buffers are added to a formulation to adjust and stabilize pH and optimize drug solubility and stability; for parenteral preparations, it is desirable that the

Table 10.3 Antioxidants in parenterals

S. no.	Antioxidant	Concentration range %
	Reducing agents	
1.	Ascorbic acid	0.02–0.1
2.	Sodium bisulfite	0.1–0.15
3.	Sodium metabisulfite	0.1–0.15
4.	Sodium formaldehyde sulfoxide	0.1–0.15
5.	Thiourea	0.005
	Blocking agents	
1.	Ascorbic acid esters	0.01–0.015
2.	Butylated hydroxytoluene (BHT)	0.005–0.02
3.	Tocopherols	0.05–0.075
	Synergist agents	
1.	Ascorbic acid	0.01–0.05
2.	Citric acid	0.005–0.01
3.	Citraconic acid	0.03–0.45
4.	Phosphoric acid	0.005–0.01
5..	Tartaric acid	0.01–0.02
	Chelating agent	
1.	Ethylenediaminetetraacetic acid salts	0.01–0.075

product pH be close to physiologic pH. Injectable products should have a pH value between 3.0 and 9.0 prior to administration. Buffer systems for parenteral consist of either a weak base and the salt of a weak base or a weak base or a weak acid and the salt of weak acid.

Sterile solutions administered intravenously in 100 mL or more quantities are known as large volume parenterals or LVPs. Typically, these solutions provide nutrients, drugs, electrolytes, and fluids to the circulation while ensuring complete rapid absorption (Hoyois et al. 2021). In many beneficial scenarios, LVPs are crucial, especially for patients who cannot swallow tablets or other nutrition (Hu et al. 2020).

LVPs are vital in clinical practice for rehydration therapy, maintaining fluid balance during surgeries, and supporting patients in intensive care who cannot eat or drink. They are also crucial in managing chronic conditions like kidney disease (Preiser et al. 2021). A key aspect of LVPs is the inclusion of additives such as vitamins, electrolytes (sodium, potassium, calcium, magnesium), trace elements, amino acids, and medications. These additives address deficiencies, enhance recovery, and support overall health by maintaining cellular function, muscle contraction, and nerve transmission.

Vitamins such as B complex, vitamin C, and vitamin D are added to LVPs to support metabolism, collagen synthesis, immune function, and calcium absorption (Anon n.d.-b). These are crucial for patients with severe deficiencies or those on long-term parenteral nutrition (Pecora et al. 2020). Trace elements like zinc, copper, manganese, and selenium support enzymatic functions, antioxidant defenses, wound healing, and immune health (Berger et al. 2021). Amino acids and proteins in parenteral nutrition solutions aid tissue repair and growth, especially in critically ill patients. Medications can be added to LVPs for rapid and precise drug delivery, essential for pain management, chemotherapy, and antibiotic therapy (Eggersdorfer et al. 2022).

10.2.4 Vitamins

Vitamins are vital for several physiological activities. They are included in large volume parenterals (LVPs) to make sure patients receive the nutrients they require, particularly in cases when oral consumption is not practical (Tubic-Grozdanis and Krämer 2015). For severely sick patients or those undergoing long-term therapy, adding vitamins to LVPs can help avoid deficits and maintain general health. Intravenous nutrition replacement therapy is a crucial intervention for patients whose diet is inadequate in providing essential nutrients since it aids in both maintaining health and rehabilitation (Tubic-Grozdanis and Krämer 2023).

10.2.4.1 Vitamin B Complex

Water-soluble vitamins, which are crucial to cell metabolism and energy generation, make up the vitamin B complex. B1 (thiamine), B2 (riboflavin), B3 (niacin), B5 (pantothenic acid), B6 (pyridoxine), B7 (biotin), B9 (folic acid), and B12 (cobalamin) are the elements that make up this complex (Serventi et al. 2023). Every vitamin in the complex performs a different role, from helping the human body convert food into energy to encouraging the production of red blood cells and cognitive function. Vitamin B complex supplementation is especially essential to patients receiving LVPs who possess greater metabolic needs, such as those recuperating from surgery or long-term medical conditions (Wijaya et al. 2015). For example, folic acid and thiamine are essential in avoiding deficits that may cause serious neurological disorders (Calder et al. 2022). Patients who are unable to consume enough nutrients orally can have their nutritional needs met by receiving a sufficient amount of these vitamins through LVPs, which promotes their overall health and faster recovery.

10.2.4.2 Vitamin C

Ascorbic acid, commonly referred to as vitamin C, is a strong antioxidant which promotes collagen formation, boosts the immune system, and improves the absorption of iron from food (Johnson and Probert 2014). It is essential for the healing of wounds and the defense of cells against oxidative damage. Vitamin C is routinely included in large volume parenterals (LVPs) to assist patients who are under intensive medical treatment or suffering from severe infections or elevated oxidative stress (Carr et al. 2017; Lee et al. 2023). Vitamin C's antioxidant properties aid in the neutralization of free radicals, minimizing cellular damage and improving recovery. This makes it especially advantageous for preserving general health while assisting the recuperation of patients who are in critical condition by ensuring they receive adequate nutritional support throughout therapies (Assouline et al. 2021).

10.2.4.3 Vitamin D

Bone condition and proper equilibrium of calcium depend on vitamin D. This helps the gastrointestinal system absorb calcium and phosphorus, which are crucial for preserving bone density and preventing diseases like osteoporosis (Anon n.d.-c). This disorders result in deficiencies of calcium. Vitamin D supplementation is especially crucial for patients receiving LVP who are immobile or have limited exposure to sunlight. Keeping these individuals' vitamin D levels adequate helps with the healthy development of their bones and overall musculoskeletal health. For patients who are unable to obtain enough vitamin D from natural sources, this is essential for promoting their recovery and overall health and minimizing issues linked to extended immobility (Anon n.d.-a).

In addition to vitamins, large volume parenteral (LVPs) may contain various other additives to meet specific medical needs. These include electrolytes like sodium, potassium, and magnesium, which are crucial for fluid balance, nerve function, and muscle contraction. Amino acids are added for protein synthesis, tissue repair, and immune function. Trace elements such as zinc, copper, and selenium are vital for enzymatic reactions and immune defense. Minerals like calcium and phosphorus are essential for bone health and metabolic processes. These additives ensure comprehensive nutritional support for patients, particularly those unable to intake nutrients orally.

10.2.5 Electrolytes

Electrolytes are electrically charged minerals found throughout body fluids that are essential to several biological processes. They support the body's hydration needs, blood pressure and acidity regulation, neuron and muscle function, and tissue repair. Electrolytes are added to large volume parenterals (LVPs) to help patients maintain the right electrolyte balance, especially when oral intake is not feasible. The important electrolytes that are added to LVPs are described below:

10.2.6 Sodium

Sodium is an important extracellular electrolyte that is necessary for neuron activity, blood pressure control, and fluid balance. It regulates the amount of water in and around cells, which is essential for healthy neuron and muscle function. Large volume parenterals (LVPs) frequently contain salt in order to avoid hyponatremia, or low sodium levels, in patients who are unable to swallow food or fluids. For those who suffer from renal disease, dehydration, or other illnesses that cause considerable fluid loss, this is especially important. Sodium chloride is frequently utilized in LVPs to promote overall cellular function and patient health by ensuring appropriate hydration and maintaining normal blood pressure levels.

10.2.7 Potassium

Potassium is an intracellular electrolyte which is crucial to healthy cell division, nerve impulse transmission, and heart contraction. It regulates the osmotic pressure and acid-base equilibrium in cells. Large volume parenterals (LVPs) usually contain potassium to avoid hypokalemia (low potassium levels), which can be caused by prolonged vomiting, diuretics, or diarrhea. Sustaining sufficient potassium levels is essential for healthy heart function, as low levels can cause arrhythmias and other heart problems. To maintain healthy patient outcomes and ensure efficient cellular function, potassium chloride is frequently utilized in LVPs.

10.2.8 Calcium

For healthy bones, blood clotting, muscular contraction, and nerve interactions, calcium is essential. It primarily accumulates as structural support in teeth and bones. Calcium is added to LVPs in order to avoid hypocalcemia, or low calcium levels, which can be caused by a number of illnesses, including pancreatitis, renal failure, and parathyroid difficulties. Enough calcium is necessary to maintain the neurological, muscular, and cardiovascular systems in optimal working order. Common types of calcium utilized in LVPs to help patients maintain appropriate calcium levels and support general physiological functions include calcium gluconate and calcium chloride.

10.2.9 Magnesium

Magnesium is a crucial electrolyte that the body uses for more than 300 metabolic processes, such as blood glucose management, muscle and neuron function, protein synthesis, and blood pressure regulation. It also contributes to the synthesis of energy and the development of bone integrity. Magnesium is necessary to add in LVPs to avoid hypomagnesemia. The magnesium deficiency is occured by long-term diuretic consumption, malnutrition, or chronic alcoholism. A specific kind of magnesium that is frequently utilized in LVPs is magnesium sulfate, which supports general metabolic processes and helps to guarantee adequate muscle and nerve function.

10.2.10 Chloride

A vital extracellular electrolyte, chloride interacts closely with sodium to maintain the body's fluid balance, osmotic pressure, and acid-base equilibrium. It is a crucial part of the hydrochloric acid in gastric juice, which is required for digestion. Chloride is included in LVPs to avoid hypochloremia, or low chloride levels, which can be brought on by prolonged vomiting, diarrhea, or dehydration. Chloride, which is often found as sodium chloride, supports general homeostasis and biological function by assisting in the maintenance of appropriate blood pressure, pH balance, and hydration.

10.2.11 Copper, Iron, and Zinc

A vital trace element involved in numerous biological processes is zinc. It is essential for wound healing, immunological response, protein synthesis, DNA synthesis, and cell division. Moreover, zinc is necessary to preserve good taste and fragrance. It helps antioxidant defense systems and metabolic activities as a cofactor for more than 300 enzymes. Zinc supplementation is essential for patients receiving LVP who have deficiencies, which can be brought on by poor nutrition, malabsorption syndromes, or higher-than-normal physiological demands from conditions including trauma, surgery, or long-term illnesses. Zinc deficiency can cause growth retardation, delayed wound healing, and reduced immunological response; consequently including zinc in LVPs is crucial for the general health and recovery of patients.

Essential mineral copper can be found as a dietary supplement or naturally occurring in some foods. It functions as a cofactor for a number of cuproenzymes, which are enzymes involved in the creation of connective tissue, energy production, iron metabolism, neuropeptide activation, and neurotransmitter synthesis (Collins 2014). The mineral iron can be found as a nutritional supplement, added to some food products, or found naturally in a variety of foods. Hemoglobin, a protein found in erythrocytes (red blood cells) that carries oxygen from the lungs to the tissues, requires iron as a necessary component (Aggett 2012). Zinc is an all-around body vitamin that supports healthy metabolism and the immune system. Your sense of smell and taste, as well as the healing of wounds, depends on zinc. Your body normally obtains enough zinc from a diverse diet. Zinc-rich foods include poultry, red meat, and breakfast cereals with added nutrients. Numerous facets of cellular metabolism involve zinc. It is necessary for hundreds of enzymes to catalyze their activities and is involved in wound healing, immune system function, protein and DNA synthesis, and cell signaling and division. In addition to being involved in taste perception, zinc also promotes healthy growth and development during pregnancy, infancy, childhood, and adolescence (MacDonald 2000).

The following range of final concentrations of copper, iron, and zinc was added to a parenteral admixture consisting of 20% intralipid (intravenous fat emulsion); it is a non-pyrogenic, highly sterile preparation for i.v. injection to deliver the required calories and fatty acids. Also, it consists of 25% Azonutril 25 (utilized to treat bacterial infections of the skin, soft dermis, respiration airways, tonsils, alveoli, ears, and throat). Additionally, enteric fever that arises due to intestinal infections, pneumonia (a viral fever), and tracheal infections brought on by coming into contact with an infected individual is treated with it. The admixture also contains 37.5% glucose-6-PO_4. The rest of the compositions were copper (0.24 mg/L), iron (0.50 mg/L), and zinc (2.00 mg/L) (Harraki et al. 1993; Desai et al. 2007; Blicharska et al. 2016). The combination, including the trace elements in the gluconate form, showed good stability, according to the pH determination. After a week of storage at both temperatures, a notable reduction in the pH of the combination containing the trace elements as chlorides was observed. The findings obtained regarding physicochemical stability were validated by flocculation kinetics. Furthermore, the kinetics showed that there is very little chance of coalescence processes in every scenario. According to B. Harraki et al. (1993), findings suggested that at a temperature of 4 and 25 °C, the two types of combinations are stable for 7 days. For extended storage, nevertheless, combinations that include trace elements in gluconate form ought to be selected. For many enzymatic processes, including those affecting iron metabolism, the development of connective tissue, the synthesis of neurotransmitters, and energy generation, copper is an essential trace metal. It is essential for preserving the neurological and cardiovascular systems' health. Superoxide dismutase, a crucial antioxidant enzyme that protects cells from injury from free radicals, contains copper. Copper deficiency may be occured due to malabsorption, and certain medical problems, which may be avoided by adding copper as supplements in LVPs. Anemia, neutropenia, anomalies in the bones, and brain diseases can all result from a copper shortage. Hence, sufficient copper supplementation in LVPs guarantees appropriate enzymatic activity and preservation of general health.

10.2.12 Manganese

Manganese is a trace element that helps create bones, coagulate blood, and lessen inflammation. It functions as a cofactor for a number of enzymes, including those that are involved in the metabolism of cholesterol, amino acids, and carbohydrates. By triggering superoxide dismutase, manganese also aids in antioxidant defense. Manganese is also necessary to add in LVPs to help individuals who have higher physiological needs or nutritional deficits. Decreased growth, skeletal abnormalities, changes in the metabolism of fats and carbohydrates, and oxidative stress can all be outcomes of a manganese deficiency. Manganese is a component of LVPs that supports appropriate enzymatic activity, bone health, and metabolic processes, all of which contribute to the overall recovery and well-being of patients.

10.2.13 Selenium

The correct operation of antioxidant enzymes, such as glutathione peroxidases, which protect cells from oxidative damage, depends on selenium, a vital trace element. It is also essential for immunological response, DNA synthesis, and thyroid hormone metabolism. Selenium is administered to LVPs to avoid deficiencies that might develop in individuals receiving parenteral nutrition for an extended period of time or who have issues with malabsorption. Thyroid dysfunction weakened immune system, and Keshan disease—a kind of cardiomyopathy—can result from a selenium deficit. Sufficient selenium supplementation in LVPs promotes immune system health, thyroid function, and antioxidant defense, all of which are beneficial to the general well-being and recuperation of patients.

10.2.14 Amino Acids

Amino acids, fundamental for protein synthesis, are vital for tissue repair, immune function, enzyme and hormone production, and various metabolic processes. They are categorized into essential (must be obtained from diet) and nonessential (synthesized by the body). In large volume parenterals (LVPs), amino acids are included to provide necessary protein substrates, particularly for patients unable to take sufficient protein orally due to severe illness, surgery, or gastrointestinal issues. This ensures patients receive essential nutrients to support vital bodily functions and promote recovery.

Large volume parenterals (LVPs) containing amino acids have several advantages. First of all, in bedridden or immobile patients in danger of muscle loss, LVPs assist in muscular development and maintenance. Maintaining muscle mass is essential for recuperation and rehabilitation, especially for patients who are severely ill or have recently undergone surgery. Second, amino acids are important for the healing of injuries. They supply the elements required for both the synthesis of new tissue and the restoration of damaged tissue. This is crucial for those recuperating after surgery or trauma since it expedites the healing process and lowers the possibility of problems. Amino acids are also essential for immunological function. They aid in the body's ability to generate white blood cells, antibodies, and other immune-supporting molecules that fight infections. This is especially important for individuals whose immune systems are weakened or who are receiving therapies like chemotherapy, which can reduce immunity.

10.2.15 Carbohydrates

The body uses carbohydrates as its main energy source, which are essential for maintaining both general health and metabolic processes. Carbohydrates are added to large volume parenterals (LVPs) to make sure patients get enough energy when oral intake is limited. They are vital in preserving steady blood glucose levels, which are necessary for normal brain function and the synthesis of all cellular energy. LVPs assist in stopping the deterioration of muscle tissue by intravenously supplying carbs, preserving proteins for their essential roles in immune response and tissue regeneration.

10.2.16 Dextrose

Glucose, or dextrose, is a basic sugar that's frequently added to LVPs to provide them an instant energy boost. It enters the bloodstream fast, which aids in raising blood glucose levels and providing energy to each of the body's cells. This stimulates metabolic processes and promotes overall healing, making it especially helpful for those recuperating from surgery, trauma, or serious sickness. Dextrose is added to LVPs to guarantee that patients get the critical carbohydrates needed to sustain energy levels and important body processes. It is essential for maintaining stable blood sugar levels, which is particularly critical for individuals who are unable to swallow food or who have higher metabolic demands. Dextrose helps to meet immediate energy needs, facilitating recovery and promoting overall health in clinical settings where nutritional support via intravenous routes is necessary.

10.2.17 Lipids

Lipids, commonly known as fats, play essential roles in the body, serving as a concentrated source of energy, contributing to cell membrane structure, and aiding in hormone production. In large volume parenterals (LVPs), lipid emulsions are included to provide necessary nutrients that support overall health and recovery, especially when oral or enteral feeding is not feasible.

Lipids in LVPs serve as a dense energy source, essential for patients needing concentrated caloric intake during extended periods without oral nutrition or increased energy demands due to illness or recovery. They also provide essential fatty acids (EFAs) like omega-3 and omega-6, vital for cell membranes, brain function, inflammation regulation, and heart health. Additionally, lipid emulsions deliver fat-soluble vitamins (A, D, E, K), critical for immune function, vision, bone health, clotting, and antioxidants, ensuring patients receive adequate levels for optimal health and healing (Table 10.4). In addition to energy and nutrient provision, lipids support immune function and aid in recovery. They help maintain immune cell integrity and function, contributing to the body's ability to fight infections and promote healing.

10.3 Role and Benefits of Additives in LVPs

10.3.1 Nutritional Support

Large volume parenteral (LVP) additives are vital for supplying vital nutrients that promote patient health and recuperation. They consist of lipids, vitamins, minerals, amino acids, and carbs, all of which work synergistically to offer a balanced diet. This is especially important for patients who are unable to consume enough nutrition orally, such as those who are having surgery, are very sick, or have digestive issues. LVPs contribute to the general well-being of patients by sustaining body functioning and promoting healing through the intravenous delivery of vital nutrients.

Table 10.4 Admixtures in parenterals

Admixture category	Examples	Functions
Vitamins	B complex	Support metabolism
	Vitamin C	Collagen synthesis
	Vitamin D	Immune function, bone health
Electrolytes	Sodium	Fluid balance
	Potassium	Nerve function
	Calcium	Muscle contraction
	Magnesium	Enzymatic reactions
Amino acids	Essential amino acids	Protein synthesis
	Nonessential amino acids	Tissue repair, immune function
Trace elements	Zinc	Enzymatic reactions
	Copper	Antioxidant defenses
	Manganese	Wound healing
	Selenium	Immune health
Carbohydrates	Dextrose	Energy source, maintain blood glucose levels
Lipids	Omega-3	Energy source, cell membrane structure
	Omega-6	Hormone production
	Fat-soluble vitamins (A, D, E, K)	Various metabolic functions

10.3.2 Role of Parentral Admixture in Nutritional Deficiencies

LVP additives are also utilized to make up for deficiencies that might result from inadequate oral intake or certain medical concerns. For example, deficits can be addressed with supplements for vitamins like B12, which is essential for nerve function, and minerals like zinc, which is critical for immunological health. This adjustment is necessary to ensure good physiological functioning and avoid difficulties, particularly in individuals who have protracted hospital stays or chronic diseases.

10.3.3 Therapeutic Benefits

Beyond basic nutrition, certain LVP additions have medicinal effects. Antioxidants such as vitamin C, for instance, can reduce oxidative stress, which is beneficial for patients receiving rigorous therapy or suffering from infections. Amino acids promote immunological response and tissue repair, which help the body heal after surgery or damage. These therapeutic benefits improve overall recovery rates, facilitate faster healing, and reduce complications, all of which improve patient outcomes.

10.3.4 Stabilization of the Solution

Additionally, additives assist in stabilizing LVP solutions, assuring their effectiveness and safety throughout administration and storage. Electrolytes, such as potassium and sodium, support the preservation of osmolarity and pH equilibrium, which are essential for cellular activity and solution stability. In order to ensure that patients receive consistent and efficient care, proper stabilization stops chemical deterioration and preserves the integrity of active components.

10.4 Preparation and Administration of Additives in LVPs

10.4.1 Guidelines for Adding Additives to LVPs

When preparing LVPs with additives, several guidelines must be followed to ensure safety, efficacy, and compliance with regulatory standards:

10.4.1.1 Compatibility and Stability

It is advisable to carry out compatibility and stability tests prior to incorporating any addition into an LVP. This involves identifying if the addition interacts with the other additives or the base solution in a way that could lead to deterioration, precipitation, or a decrease in efficacy. Testing for compatibility guarantees that the finished product is safe to use on patients and maintains its chemical stability over time.

10.4.1.2 Aseptic Techniques

To avoid contamination and preserve sterility, aseptic procedures must be meticulously followed while adding additives to LVPs. When preparing, a sterile environment should be employed, such as laminar flow hoods or isolators. Syringes, needles, and vials retaining additives are just a few examples of sterile, parenterally compatible equipment that have to be available. Using the proper personal protective equipment (PPE) can help reduce the danger of microbiological contamination in personnel handling additives.

10.4.1.3 Dosing Considerations

Additives must be applied precisely in accordance with patient needs and recognized dosage standards. This involves figuring out the appropriate concentration based on the patient's weight, health, and dietary requirements. Insufficient or excessive dosage may result in side effects or ineffective treatment results. Accuracy and patient safety depend on pharmaceutical computations being double-checked by trained personnel.

Concentration calculations ensure precise dosing for therapeutic efficacy and safety in LVPs, considering patient specifics and desired outcomes. Compatibility checks prevent interactions that could compromise solution integrity, validated through rigorous physical and chemical assessments. Ongoing training and competency assessments for personnel uphold dosing accuracy and adherence to safety protocols, ensuring consistent quality in clinical practice. Continuous education updates healthcare providers on current standards and regulatory guidelines, promoting patient safety and optimizing treatment outcomes.

10.4.1.4 Monitoring and Documentation

Thorough documentation and monitoring are essential during the preparation phase. Every stage, from choosing and preparing the additives to verifying the finished product, needs to be well documented. Batch numbers, expiration dates, concentrations, and any modifications to normal operating procedures are all covered by this. Quality control, traceability, and regulatory compliance are all supported by thorough documentation.

10.5 Regulatory Considerations in LVP Preparation

10.5.1 Guidelines and Standards by Regulatory Authorities

Regulatory authorities such as the Food and Drug Administration (FDA) in the United States and the European Medicines Agency (EMA) in Europe provide comprehensive guidelines and standards for the preparation of large volume parenterals (LVPs). These guidelines aim to ensure the safety, efficacy, and quality of intravenous therapies administered to patients.

10.5.1.1 FDA Guidelines

The FDA regulates LVPs under strict current Good Manufacturing Practice (cGMP) guidelines, ensuring facilities, personnel, equipment, and processes comply with stringent standards. Sterility assurance mandates controlled aseptic conditions to prevent microbial contamination, requiring clean room environments and validated sterilization methods. Quality control is paramount, involving rigorous testing of raw materials, in-process samples, and final products for identity, strength, purity, and sterility. Proper labeling is crucial, with LVPs requiring clear details on contents, dosage, expiration date, and storage conditions. Documentation must facilitate traceability, encompassing batch production records, testing outcomes, and adverse event reports to uphold safety and regulatory compliance.

10.5.1.2 EMA Standards

In Europe, the EMA oversees LVPs under the regulatory frameworks of the European Pharmacopoeia and EU GMP guidelines, which harmonize standards internationally for consistency and compatibility. Compliance with the European Pharmacopoeia specifications ensures that LVPs meet stringent quality and safety criteria.

The EMA mandates rigorous quality assurance akin to the FDA, including frequent inspections of manufacturing facilities to uphold validated processes and procedures. This oversight ensures that LVPs consistently adhere to high-quality standards throughout production. Comprehensive documentation is a cornerstone of

EMA regulations, encompassing detailed records of LVP manufacturing and testing. Batch records, stability studies, and validation reports are meticulously maintained to demonstrate compliance with regulatory requirements. These documentation practices support traceability and accountability, facilitating the safe and reliable production of LVPs across European markets.

10.5.1.3 Quality Control and Assurance

Quality control in the production of large volume parenterals (LVPs) involves rigorous testing and analysis of raw materials, in-process samples, and final products to verify adherence to specified standards. Tests cover aspects such as identity, strength, purity, sterility, endotoxin levels, and the presence of particulate matter, ensuring the integrity and safety of LVPs prior to patient administration. Concurrently, quality assurance oversees the entire quality management system, validating manufacturing processes, qualifying equipment, maintaining suitable facilities, and providing ongoing training to personnel to uphold rigorous standards consistently. These practices are crucial for mitigating risks, ensuring product consistency, and safeguarding the efficacy and safety of LVPs throughout their production and distribution cycles.

Stability testing is essential in the production of large volume parenterals (LVPs) to assess their shelf life and optimal storage conditions. These studies evaluate the product's integrity and potency over time, considering environmental factors like temperature, humidity, and light exposure that could affect stability. By subjecting LVPs to controlled conditions mimicking real-world storage environments, manufacturers can predict how the product will behave over its intended shelf life. This data informs decisions on storage recommendations, packaging materials, and expiration dating, ensuring that LVPs maintain their quality and efficacy until they are administered to patients. Stability testing thus plays a critical role in ensuring that healthcare providers can rely on the consistency and reliability of LVPs throughout their use in clinical settings.

10.5.1.4 Labeling and Documentation Requirements

Accurate and thorough labeling and documentation are indispensable for ensuring the safety and efficacy of large volume parenterals (LVPs). The labeling of LVPs must contain essential information such as the product name, strength, dosage form, route of administration, expiration date, and specific storage conditions. Clear and concise handling instructions, along with any necessary warnings or precautions, are also crucial to guide healthcare professionals in proper usage and patient safety. Compliance with regulatory standards is essential to ensure that labels meet all requirements, contributing to the overall quality assurance of the product.

Documentation for LVPs involves meticulous batch records that meticulously document every aspect of the preparation process, including formulation, compounding, packaging, and labeling procedures. These records serve to establish traceability and accountability throughout the product's life cycle, facilitating efficient batch recall procedures if necessary. Additionally, they include comprehensive reports of quality control tests, stability assessments, and records of adverse events, providing critical data for ongoing product monitoring and regulatory compliance. Together, accurate labeling and thorough documentation play pivotal roles in maintaining the integrity and safety of LVPs, ensuring they meet stringent regulatory standards and are administered safely in clinical settings.

10.6 Conclusion

Parenteral admixtures are utilized to increase the therapeutic effectiveness, stability of the product and maintain the nutritional and electrolyte balance in the body. It is also used to reduce adverse effect of drugs. The regulatory authorities of different countries regulate the manufacturing to assure the quality, stability, and microbiological integrity of the parenteral admixture. Preparing parenteral admixtures requires adherence to strict guidelines to prevent contamination, ensure sterility, and maintain compatibility between ingredients. Pharmacists and healthcare professionals play a crucial role in preparing these admixtures safely and accurately, following established protocols and using aseptic techniques to minimize the risk of complications for patients receiving these medications.

References

Aggett PJ (2012) Iron. In: Erdman JW, Macdonald IA, Zeisel SH (eds) Present knowledge in nutrition, 10th edn. Wiley-Blackwell, Washington, DC, pp 506–520

Ahmed TA, Khalid M, Ahmed OA, Zidan AS (2018) Sterile dosage forms loaded nanosystems for parenteral, nasal, pulmonary and ocular administration. In: Nanoscale fabrication, optimization, scale-up and biological aspects of pharmaceutical nanotechnology. William Andrew Publishing, pp 335–395

Allen LV Jr (2021) Intravenous admixture preparation considerations, part 9-A: error prevention in intravenous admixture preparation. Int J Pharm Compd 25(2):131–139

Anon (n.d.-a) A dynamic duo: the roles of calcium + vitamin D in the body. https://store.mayoclinic.com/education/the-roles-of-calcium-vitamin-d-in-the-body/. Accessed 15 Jul 2024

Anon (n.d.-b) Dietary supplements for immune function and infectious diseases—health professional fact sheet. https://ods.od.nih.gov/factsheets/ImmuneFunction-HealthProfessional/. Accessed 15 Jul 2024

Anon (n.d.-c) Vitamin D-NHS. https://www.nhs.uk/conditions/vitamins-and-minerals/vitamin-d/. Accessed 15 Jul 2024

Assouline B, Faivre A, Verissimo T, Sangla F, Berchtold L, Giraud R, Bendjelid K, Sgardello S, Elia N, Pugin J, De Seigneux S, Legouis D (2021) Thiamine, ascorbic acid, and hydrocortisone as a metabolic resuscitation cocktail in sepsis: a meta-analysis of randomized controlled trials with trial sequential analysis. Crit Care Med 49(12):2112–2120. https://doi.org/10.1097/ccm.0000000000005262

Bakrey H, Shivgotra R, Abdu A, Soni B, Shahtaghia NR, Jain SK, Thakur S (2024) Use of total parenteral nutrition (TPN) as a vehicle for drug delivery. Curr Drug Targets 25(5):306–329

Berger MM, Herter-Aeberli I, Zimmermann MB, Spieldenner J, Eggersdorfer M (2021) Strengthening the immunity of the Swiss population with micronutrients: a narrative review and call for action. Clin Nutr ESPEN 43:39–48. https://doi.org/10.1016/j.clnesp.2021.03.012

Blicharska E, Szczęsna B, Kocjan R, Gumieniczek A, Komsta Ł (2016) Analysis of dissolution profiles of iron, zinc, and manganese from complex dietary supplements by ion chromatography and chemometrics. J Liq Chromatogr Relat Technol 39(1):30–34

Boullata JI, Gilbert K, Sacks G, Labossiere RJ, Crill C, Goday P, Kumpf VJ, Mattox TW, Plogsted S, Holcombe B, American Society for Parenteral and Enteral Nutrition (2014) ASPEN clinical guidelines: parenteral nutrition ordering, order review, compounding, labeling, and dispensing. J Parenter Enter Nutr 38(3):334–377

Boullata JI, Mirtallo JM, Sacks GS, Salman G, Gura K, Canada T, Maguire A, ASPEN Parenteral Nutrition Safety Committee (2022) Parenteral nutrition compatibility and stability: a comprehensive review. J Parenter Enter Nutr 46(2):273–299

Calder PC, Ortega EF, Meydani SN, Adkins Y, Stephensen CB, Thompson B, Zwickey H (2022) Nutrition, immunosenescence, and infectious disease: an overview of the scientific evidence on micronutrients and on modulation of the gut microbiota. Adv Nutr 13(5):S1–S26. https://doi.org/10.1093/advances/nmac052

Carr AC, Rosengrave PC, Bayer S, Chambers S, Mehrtens J, Shaw GM (2017) Hypovitaminosis C and vitamin C deficiency in critically ill patients despite recommended enteral and parenteral intakes. Crit Care 21(1):300. https://doi.org/10.1186/s13054-017-1891-y

Collins JF (2014) Copper. In: Ross AC, Caballero B, Cousins RJ, Tucker KL, Ziegler TR (eds) Modern nutrition in health and disease, 11th edn. Lippincott Williams & Wilkins, Baltimore, MD, pp 206–216

Dehmel C, Braune SA, Kreymann G, Baehr M, Langebrake C, Hilgarth H, Nierhaus A, Dartsch DC, Kluge S (2011 Aug) Do centrally pre-prepared solutions achieve more reliable drug concentrations than solutions prepared on the ward? Intensive Care Med 37:1311–1316

Desai NR, Shah SM, Koczone J, Vencl-Joncic M (2007) Zinc content of commercial diluents widely used in drug admixtures prepared for intravenous infusion. Int J Pharm Compd 11(5):426

Dongare SD, Mali SS, Patrekar PV (2015) Sterile parenteral products: a narrative approach. J Drug Deliv Ther 5(1):41–48

Driscoll DF, Bhargava HN, Li L, Zaim RH, Babayan VK, Bistrian BR (1995) Physicochemical stability of total nutrient admixtures. Am J Health Syst Pharm 52:623

Driscoll DF, Bacon MN, Bistrian BR (2000) Physicochemical stability of two types of intravenous lipid emulsion as total nutrient admixtures. J Parenter Enter Nutr 24(1):15–22

Eggersdorfer M, Berger MM, Calder PC, Gombart AF, Ho E, Laviano A, Meydani SN (2022) Perspective: role of micronutrients and omega-3 long-chain polyunsaturated fatty acids for immune outcomes of relevance to infections in older adults-a narrative review and call for action. Adv Nutr (Bethesda, MD) 13(5):1415–1430. https://doi.org/10.1093/ADVANCES/NMAC058

Ezike TC, Okpala US, Onoja UL, Nwike CP, Ezeako EC, Okpara OJ et al (2023) Advances in drug delivery systems, challenges and future directions. Heliyon 9(6):e17488

Garcia J, Garg A, Song Y, Fotios A, Andersen C, Garg S (2018) Compatibility of intravenous ibuprofen with lipids and parenteral nutrition, for use as a continuous infusion. PLoS One 13(1):e0190577

Gostyńska A, Stawny M, Dettlaff K, Jelińska A (2019) The interactions between ciprofloxacin and parenteral nutrition admixtures. Pharmaceutics 12(1):27

Harraki B, Guiraud P, Rochat MH, Fussellier M, Favier A (1993) Influence of copper, iron, and zinc on the physicochemical properties of parenteral admixture. PDA J Pharm Sci Technol 47(5):199–204

Hedlund N, Beer I, Hoppe-Tichy T, Trbovich P (2017) Systematic evidence review of rates and burden of harm of intravenous admixture drug preparation errors in healthcare settings. BMJ Open 7(12):e015912

Hoyois A, Ballarin A, Thomas J, Lheureux O, Preiser JC, Coppens E, Perez-Bogerd S, Taton O, Farine S, Van Ouytsel P, Arvanitakis M (2021) Nutrition evaluation and management of critically ill patients with COVID-19 during post-intensive care rehabilitation. JPEN J Parenter Enteral Nutr 45(6):1153–1163. https://doi.org/10.1002/jpen.2101

Hu Q, Ren H, Hong Z, Wang C, Zheng T, Ren Y, Chen K, Liu S, Wang G, Guosheng G, Xiuwen W, Ren J (2020) Early enteral nutrition preserves intestinal barrier function through reducing the

formation of neutrophil extracellular traps (NETs) in critically ill surgical patients. Oxidative Med Cell Longev 2020:1. https://doi.org/10.1155/2020/8815655
Jessurun JG, Hunfeld NG, van Rosmalen J, van Dijk M, van den Bemt PM (2022) Prevalence and determinants of intravenous admixture preparation errors: a prospective observational study in a university hospital. Int J Clin Pharm 44(1):44–52
Johnson NT, Probert MR (2014) Vitamin C revisited. Crit Care 18(4):460. https://doi.org/10.1107/s2053273316095346
Kapp RW (2018) Gastrointestinal tract as a major route of pharmaceutical administration. In: Toxicology of the gastrointestinal tract, 2nd edn. CRC Press, CRC Press, pp 113–149
Khalili H, Sheikhbabayi M, Samadi N, Jamalifar H, Dalili D, Samadi N (2013) Bacterial contamination of single-and multiple-dose vials after multiple use and intravenous admixtures in three different hospitals in Iran. Iran J Pharm Res 12(1):205
Lee ZY, Ortiz-Reyes L, Lew CCH, Shahnaz Hasan M, Ke L, Patel JJ, Stoppe C, Heyland DK (2023) Intravenous vitamin C monotherapy in critically ill patients: a systematic review and meta-analysis of randomized controlled trials with trial sequential analysis. Ann Intensive Care 13(1):1–17. https://doi.org/10.1186/S13613-023-01116-X/TABLES/3
Ludwig JD (2019) Parenteral dosage forms: introduction and historical perspective. In: Parenteral medications, 4th edn. CRC Press, pp 3–9
MacDonald RS (2000) The role of zinc in growth and cell proliferation. J Nutr 130(5):1500S–1508S
Manchanda S, Das N, Chandra A, Bandyopadhyay S, Chaurasia S (2020) Fabrication of advanced parenteral drug-delivery systems. In: Drug delivery systems. Academic Press, pp 47–84
Meyer BK, Ni A, Binghua H, Shi L (2007) Antimicrobial preservative use in parenteral products: past and present. J Pharm Sci 96(12):3155–3167. https://doi.org/10.1002/jps.20976
Nema S, Brendel RJ (2016) Excipients for parenteral dosage forms: regulatory considerations and controls. In: Pharmaceutical dosage forms-parenteral medications. CRC Press, pp 123–148
Nema S, Brendel RJ, Washkuhn R (2007) Excipients: parenteral dosage forms and their role. In: Encyclopedia of pharmaceutical technology. Informa Healthcare, New York, pp 1622–1645
Nikam Nikita R, Vaishnavi A, Lalchand D (2023) Parenteral drug delivery approach: an overview. Journal of Xidian Univ 17(1):386–400
Otero-Millán L, Bea-Mascato B, Legido Soto JL, Martínez-López-De-Castro N, Lago-Rivero N (2024a) Evaluation of the stability of newborn hospital parenteral nutrition solutions. Pharmaceutics 16(3):316
Otero-Millán L, Bea-Mascato B, Legido Soto JL, Martínez-López-De-Castro N, Lago-Rivero N (2024b) Physicochemical stability of hospital parenteral nutrition solutions: effect of changes in composition and storage protocol. Pharmaceutics 16(5):572
Pecora F, Persico F, Argentiero A, Neglia C, Esposito S (2020) The role of micronutrients in support of the immune response against viral infections. Nutrients 12(10):1–45. https://doi.org/10.3390/nu12103198
Piwowarczyk L, Tomczak S, Antkowiak P, Jelińska A, Stawny M (2022) Sodium valproate incompatibility with parenteral nutrition admixtures-a risk to patient safety: an in vitro evaluation study. Pharmaceutics 14(2):371
Pramanick S, Singodia D, Chandel V (2013) Excipient selection in parenteral formulation development. Pharma Times 45(3):65–77
Preiser JC, Arabi YM, Berger MM, Casaer M, McClave S, Montejo-González JC, Peake S, Blaser AR, Van den Berghe G, van Zanten A, Wernerman J, Wischmeyer P (2021) A guide to enteral nutrition in intensive care units: 10 expert tips for the daily practice. Crit Care 25(1):1–13. https://doi.org/10.1186/S13054-021-03847-4/FIGURES/3
Rayaprolu BM, Strawser JJ, Anyarambhatla G (2018) Excipients in parenteral formulations: selection considerations and effective utilization with small molecules and biologics. Drug Dev Ind Pharm 44(10):1565–1571
Riera P, Garrido-Alejos G, Cardenete J, Moliner E, Zapico-Muñiz E, Cardona D, Garin N (2018) Physicochemical stability and sterility of standard parenteral nutrition solutions and simulated Y-site admixtures for neonates. Nutr Clin Pract 33(5):694–700

Serventi L, Keeling I, Harris A, Craig A, Adams MR, Bi Y, Ruan J, Ham EE, Broad TM, Smith GT, Blackman HL, Liu Z, Mullally SA, Mohajer M, Mehta A, Feng Z (2023) Water soluble vitamins. In: Sustainable development goals series part F2751. Springer, pp 91–106. https://doi.org/10.1007/978-3-031-12358-0_7

Stawny M, Gostyńska A, Dettlaff K, Jelińska A, Kościelniak M, Ogrodowczyk M (2019) Development, validation, and stability assessment application of RP-HPLC-DAD method for quantification of ampicillin in total parenteral nutrition admixtures. Antibiotics 8(4):268

Tomczak S, Stawny M, Jelińska A (2020) Co-administration of drugs and parenteral nutrition: in vitro compatibility studies of loop diuretics for safer clinical practice. Pharmaceutics 12(11):1092

Tomczak S, Chmielewski M, Szkudlarek J, Jelińska A (2023) Antiemetic drugs compatibility evaluation with paediatric parenteral nutrition admixtures. Pharmaceutics 15(8):2143

Tubic-Grozdanis M, Krämer I (2015) Parenteral. In: Practical pharmaceutics. Springer, pp 265–299. https://doi.org/10.1007/978-3-319-15814-3_13

Tubic-Grozdanis M, Krämer I (2023) Parenteral. In: Practical pharmaceutics: an international guideline for the preparation, care and use of medicinal products, 2nd edn. Springer, pp 473–519. https://doi.org/10.1007/978-3-031-20298-8_21

Watrobska-Swietlikowska D (2019) Compatibility of maximum inorganic and organic calcium and phosphate content in neonatal parenteral solutions. Sci Rep 9(1):10525

Wijaya CH, Wijaya W, Mehta BM (2015) General properties of minor food components. In: Handbook of food chemistry, pp 55–74. https://doi.org/10.1007/978-3-642-36605-5_40

Yeo KS, Kua SW, Teoh GS, Onsiong MK (2001) The use of thiopentone/propofol admixture for laryngeal mask airway insertion. Anaesth Intensive Care 29(1):38–42

Challenges and Advances in Pharmaceutical Development of Topical and Transdermal Dosage Forms

11

Jaspreet Kaur, Manju, Amanjot Kaur, Monu Yadav, Neha Raina, Sumit Kumar, Akhilesh Kumar, and Madhu Gupta

Abstract

This chapter explores the challenges and advances in the pharmaceutical development of topical and transdermal dosage forms. It discusses key formulation challenges, including drug stability, solubility, skin permeation, and irritation. The chapter highlights advances in skin permeation enhancement techniques, such as the use of chemical enhancers, physical methods, and vesicular systems.

J. Kaur
University Institute of Pharmaceutical Sciences, UGC Centre of Advanced Studies (UGC-CAS), Panjab University, Chandigarh, India

Rayat Institute of Pharmacy, Ropar, Punjab, India

Manju
University Institute of Pharmaceutical Sciences, UGC Centre of Advanced Studies (UGC-CAS), Panjab University, Chandigarh, India

A. Kaur
Rayat Institute of Pharmacy, Ropar, Punjab, India

M. Yadav (✉)
Amity Institute of Pharmacy, Amity University, Haryana, Amity Education Valley Gurugram, Manesar, Panchgaon, Haryana, India

N. Raina · M. Gupta (✉)
Department of Pharmaceutics, Delhi Pharmaceutical Sciences and Research University, New Delhi, India

S. Kumar
Department of Pharmaceutical Sciences, Central University Haryana, Mahendragarh, India

A. Kumar
Division of Medicine, ICAR-Indian Veterinary Research Institute, Bareilly, Uttar Pradesh, India

K. Jain, A. K. Yadav (eds.), *Advances in Pharmaceutical Product Development*,
https://doi.org/10.1007/978-981-97-9230-6_11

Additionally, it addresses the challenges associated with establishing bioequivalence and meeting regulatory requirements for generic topical and transdermal products. Furthermore, the chapter examines manufacturing challenges and the need for process optimization and quality control. On the other hand, it also outlines the advances made in the field, including the development of novel drug delivery systems, personalized medicine, combination therapies, and continuous monitoring and controlled release technologies.

Keywords

Transdermal · Topical · Bioavailability · Novel drug delivery · Controlled release

11.1 Introduction

Skin covers the major portion of the human body which consists of 15% of the total body weight with a 2 m^2 surface area. Three layers make up the skin structure, namely, epidermis, dermis, and hypodermis. The skin is a dynamic organ with numerous specialized cells and processes. It maintains homeostasis with the help of collecting sensory information from environment and plays a vital role in defense mechanism toward pathogens (Tortora and Derrickson 2018; Tapfumaneyi et al. 2022). It is also involved in maintaining the proper temperature for the body to function well. Natural and synthetic topical products are frequently used to treat the appearance of the face and abnormal condition of the skin (Domínguez-Hüttinger et al. 2013). The treatment via the transdermal method of some dermatological diseases such as alopecia, dermatitis, psoriasis, acne vulgaris, vitiligo, and skin cancer may also be used for other diseases such as diabetes, osteoporosis, rheumatoid arthritis, and Alzheimer's and Parkinson's disease (Rabiei et al. 2020).

Transdermal and topical pharmaceuticals are prescribed and administered routinely all over the world (Alany 2017). On the contrary, conventional skin formulations are overpowered that are intended for topical, regional, and transdermal drug delivery such as simple solutions, lotions, creams, ointments, gels, foams, and aerosols due to low enhancement of drug permeation and drug solubility. Over the centuries, several routes have been investigated for the administration of a variety of medicinal products. The global transdermal drug delivery market system had an estimated net worth of US $55,100.33 million in 2021 and is projected to reach US $88,422.40 million by 2030, having a compound annual growth rate of 5.0% from 2021 to 2030 (Phatale et al. 2022). However, their transport through these routes is constrained by the anatomy of the skin and physiological characteristics of cells. The skin's topmost layer, the stratum corneum (SC), serves as the primary barrier to drugs passing through the skin. It is composed of corneocytes or dead skin cells. Overcoming this barrier is the major challenge faced by researchers in formulating drugs and tools of drug delivery to the skin (Farjami et al. 2021; Peña-Juárez et al. 2021). A wide range of transdermal systems containing different drugs have been formulated, but permeation barriers in the skin limit their clinical use (Kováčik et al. 2020).

Among the different methods for drug delivery to the skin, nano-sized systems have attracted attention as drug carriers for transdermal drug delivery (Takeuchi et al. 2019). To facilitate transdermal absorption, diverse methodologies have been developed and patented (Brown et al. 2006). This has made it essential to create novel topical and transdermal drug delivery systems, such as nano-based technologies that significantly improve dermatotherapy and overcome several formulation limitations, such as drug solubility (Durga et al. 2022), resolving some of the subsequent complications with oral drug delivery, including the hepatic first-pass effect, longer dosing frequencies, and patient compliance (Ramadon et al. 2021). Such distinctively intact skin structures serve as a useful barrier for the majority of medicinal products and foreign compounds. Thus, numerous unique ways have been developed to design appropriate transdermal drug delivery systems in order to effectively administer biologically active substances via transdermal route. Some of the common and advanced investigational strategies involve the use of permeation enhancers, iontophoresis, microneedles (MNs), sonophoresis, laser ablation, thermal ablation, magnetophoresis, jet injectors, electroporation, and nanocarriers to boost skin permeation of drugs (Phatale et al. 2022). For nanocarriers, they are classified as polymer-based nanocarriers, lipid-based nanocarriers, viral nanoparticles, or inorganic nanoparticles, depending on their constituent parts. The most cutting-edge drug delivery method recently available is the lipid-based nanocarrier system which provides ease of large-scale production, biocompatibility, and biodegradability due to the nature of the materials used, low toxicity, the possibility of controlled and modified drug release, drug solubility enhancement, and the potential to incorporate hydrophilic and lipophilic drugs. The two types of lipid-based nanocarriers that might be used to carry therapeutics are vesicular and nonvesicular. Vesicular lipid-based nanocarriers include liposomes, niosomes, ethosomes, cubosomes, bilosomes, transfersomes, and glycerosomes, and nonvesicular lipid-based nanocarriers include solid lipid nanoparticles (SLNs), nanostructured lipid carriers (NLCs), and nanoemulsions (Priya et al. 2022).

In this book chapter, we will explain about strategies mediated by nanocarriers that address various transdermal drug delivery challenges, their clinical significance, and the most recent developments in research on a few of the existing and potential transdermal delivery technologies and approaches.

11.2 Formulation Challenges of Topical and Transdermal Products

There are numerous challenges associated with topical and transdermal products that are covered in the subsections below.

11.2.1 Drug Solubility

Pharmaceutical formulation has been impacted by the solubility of medicines. Polymeric micelles have gained much interest as an innovative method to overcome the poor solubility and permeability of drugs across the skin. These are amphiphilic, biocompatible, and biodegradable copolymer-based self-assembling nanocarriers. Natural polymers that are frequently utilized include gelatin and albumin, whereas synthetic polymers include polylactic acid and polyglycolic acid. They are nanocarriers that encapsulate hydrophilic, lipophilic, and charged compounds; decrease the potential for systemic side effects; and facilitate targeted drug delivery (Makhmalzade and Chavoshy 2018; Yotsumoto et al. 2018). Compared to other nanocarriers, they are smaller in size and exhibit better stability in different microenvironments (Ghezzi et al. 2021). Apart from this, ethosomes are nano-vesicular carriers suitable for poorly soluble drugs shown to increase their drug permeation (Abouhussein 2021; Apolinário et al. 2021).

11.2.2 Drug Stability

To overcome some of the limitations in topical and transdermal drug delivery, nanoemulsions have been developed to improve penetration, drug stability, and absorption of active molecules, as well as to achieve controlled release (Sarheed et al. 2020; Souto et al. 2022). Nanoemulsions are lipid-based, colloidal oil in water dispersions of finely dispersed droplets (in the nm scale) with hydrophilic and lipophilic phases (Klang et al. 2012; Sarheed et al. 2020). They improve both polar and nonpolar chemical solubility and are disposable. Compared to conventional emulsions, they have greater thermodynamic and kinetic stability and an increased bioavailability of lipophilic compounds from the addition of surfactants/emulsifier. The fluidic nature of nanoemulsions and surfactant interface promotes skin interaction. In addition, they have a high loading capacity of hydrophobic molecules and protect the active molecules from oxidation and hydrolysis, which improves bioavailability. Furthermore, a clinical study by Yamada et al. demonstrated enhanced topical drug delivery of hydrophobic drugs in a tailorable nanoemulsion using elongated microparticles that could penetrate the dermal-epidermal junction and enhance drug permeability within the epidermis (Rai et al. 2018; Yamada et al. 2018).

11.2.3 Skin Irritation

Researchers have examined the microneedles (MNs) for drug delivery by the transdermal route as well as addressing the drawbacks of the current techniques. They have created a refined method utilizing the microneedles in order to allow hydrophilic high molecular weight chemicals to penetrate the stratum corneum. Skin possesses a number of defense systems to protect itself from a variety of environmental challenges. Microneedle usage can result in mild to moderate skin irritation or allergies among people with sensitive skin. It may cause inflammation, discomfort, and redness sometimes. If the microneedles are not sterile, the holes they create in the skin can get infected. Even though the pores produced by microneedles are much smaller than those made by hypodermic needles, they exhibit less microbial invasion. These limitations are very rare and can be overcome with advanced material selection for microneedles (Waghule et al. 2019).

11.2.4 Bioequivalence Assessment and Regulatory Requirements

Bioequivalence is of utmost importance in topical and transdermal products as it ensures that generic versions of these products are therapeutically equivalent to their reference (brand-name) counterparts. Bioequivalence refers to the equivalent systemic exposure of the active pharmaceutical ingredient (API) between two products when administered in the same dosage form and route and under similar conditions. Generic substitution, safety and efficacy, regulatory requirements, formulation optimization, and post-marketing surveillance are some key reasons why bioequivalence is crucial in topical and transdermal products.

The availability of generic versions of topical and transdermal products can significantly reduce healthcare costs. Bioequivalence studies provide the scientific evidence needed to prove that a generic product can be safely and effectively substituted for the reference product. This facilitates patient access to more affordable treatment options. These studies provide assurance that generic alternatives deliver the same amount of active ingredient to the systemic circulation as the reference product. This ensures that the generic product will produce the same therapeutic effect and safety profile as the reference product. Patients can confidently switch between the generic and reference products without compromising their treatment outcomes. Regulatory authorities, such as the US Food and Drug Administration (FDA) and the European Medicines Agency (EMA), require bioequivalence data for the approval of generic topical and transdermal products. These agencies have established guidelines that specify the acceptable limits of variability in drug release and absorption profiles. Compliance with these guidelines helps manufacturers demonstrate that their product is comparable to the reference product. These studies can assist manufacturers in optimizing the formulation of their topical and transdermal products. Understanding the factors influencing drug release and absorption, such as excipient selection, formulation design, and patch or gel structure, enables manufacturers to develop products that achieve similar systemic exposure to the reference product. This helps ensure that the product delivers the desired therapeutic effect. Moreover, these studies continue to play a crucial role in post-marketing surveillance of generic topical and transdermal products. Ongoing monitoring of generic products in the market helps identify any potential variations in efficacy or safety. Bioequivalence studies can be used to compare the performance of different generic versions and ensure that they maintain the required standards.

11.2.5 Challenges in Development of Standardized Bioequivalence Studies

Developing standardized bioequivalence protocols for topical and transdermal products can be a complex and challenging process. These challenges arise due to the unique characteristics of topical and transdermal delivery systems, which require specific considerations in designing bioequivalence studies. This paragraph discusses the major challenges faced in developing standardized bioequivalence protocols for topical and transdermal products.

Topical and transdermal products present unique challenges due to their complex delivery systems. Topical products can vary in terms of formulation, such as creams, ointments, gels, or solutions, each with different properties that can affect drug release and absorption. Meanwhile, transdermal products often comprise drug-containing patches or films that adhere to the skin, requiring careful consideration of factors like patch design, adhesion, and drug release kinetics. Developing standardized bioequivalence protocols for these diverse delivery systems is challenging, as the studies need to account for these formulation variations while ensuring accuracy and reproducibility. One of the fundamental goals of bioequivalence studies is to measure drug absorption accurately. However, this becomes challenging with topical and transdermal products, as it is difficult to assess the extent of drug absorption within the skin. Unlike oral medications, where measuring drug concentrations in blood plasma provides a direct reflection of systemic exposure, bioavailability assessment for topical and transdermal products often involves assessing drug levels within the skin layers, which may require invasive techniques like microdialysis. Developing standardized protocols that can reliably measure drug absorption in a noninvasive and reproducible manner is a significant challenge.

Skin permeability varies among individuals due to factors like age, skin type, ethnicity, and underlying skin conditions. This variability can significantly affect the bioavailability of topically applied drugs, making it challenging to establish standardized protocols that account for this variation. Developing bioequivalence protocols that adequately address this variability to ensure accurate comparison of absorption between reference and generic products is a complex task. Determining a therapeutic endpoint for topical and transdermal products is another challenge in developing standardized bioequivalence protocols. Unlike oral medications, where systemic drug concentrations or pharmacodynamic effects can serve as reliable therapeutic endpoints, defining an endpoint for topically applied drugs can be more subjective. For example, in dermatologic conditions, multiple parameters such as erythema, scaling, or pruritus may need to be considered. Developing consensus on clinically relevant therapeutic endpoints and incorporating them into standardized protocols is crucial but challenging. Topical and transdermal products frequently contain various excipients and formulation differences that can affect drug release and absorption. These excipients and formulation variables are crucial in ensuring product stability, skin penetration, and drug efficacy. However, their potential influence on bioavailability and bioequivalence must be carefully considered and standardized across different products. Establishing protocols that adequately address the impact of formulation variances and excipient effects is essential but complicated, as the range of possible formulations is extensive.

Efforts to overcome these challenges involve collaboration among regulatory authorities, industry experts, and academia to establish standardized bioequivalence protocols for topical and transdermal products. This includes setting criteria for study design, selection of appropriate in vitro and in vivo methods, statistical analysis, and acceptance.

11.2.6 Regulatory Approval Process for Topical and Transdermal Dosage Forms

The regulatory approval process for topical and transdermal dosage forms involves several stages to ensure the safety, efficacy, and quality of these products. This process is overseen by regulatory authorities such as the US Food and Drug Administration (FDA) in the United States and the European Medicines Agency (EMA) in Europe. While the specific requirements may vary for different countries, the overall process remains similar. Before conducting clinical trials, the preclinical development stage focuses on evaluating the safety and efficacy of the topical or transdermal product through laboratory and animal studies. This involves conducting pharmacokinetic and toxicological studies to determine the absorption, distribution, metabolism, excretion, and potential adverse effects of the product. These studies assess the formulation, manufacturing process, stability, and compatibility of the product with the skin to ensure its integrity and safety. If the preclinical studies show promising results, an Investigational New Drug (IND) application is submitted to the regulatory authority. The IND application includes data from preclinical studies, details of the proposed clinical trials, information on the manufacturing and control of the product, and any previous human experience if available. The regulatory authority reviews the IND application to evaluate the proposed clinical trials and assess the safety of the product. The clinical trial phase involves conducting controlled studies in human subjects to assess the safety, efficacy, and bioequivalence of the topical or transdermal product. The clinical trial protocol is developed, detailing the study design, dosing regimen, inclusion and exclusion criteria, and endpoints to measure efficacy. The trials typically involve two groups—an experimental group receiving the investigational product and a control group receiving either a placebo or a reference product. The trials are conducted according to Good Clinical Practice (GCP) guidelines, ensuring the ethical conduct of the study and data integrity. Once the clinical trials are complete, the sponsor submits a New Drug Application (NDA) in the United States or a Marketing Authorization Application (MAA) in Europe to the regulatory authorities. These applications provide comprehensive data on the product, including information on its safety, efficacy, manufacturing process, quality control, labeling, and proposed intended use. The regulatory authority reviews the application and determines whether to grant marketing approval based on the evaluation of the data. After receiving marketing approval, the regulatory authority monitors the product's safety through post-marketing surveillance. This involves ongoing monitoring of adverse events, periodic safety updates, and compliance with any post-approval commitments. The regulatory authority may also conduct inspections of manufacturing facilities to ensure continued quality control and compliance with Good Manufacturing Practice (GMP) guidelines.

Throughout the regulatory approval process, it is essential to comply with specific guidelines and regulations for topical and transdermal dosage forms. These guidelines provide detailed requirements for various aspects of product development, including the selection of appropriate study designs, in vitro and in vivo methods for bioequivalence assessment, statistical analysis, and acceptance criteria. Adhering to these guidelines helps ensure standardized protocols and allows for consistent and reliable assessment of topical and transdermal products.

11.2.7 Challenges in Meeting Regulatory Guidelines

Meeting regulatory guidelines for topical and transdermal products can present various challenges. Variability in skin permeability, in vitro versus in vivo correlation, sensitivity to formulation and application factors, interchangeability of patch products, stability, and shelf life are some key challenges in meeting regulatory guidelines. The skin is a complex barrier that can vary significantly across individuals and body sites and even within the same individual over time. This variability in skin permeability poses a challenge in ensuring consistent drug delivery and absorption across different populations. Manufacturers need to carefully select appropriate test sites and consider inter- and intra-subject variability when designing bioequivalence studies (Maibach and Wester 2013). In vitro release studies are commonly employed to assess the equivalence of drug release from topical and transdermal products. However, establishing a correlation between in vitro release and in vivo drug absorption can be challenging. Factors such as skin metabolism, binding to skin components, and drug degradation within the skin can affect the in vivo bioavailability, which may not be fully predicted by in vitro studies. Therefore, manufacturers often need to conduct in vivo studies to establish bioequivalence (Barry 2012). Topical and transdermal products can be influenced by various formulation and application factors, such as the type of vehicle, pH, occlusiveness, and application technique. These factors can affect drug release, absorption, and distribution, resulting in potential differences in bioavailability between the generic and reference products. Manufacturers need to carefully control these factors to ensure bioequivalence (Ghosh et al. 2014). For transdermal patches, ensuring the interchangeability of adhesive patches can be challenging. Variations in adhesive properties, such as adhesion strength, peel force, and drug release kinetics, can impact drug absorption and efficacy. Manufacturers need to validate the interchangeability of adhesive patches through comparative bioequivalence studies to prove that generic versions are therapeutically equivalent to the reference product (Noel et al. 2017). Stability and shelf life of topical and transdermal products can impact their efficacy and quality assurance. Changes in active pharmaceutical ingredient concentration, dissolution, degradation, and physical properties can affect drug release and absorption. Manufacturers need to conduct stability studies to demonstrate that the generic product remains stable over time and is bioequivalent to the reference product throughout its stated shelf life (Sawnhey et al. 2016). By addressing these challenges, manufacturers can ensure that their topical and transdermal products meet regulatory guidelines for bioequivalence, thereby providing safe and effective alternatives to the reference products.

11.3 Manufacturing Challenges

Scaling up the manufacturing process of topical and transdermal dosage forms can be challenging due to several factors. These challenges include formulation issues, manufacturing process optimization, equipment selection and validation, regulatory compliance, and quality control.

Developing a formulation that can be scaled up without compromising the product's safety, efficacy, and stability is crucial. Formulation challenges may arise when transitioning from small-scale laboratory batches to large-scale manufacturing. For example, the viscosity and rheological properties of the formulation may change, affecting the product's spreadability and skin adhesion. Formulation adjustments may be required to ensure consistent performance throughout the manufacturing process. Scaling up the manufacturing process requires optimizing process parameters such as mixing, heating, cooling, and drying times to achieve a consistent and reproducible product. The optimization process may involve evaluating different manufacturing techniques and equipment, identifying critical process parameters, and conducting process validation studies. Selecting appropriate equipment for large-scale manufacturing is crucial to ensure consistent product quality. Equipment scale-up considerations include factors such as mixing capacity, heating and cooling rates, processing times, and the ability to maintain process control parameters. Equipment validation studies should be conducted to verify that the equipment performs as intended and consistently produces the desired product. Regulatory compliance is essential throughout the manufacturing process. Manufacturing facilities must meet Good Manufacturing Practice (GMP) guidelines to ensure the quality, safety, and efficacy of the topical or transdermal product. The facilities must be designed and maintained to prevent cross-contamination, ensure proper hygiene, and control environmental factors that can affect product quality. Adequate documentation and record-keeping are necessary to demonstrate compliance with regulatory requirements. Quality control plays a crucial role in ensuring consistent product quality during scale-up. Robust quality control measures should be in place to monitor critical quality attributes such as drug content, release rate, and physical properties of the product. Analytical testing methods should be developed and validated to ensure accurate and reliable results. Quality control checks should be performed at different stages of the manufacturing process to identify and address any deviations.

For example, let's consider the scaling up of a transdermal patch for pain management. In small-scale laboratory batches, the selected formulation effectively delivers the active ingredient across the skin barrier and provides the desired analgesic effect. However, when scaling up the manufacturing process, challenges may arise. The viscosity of the formulation may change, affecting the adhesive properties of the patch and its ability to adhere to the skin consistently. Formulation adjustments may be required to maintain the desired adhesion and drug delivery characteristics.

Additionally, the manufacturing process optimization may involve evaluating different mixing techniques, optimizing heating and cooling rates, and ensuring sufficient drying times for large-scale production. The equipment used for mixing, drying, and packaging should be validated to ensure it consistently produces patches with the desired quality attributes. Regulatory compliance is crucial in the manufacturing process of transdermal patches. The manufacturing facility must meet GMP guidelines to ensure the integrity and quality of the product. Proper documentation and record-keeping are necessary to demonstrate compliance during regulatory inspections. Quality control measures play a vital role in ensuring consistent product quality.

11.3.1 Ensuring Consistency and Reproductivity in the Manufacturing Process

Ensuring consistency and reproducibility in the manufacturing process of topical and transdermal products is essential to maintain product quality and meet regulatory requirements. Here are some key strategies to achieve this:

Establishing a standardized formulation is crucial to ensure consistent product performance. This includes defining the composition and concentrations of active ingredients, excipients, and any other components used in the formulation. This formulation should be thoroughly tested and validated to confirm its stability, efficacy, and safety. Process validation is an essential step in establishing consistency and reproducibility in manufacturing. It involves designing and executing experiments or trials to demonstrate that the manufacturing process consistently produces products that meet the predetermined quality standards. This validation process should assess critical process parameters, equipment performance, and the impact of variability on product quality. Standard Operating Procedures (SOPs) should be developed and implemented to guide operators through the manufacturing process. SOPs should include detailed instructions, specifications, and acceptance criteria for each manufacturing step. Adequate training of personnel is crucial to ensure they understand and follow the SOPs correctly. Implementing a robust quality control program is essential to monitor product quality throughout the manufacturing process. This involves conducting regular testing of both raw materials and finished products to verify their compliance with specifications. Analytical methods used for testing should be validated to ensure accuracy and reliability.

Implementing a comprehensive quality assurance program is important to oversee and control all aspects of the manufacturing process. This includes conducting regular audits, ensuring compliance with GMP guidelines, and maintaining proper documentation and record-keeping. Quality assurance personnel should regularly review batch records, test data, and other quality-related documents to identify any deviations or issues. Implementing controls to ensure batch-to-batch consistency is crucial. This can include using well-defined and controlled manufacturing processes, conducting in-process checks and testing, and implementing robust change control procedures. Any changes in raw materials, equipment, or manufacturing procedures should be thoroughly evaluated and validated before implementation. Regularly evaluating the manufacturing process for opportunities for improvement is important to achieve consistency and reproducibility. This can involve monitoring process performance metrics, conducting root cause analysis for any deviations or failures, and implementing corrective and preventive actions.

By employing these strategies, manufacturers of topical and transdermal products can ensure consistent and reproducible manufacturing processes. This, in turn, helps to maintain product quality, efficacy, and safety while also meeting regulatory requirements.

11.4 Advances in Pharmaceutical Development

11.4.1 Nanotechnology-Based Drug Delivery System in Topical and Transdermal Formulations

The latest developments in topical and transdermal drug delivery systems have made it possible to administer therapies precisely to the site of action by enhancing drug penetration through the stratum corneum and improving bioavailability. Some dermatoses currently have few treatment options available despite different technical advances because of possible side effects and difficulties developing formulations. Approximately 16% of an adult's entire body weight is made up of skin, which is said to be the body's biggest organ. As a result, it is important for maintaining homeostasis (Tortora and Derrickson 2018) and functioning as a chemical, physical, and biological barrier against external environmental hazards (Dehdashtian et al. 2018). While traditional topical and transdermal formulations such as creams and ointments are adequate for treating some dermatological diseases, the limitations and drawbacks of bioavailability and targeted drug delivery necessitate the development and advancement of technologies to improve drug permeation and increase drug delivery (Das Kurmi et al. 2017).

Other physical penetration strategies in transdermal delivery being researched are sonophoresis (Nguyen and Banga 2018; Park et al. 2019), iontophoresis (Park et al. 2019), and electroporation (Baveja 2018). They are usually regarded as secure, productive, and efficient with good medicine bioavailability. In addition, a novel method of delivering medicine called electroporation employs tiny impulses of electricity. This facilitates the passage of hydrophilic drugs into the stratum corneum (Baveja 2018). Sonophoresis is the application of various ultrasonic wave frequencies to penetrate the skin's protective layer and enhance drug diffusion through the epidermis, hence enhancing medication penetration. Compared to high-frequency ultrasound, low-frequency ultrasound improves medication penetration more (Nguyen and Banga 2018; Park et al. 2019), but in iontophoresis molecules are pushed into the skin using an electric field (the repelling forces between similarly charged molecules). For the delivery of both charged and uncharged molecules, it delivers quick drug release (Park et al. 2019). However, since the models are patient-specific, many of these potential solutions suffer challenges with cost and large-scale applicability; to evaluate the long-term effects of any of the strategies, greater depth of research and clinical trials are thus necessary.

11.4.2 Personalized Medicine Tailoring Topical and Transdermal Dosage Forms to Individual Patient Needs

The field of personalized medicine, which aims to adapt medicines to a specific individual by taking into account that person's physiology, drug action, and genetic profile, has the potential to revolutionize the healthcare industry. The primary technology driving this paradigm change from traditional "one-size-fits-all" care to personalized medicine is three-dimensional (3D) printing. A three-dimensional item is created utilizing different computer software through the process of 3D printing, one layer at a time. Using 3D printing, it is feasible to create a wide variety of pharmaceutical dosage forms with different geometries, release profiles, and drug combination. Few of the main 3D printing technical platforms being investigated in the pharmaceutical sector include stereolithography, selective laser sintering, inkjet printing, fused filament fabrication, binder jetting, and pressure-assisted microsyringe. In the future, this technology might be used in a clinical context where medications could be given based on patient needs. Creating products from three dimensions is known as three-dimensional printing, commonly referred to as additive manufacturing or 3D printing. It involves depositing layers of material on top of one another to gradually develop a solid structure. It utilizes CAD software to transmit the required signals to a 3D printer, which converts the computerized digital model into two-dimensional (2D) portions and creates solid layers to generate the required objects (Vaz and Kumar 2021).

There is great role of technology in personalized medicines. Inkjet printing, in general, refers to systems that employ pattern-generating tools to digitally direct and position tiny liquid drops on a substrate. Continuous inkjet printers, as the name suggests, continuously emit a fine mist of liquid droplets onto a substrate, even when not explicitly necessary. This process involves introducing a pressure wave into the ink supply, resulting in vibrations that break up the ink into consistently sized droplets which are subsequently expelled from the nozzle. This technique has been utilized for a considerable period of time due to its consistent emission of ink droplets. While the advantage lies in the continuous creation of droplets, preventing clogging of the nozzle, it is worth noting that drawbacks such as low resolution and costly maintenance accompany this printing method (Vaz and Kumar 2021). In a drop-on-demand inkjet printer, ink droplets are expelled only when necessary. The printer typically has 100–1000 nozzles, and each nozzle is capable of independently ejecting droplets onto the substrate. The trigger signal instructs the printhead to release the droplets precisely where needed. This technology allows for precise control and efficient use of ink. In a thermal inkjet (TIJ) printer, thermal energy is used as the trigger to release the ink droplets. Each nozzle in the printhead has a resistor that heats up when triggered by an electric charge. This rapid heating causes the ink to vaporize and form a bubble, which propels the droplet out of the nozzle onto the substrate. The resistor then cools down, allowing the nozzle to be reloaded with ink for the next droplet. The main disadvantage of thermal inkjet printers is the use of high temperatures (around 200–300 °C), which can potentially degrade thermolabile active compounds present in some inks. This limitation needs to be considered when using thermal inkjet technology for sensitive applications (Vaz and Kumar 2021). In a piezoelectric inkjet printer, a piezoelectric element or actuator is used to generate pressure and eject the ink droplets. When an electric voltage is applied to the piezoelectric material, it changes its shape, exerting pressure on the ink reservoir. This pressure forces the ink out through the nozzle, forming a droplet onto the substrate. Once the piezoelectric element returns to its original shape, the nozzle is reloaded with ink, ready to be triggered again. This offers advantages such as precise control of ink droplet size and high-quality printing. They can also handle a wider range of inks compared to thermal inkjet printers. However, they may be more expensive and have more complex mechanisms (Alomari et al. 2015; Vadodaria and Mills 2020). The ability to use less volatile and more biocompatible fluids while operating at room temperature is one of this method's primary advantages (Acosta-Vélez 2016; Acosta-Vélez and Wu 2016).

11.4.3 Benefits and Challenges of Developing Combination Therapies in Topical and Transdermal Formulations

Drug distribution via the skin is a fascinating and difficult field. It has been both an intriguing or difficult study topic. Modern technological advancements lead to the transdermal delivery of various types of medications, including traditional hydrophobic small molecules, hydrophilic medications, and macromolecules. Because of its clear advantages over alternative delivery methods, transdermal systems are a preferred method of medication administration. Patients can administer themselves conveniently and painlessly through transdermal delivery. Another advantage of transdermal drug delivery is the ability to provide a sustained release of medication. This is particularly useful for medications with a short half-life that would require frequent dosing. By continuously releasing the drug through the skin, transdermal systems can maintain a constant plasma level of the medication, resulting in more effective and consistent therapy. Additional benefit of transdermal distribution is bypassing the gastrointestinal tract that may affect the absorption of medications. The first-pass effect can be eliminated, making it safer to provide less medication to hepato-compromised patients and reducing the adverse effect. On a monthly cost basis, transdermal systems are often less expensive than other therapies since patches are made to administer medications for 1–7 days. The ability to give medications in numerous doses, on-demand, or at a variable rate is another benefit of transdermal administration. This feature enhances the advantages of traditional patch dosage forms. The market for transdermal products is expanding, which further demonstrates the extremely high level of patient acceptance of transdermal therapies. The market for transdermal medication delivery, which was valued $12.7 billion in 2005, is anticipated to grow to $32 billion by 2015 (Paudel et al. 2010).

11.4.3.1 Case Studies and Examples of Combination Therapies of Transdermal Patches and Topical Drug

Transdermal systems have predominantly focused on the patch dosage form, yet there are alternative variations such as metered sprays like Evamist®, delivering estradiol, or gel formulations like AndroGel®, delivering testosterone, which operate within the transdermal era. Technically, these newer transdermal gels can be seen as a resurgence of the nitroglycerin ointment technology that was previously approved. However, these updated formulations exhibit enhanced sensory properties and market appeal. Consequently, these technologies present a viable alternative to patch production, particularly for potent drug compounds, as they are highly effective and cost-efficient. Furthermore, for the successful distribution of medication through transdermal patches, stability is a crucial consideration. With the possibility of patches containing volatile solvents or hygroscopic materials, appropriate packaging becomes imperative (Paudel et al. 2010).

11.4.4 Continuous Monitoring and Controlled Release in Transdermal Drug Delivery System

The Internet of Things (IoT) plays a crucial role in smart medicine delivery systems (Fig. 11.1) by utilizing advanced technology to monitor and improve the effectiveness of medicinal therapy. IoT-based drug delivery systems utilize various hardware, sensors, and smart tools to monitor and evaluate a patient's health parameters in real time, leading to customized and targeted medicine therapy. These systems employ devices of different sizes and types, such as wearable and implantable devices like infusion pumps, smart pens, inhalers, and auto-injectors. However, there are a number of difficulties in developing and implementing IoT-based medicine delivery systems, including confidentiality and privacy of the data collected by these devices, strict regulations and standards, and compatibility and reliability issues.

11.4.4.1 Real-Time Monitoring and Control of Drug Release

IoT-based medication transdermal delivery systems are becoming more common in the healthcare industry as a result of their ability to improve drug delivery's precision, efficacy, and efficiency. Different techniques are used in medicine transdermal

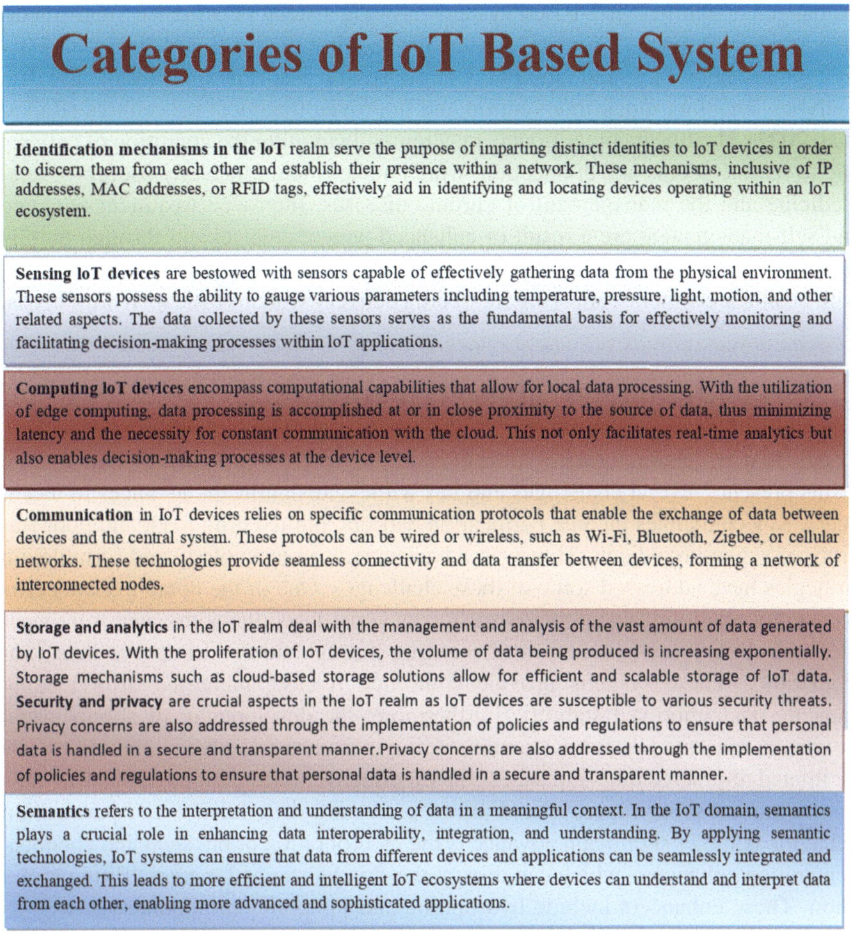

Fig. 11.1 Categories of IoT-based system

delivery system to examine data, make wise judgments, and optimize therapy. In systems like insulin administration for detecting diabetes, where data is continually monitored and evaluated to determine right dosage form, closed-loop control algorithms are utilized. Use of pharmacokinetic and pharmacodynamic algorithms can improve dosage regimes and help in forecasting medication based pharmacokinetic processes i.e. absorption, distribution, metabolism and excretion. Datasets are analyzed by automated algorithms to find patterns or correlations that guide medicine dose choices and improve treatment results. Effectiveness and safety of drug delivery systems are improved by these algorithms, which allow for customized dose recommendation and modifications (Raikar et al. 2023). Smart drug delivery systems have a great deal of promise of increasing medication adherence and reducing the load of demanding regimens. By utilizing their connection and smooth data exchange capabilities, IoT-based systems provide special benefits for smart medicine administration. IoT sensors continually monitor medication delivery parameters, patient responses, and ambient conditions with real-time monitoring and data analytics, enabling quick intervention and improved treatment procedures (Raikar et al. 2023). Healthcare practitioners may remotely monitor the condition of patients due to remote access and connection, which makes them especially useful for telemedicine and the management of chronic diseases. Improved treatment outcome and self-management are a result of enhanced patient engagement through mobile apps and wearables (Saunders et al. 2019). The infrastructure enables integration with healthcare systems, allowing smart drug delivery systems to adapt to new needs.

11.5 Conclusion and Future Perspectives

Conclusion of pharmaceutical development of transdermal and topical dosage forms presents several challenges and has witnessed significant advances in recent years. Key challenges include permeation barriers, limited drug load, and poor stability of active ingredients. However, advancements in technology and formulation strategies have addressed some of these challenges. One of the notable advances is the use of novel delivery systems such as nanoparticles, liposome, and microneedles, which increase penetration of the drug through skin and improve their bioavailability. These systems provide controlled release, increased drug loading capacity, and improved stability. Furthermore, the development of sophisticated in vitro/in vivo models for evaluation of drug permeation and skin absorption has facilitated the screening and selection of effective formulations. These models mimic human skin properties and aid in predicting drug release and permeation behavior. Another significant advance is the incorporation of physical and chemical penetration enhancers, which can disrupt the skin barrier and increase drug permeation. These enhancers include hydrophilic and lipophilic compounds, as well as techniques such as iontophoresis, sonophoresis, and microneedle-based intradermal delivery. Moreover, the introduction of solubility-enhancement strategies, such as cosolvency, complexation, and nano-emulsification, has improved the formulation of poorly soluble drugs for transdermal and topical delivery. These approaches enhance dissolution and solubility of the drugs, leading to improved drug release and permeation. Despite the notable advancements, there are still areas of research that need to be explored. Future studies should focus on developing novel strategies to overcome skin permeation barriers, enhance drug loading capacity, and improve stability of active ingredients. Additionally, the exploration of personalized medicine and the development of custom-made transdermal and topical forms based on individual patient needs hold immense potential. To conclude, the pharmaceutical development of transdermal and topical forms has gradually progressed, along with advancements in drug delivery systems, penetration enhancers, solubility-enhancement strategies, and sophisticated evaluation models. These advances have improved the efficacy and optimization of transdermal and topical system. Continued research in this field is expected to yield breakthroughs in overcoming the remaining challenges and bringing transformative changes in development of transdermal and topical delivery system.

References

Abouhussein DM (2021) Enhanced transdermal permeation of BCS class IV aprepitant using binary ethosome: optimization, characterization and ex vivo permeation. J Drug Deliv Sci Technol 61:102185

Acosta-Vélez GF (2016) 3D pharming: direct printing of personalized pharmaceutical tablets. Polym Sci 2(1):1–10

Acosta-Vélez GF, Wu BM (2016) 3D pharming: direct printing of personalized pharmaceutical tablets. Polym Sci 2(1):11

Alany R (2017) Topical and transdermal formulation and drug delivery. Pharm Dev Technol 22(4):457–457

Alomari M, Mohamed FH, Basit AW, Gaisford S (2015) Personalised dosing: printing a dose of one's own medicine. Int J Pharm 494(2):568–577

Apolinário AC, Hauschke L, Nunes JR, Lourenço FR, Lopes LB (2021) Design of multifunctional ethosomes for topical fenretinide delivery and breast cancer chemoprevention. Colloids Surf A Physicochem Eng Asp 623:126745

Barry BW (2012) Topical and transdermal drug delivery: principles and practice. Oxford University Press

Baveja S (2018) Drug delivery: steps toward replacing the pill. Berkeley Sci J 22(2). https://doi.org/10.5070/bs3222039588

Brown MB, Martin GP, Jones SA, Akomeah FK (2006) Dermal and transdermal drug delivery systems: current and future prospects. Drug Deliv 13(3):175–187

Das Kurmi B, Tekchandani P, Paliwal R, Rai Paliwal S (2017) Transdermal drug delivery: opportunities and challenges for controlled delivery of therapeutic agents using nanocarriers. Curr Drug Metab 18(5):481–495. https://doi.org/10.2174/1389200218666170222150555

Dehdashtian A, Stringer TP, Warren AJ, Mu EW, Amirlak B, Shahabi L (2018) Anatomy and physiology of the skin. In: Riker A (ed) Melanoma. Springer, Cham, pp 15–26

Domínguez-Hüttinger E, Ono M, Barahona M, Tanaka RJ (2013) Risk factor-dependent dynamics of atopic dermatitis: modelling multi-scale regulation of epithelium homeostasis. Interface Focus 3(2):20120090

Durga KN, Bhuvaneswari P, Hemalatha B, Padmalatha K (2022) A review on transdermal drug delivery system. Asian J Pharm Technol 12(2):159–166

Farjami A, Salatin S, Jafari S, Mahmoudian M, Jelvehgari M (2021) The factors determining the skin penetration and cellular uptake of nanocarriers: new hope for clinical development. Curr Pharm Des 27(42):4315–4329

Ghezzi M, Pescina S, Padula C, Santi P, Del Favero E, Cantù L, Nicoli S (2021) Polymeric micelles in drug delivery: an insight of the techniques for their characterization and assessment in biorelevant conditions. J Control Release 332:312–336

Ghosh TK, Ghosh A, Bishop M (2014) Topical dermatological drug delivery. Springer

Klang V, Matsko NB, Valenta C, Hofer F (2012) Electron microscopy of nanoemulsions: an essential tool for characterisation and stability assessment. Micron 43(23):85–103

Kováčik A, Kopečná M, Vávrová K (2020) Permeation enhancers in transdermal drug delivery: benefits and limitations. Expert Opin Drug Deliv 17(2):145–155

Maibach H, Wester RC (2013) Skin bioavailability and bioequivalence. J Control Release 171(3):605–608

Makhmalzade BS, Chavoshy F (2018) Polymeric micelles as cutaneous drug delivery system in normal skin and dermatological disorders. J Adv Pharm Technol Res 9(1):2

Nguyen HX, Banga AK (2018) Electrically and ultrasonically enhanced transdermal delivery of methotrexate. Pharmaceutics 10(3):117. https://doi.org/10.3390/pharmaceutics10030117

Noel R, Finnin BC, Patel RR (2017) Adhesive transdermal delivery systems. In: Handbook of non-invasive drug delivery systems: science and technology, vol 2. Elsevier, pp 359–385

Park J, Lee H, Lim G-S, Kim N, Kim D, Kim Y-C et al (2019) Enhanced transdermal drug delivery by sonophoresis and simultaneous application of sonophoresis and iontophoresis. AAPS PharmSciTech 20(3):96. https://doi.org/10.1208/s12249-019-1309-z

Paudel KS, Milewski M, Swadley CL, Brogden NK, Ghosh P, Stinchcomb AL (2010) Challenges and opportunities in dermal/transdermal delivery. Ther Deliv 1(1):109–131

Peña-Juárez MC, Guadarrama-Escobar OR, Escobar-Chávez JJ (2021) Transdermal delivery systems for biomolecules. J Pharm Innov 17:1–14

Phatale V, Vaiphei KK, Jha S, Patil D, Agrawal M, Alexander A (2022) Overcoming skin barriers through advanced transdermal drug delivery approaches. J Control Release 351:361–380

Priya S, Desai VM, Singhvi G (2022) Surface modification of lipid-based nanocarriers: A potential approach to enhance targeted drug delivery. ACS Omega 8(1):74–86

Rabiei M, Kashanian S, Samavati SS, Jamasb S, McInnes SJ (2020) Nanomaterial and advanced technologies in transdermal drug delivery. J Drug Target 28(4):356–367

Rai VK, Mishra N, Yadav KS, Yadav NP (2018) Nanoemulsion as pharmaceutical carrier for dermal and transdermal drug delivery: formulation development, stability issues, basic considerations and applications. J Control Release 270:203–225

Raikar AS, Kumar P, Raikar GV, Somnache SN (2023) Advances and challenges in IoT-based smart drug delivery systems: a comprehensive review. Appl Syst Innov 6(4):62

Ramadon D, McCrudden MT, Courtenay AJ, Donnelly RF (2021) Enhancement strategies for transdermal drug delivery systems: current trends and applications. Drug Deliv Transl Res 12:1–34

Sarheed O, Shouqair D, Ramesh KVRNS, Khaleel T, Amin M, Boateng J, Drechsler M (2020) Formation of stable nanoemulsions by ultrasound-assisted two-step emulsification process for topical drug delivery: effect of oil phase composition and surfactant concentration and loratadine as ripening inhibitor. Int J Pharm 576:118952

Saunders A, Messer LH, Forlenza GP (2019) MiniMed 670G hybrid closed loop artificial pancreas system for the treatment of type 1 diabetes mellitus: overview of its safety and efficacy. Expert Rev Med Devices 16(10):845–853

Sawnhey RR, Bahl D, Kansara K (2016) Stability issues in topical medications: a review. J Adv Pharm Technol Res 7(2):44–53

Souto EB, Cano A, Martins-Gomes C, Coutinho TE, Zielińska A, Silva AM (2022) Microemulsions and nanoemulsions in skin drug delivery. Bioengineering 9(4):158

Takeuchi I, Shimamura Y, Kakami Y, Kameda T, Hattori K, Miura S, Shirai H, Okumura M, Inagi T, Terada H, Makino K (2019) Transdermal delivery of 40-nm silk fibroin nanoparticles. Colloids Surf B Biointerfaces 175:564–568

Tapfumaneyi P, Imran M, Mohammed Y, Roberts MS (2022) Recent advances and future prospective of topical and transdermal delivery systems. Front Drug Deliv 2:957732

Tortora GJ, Derrickson BH (2018) Principles of anatomy and physiology. John Wiley & Sons

Vadodaria S, Mills T (2020) Jetting-based 3D printing of edible materials. Food Hydrocoll 106:105857

Vaz VM, Kumar L (2021) 3D printing as a promising tool in personalized medicine. AAPS PharmSciTech 22:1–20

Waghule T, Singhvi G, Dubey SK, Pandey MM, Gupta G, Singh M, Dua K (2019) Microneedles: A smart approach and increasing potential for transdermal drug delivery system. Biomed Pharmacother 109:1249–1258

Yamada M, Tayeb H, Wang H, Dang N, Mohammed YH, Osseiran S, Belt PJ, Roberts MS, Evans CL, Sainsbury F, Prow TW (2018) Using elongated microparticles to enhance tailorable nano-emulsion delivery in excised human skin and volunteers. J Control Release 288:264–276

Yotsumoto K, Ishii K, Kokubo M, Yasuoka S (2018) Improvement of the skin penetration of hydrophobic drugs by polymeric micelles. Int J Pharm 553(1–2):132–140

Advances in Ophthalmic Formulation Development

12

Rohit Bhawale, Vaibhavi Srivastava, and Neelesh Kumar Mehra

Abstract

Recent advancement in ophthalmic drug delivery is mostly concentrated on medication administering in the form of simple eye drops. Novel, innovative ophthalmic dosage forms possess advancement such as sustained release, less dosing frequency and improved ocular bioavailability and permeability. The present chapter will give an overview of the recent progress and challenges related to advanced ophthalmic dosage forms in addition to anatomy of the human eye and barriers which will be under consideration while developing ophthalmic drug delivery systems. Application of 3DP (three-dimensional printing) in advancement of ocular delivery systems will also be discussed in detail along with the evaluation tests and their regulatory considerations.

Keywords

Ophthalmic · Nanomedicine · Liposomes · Nanoparticles · Regulatory

12.1 Introduction

The eye is a complex but essential sensory organ playing a characteristic role in providing vision to living organisms. The eye is a delicate organ with complex anatomy and physiology (Fig. 12.1). The eye's three adjacent tissue layers—the centre pigmented layer, the inner neurosensory layer and the outermost collagenous

R. Bhawale · V. Srivastava · N. K. Mehra (✉)
Pharmaceutical Nanotechnology Research Laboratory, Department of Pharmaceutics, National Institute of Pharmaceutical Education and Research (NIPER), Ministry of Chemical and Family Welfare, Hyderabad, Telangana, India
e-mail: neelesh@niperhyd.ac.in

K. Jain, A. K. Yadav (eds.), *Advances in Pharmaceutical Product Development*, https://doi.org/10.1007/978-981-97-9230-6_12

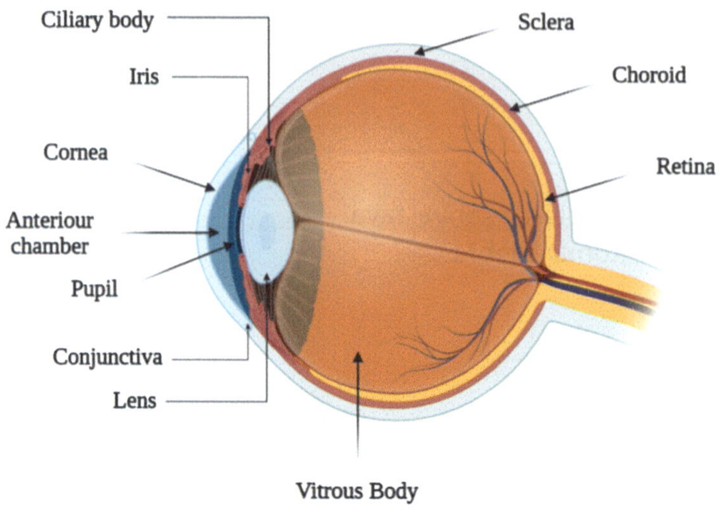

Fig. 12.1 The anatomy of the human eye

layer—are anatomically arranged in this order. The two regions of this physically and physiologically distinct organ that enable vision are (Irsch and Guyton 2009) the front or anterior segment (which makes up one-third of the eye) including the cornea, aqueous humour, conjunctiva, ciliary body, lens and iris and (Patel 2013) the back or posterior segment (which makes up the remaining two-thirds of the eye), which is made up of the sclera, retina, vitreous humour, choroid and optic nerve (Fig. 12.2).

To protect and keep this immune-privileged organ apart from the systemic circulation, each eye tissue has a unique structure (Cunha-Vaz 1979). In the anterior region, the collagenous layer that gives mechanical strength is known as the cornea. It is translucent in nature and oversees the focusing light on the retina. The fibrous collagenous posterior portion is opaque in nature, called sclera. The middle layer, or uvea, contains the ciliary body and iris. Aqueous humour, which is secreted by the ciliary body's smooth muscles and serves as both food for the anterior segment's avascular tissues and a regulator of intraocular pressure, is also produced by the ciliary body. The ciliary body also carries out debris drainage from the cornea and lens. The intermediate layer in the posterior part is made up of a huge network of capillaries known as the vascular choroid, which provides the retina with all its necessary nutrients. The retina, which is the innermost layer, aids in transmitting light signals to the brain. The lens refracts light entering the eye and is located between the two chambers (Kim et al. 2014). The eye, while having a well-defined physiological structure and makeup, is a very delicate organ that is prone to a number of disorders, most often uveitis, which encompasses any inflammation of the iris, choroid and ciliary body together known as the uvea (Yellepeddi and Palakurthi 2016). In many circumstances, uveal inflammation affects the posterior section as well.

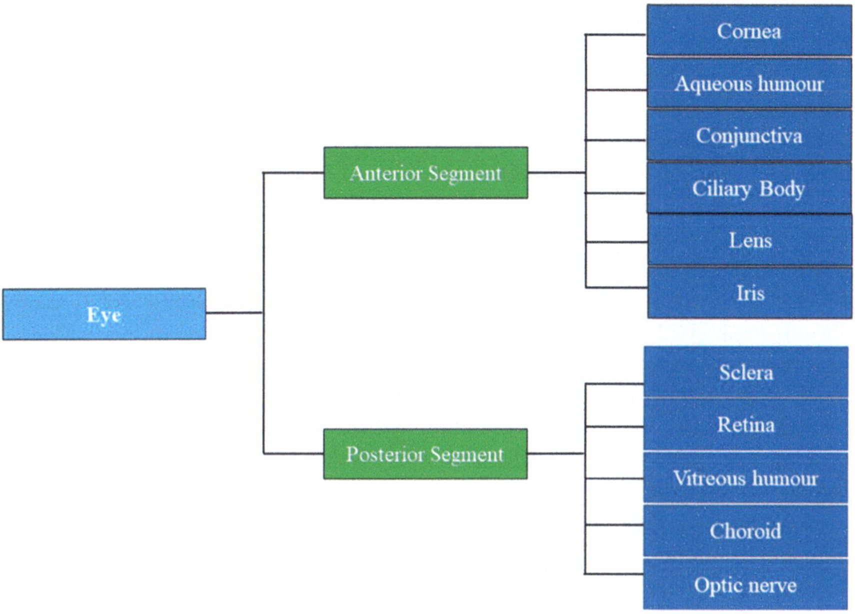

Fig. 12.2 Physiological classification of the human eye

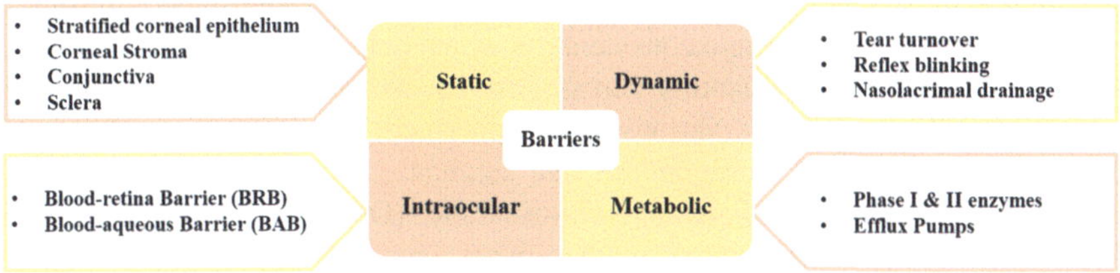

Fig. 12.3 Various ophthalmic barriers in the human eye

To preserve the visual accuracy and vision ability of the eye, several ocular tissues, membranes and mechanisms are working, but these defence mechanisms are creating hurdles for effective ocular drug delivery, thus commonly known as ocular barriers (Srivastava et al. 2023; Zulliger et al. 2015). The ocular barriers are mainly classified into two types, i.e. static and dynamic. The cornea, retina, sclera and blood-retinal barriers come under static barriers, whereas tear dilution, lymphatic clearance and conjunctival blood flow are considered as dynamic barriers. Delivering a drug across the cornea is a major restriction for drug delivery systems; thus the ocular bioavailability of drugs depends on how easily the drug will cross the above-mentioned barriers. The existence of cellular cornea and tear film acts as amphiphilic barriers for drug molecules (Agarwal et al. 2016). The different types of ocular barriers are enlisted in Fig. 12.3.

The conventional method of drug administration, i.e. simple eye drops and solution, is used as a preferred and easier ophthalmic dosage form; however poor ocular bioavailability is associated with them (Wu et al. 2019). High tear fluid turnover washing off 90–95% of the total drug administered produces the most challenging event of ocular drug delivery. Advancement in ocular delivery is very important to improve or prolong the ocular residence time of drug. Undesirable loss of drug molecules mostly occurs through nasolacrimal drainage, systemic absorption, blood-retinal barrier, enzymatic degradation and protein binding (Zhu et al. 2018; Bisht et al. 2018). Recent development in ophthalmic preparations has made it possible for prescribers to choose the best dosage forms for certain eye diseases (Bhosale et al. 2022; Maulvi et al. 2021; Rakhmetova 2020).

Even having some disadvantages and limitations like the administration of formulation and high drug loss due to precorneal space, the topical route is the most accepted route for ophthalmic drug administration. Systemic administration consists high amount of drugs which can cause various side effects. Drug delivery to the posterior structure of the eye by injection is the effective way but has lots of dangerous disadvantages like retinal detachment and haemorrhage. Considering these points the development of ophthalmic preparations focuses on the topical route of administration with prolonged residence time, site-specific delivery, sustained release profile and better corneal permeability (Baranowski et al. 2014). In the past few years, various drug delivery systems came into research with the help of using permeation enhancers, in situ gel, implants, stimuli-responsive and prodrugs, and also some nanotechnologies like nanoparticles, liposomes, dendrimers, niosomes and microneedles are taken into consideration which helped to increase ocular residence time and drug permeability across different ocular barriers and helped to improve drug bioavailability. Various natural polymers are used to formulate sustained-release ophthalmic formulations like xanthan gum, hyaluronic acid, cellulose derivative, gellan gum, alginate, chitosan, etc. (Dubashynskaya et al. 2019; Pathak et al. 2023). This chapter included the recent advancement in ophthalmic formulation till now and also discussed about different technologies that contribute to the ophthalmic field to deliver drugs.

12.2 Ideal Properties and Types of Ophthalmic Preparation

12.2.1 Ideal Properties for Ophthalmic Preparation

12.2.1.1 Sterility (Absence of Pyrogen)

Sterility is one of the essential properties of ophthalmic preparations as the eye contains favourable conditions for growth of microorganisms. Nutrient-rich tear fluid, humidity, mucin and optimum temperature are perfect environment for microorganism flourishment. Presence of microorganisms in ophthalmic preparations may cause several severe corneal infections. To avoid such horrifying conditions and protect the eye, ophthalmic products should be sterile and free from pyrogens or its byproducts. For multi-dose containers, preservatives are very essentials in order to maintain sterility. The preservative should not be poisonous, irritating or incompatible with medications (Hasegawa et al. 2021).

12.2.1.2 Absence of Foreign Particle

As explained earlier the eye is a very sensitive and delicate organ; thus it needs to be taken extra care during administration of the dosage form. Even a small foreign particle may cause severe irritation, redness or allergic conditions to the eye. To avoid entrance of foreign particles inside the eye, several barriers are present such as eye lid, eye lashes, tear fluid turnover, blinking and a submucosal secretory immunoglobulin system. In the same order, ophthalmic products should be clear, non-hazy and devoid of impurities or particles. To achieve the foreign body-free product, ophthalmic preparations should pass through an appropriate membrane and sintered glass filters, and check for clarity test (Paradkar and Parmar 2017).

12.2.1.3 Corneal Tissue Compatibility

To penetrate the cornea and ocular tissues, the ophthalmic preparation should have the vehicle, which shows good wettability and penetrability. Advanced ocular products were developed to achieve prolonged residence time and enhanced contact time with corneal tissue; thus the product should be compatible with ocular tissues. The excipients and vehicles of formulation should be tested for biocompatibility study with corneal tissues to avoid irritation and toxicity (Patel 2013).

12.2.1.4 Isotonicity

Isotonicity is again an important characteristic of ocular drug delivery systems. To avoid the irritation and discomfort of the eye, the formulation should be isotonic with lachrymal secretions. Osmolarity of tear fluid ranges from 290 to 310 mOsmol/kg that will be isotonic to 0.9% NaCl solution, i.e. normal saline. Various isotonic vehicles are used to prepare ophthalmic formulations such as sodium acid phosphate buffer and 1.9% w/v boric acid (Patel 2013; Pramar 2009). Formulation should be isotonic to physiological; fluid is necessary to avoid irritation and discomfort to the eyes.

12.2.1.5 Optimum pH of the Solution

The stability, therapeutic activity, solubility and patient convenience of ophthalmic formulations mostly depend on pH which should be almost near to 7.4, i.e. pH of physiological fluid. The eye can handle solutions with a wide pH range as long as they are not heavily buffered since the tear will quickly return the eye's pH to normal (Lim et al. 2014).

12.2.1.6 Viscosity (Appropriate Rheological Properties)

Conventional ocular products are of liquid physical state, possessing very less viscosity, but the advanced ocular products have variable rheological properties in order to improve residence time. Sol-gel transitions are the most fascinating approach for advancing ophthalmic products, but they need to maintain optimum viscosity as highly viscous product may cause irritation and discomfort to the eye.

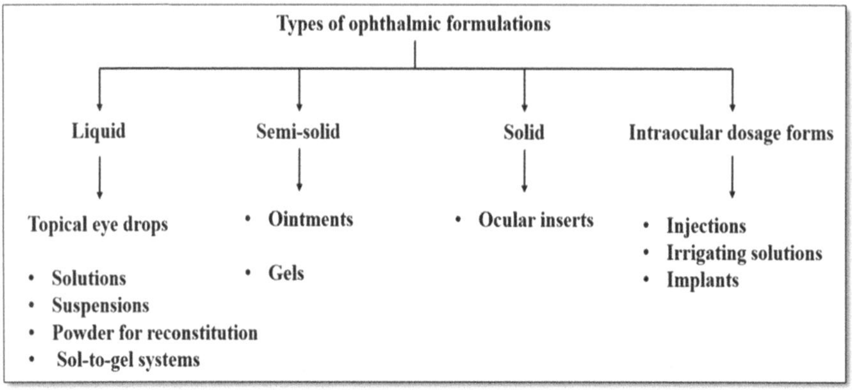

Fig. 12.4 Classification of ophthalmic dosage forms

Optimum viscosity for ophthalmic products ranges from 25 to 50 cps. Thickening agents are added in ophthalmic preparation to improve and prolong the contact time of drug in the eye (Lievens et al. 2019). The preparations should be viscous enough to stick to the corneal surface but liquid enough to pass through filters at the time of sterilization.

12.2.2 Types of Ophthalmic Preparations

The various ophthalmic formulations are classified based on the state of the final formulations, viz. liquid, solid and semi-solid (Baranowski et al. 2014). The classification of the various ophthalmic formulations is shown in Fig. 12.4.

12.3 Novel Drug Delivery Systems (NDDS) for Ophthalmic Formulations

12.3.1 In Situ Gelling System

Drug-laden polymeric solutions are intended to have low viscosity for simple instillation; nevertheless, when administered in cul-de-sac, it alters into pseudo-plastic gels that prolong the drug contact duration with the ocular tissues (Sokolowski 2018). In situ gelling can be achieved by various stimulations such as ionic alteration, thermo-stimulation, pH responsive or change in microenvironment (Ranch et al. 2021).

The antibacterial activity against methicillin-resistant *Staphylococcus aureus* (MRSA) in rabbits was increased by another pH-sensitive in situ gel containing vancomycin that provided a sustained release, releasing 39.2% after 24 h. The lipid film hydration process was utilized to formulate vancomycin-loaded niosomes with an average diameter of 53.7 nm (Allam et al. 2019). When compared to a regular eye drop solution, a pH-sensitive in situ gel containing bimatoprost (BIM) solid-lipid nanoparticles (SLN) was safe and increased the precorneal drug residence time. Rita et al. prepared, characterized and evaluated the BIM-SLN for the treatment of glaucoma. The high shear homogenization prepares SLN by following the probe sonication and later on incorporates into the pH-sensitive gel to form in situ gel containing BIM. The prepared formulation results in prolonged release, and HET-CAM test results show there is no evidence of eye irritation by formulation. On the same side, histopathological study proves there is no tissue damage due to application of the prepared formulation (Wadetwar et al. 2020).

12.3.2 Mucoadhesives

To improve contact time and reduce the clearance of drug through tear turnover, advancement in drug delivery system is introduced by utilizing the mucous layer of the eye. The contact of a substance, often a polymer, whether natural or synthetic, with the mucosa or any accompanying mucus is known as mucoadhesion (Bayer 2022). The interaction of the dosage form with mucus-coated mucosal epithelial membranes is the key to the administration of mucoadhesive drugs. Because of this interaction, residence duration rises, giving the drug more opportunity to enter its designated site of action (Grassiri et al. 2021). The various theories are intended to explain the mucoadhesion of dosage form to the ocular surface (Shaikh et al. 2011).

- **Electronic theory**: Electron transport between the dosage form and the mucosal surface causes the interaction.
- **Wetting theory**: It generally applies to liquid mucoadhesive where the polymer and mucin have structural similarities. As a result, the mucoadhesive polymer may spread over the mucosal surface, and surface tension is reduced.
- **Adsorption theory**: Electrostatic effects, hydrogen bonds and van der Waals forces are used as the attraction processes at this place. Additionally, hydrophobic effects are involved, especially when amphiphilic mucoadhesive polymers are used and mucins and other particular polymers may form covalent bonds.
- **Diffusion theory**: It takes into account how soluble mucins diffuse into the mucoadhesive and how polymers interact with mucus.

Abovementioned theories do not take place as they explain straightforward. But the mechanism of adhesion appears according to the abovementioned theories. We can't say any theory via adhesion takes place, but the part of theory might involve at the time of adhesion (Smart 2005; Khutoryanskiy 2011). If we consider the series of events in how this theory participates in the mucoadhesive mechanism, initially wetting is followed by electronic, adsorption and diffusion. Again, clectronic like this cycle repeats (Khutoryanskiy 2011; Ravindran et al. 2012).

Li et al. prepared drug-loaded chitosan film for ocular drug delivery to treat glaucoma. In order to prepare a chitosan-brimonidine tartrate (BT) film, the anti-glaucoma drug BT was dissolved in a solution based on ordinary water. The resultant film is transparent, structurally sound and mucoadhesive. Crystals of the anti-glaucoma BT medication are in micron-sized form and are evenly scattered in the chitosan film. High corneal permeability and quick drug release kinetics are made possible by the chitosan-BT film, which may one day be used to deliver drugs to the eyes (Li et al. 2020).

12.3.3 Ophthalmic Micro- and Nano-Emulsions

In this biphasic system, a non-ionic surfactant or stabilizer is used to disperse a non-aqueous liquid into aqueous vehicles (Koli et al. 2021). We can deliver the hydrophobic drug incorporated in oil droplets as an emulsion that will improve aqueous solubility of cargo and thus finally available in the form of eye drops. Comparing the aqueous external phase-containing emulsion to the non-aqueous external phase, it was discovered to be less irritating and more tolerated by the eye (Maulvi et al. 2020a; Pandey et al. 2020). The erythromycin-loaded microemulsion was prepared by using ethanol, oleic acid and Tween® 80 to improve permeation in the goat model. Microemulsion also appeared as a transparent, and globule size was found to be a minimum of ~12.1 nm (Mukhopadhyay and Butola 2020). When the drug was delivered to the posterior region of the eye using a microemulsion loaded with fluocinolone acetonide and cow ghee (CG) as a permeation enhancer, there were increased penetration and improved in vivo drug kinetic characteristics. Water titration methods were used to prepare microemulsion, and CG, Labrasol (surfactant) and Transcutol (co-surfactant) act as corneal penetration enhancer. The transcorneal drug permeability got improved because of micellar formation. Drugs' ocular permeability is improved by the fatty acids in CG, which serve as an effective carrier (Gupta et al. 2019).

Another study involves cannabidiol-loaded nano-emulsion for effective ocular delivery. The nano-emulsion was prepared by using glycerin (tonicity adjuster) and Carbopol® 940 NF (mucoadhesive agent) in the rabbit model. The investigation's primary goal was to fully assess CBD's effects on the IOP (intraocular pressure) in the model. Utilizing hot homogenization and probe sonication, the nano-emulsion was formulated. At various temperatures, the formulation was tested for its stability study, and then the effect of formulation on IOP of the rabbit model was observed. IOP was decreased in the animal model under study when CBD was applied topically and prepared as a mucoadhesive nano-emulsion (Senapati et al. 2022).

12.3.4 Ophthalmic Nano-Suspensions

The water-insoluble drug dispersed in aqueous vehicles to form ophthalmic suspensions. Ophthalmic suspensions sometimes have longer residence time because drug particles have a propensity to accumulate in the conjunctival cul-de-sac (Peters et al. 2020). Particle size, dissolution rate of drug in tear fluids and intrinsic solubility are considered as a critical factor during the production of ophthalmic suspensions (Schoenwald and Stewart 1980). The particle size of < 10 μm shows poor drug retention time on the ocular surface but also shows high solubility and dissolution rate. On the other hand, a particle size of > 10 μm shows ocular irritation and rapid drainage from the ocular surface because of excessive tearing (Li et al. 2014). The stability of ophthalmic suspension is one of the major issues faced by the ophthalmologist. If a suspension is stored in a frozen condition will provide the problem of re-dispersion since in cold condition particles will clog together to produce a cack (Joshi 1994). To avoid the side effects associated with frequent dosage, a high-pressure homogenizer was used to manufacture a stable long retentive posaconazole ocular solution utilizing a polymer system (Carbopol 974P and xanthan gum). The zeta potential of −47.3 mV indicates an electrically stable suspension that is resistant to flocculation. Posaconazole suspension proved effective against *Candida albicans* fungal infection (10^{-8} CFU/mL). In accelerated stability studies, the product (which has a shelf life of 24 months) was stable for 6 months (Simta et al. 2020). During storage, changes in the crystal structure may occur, causing particle size to grow or decrease, influencing medication solubility and bioavailability. The result of olopatadine-loaded Kollidon® SR nanosuspension via spray once a day shows sustained release and relatively narrow particle size. The spray drying method was employed to prepare nanosuspension along with particle size ranging from 75 to 145 nm. The finding of in vivo experiment performed on Anatolian Merino sheep shows improved drug retention time (Güven and Yenilmez 2019).

12.3.5 Nanotechnologies Enabled Drug Delivery System in Ophthalmic Formulation Development

In the past few years, various drug delivery systems came into research with the help of using permeation enhancers, in situ gel, implants, stimuli-responsive and prodrugs, and also some nanotechnologies like nanoparticles, liposomes, dendrimers, niosomes and microneedles are taken into consideration which helped to increase ocular residence time and drug permeability across different ocular barriers and also helped to improve drug bioavailability. Qiao et al. developed a liposomal formulation with high stability, good patient compliance and a sustained release to treat dry eye disease. Rats were used to assess the retention, tolerance and therapeutic efficacy of rebamipide (RBM) liposomes. In vivo evaluation of RBM's ocular distribution was done on rabbit eyes. In comparison to suspensions, RBM liposomes exhibit greater corneal retention. RBM liposomes maintain the same effectiveness and enhance the concentration in the cornea and aqueous humour while lowering the dose of the drug and causing ocular surface irritation. RBM liposomes offered an additional use in ophthalmic medicine, particularly for the treatment of dry eye disease (DED) (Qiao et al. 2022). Mitra et al. formulated a 0.1% cyclosporine A (CsA) nanomicelle system using a 0.05:1.0% mixture of octoxynol-40 and hydrogenated castor oil-40, which has extremely low critical micelle concentrations (0.00707%). The rabbit's cornea, conjunctiva and sclera received therapeutically

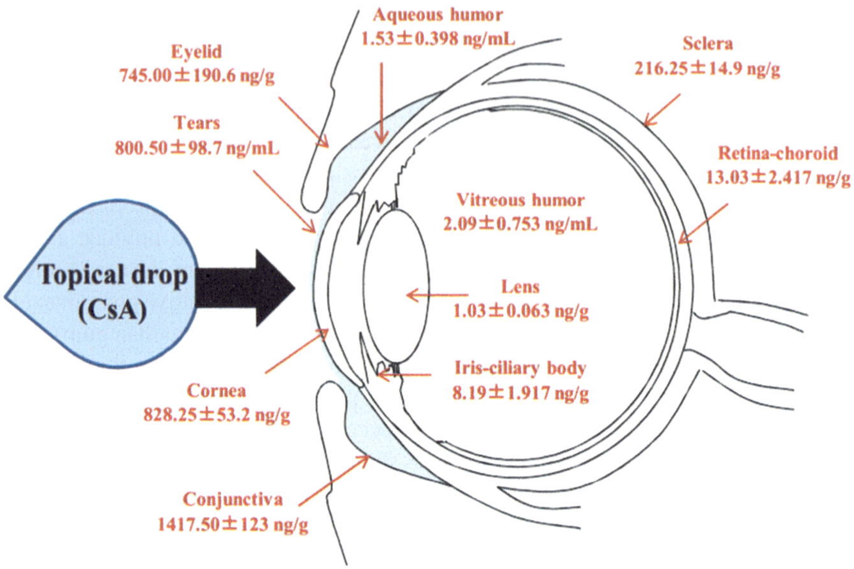

Fig. 12.5 The CsA concentration after single installation in each ocular tissue of rabbit (Nagai and Otake 2022)

substantial doses of CsA from the 0.1% CsA nanomicelle preparation (Cholkar et al. 2015). The CsA concentration after a single installation in each ocular tissue of rabbits is shown in Fig. 12.5.

Additionally, CsA concentrations in all ocular tissues—aside from the tear and upper eyelid—were markedly increased compared to Restasis® when repeatedly administered with a 0.05% CsA nanomicelle formulation (OTX-101) (Weiss and Kramer 2019). These results could be the consequence of better interaction between the tear's outer oil layer and the oil-based vehicle. In adult patient for treatment of DED, the US FDA approved OTX-101 in 2018 as Cequa® (OTX-101, 0.09%) ophthalmic solution (Mandal et al. 2019). Therefore, effective medication therapy for DED may be made possible by surfactant-based nanomicellar systems. Allam et al. sought to extend the pharmacological effect of vancomycin by inserting drug-loaded niosomes into a pH-sensitive in situ forming gel. Unexpectedly, the formulation showed physicochemical stability. Additionally, greater antibacterial activity was seen in vitro and in vivo as compared to vancomycin hydrochloride-free medication solution, and EE% of niosomes was greater than 46%. Additionally, niosomes had shown sustained release for longer than 24 h (Allam et al. 2019).

12.3.6 Therapeutic Contact Lenses

The contact lenses loaded with drug which are able to show therapeutic effect are known as therapeutic contact lenses. Therapeutic contact lenses may increase the bioavailability of drug by at least 50% by causing the high residence time and come into close contact with the cornea (Desai et al. 2020). Drugs released through contact lenses have a residence time of more than 30 min in the pre- and post-tear film, compared to eye drops which are only 1–2 min (Tieppo et al. 2012; Hui and Willcox 2016). The use of therapeutic contact lenses can lower the total drug dose required, the frequency of dosing and the quantity of drug lost through systemic absorption (Hiratani et al. 2005; Hsu et al. 2015). In general, the lenses are well tolerated (Maulvi et al. 2020b). Soaking, ion ligation, molecular imprinting and the use of colloidal polymeric nanoparticles or supercritical fluids are some of the methods developed to laden contact lenses (Maulvi et al. 2014, 2016, 2019, 2020c). When compared to routinely inserted eye drops, dexamethasone kept in lenses using a film resulted in 200× larger quantities of the medication in the retina (Ross et al. 2019). Coating the surface of the ACUVUE® TrueEye™ contact lenses with phomopsidione nanoparticles (average size = 77.45 nm) produced a sustained release of the drug for 48 h. Keratitis might be treated using this lens. The test microorganisms employed in this investigation were methicillin-resistant *Staphylococcus aureus* (MRSA), *Candida albicans*, *Proteus mirabilis*, *Pseudomonas aeruginosa*, *Serratia marcescens*, and *Candida utilis* (Bin Sahadan et al. 2019). Dexamethasone-laden contact lenses are able to prevent anterior uveitis caused by lipopolysaccharide for 5 days and suture-induced corneal neo-vascularization for 7 days. The PLGA encapsulation process was used to prepare film and merged it into a hydrogel contact lens (Bengani et al. 2020).

12.3.7 Ocular Inserts

These are intended to deliver the drug based on diffusion at zero order rate constant (Rathore and Nema 2009). Ocular insert can prolong drug delivery, increase drug residence durations, improve bioavailability, and prevent peak and valley release patterns to get around the side effects that go along with those features (Wolters Kluwer 2023). Ketorolac-laden polymethyl methacrylate (PMMA) microspheres can be used as ocular coils in the lower conjunctival fornix to administer drugs over several months. By using the o/w emulsification method, the microspheres were prepared. With the aid of technology based on funnel volume, the ocular coils were filled with PMMA microspheres in their lumen. Voriconazole was successfully used in antibacterial investigations by porous rod-shaped hydrogel polymer that provided a regulated release of the medication for 12 h in *Aspergillus* and *Fusarium*. The rods, however, induced more tear production and were removed from the eye (Rakhmetova 2020).

12.4 Implementations of 3DP (Three-Dimensional Printing) in Ophthalmology and Its Regulatory Consideration

Three-dimensional printing (3DP) is not a new invention, but its use in the healthcare sector is a new approach which is helpful to solve the recent problems and clinical challenges in ophthalmic drug delivery as well as in other drug delivery too. This technique is cost-efficient and requires a biocompatible material for printing that should match to human physiology and thus helpful in organ transplantation, surgery as well as in the development of new drug delivery systems. Additionally, 3D printed medical implants and gadgets may be personalized for every patient to offer a more customized treatment strategy. Patients with various eye problems can benefit from the advantages and uses of 3DP, and recent developments in 3D bioprinting have opened up new therapeutic possibilities in ophthalmology. 3DP has the ability to offer more flexibility for patient-specific customization is a key aspect (Bhawale et al. 2023; Fan et al. 2020).

The future and present outlook for treating patients with a variety of eye problems may be greatly altered by the prospective applications of 3DP. A solution to the serious corneal donor scarcity may have been found with the development of 3DP cornea (Zhang et al. 2019). The 3DP technology also has the potential to customize a design to the demands of each patient, which can lead to a better anatomical fit. Traditional manufacturing techniques have been used to create a large number of ocular implants likely laser beam machining (LBM) and computer numerical control (CNC) machining which are subtractive manufacturing technique types via casting and forging (Davis et al. 2022). By using an effective 3DP medication delivery system in place of the current medical therapy, 3DP can be helpful in ophthalmology. Utilizing properly chosen biomaterials, the 3DP technology's adaptability and personalization may potentially be utilized to design glaucoma and cataract implants that are unique to each patient. The future of 3DP can help many individuals who have problems after eye surgery and from the side effects of medical treatments, even if there are still many challenges in choosing biocompatible materials and their printability under various 3DP techniques.

12.4.1 Corneal Tissue Bioprinting

To design 3D bioprints, the corneal tissue is viable to combine biomaterials with human proteins (Table 12.1). Corneal tissues can be rebuilt by using biocompatible materials, but at the time of selection, the criteria should be followed carefully for its in vitro settings. A stromal structure made of human adipose tissue-derived stem cells (hASCs) was able to mimic the characteristics of native corneal stroma with high cell survival rates using 3D bioprinting (Sorkio et al. 2018). The form and biological activities of limbal epithelium produced from human embryonic stem cells that were employed in different bioprinting were likewise those of the corneal epithelium. The interaction of the 3DP stroma integrating hASCs with a porcine cornea revealed its preliminary biocompatibility to integrate with the host tissue.

Table 12.1 Types of bio-inks made up of various types of biomaterials

Bioinks	Biomaterials	References
1. Bioink consist of hydrogels and particles	• PLGA-PEG (poly(lactic-co-glycolic acid-polyethylene glycol) with cell-laden CMC • PLGA/gelatin combines with methacryloyl (GelMa)	Ashammakhi et al. (2019)
2. Bioink made up of synthetic and natural biomaterials	• GelMa	
3. Bioink made up of synthetic biomaterials	• PEG-diacrylate (PEGDA) • PEGDA with alginate	
4. Bioink made up of natural biomaterials	• Chitosan with gelatin • Cellulose with alginate	

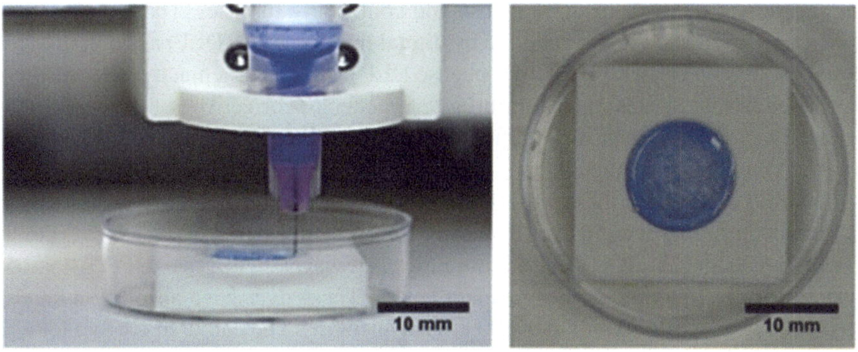

Fig. 12.6 The image showing the 3DP process of printing corneal structure and its outcome before the incubation (Isaacson et al. 2018)

The various bioprinting technologies are available by which we can produce high-resolution medical devices with any type of stem cells. The laser-assisted bioprinting (LaBP) technology is one of the most precise, flexible and powerful techniques for generating corneal tissues from stem cells. First and foremost, bioprints do not affect their biological and functional properties. The most interesting thing is that the artificial corneas are able to maintain the ideal systemic curved shape and distinctive arrangement of collagen lamellae. Isaacson et al. used a pneumatic 3D extrusion bioprinter in 2018 to develop 3D bioprinted corneas (Fig. 12.6) (Isaacson et al. 2018).

3D model of human cornea is constructed by collecting topographical data from matured human cornea. Different types of bio-ink combination with a low viscosity like collagen and alginate were tested while doing these experiments. In result they found the enhancement in mechanical stability of models which are printed with collagen-1 bio-ink combined with alginate. The corneal keratocytes' remarkable vitality persisted for 7 days following printing and cell insertion. Therefore, the preservation of high viability keratocytes implies that composite bio-ink made up of collagen and alginate is able to use for 3DP corneas. Despite innovative ways for replicating corneal structures, like magnetism or electrospinning, the findings demonstrate poor transparency. Shear-induced fibres may be produced with the use of 3DP technology. As collagen fibres differ from collagen fibrils and can even impact corneal keratocytes, thin collagen fibrils generated from decellularized corneal tissues can be employed (Muthusubramaniam et al. 2012; Kim et al. 2019).

12.4.2 Contact Lens

The introduction of hydrogel soft lenses was a key advancement in contact lens development. Contact lens manufacturing is still a complex procedure as it involved many steps and has limited flexibility for design. The utilization of newer techniques like 3DP is used to develop different types of contact lenses such as smart contact lenses which helps to diagnose and control eye diseases. The light-curing-based polymerization is mainly done by digital light printing (DLP) as compared to fused deposition modelling (FDM) because DLP printing has a much higher resolution (Bandari et al. 2021). DLP printing can generate corrective contact lenses by using Asiga Dental Clear resin which is mostly used in the dental industry (Alam et al. 2021). Fahad Alam et al. prepared and customized contact lenses for the management of colour blindness with the help of 3D printing. They used wavelength-selective filtering dyes and a resin made up of in situ synthesized poly(2-hydroxyethyl methacrylate, or pHEMA). To filter out unwanted wavelengths, two dyes (Atto 565 and 488) were functionalized in contact lenses. These dyes have absorption bands in the range of 550–580 and 480–500 nm wavelengths. The coloured contact lenses efficiently blocked 80–90% of the unwanted wavelengths, where colour vision-deficient patients have difficulty discriminating between shades of specific colours. Using the Ishihara test, contact lenses were evaluated on colourblind patients, and the results showed a clear improvement in the participants' sense of colour (Alam et al. 2022).

12.4.3 Drug Delivery

Hydrogel-based formulations for 3D printed patches can also enable the drug release effectively in the eye, such as the conjunctiva, without impairing vision or making the user uncomfortable blinking (Table 12.1). Tagami et al. developed lyophilized ophthalmic patches capable of manufacturing unique doses that may be tailored to hospital patients (Tagami et al. 2022). The antibiotic levofloxacin was incorporated in the drug release patches. The patch is printed by using different hydrogel-based bioink containing mannitol, HPMC, xylitol and drug. Most crucially, the patches can contain numerous active medicinal components and be designed to release varied amounts based on the needs of the patients. These eye patches may be made to administer various medication formulations, such as eye drops for those having cataract surgery (Gonzalez-Gonzalez et al. 2014) and mini tablets to treat inferior conjunctival fornix (Moosa et al. 2014). A tailored ophthalmic patch for each patient may be built using the chemical and physical properties of the materials used in the drug-release patch. For individuals who have had cataracts and glaucoma surgery, the biocompatible drug-releasing patch can assist reduce the need for frequent eye drop administration.

12.4.4 Regulatory Considerations of 3DP in Ophthalmology

Despite the promising future of 3DP applications in the medical industry, the legal framework for 3DP technology in pharmaceutical goods is not yet complete. Although the FDA authorized the 3DP medication Spritam in 2015, the application guidelines for 3DP were not available until 2017 (Tsui et al. 2022). Furthermore, the FDA has not published any restrictions regarding 3DP technology (Mohammed et al. 2021). It is a hindrance to the application of 3DP medical devices and restricts the clinical translation of 3DP goods to patients. As a result, it is hard to specify who will bear legal liability for disputes over the safety of 3DP items. However, the regulatory bodies have just recently designated teams and started programs to begin developing the 3DP standard requirements (Mohammed et al. 2021). Following the FDA's early attempts in 2019, the Medicines and Healthcare Products Regulatory Agency (MHRA), Therapeutic Goods Administration (TGA, Australia), European Medicines Agency (EMA) and other national regulatory bodies have begun to explore laws enabling innovative 3DP uses in medicine (Tsui et al. 2022).

12.5 Challenges in the Development of Ophthalmic Formulation

Although eye drops or ointments can be used to treat some viral or inflammatory eye conditions. The various anatomical and physiological obstacles that the eye has mean that achieving the required therapeutic effectiveness as well as ocular bioavailability remains challenging. Despite enormous effort in developing an effective ocular delivery method, several innovations have yet to be successfully turned into a marketable product. The majority of technologies that have undergone clinical trials have proven to be suitable effective and safe in animal models. Promising findings in animal models, however, may not always transfer into the same results in human trials. Any animal model showing illnesses and conditions cannot be the same as the physiological reactions in humans, particularly for ocular disorders. Although rodent models are well known, inexpensive and simple to use, rodent eyes are smaller and have a proportionately bigger lens and corneal surface than human eyes; rodent eyes also contain nictitating membranes (Mann et al. 2018; Rodrigues et al. 2018).

Rabbits are widely utilized to investigate ophthalmic formulations because their eyes are more similar in size to human eyes than rodent eyes. However, as compared to human eyes, rabbit eyes create more mucus, have fewer tears and blink less

(Mann et al. 2018; Ahn et al. 2016). These variations in anatomy and physiology have an impact on PK, highlighting how challenging it is to extrapolate from promising preclinical findings to clinical success (Rodrigues et al. 2018). There are genetically modified animal models available for testing medicinal effectiveness. Although naturally occurring disease models are more likely to describe correct pathophysiology, the practicalities of having animal models available on demand are hard (Mann et al. 2018). As previously stated, good preclinical results do not correctly predict comparable outcomes in humans, particularly in research related to nanoformulations. The effective clinical translation of nanomedicines is predicted to enhance medication therapy for human illnesses; yet nanomedicines have had limited success in reaching the clinic and have remained as "possible" remedies for the previous few decades (Satalkar et al. 2016; Park 2017).

Because of several anatomical and physiological obstacles, drug delivery experts and ophthalmologists have long had difficulty delivering medications into the eye (Fig. 12.7). Invincible static and dynamic ocular barriers prevent xenobiotics from entering the eye and inhibit therapeutic drugs from being actively absorbed. The intention behind developing ideal drug delivery is to improve the bioavailability and controlled release of the drug at the site of action which helps to cross different ocular barriers. Current advancements for anterior ocular segment disorders, such as punctum plugs, drug-eluting contact lenses and ocular implants, represent state-of-the-art drug release technologies. Parallel attempts for ocular delivery technology for back of the eye ailments have led in the approval of different intravitreal implants. Nanoparticles, nanowafers, dendrimers, nano micelles, liposomes and microneedles are among the novel drug-delivery methods being investigated for

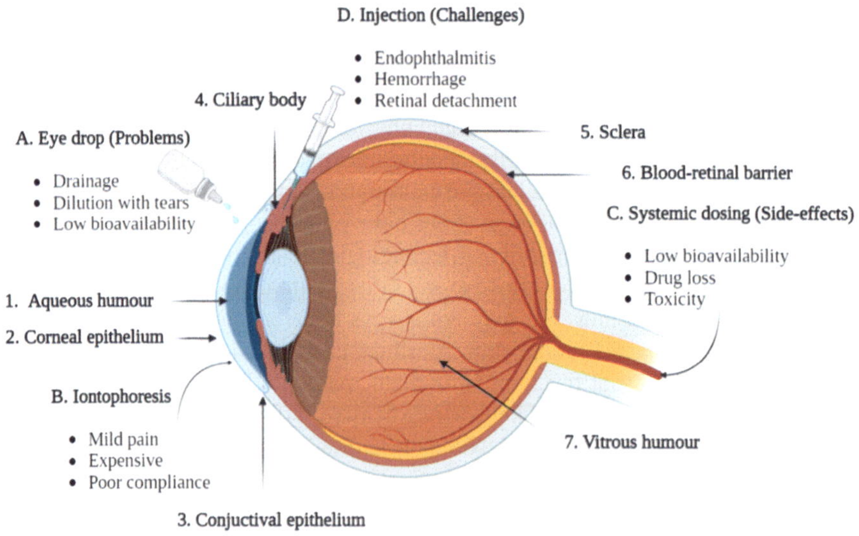

Fig. 12.7 The anatomical diagram of the eye shows various ocular barriers and drug delivery sites and challenges/problems/side effects, respectively

anterior and posterior disorders. To improve patient compliance for back of the eye problems, innovative techniques for the non-invasive administration of effective treatment drugs are being developed (Cabrera et al. 2019). As a result, the primary issue in the field of ocular drug delivery is to overcome these obstacles to obtain therapeutically appropriate drug molecule concentrations with increased sustained bioavailability in the target ocular location.

12.6 Evaluations Parameters/Test of the Ophthalmic Formulations

The eye is a delicate organ, and thus ophthalmic drug delivery systems need to evaluate on different parameters to confirm its safety and compatibility to ocular tissues. Unlike other delivery systems, ophthalmic delivery systems used to evaluate in terms of its stability, compatibility with ocular tissue, along with physiochemical properties such as viscosity, osmolarity, pH, etc.

There are two kinds of testing protocols and acceptability standards for ophthalmic products: (1) those that evaluate general quality features, such as identity, impurities, potency, particulate matter and sterility and (2) those that evaluate in vitro product performance, such as dissolution or release of the drug ingredient from the drug product. This section addresses the quality tests for ophthalmic products. The performance tests (dissolution/drug release) are addressed by Ophthalmic Products-Performance Tests <1771>. All the tests described below are according to the United States Pharmacopoeia (USP) 43 NF 38.

12.6.1 Physical Appearance

A qualitative description of the drug product should be provided. The final qualifying criteria should include a clear appeared formulation, packaging, colour, etc. A quantitative technique can be necessary if the hue varies while being stored. This is not a standard test, but it is part of the drug's manufacturer's specification.

12.6.2 Identification

In order to distinguish between compounds with potentially present closely related structural similarities, identification tests must first identify the drug or drugs' present moiety. Identity testing should be drug-specific [e.g., infrared (IR) spectroscopy]. If the test is nonspecific, at least two orthogonal nonspecific tests should be used. Near-infrared (NIR) or Raman spectrophotometric methods could be acceptable for the identification of the drug product. The chromatographic identification of the products is considered under the <621> and <201> in USP 43 NF 38.

12.6.3 Assay

The strength (content) of the medicinal product should be determined using a particular and stability-indicating test. When using a nonspecific assay test, it is important to apply additional supporting analytical techniques to obtain overall specificity. When there is indication that an excipient is interfering with a nonspecific assay test, a specific approach should be utilized. Antibiotics-Microbial experiments <81>, Ion Chromatography <1065> and <621> all have further details on individual experiments.

12.6.4 Impurities

The drug ingredient and excipients utilized in the production of the drug product may contain process impurities, synthetic byproducts and additional inorganic and organic contaminants. We can avoid respective impurities with the help of monographs of drug ingredients and excipients mentioned in the pharmacopeia. It's important to keep an eye on organic impurities that develop during the production process of a drug product as well as those that result from the degradation of the active ingredient in the drug. All articles fulfill the standards of Elemental Impurities-Limits <232> and Residual Solvents <467>.

12.6.5 Particulate and Foreign Matter

All ophthalmic items should be scrutinized for package integrity and the presence of visible foreign and particulate materials, to the greatest degree practicable. Extrinsic, or external matter, and intrinsic, or stuff associated to the product, are the two origins of these undesirable particles. Extrinsic material cannot be connected to the product or process. Intrinsic particles are introduced during product manufacturing or as a result of a change over time. One hundred percent inspection of injectable products in clear packages is required to remove final packages with visible particles. When 100% inspection of the injectable product in the final package is difficult, such as in opaque containers, alternative methods may be used to determine acceptability. Visible Particulates in Injections <790> provides further guidance for general inspection methods and a definition of "essentially free of visible particles" for batch compliance. Products intended for intraocular use must abide by Particulate Matter in Ophthalmic Solutions <789>. Products intended for external use must adhere to Particulate Matter in Injections <788>. Considerations for product evaluation and the background for both subvisible methods are found in Methods for the Determination of Particulate Matter in Injections and Ophthalmic Solutions <1788>.

12.6.6 Antimicrobial Preservatives

Products that are stored in containers that permit the extraction or administration of numerous doses require the addition of antimicrobial agents unless one of the following situations occurs:

1. Each monograph contains a separate set of instructions.
2. A radionuclide in the material has a physical half-life of <24 h.
3. The product satisfies the criteria of the Antimicrobial Effectiveness Testing <51>, even without the use of additional agents.

Antimicrobial Agents-Content <341> and <51> specifications must be followed by substances. It is important to set acceptance standards for antimicrobial preservatives used in multi-unit products.

12.6.7 Bacterial Endotoxins

All injectable ophthalmic medication products must be manufactured in a way that minimizes bacterial endotoxins, as specified in the Bacterial Endotoxins Test <85>. For ocular irrigation products, the limit is NMT 0.5 EU/mL, and for injectable or implanted medication products, the limit is NMT 2.0 EU/dose/eye. It is not applicable to topical ophthalmic products.

12.6.8 Uniformity of Dosage Units

Dosage forms supplied in single-unit containers are suitable for this test. It takes into account both the dosage form's bulk and the amount of drug substance(s) contained within. The test can be carried out using either content uniformity or weight variation (see Uniformity of Dosage Units <905>).

12.6.9 Sterility Test

Ophthalmic dosage formulations must fulfill the Sterility Tests <71> criteria. If the formulation's individual constituents do not lend themselves to conventional sterilizing processes, substances that fulfill the sterility standards mentioned in <71>, as well as aseptic production, may be employed. At the moment of filling and sealing, the immediate container for ophthalmic products must be sterile. The sterility test according to various official pharmacopeias is shown below in the diagram (Fig. 12.8a, b) (Paradkar and Parmar 2017).

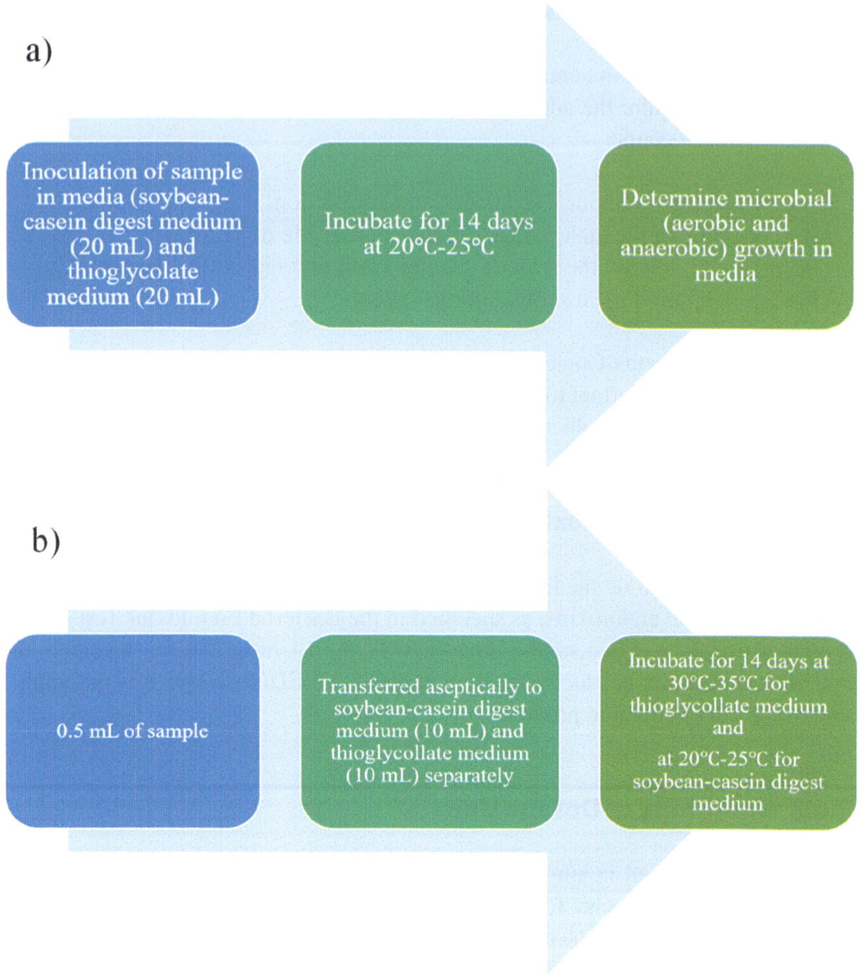

Fig. 12.8 Schematic presentation of ophthalmic formulation sterility test according to (**a**) US Pharmacopeia (USP) (2020) and British Pharmacopoeia (BP) (2022), (**b**) Indian Pharmacopoeia (IP) (2022)

12.6.10 Osmolarity

Ophthalmic products may be tolerated over a wide range of tonicity (0.5–5% sodium chloride, equivalent to about 171–1711 mOsm/kg). Hypotonic solutions are better tolerated than hypertonic solutions. Precautions should be taken to ensure that the product maintains its osmolarity during shelf life. Any possible contributions or interferences from the packaging system should be considered <771>.

Human tears have an osmolarity of around 310 mOsm/kg and a tonicity comparable to 0.9% w/v NaCl solution. The presence of sodium, chorine and bicarbonate transport helps to keep osmotic pressure in check. The physical characteristics of drug also have a role in the permeability of the blood-aqueous barrier. The pH of the formulation must be the same as the tear which is around 7.4. Various hydrophilic ointments are easy to wash, but their use is restricted by consideration of the osmotic effect (Dubald et al. 2018). The osmolarity of the prepared formulation is calculated by the following equation:

$$\mathrm{mOsm}/\mathrm{L} = \frac{\text{Concentration}\,(\mathrm{g}/\mathrm{L})}{\text{Molecular weight}} \times 100$$

If the prepared formulation is found to be hypotonic, we must add some specific amount of substance which will adjust the tonicity of the prepared formulation. Sodium chloride (NaCl) is mainly added as an adjustable substance. To calculate the required amount of sodium chloride (NaCl), the following equation is useful:

$$\text{Weight of NaCl required} = \frac{0.52 - \text{Freezing point of unadjustable solution}}{\text{Freezing point depression of 1\% solution of drug}}$$

12.6.11 Ocular Irritation

Ocular irritation is one of the most important parameters to evaluate and check the irritation caused by the prepared formulation. It promises safety and valuable assessment for local compatibility. The Draize rabbit eye test developed by John H. Draize in 1944 is accepted worldwide and accepted by FDA which is used to check eye irritation after applying the formulation to the eyes (Luechtefeld 2016). Three separate eye tissues—the iris, conjunctiva and cornea—are examined as part of the examination. The albino rabbit was chosen as the test species. To evaluate the irritancy of the prepared formulation, either 0.1 mL or 0.1 g of the formulation which may be an ointment, liquid, solid or paste is installed in different tissue like the cul-de-sac of the conjunctiva and cornea of the one eye. Another eye should be untreated or treated by only the excipient or vehicle. The evaluation is done at 1, 4, 24, 48 and 71 h on exposure to the material. If needed time may extend up to 4, 7 and 21 days. Record the irritation behaviour based on observation (Kolle et al. 2015).

12.6.12 Isotonicity Evaluation

Eye irritation and damage of eye tissues are the main reasons of ophthalmic preparation whose isotonicity is not maintained according to standard. The prepared formulation was mixed with a small quantity of blood to evaluate and observe the changes in blood cell shapes under 40× magnification. This change in the shape of blood cells is compared with the shape of blood cells kept in an isotonic solution

Table 12.2 Stability test conditions for ophthalmic formulations

Temperature	Stored for
30–35 °C	30 days
2–8 °C	30 days

(0.9% NaCl). If the formulation is hypertonic, cells will shrink, and if the formulation is hypotonic, cells will get swell or might burst (Maiti et al. 2011).

12.6.13 Stability Study

As per ICH guidelines, keep the formulation at different conditions mentioned below in Table 12.2. Later, evaluate the formulation for its particle size, viscosity, drug content and % entrapment efficiency (Paradkar and Parmar 2017).

12.6.14 pH

The pH of normal tears is around 7.4. The eye can withstand items with pH levels ranging from 3.0 to 8.6, depending on the formulation's buffering ability. The pH value of the formulation should be the one where the drug product is the most stable. Patients will tolerate formulations that aim for the extremes of the permissible pH range more readily if they have a limited buffering capacity <791> (Lim et al. 2014).

12.6.15 Viscosity

The residence time of ophthalmic formulation in the eye is an important consideration which totally depends on the viscosity of prepared formulation. Rheometer is used to determine the viscosity of prepared formulation with the 101–1800 rpm shear rate (Lievens et al. 2019).

12.7 Regulatory and Future Consideration

Due to a lack of stringent laws and appropriate regulatory requirements for ocular delivery, developing fresh administration routes and innovative formulations faces unsurmountable difficulties when it comes to constructing nonclinical programs, which are essential to accurately assessing risk for people. Regulations pertaining to nonclinical testing of ocular formulations are also not well defined due to concurrent regional variances. Therefore, these complexities offer new prospects to NDA and ANDA applicants in ophthalmic drug development and commercialization. This section includes a brief summary of the regulatory perspective which is helpful in the development of a safe, effective and quality final product. A further requirement of 21 CFR 211.167, which regulates foreign, harsh or abrasive particles in formulations, is a requirement of USFDA federal law that must be met by ophthalmic products. According to 21 CFR 200.50, ophthalmic equipment and goods must be sterile and devoid of pyrogens (US FDA 2023). Regulatory authorities sternly demand that formulations be isotonic and compatible with eye tissues. In order to reduce or avoid the threat of microbiological infections, 21 CFR 200.50 requests preservatives in multi-dose ophthalmic complex solutions (eCFR 2023). There is still confusion about drug delivery regarding ophthalmic products. For now, it depends on the use; if the contact lens is used for drug delivery purpose, then it acts as a drug, and if in addition it is given by correcting refractive index, then it is considered as a device (Srivastava et al. 2023).

As per the Drugs and Cosmetics Act (1940) and Rules (1945) in India, the standards for ophthalmic formulations (OFs) come under schedule FF (Rule 126a). Approval for a new drug ophthalmic product for a large number of people in the country is given by the Central Drug Standards Control Organization (CDSCO), which makes sure the availability of quality ophthalmic formulations at low costs will be affordable for a large number of people. On the other hand, the responsibility to set the standards for ocular formulations is of Indian Pharmacopoeia Commission (IPC). Therefore, the manufacturing, regulation and rational prescription of an ophthalmic formulation are standardized by IP standards and NFI prescribing information.

In the development of ocular drug delivery for both an anterior and posterior segment of the excipient, raw material will be the main consideration, and they should not affect the formulation stability and bioavailability of API in any aspect. The colouring agent is prohibited in the development of ophthalmic formulation. The development of nanocarriers for the ocular purpose used various ingredients, some of them shown in Table 12.3. Aseptic production processes and Good Manufacturing Practices (GMP), particularly with regard to cross-contamination, are essential because ocular formulations must be sterile. To ensure the efficacy of the formulation, the processes should be evaluated and monitored at every stage of manufacturing. The USFDA has intervened in several instances when the standards for ophthalmic product production have not been satisfied, which calls for strict planning and design of ocular products. As per IP, quality standards for the ophthalmic formulation mainly include identification, excipients, pH value, morphology, assay, water content, drug-carrier compatibility, irritation study, stability test and sterility test bacterial endotoxins and also container type and labelling also stated. Even if the transport of ocular drugs has altered due to nanotechnology, several aspects of the design and development of nanocarriers still need attention. The preparation of ophthalmic formulations without preservatives is now necessary since the use of preservatives and dispersants can result in sterile endophthalmitis or vision loss, and excessive surfactant concentrations can be hazardous. Vitreous clouding might cause due to delivery of microparticles/nanoparticles/liposomes/dendrimers via intraocular and periocular routes. The aggregation of nanoparticles remains a major challenge. The destiny of medications encapsulated in nanocarriers and their toxicity in the body must be determined via more research because there is a lack of regulatory guidelines for ocular pharmaceuticals and formulations (Gorantla et al. 2020; Mehra et al. 2016).

Table 12.3 A list of compounds used in the production of various nanomedicines for ocular DDS

Sr. no.	Carrier	Key ingredients	API/drug used	Ex vivo/cell lines/in vivo studies outcomes	References
1.	Nanospheres and nanocapsules	PLGA and poloxamer	Lactoferrin	Increase in surface bioadhesion and permanence/consistency	Varela-Fernández et al. (2022)
2.	SLN and NLC	Precirol® ATO 5 (P ATO5), oleic acid (OA) and Pluronic® F68 (P F68)	Loteprednol etabonate	The reduction in IL-1 and IL-6 levels counted by the ELISA test	Uner et al. (2023)
3.	Liposomes	L-α-Phosphatidylcholine (PC), cholesterol (CHO) and PEGylated lipids	Ciprofloxacin hydrochloride	The conjunctival (bovine) retention of the lipid vesicles was found more as compared to on the cornea	Moiseev et al. (2022)
4.	Niosomes	Cholesterol, Tween, 1,2-di-*O*-octadecenyl-3-trimethylammonium propane (DOTMA)	Epalrestat	The heat map shows the permeation of the drug through different tissues of the eye. Evaluation done with the help of IR-Raman	Kattar et al. (2023)
5.	Nanomicelles	Soluplus®, [grafted polymer of polyvinyl caprolactam-polyvinylalcohol-polyethyleneglycol (PVCL-PVA-PEG)]	Everolimus	Corneal penetration of nanomicelles of a drug is more as compared to the suspension of a drug	Mehra et al. (2021)
6.	Cubosomes	Myverol® 18–99 K, Poloxamer 407 (P 407)	Fluconazole	The cubosomal preparation of the drug found safer and more efficacious compared to the normal drug solution	Nasr et al. (2020)

Sr. no.	Carrier	Key ingredients	API/drug used	Ex vivo/cell lines/in vivo studies outcomes	References
7.	Dendrimers	DAB-core PAMAM dendrimer generation 5 (G5), Span80, Tween80	Brimonidine tartrate and timolol maleate	The nDHP (nano-in-nano dendrimer gel particles) found more superior compared to other formulation μDHP3 and μDHP10 in terms of cytocompatibility, degradability, drug release kinetics and corneal permeability	Wang et al. (2021)
8.	Gelling system	PEOz-PCL-PEOz gel, silk hydrogels, PCM–HEMA-based in situ gel	Sparfloxacin	The corneal permeation was enhanced and formulation also found non-irritant, tolerable and antimicrobial with the help of ocular tolerance test and histopathological study	Khan et al. (2015)
9.	Nanowafers	Poly (vinyl alcohol), silicone CMC polymer	Axitinib	Nanowafers are found to be more advantageous over conventional eye drop in case of penetration of the cornea and to treat the eye from the inside	Yuan et al. (2015)
10.	Implants/ inserts	Eudragit, PEO 2000, chitosan-thiolated PAA sodium alginate, polyvinyl acetate	Dexamethasone	Examethasone polyurethane dispersions (DX-PUD) downregulated proinflammatory cytokines/chemokines (IL-1b, IL-6) and inducible nitric oxidesynthase (iNOS) and upregulated IL-10 anti-inflammatory cytokine	Barbosa Saliba et al. (2016)

12.8 Conclusion

The complex ophthalmic dosage form is an important formulation consideration to overcome the conventional method and dosage form limitations. The eye drop is the easiest way and technique to deliver the drug into the eyes but has disadvantages like low retention time, low bioavailability, etc. In this chapter overall, we tried to include the historical drug delivery system, and the newest drug delivery system which we can use in ophthalmic after fewer modifications will help to overcome most of the convention method limitations. Delivering a drug across the protective barriers and elimination mechanism is challenging. The most challenging thing in ophthalmic preparations is to improve or prolong the ocular residence time. Overall, it gives an idea about advancements in the formulation of ophthalmic formulation till now and also by using different technologies how we can contribute to the ophthalmic field to deliver drugs, which will be helpful in the delivery of medicine in different eye diseases.

Acknowledgment Authors would like to acknowledge National Institute of Pharmaceutical Education and Research (NIPER) and Hyderabad (Department of Pharmaceutical, Ministry of Chemical and Fertilizers, India) for providing extending facilities during this manuscript writing (manuscript communication no. NIPER-HYD/2023/xxx). The principal investigator (Dr. Neelesh Kumar Mehra) would like to thank the Department of Science and Technology (DST), New Delhi, Government of India (Grant No. DST/NM/NS/2021/405) for extending financial support.

Data Availability Not applicable.

Conflict of Interest The authors declare no competing financial interest.

References

Agarwal R, Iezhitsa I, Agarwal P, Abdul Nasir NA, Razali N, Alyautdin R et al (2016) Liposomes in topical ophthalmic drug delivery: an update. Drug Deliv 23(4):1075–1091

Ahn SJ, Hong HK, Na YM, Park SJ, Ahn J, Oh J et al (2016) Use of rabbit eyes in pharmacokinetic studies of intraocular drugs. J Vis Exp (113):53878

Alam F, Elsherif M, AlQattan B, Salih A, Lee SM, Yetisen AK et al (2021) 3D printed contact lenses. ACS Biomater Sci Eng 7(2):794–803

Alam F, Salih AE, Elsherif M, Yetisen AK, Butt H (2022) 3D printed contact lenses for the management of color blindness. Addit Manuf 49:102464

Allam A, El-Mokhtar MA, Elsabahy M (2019) Vancomycin-loaded niosomes integrated within pH-sensitive in-situ forming gel for treatment of ocular infections while minimizing drug irritation. J Pharm Pharmacol 71(8):1209–1221

Ashammakhi N, Ahadian S, Xu C, Montazerian H, Ko H, Nasiri R et al (2019) Bioinks and bioprinting technologies to make heterogeneous and biomimetic tissue constructs. Mater Today Bio 1:100008

Bandari S, Nyavanandi D, Dumpa N, Repka MA (2021) Coupling hot melt extrusion and fused deposition modeling: critical properties for successful performance. Adv Drug Deliv Rev 172:52–63

Baranowski P, Karolewicz B, Gajda M, Pluta J (2014) Ophthalmic drug dosage forms: characterisation and research methods. Sci World J 2014:1–14

Barbosa Saliba J, Vieira L, Fernandes-Cunha GM, Rodrigues Da Silva G, Ligório Fialho S, Silva-Cunha A et al (2016) Anti-inflammatory effect of dexamethasone controlled released from anterior suprachoroidal polyurethane implants on endotoxin-induced uveitis in rats. Invest Opthalmol Vis Sci 57(4):1671

Bayer IS (2022) Recent advances in mucoadhesive interface materials, mucoadhesion characterization, and technologies. Adv Mater Interfaces 9(18):2200211

Bengani LC, Kobashi H, Ross AE, Zhai H, Salvador-Culla B, Tulsan R et al (2020) Steroid-eluting contact lenses for corneal and intraocular inflammation. Acta Biomater 116:149–161

Bhawale R, Suryavanshi P, Banerjee S (2023) Three-dimensional (3D) printing of oral dental films (ODFs) using blended Compactcel® polymers through semi-solid extrusion (SSE) bioprinter. Bioprinting 33:e00287

Bhosale VA, Srivastava V, Valamla B, Yadav R, Singh SB, Mehra NK (2022) Preparation and evaluation of modified chitosan nanoparticles using anionic sodium alginate polymer for treatment of ocular disease. Pharmaceutics 14(12):2802

Bin Sahadan MY, Tong WY, Tan WN, Leong CR, Bin Misri MN, Chan M et al (2019) Phomopsidione nanoparticles coated contact lenses reduce microbial keratitis causing pathogens. Exp Eye Res 178:10–14

Bisht R, Mandal A, Jaiswal JK, Rupenthal ID (2018) Nanocarrier mediated retinal drug delivery: overcoming ocular barriers to treat posterior eye diseases. WIREs Nanomed Nanobiotechnol 10(2). https://doi.org/10.1002/wnan.1473

Cabrera FJ, Wang DC, Reddy K, Acharya G, Shin CS (2019) Challenges and opportunities for drug delivery to the posterior of the eye. Drug Discov Today 24(8):1679–1684

Cholkar K, Gilger BC, Mitra AK (2015) Topical, aqueous, clear cyclosporine formulation design for anterior and posterior ocular delivery. Transl Vis Sci Technol 4(3):1

Cunha-Vaz J (1979) The blood-ocular barriers. Surv Ophthalmol 23(5):279–296

Davis R, Singh A, Jackson MJ, Coelho RT, Prakash D, Charalambous CP et al (2022) A comprehensive review on metallic implant biomaterials and their subtractive manufacturing. Int J Adv Manuf Technol 120(3–4):1473–1530

Desai AR, Maulvi FA, Desai DM, Shukla MR, Ranch KM, Vyas BA et al (2020) Multiple drug delivery from the drug-implants-laden silicone contact lens: addressing the issue of burst drug release. Mater Sci Eng C 112:110885

Dubald M, Bourgeois S, Andrieu V, Fessi H (2018) Ophthalmic drug delivery systems for antibiotherapy—a review. Pharmaceutics 10(1):10

Dubashynskaya N, Poshina D, Raik S, Urtti A, Skorik YA (2019) Polysaccharides in ocular drug delivery. Pharmaceutics 12(1):22

eCFR (2023) 21 CFR 200.50—Ophthalmic preparations and dispensers [cited 2023 Aug 2]. https://www.ecfr.gov/current/title-21/chapter-I/subchapter-C/part-200/subpart-C/section-200.50

Fan D, Li Y, Wang X, Zhu T, Wang Q, Cai H et al (2020) Progressive 3D printing technology and its application in medical materials. Front Pharmacol 11:122

Gonzalez-Gonzalez L, Grob S, Daly M (2014) Management of mydriasis and pain in cataract and intraocular lens surgery: review of current medications and future directions. Clin Ophthalmol 8:1281

Gorantla S, Rapalli VK, Waghule T, Singh PP, Dubey SK, Saha RN et al (2020) Nanocarriers for ocular drug delivery: current status and translational opportunity. RSC Adv 10(46):27835–27855

Grassiri B, Zambito Y, Bernkop-Schnürch A (2021) Strategies to prolong the residence time of drug delivery systems on ocular surface. Adv Colloid Interf Sci 288:102342

Gupta A, Nayak K, Misra M (2019) Cow ghee fortified ocular topical microemulsion; *in vitro, ex vivo*, and *in vivo* evaluation. J Microencapsul 36(7):603–621

Güven UM, Yenilmez E (2019) Olopatadine hydrochloride loaded Kollidon® SR nanoparticles for ocular delivery: nanosuspension formulation and in vitro–in vivo evaluation. J Drug Deliv Sci Technol 51:506–512

Hasegawa A, Gulmezian-Sefer M, Cheng Y, Srikumar R (2021) Microbiological considerations for ophthalmic products: sterility, endotoxin limits, and preservatives. In: Ophthalmic product development. Springer, pp 199–227

Hiratani H, Fujiwara A, Tamiya Y, Mizutani Y, Alvarez-Lorenzo C (2005) Ocular release of timolol from molecularly imprinted soft contact lenses. Biomaterials 26(11):1293–1298

Hsu KH, Carbia BE, Plummer C, Chauhan A (2015) Dual drug delivery from vitamin E loaded contact lenses for glaucoma therapy. Eur J Pharm Biopharm 94:312–321

Hui A, Willcox M (2016) In vivo studies evaluating the use of contact lenses for drug delivery. Optom Vis Sci 93(4):367–376

Irsch K, Guyton DL (2009) Anatomy of eyes. In: Encyclopedia of biometrics. Springer, Boston, MA, pp 11–16

Isaacson A, Swioklo S, Connon CJ (2018) 3D bioprinting of a corneal stroma equivalent. Exp Eye Res 173:188–193

Joshi A (1994) Microparticulates for ophthalmic drug delivery. J Ocul Pharmacol Ther 10(1):29–45

Kattar A, Quelle-Regaldie A, Sánchez L, Concheiro A, Alvarez-Lorenzo C (2023) Formulation and characterization of epalrestat-loaded polysorbate 60 cationic niosomes for ocular delivery. Pharmaceutics 15(4):1247

Khan N, Aqil M, Imam SS, Ali A (2015) Development and evaluation of a novel *in situ* gel of sparfloxacin for sustained ocular drug delivery: *in vitro* and *ex vivo* characterization. Pharm Dev Technol 20(6):662–669

Khutoryanskiy VV (2011) Advances in mucoadhesion and mucoadhesive polymers. Macromol Biosci 11(6):748–764

Kim YC, Chiang B, Wu X, Prausnitz MR (2014) Ocular delivery of macromolecules. J Control Release 190:172–181

Kim H, Park MN, Kim J, Jang J, Kim HK, Cho DW (2019) Characterization of cornea-specific bioink: high transparency, improved in vivo safety. J Tissue Eng 10:204173141882338

Koli AR, Ranch KM, Patel HP, Parikh RK, Shah DO, Maulvi FA (2021) Oral bioavailability improvement of felodipine using tailored microemulsion: surface science, ex vivo and in vivo studies. Int J Pharm 596:120202

Kolle SN, Moreno MCR, Mayer W, van Cott A, van Ravenzwaay B, Landsiedel R (2015) The EpiOcular™ eye irritation test is the method of choice for the *in vitro* eye irritation testing of agrochemical formulations: correlation analysis of EpiOcular eye irritation test and BCOP test data according to the UN GHS, US EPA and Brazil ANVISA classification schemes. Altern Lab Anim 43(3):181–198

Li Q, Li Z, Zeng W, Ge S, Lu H, Wu C et al (2014) Proniosome-derived niosomes for tacrolimus topical ocular delivery: in vitro cornea permeation, ocular irritation, and in vivo anti-allograft rejection. Eur J Pharm Sci 62:115–123

Li B, Wang J, Gui Q, Yang H (2020) Drug-loaded chitosan film prepared via facile solution casting and air-drying of plain water-based chitosan solution for ocular drug delivery. Bioact Mater 5(3):577–583

Lievens C, Berdy G, Douglass D, Montaquila S, Lin H, Simmons P et al (2019) Evaluation of an enhanced viscosity artificial tear for moderate to severe dry eye disease: a multicenter, double-masked, randomized 30-day study. Contact Lens Anterior Eye 42(4):443–449

Lim LT, Ah-Kee EY, Collins CE (2014) Common eye drops and their implications for pH measurements in the management of chemical eye injuries. Int J Ophthalmol 7:1067 [cited 2023 May 4]. https://doi.org/10.3980/j.issn.2222-3959.2014.06.29

Luechtefeld T (2016) Analysis of Draize eye irritation testing and its prediction by mining publicly available 2008-2014 REACH data. ALTEX 33:123

Maiti S, Paul S, Mondol R, Ray S, Sa B (2011) Nanovesicular formulation of brimonidine tartrate for the management of glaucoma: in vitro and in vivo evaluation. AAPS PharmSciTech 12(2):755–763

Mandal A, Gote V, Pal D, Ogundele A, Mitra AK (2019) Ocular pharmacokinetics of a topical ophthalmic nanomicellar solution of cyclosporine (Cequa®) for dry eye disease. Pharm Res 36(2):36

Mann BK, Stirland DL, Lee HK, Wirostko BM (2018) Ocular translational science: a review of development steps and paths. Adv Drug Deliv Rev 126:195–203

Maulvi FA, Soni TG, Shah DO (2014) Effect of timolol maleate concentration on uptake and release from hydrogel contact lenses using soaking method. J Pharm Appl Sci 1(1):17
Maulvi FA, Soni TG, Shah DO (2016) A review on therapeutic contact lenses for ocular drug delivery. Drug Deliv 23(8):3017–3026
Maulvi FA, Parmar RJ, Shukla MR, Desai AR, Desai DT, Ranch KM et al (2019) Plackett-Burman design for screening of critical variables and their effects on the optical transparency and swelling of gatifloxacin-pluronic-loaded contact lens. Int J Pharm 566:513–519
Maulvi FA, Pillai LV, Patel KP, Desai AR, Shukla MR, Desai DT et al (2020a) Lidocaine tripotassium phosphate complex laden microemulsion for prolonged local anaesthesia: in vitro and in vivo studies. Colloids Surf B Biointerfaces 185:110632
Maulvi FA, Patel PJ, Soni PD, Desai AR, Desai DT, Shukla MR et al (2020b) Novel poly(vinylpyrrolidone)-coated silicone contact lenses to improve tear volume during lens wear: in vitro and in vivo studies. ACS Omega 5(29):18148–18154
Maulvi FA, Parmar RJ, Desai AR, Desai DM, Shukla MR, Ranch KM et al (2020c) Tailored gatifloxacin Pluronic® F-68-loaded contact lens: addressing the issue of transmittance and swelling. Int J Pharm 581:119279
Maulvi FA, Ranch KM, Desai AR, Desai DT, Shukla MR (2021) Ophthalmic preparations. In: Remington. Elsevier, pp 565–575
Mehra NK, Cai D, Kuo L, Hein T, Palakurthi S (2016) Safety and toxicity of nanomaterials for ocular drug delivery applications. Nanotoxicology 10(7):836–860
Mehra N, Mohd A, Sultana Y (2021) A grafted copolymer-based nanomicelles for topical ocular delivery of everolimus: formulation, characterization, ex-vivo permeation, in-vitro ocular toxicity, and stability study. Eur J Pharm Sci 159:105735
Mohammed AA, Algahtani MS, Ahmad MZ, Ahmad J, Kotta S (2021) 3D printing in medicine: technology overview and drug delivery applications. Ann 3D Print Med 4:100037
Moiseev RV, Kaldybekov DB, Filippov SK, Radulescu A, Khutoryanskiy VV (2022) Maleimide-decorated PEGylated mucoadhesive liposomes for ocular drug delivery. Langmuir 38(45):13870–13879
Moosa RM, Choonara YE, du Toit LC, Kumar P, Carmichael T, Tomar LK et al (2014) A review of topically administered mini-tablets for drug delivery to the anterior segment of the eye. J Pharm Pharmacol 66(4):490–506
Mukhopadhyay S, Butola M (2020) Preparation and evaluation of erythromycin microemulsion for opthalmic drug delivery. J Adv Sci Res 11(1):22–6 [cited 2023 May 8]. https://www.sciensage.info/index.php/JASR/article/view/404
Muthusubramaniam L, Peng L, Zaitseva T, Paukshto M, Martin GR, Desai TA (2012) Collagen fibril diameter and alignment promote the quiescent keratocyte phenotype. J Biomed Mater Res A 100A(3):613–621
Nagai N, Otake H (2022) Novel drug delivery systems for the management of dry eye. Adv Drug Deliv Rev 191:114582
Nasr M, Teiama M, Ismail A, Ebada A, Saber S (2020) In vitro and in vivo evaluation of cubosomal nanoparticles as an ocular delivery system for fluconazole in treatment of keratomycosis. Drug Deliv Transl Res 10(6):1841–1852
Pandey SS, Patel MA, Desai DT, Patel HP, Gupta AR, Joshi SV et al (2020) Bioavailability enhancement of repaglinide from transdermally applied nanostructured lipid carrier gel: optimization, in vitro and in vivo studies. J Drug Deliv Sci Technol 57:101731
Paradkar MU, Parmar M (2017) Formulation development and evaluation of natamycin niosomal in-situ gel for ophthalmic drug delivery. J Drug Deliv Sci Technol 39:113–122
Park K (2017) The drug delivery field at the inflection point: time to fight its way out of the egg. J Control Release 267:2–14
Patel A (2013) Ocular drug delivery systems: an overview. World J Pharmacol 2(2):47
Pathak A, Jain K et al (2023) "Dendrimer-drug conjugates" in International Book "Polymer-drug conjugates: linker chemistry, protocols and applications". Elsevier, ISBN: 9780323916639. https://doi.org/10.1016/B978-0-323-91663-9.00005-9

Peters MCC, dos Santos Neto E, Monteiro LM, Yukuyama MN, Machado MGM, de Oliveira IF et al (2020) Advances in ophthalmic preparation: the role of drug nanocrystals and lipid-based nanosystems. J Drug Target 28(3):259–270

Pramar Y (2009) Compounding ophthalmic liquids

Qiao H, Xu Z, Sun M, Fu S, Zhao F, Wang D et al (2022) Rebamipide liposome as an effective ocular delivery system for the management of dry eye disease. J Drug Deliv Sci Technol 75:103654

Rakhmetova A (2020) Development of a new polymer ocular insert to treat fungal infections threatening the cornea. Nazarbayev University School of Engineering Digital Science [cited 2023 May 4]. http://nur.nu.edu.kz/handle/123456789/4650

Ranch KM, Maulvi FA, Koli AR, Desai DT, Parikh RK, Shah DO (2021) Tailored doxycycline hyclate loaded in situ gel for the treatment of periodontitis: optimization, in vitro characterization, and antimicrobial studies. AAPS PharmSciTech 22(3):77

Rathore KS, Nema RK (2009) Review on ocular inserts. Int J PharmTech Res CODEN 1(2):164–169

Ravindran V, Reddy R, Sandhya S, Banji D, Reddy V (2012) Critical review on mucoadhesive drug delivery systems. Hygeia J Drug Med 4:7–28

Rodrigues GA, Lutz D, Shen J, Yuan X, Shen H, Cunningham J et al (2018) Topical drug delivery to the posterior segment of the eye: addressing the challenge of preclinical to clinical translation. Pharm Res 35(12):245

Ross AE, Bengani LC, Tulsan R, Maidana DE, Salvador-Culla B, Kobashi H et al (2019) Topical sustained drug delivery to the retina with a drug-eluting contact lens. Biomaterials 217:119285

Satalkar P, Elger BS, Hunziker P, Shaw D (2016) Challenges of clinical translation in nanomedicine: a qualitative study. Nanomedicine 12(4):893–900

Schoenwald RD, Stewart P (1980) Effect of particle size on ophthalmic bioavailability of dexamethasone suspensions in rabbits. J Pharm Sci 69(4):391–394

Senapati S, Youssef AAA, Sweeney C, Cai C, Dudhipala N, Majumdar S (2022) Cannabidiol loaded topical ophthalmic nanoemulsion lowers intraocular pressure in normotensive Dutch-belted rabbits. Pharmaceutics 14(12):2585

Shaikh R, Raj Singh T, Garland M, Woolfson A, Donnelly R (2011) Mucoadhesive drug delivery systems. J Pharm Bioallied Sci 3(1):89

Simta J, Kavita I, Milind B, Adieboye Oforibika G, Adokiye Oforibika D, Atieno Ochung A et al (2020) Novel long retentive posaconazole ophthalmic suspension. Pharm Sci Technol 4(1):1–10 [cited 2023 May 8]. https://repository.maseno.ac.ke/handle/123456789/1496

Smart J (2005) The basics and underlying mechanisms of mucoadhesion. Adv Drug Deliv Rev 57(11):1556–1568

Sokolowski R (ed) (2018) Edmund Husserl and the phenomenological tradition. Catholic University of America Press

Sorkio A, Koch L, Koivusalo L, Deiwick A, Miettinen S, Chichkov B et al (2018) Human stem cell based corneal tissue mimicking structures using laser-assisted 3D bioprinting and functional bioinks. Biomaterials 171:57–71

Srivastava V, Chary PS, Rajana N, Pardhi ER, Singh V, Khatri D et al (2023) Complex ophthalmic formulation technologies: advancement and future perspectives. J Drug Deliv Sci Technol 82:104317

Tagami T, Goto E, Kida R, Hirose K, Noda T, Ozeki T (2022) Lyophilized ophthalmologic patches as novel corneal drug formulations using a semi-solid extrusion 3D printer. Int J Pharm 617:121448

The British Pharmacopoeia (2022) Appendix XVI A T for SS (ph. Eur method 2.6.1), vol 5

The Indian Pharmacopoeia (IP) (2022) 2.2.11. Sterility, p 1

The United States Pharmacopeia (2020) USP 43 NF 38 ST <71>. The National Formulary, vol IV

Tieppo A, White CJ, Paine AC, Voyles ML, McBride MK, Byrne ME (2012) Sustained in vivo release from imprinted therapeutic contact lenses. J Control Release 157(3):391–397

Tsui JKS, Bell S, da Cruz L, Dick AD, Sagoo MS (2022) Applications of three-dimensional printing in ophthalmology. Surv Ophthalmol 67(4):1287–1310

Uner B, Ozdemir S, Yildirim E, Yaba A, Tas C, Uner M et al (2023) Loteprednol loaded nanoformulations for corneal delivery: ex-vivo permeation study, ocular safety assessment and stability studies. J Drug Deliv Sci Technol 81:104252

US FDA (2023) CFR—Code of Federal Regulations Title 21 [cited 2023 Aug 2]. https://www.accessdata.fda.gov/scripts/cdrh/cfdocs/cfcfr/cfrsearch.cfm

Varela-Fernández R, García-Otero X, Díaz-Tomé V, Regueiro U, López-López M, González-Barcia M et al (2022) Mucoadhesive PLGA nanospheres and nanocapsules for lactoferrin controlled ocular delivery. Pharmaceutics 14(4):799

Wadetwar RN, Agrawal AR, Kanojiya PS (2020) In situ gel containing Bimatoprost solid lipid nanoparticles for ocular delivery: in-vitro and ex-vivo evaluation. J Drug Deliv Sci Technol 56:101575

Wang J, Li B, Huang D, Norat P, Grannonico M, Cooper RC et al (2021) Nano-in-Nano dendrimer gel particles for efficient topical delivery of antiglaucoma drugs into the eye. Chem Eng J 425:130498

Weiss SL, Kramer WG (2019) Ocular distribution of cyclosporine following topical administration of OTX-101 in New Zealand white rabbits. J Ocul Pharmacol Ther 35(7):395–402

Wolters Kluwer (2023) Lippincott Medicine [cited 2023 May 8]. https://www.wolterskluwer.com/en/solutions/lippincott-medicine

Wu Y, Liu Y, Li X, Kebebe D, Zhang B, Ren J et al (2019) Research progress of in-situ gelling ophthalmic drug delivery system. Asian J Pharm Sci 14(1):1–15

Yellepeddi VK, Palakurthi S (2016) Recent advances in topical ocular drug delivery. J Ocul Pharmacol Ther 32(2):67–82

Yuan X, Marcano DC, Shin CS, Hua X, Isenhart LC, Pflugfelder SC et al (2015) Ocular drug delivery nanowafer with enhanced therapeutic efficacy. ACS Nano 9(2):1749–1758

Zhang B, Xue Q, Li J, Ma L, Yao Y, Ye H et al (2019) 3D bioprinting for artificial cornea: challenges and perspectives. Med Eng Phys 71:68–78

Zhu M, Wang J, Li N (2018) A novel thermo-sensitive hydrogel-based on poly(*N*-isopropylacrylamide)/hyaluronic acid of ketoconazole for ophthalmic delivery. Artif Cell Nanomed Biotechnol 46(6):1282–1287

Zulliger R, Conley SM, Naash MI (2015) Non-viral therapeutic approaches to ocular diseases: an overview and future directions. J Control Release 219:471–487

Advances and Developments in Formulation of Drug Nanocrystals

13

Manshi, Sonal Setya, and Sushama Talegaonkar

Abstract

The field of pharmaceutical sciences has witnessed remarkable progress in recent years, particularly in the formulation of drug nanocrystals. Nanocrystals represent a promising approach to address challenges associated with the solubility and bioavailability of poorly water-soluble drugs. Nanocrystals are microscopic particles with dimensions typically in the nanometre range, from 1 to 100 nm in size. They are a type of nanomaterial characterized by their small size and large surface area relative to their volume. Nanocrystals can be composed of various materials, including metals, semiconductors, or organic compounds. Their enhanced surface area, a consequence of being in the nanometre range, contributes to improved reactivity and dissolution rates. This chapter comprehensively covers the fundamental principles and innovative techniques employed in the production of drug nanocrystals, emphasizing methodologies such as high-pressure homogenization, wet milling, and precipitation methods. It has also given considerable information about the selection and design of stabilizers and surfactants, crucial for ensuring the stability and performance of nanocrystals. It explores the implications of nanocrystal technology on drug delivery systems, elucidating how the enhanced surface area and dissolution rates of nanocrystals contribute to improved bioavailability. This discussion extends to the incorporation of nanocrystals into various dosage forms, including oral tablets, injectables, and topical formulations, highlighting their versatility and potential for personalized medicine. In the commercial landscape, various pharmaceutical companies

Manshi · S. Setya
Department of Pharmaceutics and Pharmacy Practice, SGT College of Pharmacy, SGT University, Gurugram, Haryana, India

S. Talegaonkar (✉)
Department of Pharmaceutics, School of Pharma Sciences, DPSRU, New Delhi, Delhi, India

K. Jain, A. K. Yadav (eds.), *Advances in Pharmaceutical Product Development*, https://doi.org/10.1007/978-981-97-9230-6_13

have utilized the potential of nanocrystals in drug formulations. Products such as TRIGLIDE, EMEND, and MEGACE ES incorporating nanocrystals have been developed to improve the delivery and performance of drugs with low solubility, addressing a critical need in the industry. These commercial products underscore the successful translation of nanocrystal research into practical solutions, marking a significant stride forward in the field of pharmaceutical sciences.

Keywords

Nanocrystals · Nanosuspension · Poorly soluble drugs · Drug delivery · Milling · Nanotechnology · Bioavailability

Abbreviations

ABZ	Albendazole
AFM	Atomic force microscopy
API	Active pharmaceutical ingredient
ATR	Microattenuated total reflectance
AUC	Area under curve
BCS	Biopharmaceutical classification system
BET	Brunauer-Emmett-Teller
BRA	Brinzolamide
CT	Combination technology
Cur/GMs/Coll-CNCs	Porous collagen cellulose nanocarrier
DSC	Differential scanning calorimetry
DTA	Differential thermal analysis
DTG	Derivative thermogravimetry
FF	Fenofibrate
FTIR	Fourier-transform infrared spectroscopy
HPH	High pressure homogenization
LAS	Liquid antisolvent
MDIs	Metred-dose inhalers
MM	Media milling
NCs	Nanocrystals
PCS	Photon correlation spectroscopy
PDI	Poly dispersity index
PG-NCs	Progesterone-loaded nanocarriers
RESOLV	Rapid expansion of a supercritical solution into liquid solvent
RESS	Rapid expansion of supercritical solution
SAS	Supercritical antisolvent
SCF	Super critical fluid

SDNCS	Self-dispersible nanocrystals
SDS	Sodium dodecyl sulphate
SEM	Scanning electron microscopy
SLS	Sodium lauryl sulfate
TEM	Transmission electron microscopy
TGA	Thermogravimetric analysis
TPGS	D-α-tocopherol polyethylene glycol 1000 succinate
XRD	X-ray diffraction

13.1 Introduction

Nanocrystals are nanometre-sized crystals, which implies they are nanoparticles having crystalline properties because of the nanosizing and employing nanocrystalline technology leads to an improved ratio of surface area to volume and improved dissolution rate. This, in turn, enhances the solubility of aquaphobic pharmaceuticals (Jarvis et al. 2018). The difference between them emerges from the fact that nanocrystals are wholly formed as a therapeutic agent or payload, preventing the need for a carrier. The dispersion of drug nanocrystals in fluid substances results in what is known as "nanosuspensions," with surface-active agents or stabilizers frequently employed to stabilize crystal dispersal in liquid substances (Oliveira et al. 2022). The dispersal media may consist of non-aqueous solutions or aqueous solutions. They are widely used to deliver drugs of BCS classification system, class II and IV drugs (low solubility drugs). Nanocrystals are widely used because of their great loading capacity and surface area; nanocrystals have intriguing features, notably their significant adhesiveness, which leads to a high retention period at the target location (Viswanathan et al. 2017). Nanocrystals are commonly discovered in individual or polycrystalline forms containing rutin (flavonoids) as an active component. Furthermore, they have the necessary characteristics to be acceptable for cutaneous applications involving poorly soluble compounds (Ran et al. 2022).

The primary benefits of nanocrystals in improving oral bioavailability encompass (1) elevated saturation solubility and faster dissolution rates for drugs with low water solubility; (2) improved adherence to biological membranes, leading to prolonged retention in the gastrointestinal tract; and (3) nanocrystals that can be absorbed through intestinal epithelial cells and the lymphatic pathway while maintaining the integrity of crystalline particles (Guo et al. 2023).

Nanocrystals, with their diminutive size in the nanometre range, represent a fascinating field in drug delivery technology. Unlike nanoparticles that may require a carrier, nanocrystals are themselves therapeutic agents or payloads, eliminating the need for an additional component. The integration of nanocrystals into drug delivery systems results in nanosuspensions, where stabilizers or surfactants are commonly employed to maintain the stability of crystalline dispersions in liquid substances (Patra et al. 2018).

Nanocrystals typically have dimensions smaller than 1 μm, often being smaller than 500 nm for practical reasons. Following nanonization, these API nanocrystals are frequently incorporated within traditional dosage unit forms like tablets, pellets, and capsules along with intravenous suspensions. The advantages of nanocrystals include their exceptional reproducibility and applicability to a broad spectrum of drugs with varying solubility profiles. However, the nanocrystal approach can entail high-energy input, leading to increased production costs. The physicochemical attributes of nanocrystals are influenced by the type of stabilizer employed, whether non-ionic surfactants (such as polysorbates, poloxamers, and poloxamer) or ionic surfactants (including bile salts and alkyl-sulfonates) (Attama et al. 2016). In comparison with matrix nanoparticles made of lipid or polymeric matrices, nanocrystals exhibit a significant drug loading, closer to 100%.Thus, the fundamental benefit of nanocrystals is the minimum quantity of formulation component (excipients) utilized, which allows for achieving a substantial concentration of drug at the targeted site.

Nanocrystals represent a promising avenue in drug delivery technology. Their unique properties, including size, loading capacity, and surface area, make them well-suited for addressing challenges associated with poorly soluble drugs. While the energy input during production and the choice of stabilizers pose considerations, the benefits, including high drug loading and minimal excipient use, underscore the potential of nanocrystals in advancing drug delivery across various therapeutic domains (Naduparambath et al. 2018).

The production of nanocrystals involves the utilization of drugs with low solubility, as those with high water solubility cannot be configured into nanocrystals, especially within a water-based dispersion medium. Developing formulations for poorly soluble APIs into nanocrystals serves to address various challenges associated with their biopharmaceutical delivery. These challenges include issues such as inadequate bioavailability post oral administration, limited skin penetration resulting in low dermal bioavailability, the requirement for excessive injection volumes in intravenous administration, and the occurrence of undesirable adverse effects following IV injection when conventional formulations, like solutions with dissolved drugs, are employed. The present innovative approach of employing nanocrystals holds promise in overcoming these obstacles and improving the efficiency of drug delivery systems (Müller et al. 2011).

13.1.1 Properties of Nanocrystals

Nanocrystals have different properties from larger crystals because of their small size and large surface area. Some of these properties are the following:

- The utilization of nanocrystals for poorly soluble pharmaceuticals can enhance their therapeutic efficacy by augmenting their solubility rate and absorption in the body.
- They should have a size below 1 μm.
- These are characterized by their extremely small size at the nanometre scale and exhibit enhanced dissolution behaviour, leading to a higher saturation solubility

compared to larger particles. This heightened solubility is advantageous, especially for poorly soluble drugs, as it can contribute to improved bioavailability and therapeutic efficacy.
- The nanocrystals can be present in two forms: crystalline or amorphous structure.
- They provide a safe and efficient direction through the skin.
- They have high penetration power.
- They can enhance membrane penetration, permeability, and adhesion.
- The nano-crystallization of inadequately soluble drugs enhances both their physicochemical stability and bioavailability (Gu et al. 2023).

13.1.2 Advantages of Nanocrystals

Improving the dissolving rate and saturation solubility presents a valuable opportunity to enhance the oral bioavailability and efficacy of medications characterized by poor solubility. The use of nanocrystals allows for drug formulations without the need for carriers, ensuring that 100% of the formulation consists solely of the active drug substance. Additionally, nanocrystals contribute to improved gastrointestinal mucosal adhesion, leading to extended residency in the gastrointestinal tract and, consequently, enhanced absorption throughout the gastrointestinal system. Notably, nanocrystals demonstrate limited toxicity, enhancing their safety profile across various applications. Furthermore, their versatility is highlighted by the possibility of administering drugs through different routes, providing flexibility in drug delivery methods (Jahangir et al. 2022; Tian et al. 2021). Figure 13.1 illustrates the various advantages of nanocrystals.

13.1.3 Disadvantages of Nanocrytals (Jahangir et al. 2022; Moroz et al. 2018)

Nanocrystals are not without drawbacks. The following are some of the issues and limits of nanocrystals:

- Stability issues: Because of their high surface energy and reactivity, nanocrystals are prone to aggregation, oxidation, degradation, and phase transformation. Stability depends on molecular structure of drug.
- Casting of high-value equipment: Need of expensive equipments.
- Release issue: The key drawback is that being crystalline in nature they are thermodynamically stable but do not release medication fast. Solid dispersions are formed rather than amorphous solid dispersions, which were more thermodynamically stable but did not release the medication as fast.
- Difficulties in handling and transport: Nanocrystals require handling and transport equipment to minimize exposure, contamination, and damage during processing and storage.

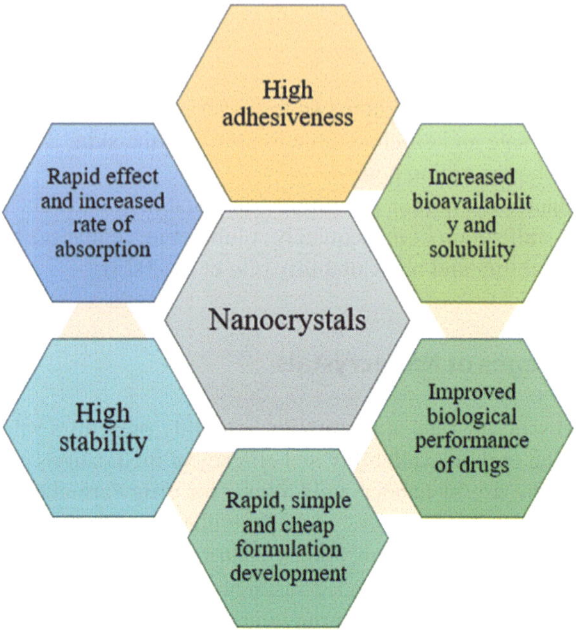

Fig. 13.1 Schematic diagram of advantages of nanocrystals

- Dose accuracy issues: Because nanocrystals have an uneven distribution and poor flowability, it is difficult to quantify and administer accurate medication doses.
- Toxicity concerns: Because of their potential accumulation, penetration, and interaction with biological systems, nanocrystals may have negative consequences on the environment and human health.
- Applicability: Only certain categories of drugs are suitable.

13.2 Performance Attributes of Nanocrystals

Nanosized crystals possess unique characteristics that enable them to improve the rate of dissolution and the bioavailability of BCS class 2 and 4 drugs. The subsequent sections outline and briefly explore the specific characteristics of nanocrystals that contribute to their enhanced functionality.

13.2.1 Improved Dissolution Rate by Surface Area Enlargement

The Noyes-Whitney equation describes the key features of nanocrystals and their impact on drug dissolution.

$$dC / dt = DA / h(C_s - C)$$

where *dC/dt* is the rate of dissolution of drug particles, *D* is the diffusion coefficient of drug in the GI media, *A* is the effective surface area of the drug particles, *h* is the thickness of the diffusion layer around drug particles, C_s is the saturation solubility of drug in solution, and *C* is concentration of the drug. This equation focused on factors such as surface area and solubility and provides insights into the speed at which a drug dissolves. The reduction in the size of a drug results in an expanded surface area, contributing to an augmented dissolution rate. Micronization proves to be a viable strategy for drugs where dissolution rate is crucial, effectively enhancing bioavailability. Progressing from micronization to nanonization further increases particle surface, leading to a more pronounced enhancement in dissolution rate. The heightened surface energy in nanocrystals holds potential for influencing their solubility and reactivity (Murdande et al. 2015). The Noyes-Whitney equation explains the improved rate of dissolution of nanocrystals, where heightened saturation solubility of drug (C_s) and increased surface area (A) result in an elevated dissolution rate for drug nanocrystals (Chogale et al. 2016).

13.2.2 Increase in Saturation Solubility

The Ostwald-Freundlich equation, as expressed in Eq. 2 explicitly defines the relationship between the saturation solubility of drug and its particulate size.

$$\log(C_s / C_\alpha) = 2pV / 2.303RT\rho r$$

where C_s represents saturation solubility of drug, C_α denotes the solubility of the solid with larger particles, *r* signifies the interfacial (surface) tension of the compound, *V* is the molar volume of the particulate substance, *R* stands for the ideal gas constant, *T* represents the absolute temperature of substance, *ρ* is the density of the solid, and *r* denotes the radius of the particulate size. The solubility of nanocrystals may deviate from that of bulk materials due to size and surface properties, and the equation aids in comprehending the impact of these factors on nanocrystal solubility. The radius of nanocrystals (*r*) emerges as a pivotal parameter, given the small size of nanocrystals, influencing their solubility and interactions with the solvent (Shegokar and Müller 2010). Acknowledging the role of temperature in nanocrystal solubility, the equation incorporates temperature considerations, recognizing that alterations in temperature can impact the dissolution and stability of nanocrystals (Junyaprasert and Morakul 2015).

13.2.3 Increase in Adhesiveness

Nanocrystals of drugs exhibit enhanced affinity to tissues, contributing to an enhanced oral absorption of poorly soluble drugs, in addition to the improvements in the dissolution rate and saturation solubility of the drug. On the surface of the gastrointestinal system, there is a permeable mucus layer. The nanosized particles allow it to quickly infiltrate and attach to the pore channels of the mucus layer. So, it prolongs the time of drug in GIT, hence enhancing bioavailability and metabolism behaviour of drugs (Junghanns and Müller 2008).

13.2.4 Increase in Permeability

Nanocrystals (NCs) within the range of 200–300 nm provide benefits in terms of increased permeability through the skin and mucosal membranes. This challenge becomes more pronounced for drugs with poor aqueous solubility. Despite the improved permeation properties attributed to the lipophilicity of poorly water-soluble drugs, the rate of the drug release remains a restricting factor. The improved penetrability of drug nanocrystals is dependent on the size of the particle; the particulates smaller than 40 nm were observed to penetrate the skin via the follicular route, whereas bigger particles faced limited penetration due to the dense system of Langerhans cells in the epidermis. Different research suggests that particles with a size exceeding 5 μm demonstrated minimal permeation through the stratum corneum. Conversely, particles within the range of 500–750 nm showed enhanced permeation into the hair follicles of human skin (Patel et al. 2018).

13.3 Preparation Techniques of Nanocrystals

There are several approaches available for producing nanocrystals with specific shapes and sizes. Essentially, three fundamental principles can be employed: milling, precipitation methods, homogenization methods, or a combination of these techniques. The primary methods utilized in the industry involve top-down technologies, which entail the reduction of the size of a large drug powder. In contrast, bottom-up technologies, starting from a dissolved molecule and involving precipitation, are not currently utilized in the production of commercial products, to the best of our knowledge. This can be ascribed to difficulties such as the requirement for removing solvents, challenges in controlling the process, and the observation that numerous poorly soluble drugs display low solubility not only in water but also in organic substances. Figure 13.2 illustrates nanocrystal preparation techniques and schematically shows top-down and bottom-up approaches.

Various methods for producing nanocrystals are available, depending on the nature and size of the substance. Nanocrystals can be produced using top-down methods, bottom-up techniques, or a combination of both approaches.

- **Top-down methods**: This involves employing mechanical forces to reduce larger crystals into smaller ones, using methods like milling, sonication, or high-pressure homogenization. The top-down approach is suitable for drugs that are

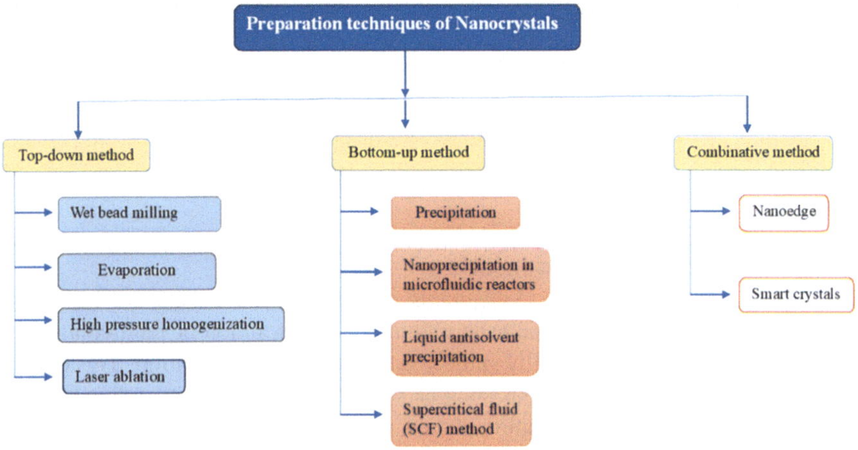

Fig. 13.2 Schematic diagram of preparation techniques of nanocrystals

poorly soluble in both water and organic solvents. It operates quickly and is commonly utilized in the production of commercially available drugs nanocrystals.

- **Bottom-up methods**: These include the formation of nanocrystals from smaller molecules or atoms by chemical processes such as precipitation, sol-gel, hydrothermal, or solvothermal synthesis.
- **Combinative techniques**: These methods combine top-down and bottom-up approaches to obtain ideal nanocrystal size, shape, and stability (Mengge et al. 2022; Kulkarni and Myerson 2017).

13.3.1 Top-Down Approaches

By exerting mechanical pressure (top-down method), bigger crystals are divided into smaller ones in the top-down process of creating nanocrystals. This strategy employs several techniques, including the following:

13.3.1.1 Wet Bead Milling

This process involves grinding a solid substance in a rotating container with balls or beads to create minuscule particles. The nanocrystal particle size is mainly influenced by factors such as the milling bead size (typically 0.1–20 nm), the drug's properties, and the setup parameters. Wet bead milling is particularly effective for producing nanocrystals of thermally unstable drugs because the absolute temperature can be regulated during the manufacturing process. It is user-friendly and yields a consistent result. Nevertheless, the addition of wetting agents and stabilizers is required, and multiple rotations are essential to reach the intended particle size (Malamatari et al. 2018). The media mill (MM) comprises essential components such as a milling chamber, milling media, recirculating chamber, motor, and coolant. The processing chamber is loaded with a crude mixture of drugs, water, and stabilizers. During the milling process, the motor stirs an unrefined mixture of drugs, stabilizers, and water in the milling chamber. In the end, the slurry makes up 2–30% (w/v) of the milling chamber capacity, and the milling media comprises 10–50% (w/v) of the slurry. Reduction in size is achieved through mechanical shear and attrition, arising from collisions between a drug molecule and the milling chamber components or processing media, as well as between two drug particles (Salazar et al. 2013).

13.3.1.2 Evaporation/Condensation

This involves heating a metal to evaporate it and then condensing the vapour to obtain nano powders. During the evaporation phase, a solution or melt containing the material of interest is carefully subjected to controlled evaporation. This can be achieved through various methods such as solvent evaporation or cooling. As the solvent evaporates, the concentration of the solute increases, leading to supersaturation. At a critical point, nucleation occurs, initiating the formation of nanosized particles.

The subsequent condensation step is crucial in shaping the nanocrystals. The vapour generated during evaporation undergoes controlled cooling or condensation, promoting the assembly of nanoparticles. The condensation conditions, including temperature and pressure, play an essential role in examining the final characteristics of nanocrystals (Harish et al. 2022).

13.3.1.3 High-Pressure Homogenization

High-pressure homogenization (HPH) is an additional top-down process wherein drug molecule size reduction is achieved through shearing forces, cavitational forces, and drug molecule colloidal system, all supported by pressurized airflow conditions. There are commonly two types of HPH procedures: microfluidization and piston gap homogenization.

In HPH process, drug suspension undergoes high-pressure homogenization by passing through a constrained pathway in a sudden burst under elevated pressure. The two types of HPH procedures are described as follows:

1. Microfluidization, alternatively called air-jet milling or jet stream homogenization, encompasses the breaking down of particles within a pressurized airflow formed by the merging of two fluid streams at a pressure of 1700 bar.
2. The process of piston-gap homogenization employs elevated pressure to propel a fluid suspension through a restricted channel or a narrow gap in a pipeline. This process typically consists of three stages:
   (a) Distribution of powdered medication in either a pure solution or a solution that includes a stabilizing agent.
   (b) Minimizing the size of molecules by either rapidly shearing them or using low-pressure homogenization.
   (c) Application of pressurized homogenization to attain the desired particle size and size distribution.

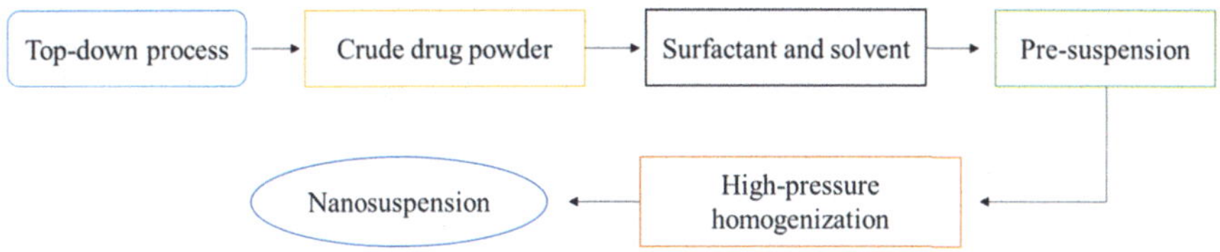

Fig. 13.3 Schematic representation of preparation technique of Dissocube

In summary, HPH is a top-down technique employing high-pressure conditions, and its variations, microfluidization and piston-gap homogenization, involve specific processes to reduce drug particle size effectively.

High-pressure homogenization in water, referred to as Dissocubes®, is a method that yields nanosuspensions with exceptional dissolution and disintegration properties. The particles take on a cuboid shape, and the trade name for these prepared nanosuspensions is Dissocubes®. This innovative process avoids material erosion and contamination from production/processing equipment, maintaining levels typically below 1 ppm, falling within an acceptable range. The technique allows for the preparation of highly concentrated or diluted nanosuspensions, handling drug amounts ranging from 1 to 400 mg/mL. This is particularly relevant for drugs with low water solubility in both aqueous and non-aqueous medium, enabling aseptic manufacturing of nanosuspensions for subcutaneous (parenteral) administration (Sun and Yeo 2012).

In contrast, homogenization in water-free media and mixtures of water, known as Nanopure, prove to be more appropriate for specific administration routes. In the development of the second generation of drug nanocrystals (Nanopure), drug suspensions in organic medium, like propylene glycol, undergo homogenization (Müller et al. 2001). Figure 13.3 illustrates the preparation method of Disso cubes.

13.3.1.4 Laser Ablation

Laser ablation involves using a laser to heat a gas mixture containing a metal precursor and then cooling it to form nanoparticles. There are three categories of laser processing durations: picosecond, nanosecond, and femtosecond. Included these, nanoscale particles can be produced more abundantly. Particle size is influenced by factors such as laser intensity, scanner velocity, suspension characteristics, and other related factors. Although there are no organic solvents used in this procedure, a tiny amount of the medicine might experience oxidative degradation and crystal state alterations because of using too much power (Sylvestre et al. 2011).

13.3.1.5 Ultrasound

Ultrasound proves to be an effective method for breaking down medication particles into tiny ones through the vibrations of auditory waves. By inducing acoustic cavitational forces into a solution and quickly dispersing the drug solution, ultrasound has been demonstrated to enhance formation of nuclei. Its ease of use in the laboratory and reliable reproducibility often lead to its integration with other

Fig. 13.4 Schematic illustration of the precipitation technique of bottom-up approaches

processes. Ultrasound-assisted precipitation of nanoparticles influences key processes like mixing, nucleation, growth, and agglomeration. Factors such as the length of the horn, immersion depth of the horn, cavitation depth, and the intensity of ultrasonic treatment all play an essential role in examining the size of the nanocrystals (Waard et al. 2011; Baig et al. 2021).

13.3.2 Bottom-Up Approaches

The bottom-up technique for making nanocrystals includes creating larger molecules or atoms from smaller ones through chemical processes. These are a few of the procedures this technique employs:

13.3.2.1 Precipitation

In the bottom-up technique of nanocrystal synthesis, precipitation serves a critical function in the formation of nanocrystals. This technique involves the controlled and gradual reduction of solutes from a solution, leading to the spontaneous nucleation and growth of nanoscale particles. Typically, a supersaturated solution of the desired material is prepared, and then, through the introduction of a precipitating agent or by altering the environmental conditions, the solutes reach a point of saturation, triggering their aggregation into nanocrystals. The selection of appropriate precipitation conditions, such as temperature, solvent, and concentration, is vital in determining the size, shape, and uniformity of the resulting nanocrystals (Sinha et al. 2013). This method allows for a high degree of control over the particle size distribution and is commonly used in the manufacturing of nanocrystals and various applications, including drug delivery, catalysis, and advanced materials. Precipitation in the bottom-up approach offers advantages in terms of scalability, reproducibility, and the ability to tailor nanocrystal properties, making it a versatile and widely utilized method in the field of nanotechnology (Tran et al. 2016). The preparation method of precipitation is shown in Fig. 13.4.

13.3.2.2 Sol-Gel

In the bottom-up technique of nanocrystal synthesis, the sol-gel method serves as a versatile and effective approach. The sol-gel process involves the transformation of a precursor sol into a gel-like network and subsequent solid material. In the context of nanocrystal formation, this technique starts with a solution containing precursor molecules, often metal alkoxides or other chemical precursors. Through controlled hydrolysis and condensation reactions, the sol evolves into a three-dimensional gel structure with nanoscale dimensions. The gel is then subjected to further processing, such as drying and calcination, resulting in the formation of nanocrystals (Yarbrough et al. 2020). The sol-gel method provides precise control over particle size, composition, and morphology by manipulating precursor chemistry and reaction conditions. This approach is particularly advantageous for producing nanocrystals of various materials, including metal oxides, semiconductors, and hybrid organic-inorganic compounds. The versatility and tunability of the sol-gel process make it a widely employed technique in the bottom-up synthesis of nanocrystals for applications spanning from catalysis and sensing to optoelectronics and advanced materials (Kumar et al. 2017).

13.3.2.3 Nanoprecipitation in Microfluidic Reactors

Nanoprecipitation in microfluidic reactors represents an innovative approach within the bottom-up technique for preparing nanocrystals. This method harnesses the advantages of microfluidic technology to precisely control the nanoprecipitation process. In this technique, a continuous flow of a solvent containing a dissolved precursor or drug is introduced into a microfluidic channel. Simultaneously, another continuous flow of a non-solvent, which is capable of being mixed with the solvent but incompatible with the precursor, is introduced. The controlled mixing of these two streams induces rapid supersaturation and leads to the spontaneous nucleation and formation of nanocrystals. The confined and precisely regulated environment of the microfluidic reactor allows for superior control over reaction parameters, including mixing ratios, flow rates, and residence times (Ali et al. 2009). This results in nanocrystals with well-defined sizes, shapes, and narrow size distributions. The nanoprecipitation in microfluidic reactors method is particularly advantageous for producing nanocrystals of materials with varying solubilities. This bottom-up technique offers scalability, reproducibility, and the ability to tailor nanocrystal properties, making it a promising method for applications such as drug delivery, imaging, and advanced materials (Niculescu et al. 2021).

13.3.2.4 Liquid Antisolvent Precipitation

An insoluble drug-containing solution stream (organic phase) is combined with an aqueous antisolvent to create liquid antisolvent (LAS) precipitation, which produces nanocrystals. The most often described technique of nanoprecipitation is solution-antisolvent. This procedure is straightforward and economical because it simply calls for the nucleation and growth processes. Two steps can be used to prepare the optimized nanocrystals. Nanocrystals are precipitated and aggregated because of this process, which also recrystallizes unstable crystal particles. Medications that do not dissolve in water or are not insoluble in non-aqueous solvents should not be formulated with organic solvents due to the issue of solvent residues (Bajaj et al. 2012).

13.3.2.5 Precipitation Assisted by Acid-Base Method

In the CO2-assisted precipitation technique, acid-base reactions are commonly utilized. This includes dispersing the medication in a mildly acidic solution for the acid phase and in a weakly basic solution containing a stabilizer for the base phase.

Nanocrystals of the drug are formed through vapour effervescence by gradually introducing the acid phase into the base phase, generating carbon dioxide. This method is environmentally friendlier as it does not require the use of carbon-based solvents, although it is applicable only to poorly soluble medications whose solubility depends on pH and remains stable during acid-base interactions (Pınar et al. 2023). This includes dissolving the drug in a mild acid solution for the acid phase and a weak base solution containing a stabilizer for the base phase. This method is environmentally friendly as it eliminates the need for organic solvents. However, it is only suitable for medications that are insoluble and exhibit solubility dependence on pH, while also remaining stable during acid-base interactions.

13.3.2.6 High Gravity-Controlled Precipitation

This method enhances the uniformity and density of drug nanocrystals through advancements in the gravity-controlled precipitation process. The primary factors influencing particle size include the concentration of reactants, rotation speed, and volumetric flow rate. Continuous mixing and reaction of the drug solution within the apparatus are facilitated by this technique. Nevertheless, the practical implementation of this approach is limited due to persistent nucleation caused by localized oversaturation near the turbulent edge during mixing (Pardhi et al. 2018).

13.3.2.7 Supercritical Fluid (SCF) Method

This method entails the dissolution of pharmaceuticals in a supercritical fluid, such as CO_2, and the formation of nanocrystals through the quick evaporation of the supercritical fluid. This evaporation occurs as the fluid is sprayed through a nozzle with a narrow aperture under lowered pressure. Two techniques, rapid expansion of supercritical solution (RESS) and supercritical antisolvent (SAS), utilize supercritical fluids in line with their role in the crystallization process. An enhancement of the rapid expansion of supercritical fluid method has been developed, resulting in the rapid expansion of a supercritical solution into a liquid solvent (RESOLV) approach. In RESOLV, the propeller is positioned in the air for the former and in an aqueous solution for the latter, depending on distinct nozzle placements (Ahmadi et al. 2019). The preparation method of supercritical fluid method is shown in Fig. 13.5.

13.3.2.8 Emulsion Polymerization Method

An oil-in-water (O/W) emulsion is generated by dissolving the active pharmaceutical ingredient (API) in organic solvents with high volatility or solvents that are partially blended with water as the dispersion phase. Subsequently, the carbon-based (organic) solvent is homogenized drop by drop into liquid state, typically incorporating emulsifiers to facilitate easy control of the emulsion droplet size. The process of obtaining drug nanocrystals involves evaporating, mixing, and extracting the emulsions. The final product's quality is significantly affected by factors such as the emulsifier, shearing rate, vaporization rate, temperature gradient, and pH scale. While this emulsion polymerization process is well-suited for laboratory operations, it may not be practical for large-scale pilot production without the need of ultrasound or homogenization due to its requirements (Boles et al. 2016).

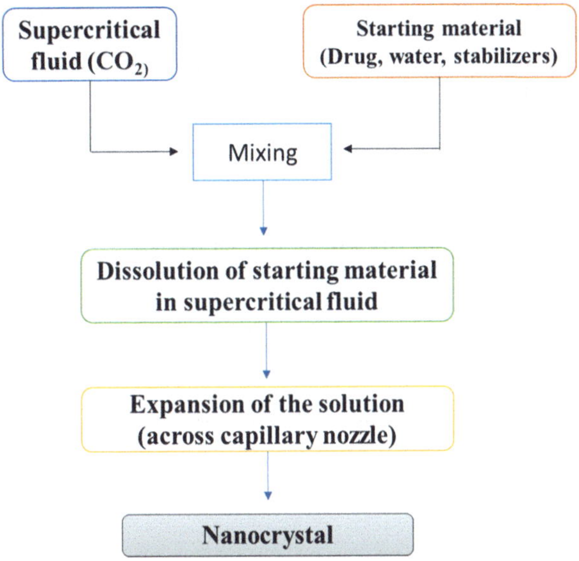

Fig. 13.5 Schematic representation of the supercritical fluid technology

The key benefit of top-down approaches as compared to bottom-up techniques is the ability to create nanosuspensions with higher loading of drug. Furthermore, these methods avoid the use of severe organic solvents, because the solvent in which the medication is distributed but not dissolvable in water for most weakly aqueous-soluble medicines, forming top-down approaches environmentally acceptable. This enables the production of a wide spectrum of weakly soluble APIs, referred to as "brick dust."

13.3.3 Combinative Technology

13.3.3.1 Nano Edge Technology

The initial combined technique developed for reducing particle size in the manufacture of nanodrugs is known as nano edge technology. This method utilizes the high-pressure homogenization (HPH) approach, complemented by the precipitation method. In this process, precipitation is employed to form initial crystal particles, and reducing slit obstruction in the high-pressurized airflows is employed to improve the effectiveness of reducing the particle size during the approach of homogenization. Following this, the HPH procedure is utilized to further compress the drug molecules, prohibiting secondary growth and addressing issues like Oswald ripening and uneven molecule size distribution and associated with the precipitation method. This, in turn, enhances the structural stability of the nanocrystal fragments. Additionally, for the high-energy procedure, alternative methods such as ultrasound

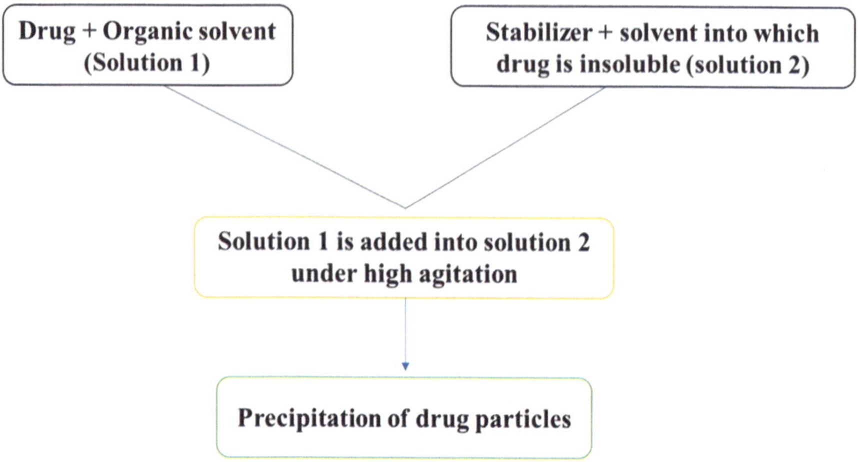

Fig. 13.6 Method of preparation of nanoedge

or micro fluidization can be applied (Zhou et al. 2019). The preparation method of nanoedge is shown in Fig. 13.6.

13.3.3.2 Smart Crystal Technology

Smart Crystal technology comprises a primary stage of pre-treatment, succeeded by a micronization (HPH) process. Preliminary treatment procedures, such as bead milling, spray drying, atomization drying, or sedimentation, may precede HPH. Acknowledged as a second-generation technique for generating nanocrystals, Smart Crystal technology embraces a comprehensive strategy termed "collection." This collection functions as a set of tools to enhance technology in the creation of drug nanocrystals and encompasses diverse methodologies, including H69, H42, H96, and combination technology (CT) (Sun et al. 2018).

13.4 Characterization of Nanocrystals

Characterizing nanocrystals is essential to understand their properties and potential applications. Characterization methods help researchers assess the size, shape, composition, crystallinity, and surface characteristics of nanocrystals. The following are typical methods employed for the characterization of nanocrystals:

13.4.1 Particle Size and Size Distribution

The mean particle sizes and polydispersity indexes (PIs) of the nanocrystals are determined using photon correlation spectroscopy (PCS) with the Malvern Zetasizer 3000HS (Malvern Instrument, Malvern, UK). The PI reflects the variability in the particle size distribution, with lower PI values indicating greater uniformity among particles. A PI exceeding 0.7 signals polydispersity in the suspended particles. To achieve an appropriate concentration for PCS measurements, a portion of the fresh nanosuspensions is diluted with a saturated solution containing approximately 0.1 wt% of stabilizer. Before size measurements, the nanosuspensions undergo sonication for 4 min. The analyses utilize a dispersant refractive index of 1.33, and each sample undergoes three measurements for accuracy (Yue et al. 2022).

13.4.2 SEM

SEM is an alternative microscopy approach that provides surface imaging of nanocrystals. While it offers lower resolution than TEM, it is useful for studying the overall morphology and size distribution of nanocrystals (Kachhadiya et al. 2022). When transforming formulated nanosuspensions into a dried powder, either through processes like lyophilization or spray drying, it is essential to conduct SEM analysis. This analysis helps monitor changes in particle size and shape both before and after the removal of water. Typically, clustering may occur post water removal, leading to maximization in particle size. SEM enables the observation of such alterations. To mitigate the extent of particle size increases, certain inactive ingredients are added as "protectants." Mannitol is frequently employed as a cryoprotectant in the lyophilization process. It undergoes recrystallization around the nanocrystals as water is removed, preventing particle interaction and agglomeration (Hecq et al. 2005).

While a certain level of agglomeration is acceptable if the final particle size remains within an acceptable range, it is crucial that the dried powder can be easily redispersed into stable nanosuspensions. Additionally, the shape of the drug crystals is influenced by their crystalline structure (Thakur et al. 2020).

13.4.3 TEM

TEM is a powerful method for visualizing the size, shape, and morphology of nanocrystals at the nanoscale. It provides high-resolution images that reveal details about individual nanoparticles and their distribution (Rana et al. 2021). TEM (transmission electron microscopy) captures images by directing a beam of electrons through a thin specimen, providing a detailed view of the interior of the sample. Widely utilized in nanomedical research, this microscopy technique offers high resolution, enabling the detailed examination of interactions between nanoparticles and cell/tissue components. The electron beams' remarkably short wavelength, approximately 100,000 times shorter than photons in the visible spectrum, allows for sub-nanometre resolution, reaching about 0.2 nm in conventional TEM (Malatesta 2021).

13.4.4 AFM

The morphology and topography of the CNCs are examined through AFM (atomic force microscopy). An aqueous CNC suspension is initially dried on a glass slide for imaging. Subsequently, scans are performed in semi-contact mode in the air using a Nanotec Electronica scanning probe instrument based in Madrid, Spain. The AFM imaging system allows environmental control within the chamber, allowing adjustments to either ambient conditions with approximately 30% relative humidity (RH) or to around 0.1% RH by introducing N_2 gas from liquid nitrogen boil-off into the chamber. The AFM images are then processed using WSxM software (version 12.0) (Bushell et al. 2021; Chen et al. 2021).

13.4.5 Surface Area and Pore Size Analysis

Techniques like BET (Brunauer-Emmett-Teller) evaluation can be used to determine the defined pore size distribution and surface area of nanocrystals, which is important for understanding their surface properties and reactivity (Uke et al. 2020).

13.4.6 Zeta Potential

The physical stability of nanocrystals is influenced by the surface charge they carry. Greater physical stability is observed when particles have elevated equal charges, leading to heightened electrostatic repulsion. The measurement of particle surface charge is best represented by "zeta potential," which is determined by analysing the electrophoretic mobility of particles in an electric field. The particle charge can be quantified in surface charge per unit through colloid titration (Dhibar et al. 2023). Particle aggregation is less probable when particles possess a suitable zeta potential, guaranteeing efficient electric repulsion, or when there is an ample steric barrier, ensuring appropriate steric repulsion between them. According to the literature, a zeta potential of at least −30 mV is advised for electrostatically stabilized systems and −20 mV for sterically stabilized systems, to attain physical stability in nanocrystal suspensions (Prasanna and Mitra 2020). The upper limit of the zeta potential is set at +30 mV.

13.4.7 DSC

Differential scanning calorimetry (DSC) is a commonly used technique to investigate the thermal properties of drugs and drug nanocrystals. DSC studies are carried out to evaluate the crystalline nature of the drug and the interactions between excipients and the drug after the production of nanocrystals. This is especially important for drugs that exist in various polymorphic forms. Additionally, certain top-down techniques, such as high-pressure homogenization, can lead to particles containing an amorphous fraction, thereby improving saturation solubility. DSC analysis involves examining pure drug, a physical mixture of the drug and excipients (stabilizer), and the final formulation, which may be in dried form. DSC can be classified based on the operating mechanism into two categories: heat flux DSC and power-compensated DSC. In heat flux DSC, two pans are placed on a thermoelectric disk within a furnace containing the sample and an empty reference pan. The furnace is heated at a linear rate, and heat is transferred to the sample and reference pan through the thermoelectric disk (Kang and Kim 2023; Wadhawan et al. 2021).

13.4.8 XRD

The evaluation of nanocrystals' morphological and polymorphic changes involves an examination of their morphology and crystalline state (Nath et al. 2020). Typically, the X-ray diffraction (XRD) method is employed to assess the crystallinity form of the drug. Confirming the formulation of the nanocrystal is achieved by observing changes in the polymorphic state. In this process, the X-ray diffraction pattern of the crystalline substances is compared with that of the pure sample. Each crystalline substance produces a unique pattern, and the amalgamation of various substances presents its distinct pattern. The X-ray diffraction pattern of a particle serves as a unique fingerprint for identifying the substance. The specimen is positioned on a level aluminium sample holder. Information is gathered by scanning from 5° to 40° with 0.02° increments, and each step takes 0.5 s for measurement (Koneti et al. 2014).

13.4.9 FTIR

The FTIR technique is employed to assess the chemical characteristics of a drug and its interactions with various excipients used in the formulation (Aguayo et al. 2018). The samples' spectra are captured within the range of 4000–400 cm^{-1} using an FTIR spectrophotometer (Vertex 70, Bruker Optik, Ettlingen, Germany) equipped with a micro attenuated total reflectance (ATR) accessory. The ATR unit incorporates a disc of diamond as the internal reflection element, and the sample penetration depth ranges from 0.1 to 2 μm. To acquire FTIR spectra at room temperature, 2 μL of the samples are applied twice and dried using a triple dental syringe. The air spectrum is utilized as a background in the FTIR analysis. The analysis of sample spectra is performed with a resolution of 2 cm^{-1} and 34 scans (Valle et al. 2021).

13.4.10 Raman Spectroscopy

The Raman spectroscopy technique functions based on the principle of inelastic scattering of monochromatic light emitted by a laser source. In this process, the frequency of photons in the monochromatic light undergoes changes due to interaction with the sample, a phenomenon known as inelastic scattering. The sample absorbs and then re-emits the laser photon light, resulting in a shift in the remission frequency either upwards or downwards compared to the original monochromatic light frequency. This occurrence is referred to as the Raman effect, offering insights into rotational, vibrational, and other low-frequency transitions within molecules (Testa-Anta et al. 2019). During sample preparation, a minimal quantity of water is used to mitigate water interference with Raman spectra. This spectroscopy serves as a valuable tool for characterizing phase transitions and identifying the crystalline or amorphous nature of various nanostructured materials, including nanoparticles and nanocrystals. Additionally, it is employed to assess defects in nanomaterials, determine the size and shape of nanomaterials, and analyse the distribution of nanostructured materials as either homogenous or heterogeneous (Bala et al. 2020).

13.4.11 TGA

Different thermal analyses, such as thermogravimetry (TG), derivative thermogravimetry (DTG), and differential thermal analysis (DTA), are utilized to evaluate the thermal stability, crystalline properties, and pyrolysis behaviour of polymers containing nanocrystals (Rana et al. 2023). Thermogravimetric analysis combines microscopy and thermal analysis to examine and characterize materials concerning temperature and time. Hot stage microscopy is useful for both screening and characterizing polymorphs, as well as discerning the crystalline and amorphous regions of nanocrystals (Pandi et al. 2021).

13.4.12 Permeation Study

Permeation studies are commonly carried out utilizing the Franz diffusion cell apparatus, utilizing skin samples such as human cadaver skin, pig ear skin, rat skin, or pig skin. The enhanced saturation solubility and dissolution of nanocrystals also indicate heightened adherence to the skin, thus aiding in the transportation of the drug through the skin membrane (Imono et al. 2020; Patzelt et al. 2011).

Characterizing nanocrystals often requires an integration of these methods to attain a thorough recognition of their physical, chemical, and structural properties. The preference of characterization methods relies on the properties and applications of the nanocrystals under investigation.

13.5 Applications of Nanocrystals

13.5.1 Oral Delivery

The advantages of orally administering drugs are significantly enhanced using drug nanocrystals (NCs), especially in improving the bioavailability of poorly soluble medications. When there is a need for a quick onset of action for such drugs, employing drug NCs proves beneficial, as observed in the case of analgesics. For example, formulating the analgesic Naproxen as a nanosuspension resulted in a reduced time to reach maximum concentration (t_{max}) and an approximately threefold increase in the area under the curve (AUC) compared to a regular suspension (Naprosyn®) (Jacobs et al. 2000). This formulation not only minimizes gastric irritability but also takes advantage of the adhesive properties and prolonged residence time inherent in drug NCs. The mucoadhesiveness of these formulations can be further enhanced by incorporating mucoadhesive polymers into the distribution medium, providing an additional benefit in manipulating bioavailability (Müller and Jacobs 2002).

13.5.2 Parenteral Administration

Administering poorly soluble drugs parenterally, especially through intravenous (IV) injection for practically water insoluble substances using cyclodextrins, liposomes, surfactants, or cosolvents, often results in either toxic side effects or large injection volumes. On the other side, nanosuspensions without carriers have the potential to provide greater loading capacity than alternative parenteral application system. When a drug is delivered as a nanosuspension, the rapid dissolution of the nanocrystals (NCs) mimics the plasma concentration profile of a solution. The nanosuspension showed significantly increased tolerance, resulting in an around twofold increase in the LD50 value (Sarnes et al. 2013; Testa-Anta et al. 2019).

13.5.3 Pulmonary Drug Delivery

Numerous essential APIs designed for pulmonary delivery face challenges due to poor solubility in both non-aqueous media and water. This is evident in widely used corticosteroids like beclomethasone dipropionate and budesonide. Nanosuspensions provide an effective solution to overcome these difficulties. Nebulizing nanosuspensions generates aerosol droplets of the desired size, carrying a substantial quantity of drug nanocrystals. The use of these nebulized nano-suspensions significantly enhances the respirable fraction compared to conventional metred-dose inhalers (MDIs). Furthermore, drug nanocrystals exhibit improved mucoadhesiveness, extending their time of residence on the mucosal surface of the lungs (Pardhi and Jain 2021; Yue et al. 2022; Tuomela et al. 2016).

13.5.4 Ocular Drug Delivery

Delivering medication to the eyes poses challenges and holds promise as an exciting area of research. Consequently, several topical formulations are under development to address ocular disorders. However, achieving effective ocular drug delivery has been a persistent challenge, marked by significantly low bioavailability. Only about 5% of locally administered drugs reached to ocular tissues due to factors such as rapid corneal drug elimination, the presence of the ocular surface epithelial barrier, and confined absorption from the ocular fornix. Tuomela et al. developed nanocrystals (NCs) of brinzolamide (BRA) to enhance the solubility of the medication and reduce intraocular (eye) pressure. The BRA-NCs surpassed the performance of the marketed product in terms of solubility, intraocular pressure reduction in an ocular hypertension model, and cytotoxicity to corneal epithelial cells (ocular) (Parmar et al. 2021).

13.5.5 Topical Drug Delivery

Nanocrystals (NCs) exhibit characteristics like improved permeation, bioadhesiveness, and heightened membrane penetration. However, sustained focus on harnessing adhesion, rapid dissolution, and enhanced penetration for dermal and mucosal applications was lacking for an extended period. This became advantageous when formulating poorly soluble antioxidants such as apigenin, hesperidin, and rutin as nanosuspensions for use in anti-aging and skin protective cosmetic products. The nanocrystals are simply incorporated into the water phase of dermal creams and oil-in-water (O/W) lotions. Despite containing nanosized crystals, these products do not meet the criteria for nano products under the new European cosmetics regulations. This is because the size of the NCs exceeds 100 nm, and the particles are required to be biodegradable rather than bio-persistent (Sun et al. 2021).

The fundamental mechanism involves the NCs enhancing the solubility of poorly water-soluble actives in the aqueous phase, resulting in a higher concentration gradient. This principle is applicable to pharmaceutical dermal formulations as well. In the case of diclofenac sodium nanosuspension for transdermal delivery, there was a notable increase in the permeability flux of the drug across the skin, up to 3.8 times higher compared to the control, as demonstrated in testing with the Yucatan micropig (YMP) skin model (Patel et al. 2018).

13.5.6 Targeted Drug Delivery

The desire to targeting medications to particular locations using nanocrystals is steadily growing, driven by both economic and therapeutic considerations. Nanosuspensions provide a practical option for targeted (precise) delivery because their external characteristics and in vivo performances can be conveniently manipulated. For example, in treating pulmonary aspergillosis, one can opt for amphotericin B nanosuspensions alternative of choosing pegylated liposomes. Achieving effective intravenous targeting requires modifying the surface properties of the nanocrystals (NCs). These surface properties play a crucial role in determining the quantitative and qualitative composition of the blood protein pattern of adsorption. By adjusting the surface properties, the particles can selectively adsorb to blood proteins responsible for enrichment at the desired target site (Joseph and Singhvi 2019; Bansal and Kumria 2012; Bajwa et al. 2016).

To explore the intricacies of these application of nanocrystals, Table 13.1 presents details on several drug nanocrystals, including examples such as albendazole, celecoxib, and ibuprofen, along with their respective preparation methods.

Table 13.1 Formulations of drug nanocrystals

API	Stabilizers	Aqueous phase	Dosage form	Method of prep.	Route	Reference
Flubendazole	Poloxamer 188	Water	Nanocrystal	Nanoprecipitation process	Pulmonary	Miyagi et al. (2023)
Valsartan	Poloxamer 188	Purified water	Nanosuspension	Antisolvent precipitation	Oral	Sreeharsha et al. (2022)
Albendazole	Tween 80, P127, or PVA screened as stabilizer	Water	Nanosuspension	Wet media milling technique	Parenteral	Permana et al. (2021)
Curcumin	D-α-Tocopherol polyethylene glycol 1000 succinate (TPGS)	Water	Nanosuspension	Wet milling method	Topical	Pelikh et al. (2021)
Quercetin	Poloxamer 188 (mg), Tween 80	Water	Nanosuspension	Wet milling technique	Topical	Manca et al. (2020)
Resveratrol	Five different stabilizers, such as HPMC, D-α-tocopheryl polyethylene glycol 1000 succinate (TPGS), Poloxamer 407, PVP K90, and Tween 80	Acetone, water	Nanocrystal	Antisolvent precipitation technique	Oral	Xiong et al. (2020)
Valero-fenbendazole	HPMC, PVP, lactose, mannitol, P407 or P188	Ultrapure water	Nanosuspension	Wet bead milling	Oral	Melian et al. (2020)
Ibuprofen	Lutrol (poloxamer) F68 and Solu plus, Natrosol 250, Methocel K4M, Pharma coat 603, 606, 615 as suspension stabilizers	Distilled water	Nanosuspension	Wet media milling technique	Oral	Ouranidis et al. (2020)
Methotrexate	Citric acid and PVA 10 K	Ultrapure water	Nano suspension	Acid-base neutralization precipitation method	Intradermal injection	Tekko et al. (2020)

(continued)

Table 13.1 (continued)

API	Stabilizers	Aqueous phase	Dosage form	Method of prep.	Route	Reference
Irbesartan	Poloxamer 407 as a stabilizer	Purified water	Nanocrystalline suspension	Wet milling technique	Oral	Meruva et al. (2019)
Celecoxib	Hydroxypropyl methylcellulose acetate succinate and hydroxypropyl cellulose, polyvinylpyrrolidone as primary stabilizer and sodium dodecyl sulfate (SDS), used as secondary stabilizer	Ultra purified Milli-Q water	Nanosuspension	Wet medium milling technique	Oral	Ding et al. (2019)
Silybin	PEG-chitosan as stabilizer	Ethanol	Nanocrystal	Antisolvent precipitation method	Oral	Liu et al. (2019)
Domperidone	PVP, HPMC, PVA, Eudragit, Pluronic F127, and ethyl cellulose	Water	Nanosuspension	Antisolvent-precipitation method	Oral	Ndlovu et al. (2019)
Fenofibrate	HPMC, PVP, PVA, and sodium dodecyl sulfate (SDS)	Purified water	Nanoparticles	Evaporation assisted solvent-antisolvent interaction	Oral	Kumar and Siril (2018)
Budesonide	Mannitol as stabilizer	Methanol	Nanoparticles	Liquid antisolvent precipitation	Oral or pulmonary delivery	Hu et al. (2018)
Lutein	Soy phosphatidylcholine, mannitol, and PVP K30	Double-distilled water	Nanosuspension	Antisolvent nanoprecipitation method	Oral	Chang et al. (2018)
Acelofenac	Sodium lauryl sulfate (SLS), hydroxypropyl methylcellulose, PVP K30	Ethanol	Nanosuspension	Precipitation-ultrasonication approach	Oral	Rahim et al. (2017)

API	Stabilizers	Aqueous phase	Dosage form	Method of prep.	Route	Reference
Diclofenac	Poloxamer 188	Bi-distilled 11 water	Nanosuspension	Wet media milling technique	Topical	Pireddu et al. (2016)
Spironolactone	Poloxamers 407 and 188 (Pluronic F127 and F68), sodium hydroxy cholate, hydroxypropyl methylcellulose	Deionized water	Nanosuspension	Wet milling technique	Oral	Mu et al. (2016)
Indomethacin and itraconazole	Poloxamer 188, poloxamer 407, polysorbate 80 (Tween 80), and propylene glycol	Water	Nanosuspension	Wet milling technique	Oral	Liu et al. (2011)

13.6 Marketed Products of Nanocrystals

Traditional treatments and certain innovative carrier systems face challenges related to efficacy, stability, toxicity, and bioavailability. Consequently, certain drugs are developed in the form of nanocarriers and introduced to the market to improve the efficacy of treatment. Details about several marketed nanocarriers and their preparation methods can be found in the accompanying Table 13.2 (Jia et al. 2023; Joshi et al. 2019; Junghanns and Müller 2008).

13.7 Conclusion

The advances and developments in the formulation of drug nanocrystals represent a pivotal strategy for addressing the challenges associated with poorly soluble drugs in the twenty-first century. The adaptability of this technology is remarkable, as it can be effortlessly employed for a broad spectrum of medications, making it a universal formulation principle analogous to micronization but specifically beneficial for BCS class II drugs. One of the key advantages lies in the ability to employ nanocrystal formulations for various routes of administration, including oral, parenteral (particularly intravenous), dermal, nasal, ophthalmic, vaginal, and pulmonary delivery. This broad applicability underscores the potential of nanocrystals in tailoring drug delivery systems to specific therapeutic needs.

Despite the widespread applications of nanocrystals, certain limitations exist, for example, how the size of the end product affects the hardness of the original crystals and the possibility of contamination. Overcoming these challenges, particularly with the introduction of bottom-up techniques, has expanded the scope of nanocrystal technology. Notably, bottom-up methods like fast precipitation have been enhanced to control particle size, and innovations such as binding organic molecules in a network have mitigated issues related to excessive growth or size enlargement. Recent developments in bottom-up approaches, specifically those involving seeding and mechanical pressure, have paved the way to organic nanocrystals with dimensions spanning from 100 to 150 nm across a variety of pharmaceuticals. It is crucial to acknowledge that while these advancements provide promising solutions, the overarching consensus from reviews and studies affirms that nanocrystal technology effectively addresses solubility and bioavailability challenges in drug formulations. As we continue to refine and expand these methodologies, the formulation of drug nanocrystals stands as a transformative and indispensable facet in the quest for improved drug delivery and therapeutic outcomes.

Table 13.2 Overview of current state of development of drugs using the Nanocrystal® technology

Trade name	API	Category	Company	Preparation method	Route
Zanaflex® capsules	Tizanidine hydrochloride	Muscle relaxant	Acorda	Ball milling	Oral
Gris-PEG®	Griseofulvin	Antifungal	Recro Gainesville LLC	Precipitation	Oral
Invega™ Hafyera	Paliperidone palmitate	Antipsychotics	Janssen Pharmaceuticals Inc.	Media milling	Intramuscular
Apretude	Cabotegravir	Antiretrovirals	ViiV Healthcare	Media milling	Intramuscular
Anjeso®	Meloxicam	Anti-inflammatory	Baudax Bio	Media milling	Intravenous
Theo-Dur®	Theophylline	Bronchodilator	Mitsubishi	Ball milling	Oral
Tricor®	Fenofibrate	Hypercholesterolaemia	Abbott	Ball milling	Oral
Avinza®	Morphine sulfate	Analgesic	King Pharmaceutical	Ball milling	Oral
Ritalin®	Methylphenidate hydrochloride	ADHD	Acorda	Ball milling	Oral
Cabenuva	Cabotegravir and rilpivirine	Antiviral combination	ViiV Healthcare	Media milling	Intravenous
Herbesser®	Diltiazem	Antihypertensive	Mitsubishi	Ball milling	Oral
Aristada™	Aripiprazole lauroxil	Atypical antipsychotic	Alkermes	High pressure homogenization	Intravenous
Invega Trinza®	Paliperidone palmitate	Schizophrenia	Janssen Pharmaceuticals Inc.	Ball milling	Intramuscular
Vitoss®	β-Tricalcium phosphate	Bone-grafting material	Orthovita Inc.	HPH	Orthopaedic
Verelan®	Verapamil HCL	Antihypertensive	Schwarz Pharma	Ball milling	Oral
Triglide®	Fenofibrate	Hypercholesterolaemia	Sciele Pharma Inc.	Microfluidizer	Oral
Azopt®	Brinzolamide	Ocular hypertension	Alcon	Ball milling	Ocular

(continued)

Table 13.2 (continued)

Trade name	API	Category	Company	Preparation method	Route
Focalin®	Dexmethylphenidate hydrochloride	Attention-deficit/ hyperactivity disorder (ADHD)	Novartis	Ball milling	Oral
Invega Sustenna®	Paliperidone palmitate	Antipsychotic	Johnson-Johnson	Ball milling	Intramuscular
Cesamet®	Nabilone	Antiemetic	Lilly	Precipitation	Oral
Megace ES®	Megestrol	Antianorexic	Par Pharmaceutical Companies, Inc.	Ball milling	Oral
Rapamune®	Sirolimus	Immunosuppressive	Wyeth	Ball milling	Oral
Emend®	Aprepitant	Antiemetic	Merck	Ball milling	Oral
Naprelan	Naproxen sodium	Anti-inflammatory	Wyeth	Ball milling	Oral

References

Aguayo MG, Pérez AF, Reyes G, Oviedo C, Gacitúa W, Gonzalez R, Uyarte O (2018) Isolation and characterization of cellulose nanocrystals from rejected fibers originated in the Kraft pulping process. Polymers 10(10):1145. https://doi.org/10.3390/polym10101145

Ahmadi TA, Omranpoor MM, Vatanara A, Seyedabadi M, Ramezani V (2019) Formation of nanosuspensions in bottom-up approach: theories and optimization. Daru 27(1):451–473. https://doi.org/10.1007/s40199-018-00235-2

Ali HS, York P, Blagden N (2009) Preparation of hydrocortisone nanosuspension through a bottom-up nanoprecipitation technique using microfluidic reactors. Int J Pharm 375(1–2):107–113. https://doi.org/10.1016/j.ijpharm.2009.03.029

Attama AA, Reginald-Opara JN, Uronnachi EM, Onuigbo EB (2016) Nanomedicines for the eye. In: Nanoscience in dermatology. Academic Press, pp 323–336. https://doi.org/10.1016/b978-0-12-802926-8.00025-2

Baig N, Kammakakam I, Falath W (2021) Nanomaterials: a review of synthesis methods properties recent progress and challenges. Mater Adv 2:1821–1871. https://doi.org/10.1039/D0MA00807A

Bajaj A, Rao MR, Pardeshi A, Sali D (2012) Nano crystallization by evaporative antisolvent technique for solubility and bioavailability enhancement of telmisartan. AAPS PharmSciTech 13(4):1331–1340. https://doi.org/10.1208/s12249-012-9860-x

Bajwa N, Mehra NK, Jain K, Jain NK (2016) Targeted anticancer drug delivery through anthracycline antibiotic bearing functionalized quantum dots. Artif Cells Nanomed Biotechnol 44(7):1774–1782. https://doi.org/10.3109/21691401.2015.1102740

Bala N, Singh HK, Verma S, Rath S (2020) Magnetic-order induced effects in nanocrystalline NiO probed by Raman spectroscopy. Phys Rev B 102(2):024423

Bansal S, Kumria R (2012) Nanocrystals: current strategies and trends. Int J Res Pharmaceut Biomed Sci 3(1):407–419

Boles M, Ling D, Hyeon T et al (2016) The surface science of nanocrystals. Nat Mater 15:141–153. https://doi.org/10.1038/nmat4526

Bushell M, Meija J, Chen M, Batchelor W, Browne C, Cho JY, Clifford CA, Al-Rekabi Z, Vanderfleet OM, Cranston ED, Lawn M (2021) Particle size distributions for cellulose nanocrystals measured by atomic force microscopy: an interlaboratory comparison. Cellulose 28:1387–1403

Chang D, Ma Y, Cao G, Wang J, Zhang X, Feng J, Wang W (2018) Improved oral bioavailability for lutein by nanocrystal technology: formulation development, in vitro and in vivo evaluation. Artif Cells Nanomed Biotechnol 46(5):1018–1024

Chen M, Parot J, Hackley VA, Zou S, Johnston LJ (2021) AFM characterization of cellulose nanocrystal height and width using internal calibration standards. Cellulose 28:1933–1946

Chogale MM, Ghodake VN, Patravale VB (2016) Performance parameters and characterizations of nanocrystals: a brief review. Pharmaceutics 8(3):26

Dhibar M, Chakraborty S, Kundu A, Laha P (2023) Chemistry characterization and application of nanocrystals-based drug delivery system: present to future perspective. Pharm Nanotechnol 11(3):265–275. https://doi.org/10.2174/2211738511666230109151410

Ding Z, Wang L, Xing Y, Zhao Y, Wang Z, Han J (2019) Enhanced oral bioavailability of celecoxib nanocrystalline solid dispersion based on wet media milling technique: formulation, optimization and in vitro/in vivo evaluation. Pharmaceutics 11(7):328

Gu W, Zeng Y, Deng Y, Huang P, Jin G, Liu F, Wei J, Li H (2023) Colloidal synthesis and optical properties of Cs2CuCl4 nanocrystals. Crystals 13(6):864. https://doi.org/10.3390/cryst13060864

Guo M, Qin S, Wang S, Sun M, Yang H, Wang X, Fan P, Jin Z (2023) Herbal medicine nanocrystals: a potential novel therapeutic strategy. Molecules 28(17):6370. https://doi.org/10.3390/molecules28176370

Harish V, Ansari MM, Tewari D, Gaur M, Yadav AB, García-Betancourt ML, Abdel-Haleem FM, Bechelany M, Barhoum A (2022) Nanoparticle and nanostructure synthesis and controlled growth methods. Nanomaterials (Basel, Switzerland) 12(18):3226. https://doi.org/10.3390/nano12183226
Hecq J, Deleers M, Fanara D, Vranckx H, Amighi K (2005) Preparation, and characterization of nanocrystals for solubility and dissolution rate enhancement of nifedipine. Int J Pharm 299(1–2):167–177
Hu J, Dong Y, Ng WK, Pastorin G (2018) Preparation of drug nanocrystals embedded in mannitol microcrystals via liquid antisolvent precipitation followed by immediate (on-line) spray drying. Adv Powder Technol 29(4):957–963
Imono M, Uchiyama H, Yoshida S, Miyazaki S, Tamura N, Tsutsumimoto H, Kadota K, Tozuka Y (2020) The elucidation of key factors for oral absorption enhancement of nanocrystal formulations: in vitro–in vivo correlation of nanocrystals. Eur J Pharm Biopharm 146:84–92
Jacobs C, Kayser O, Müller RH (2000) Nanosuspensions as a new approach for the formulation for the poorly soluble drug tarazepide. Int J Pharm 196(2):161–164
Jahangir MA, Imam SS, Muheem A et al (2022) Nanocrystals: characterization overview, applications in drug delivery, and their toxicity concerns. J Pharm Innov 17:237–248. https://doi.org/10.1007/s12247-020-09499-1
Jarvis M, Krishnan V, Mitragotri S (2018) Nanocrystals: a perspective on translational research and clinical studies. Bioeng Transl Med 4(1):5–16. https://doi.org/10.1002/btm2.10122
Jia Y, Jiang Y, He Y, Zhang W, Zou J, Magar KT, Boucetta H, Teng C, He W (2023) Approved nanomedicine against diseases. Pharmaceutics 15(3):774
Joseph E, Singhvi G (2019) Multifunctional nanocrystals for cancer therapy: a potential nanocarrier. In: Nanomaterials for drug delivery and therapy. Elsevier, pp 91–116. https://doi.org/10.1016/b978-0-12-816505-8.00007-2
Joshi K, Chandra A, Jain K, Talegaonkar S (2019) Nanocrystallization: an emerging technology to enhance the bioavailability of poorly soluble drugs. Pharm Nanotechnol 7(4):259–278. https://doi.org/10.2174/2211738507666190405182524
Junghanns JU, Müller RH (2008) Nanocrystal technology, drug delivery and clinical applications. Int J Nanomedicine 3(3):295–310
Junyaprasert VB, Morakul B (2015) Nanocrystals for enhancement of oral bioavailability of poorly water-soluble drugs. Asian J Pharm Sci 10(1):13–23. https://doi.org/10.1016/j.ajps.2014.08.005
Kachhadiya DD, Sonawane AV, Murthy ZVP (2022) Metal-organic frameworks (MOFs) characterization using different analytical methods. In: Gulati S (ed) Metal-organic frameworks (MOFs) as catalysts. Springer, Singapore. https://doi.org/10.1007/978-981-16-7959-9_6
Kang H, Kim DS (2023) A study on the crystallization and melting of PLA nanocomposites with cellulose nanocrystals by DSC. Polym Compos 44(11):7727–7736
Koneti V, Singh SK, Gulati M (2014) A comparative study of top-down and bottom-up approaches for the preparation of nanosuspensions of glipizide. Powder Technol 256:436–449
Kulkarni SA, Myerson AS (2017) Methods for nano-crystals preparation. In: Roberts K, Docherty R, Tamura R (eds) Engineering crystallography: from molecule to crystal to functional form, NATO science for peace and security series A: chemistry and biology. Springer, Dordrecht. https://doi.org/10.1007/978-94-024-1117-1_16
Kumar R, Siril PF (2018) Enhancing the solubility of fenofibrate by nanocrystal formation and encapsulation. AAPS PharmSciTech 19:284–292
Kumar S, Bhushan P, Bhattacharya S (2017) Fabrication of nanostructures with bottom-up approach and their utility in diagnostics, therapeutics, and others. Environ Chem Med Sensors:167–198. https://doi.org/10.1007/978-981-10-7751-7_8
Liu P, Rong X, Laru J, van Veen B, Kiesvaara J, Hirvonen J et al (2011) Nanosuspensions of poorly soluble drugs: preparation and development by wet milling. Int J Pharm 411(1–2):215–222. https://doi.org/10.1016/j.ijpharm.2011.03.05

Liu Y, Wang Y, Zhao J (2019) Design, optimization and in vitro-in vivo evaluation of smart nanocaged carrier delivery of multifunctional PEG-chitosan stabilized silybin nanocrystals. Int J Biol Macromol 124:667–680

Malamatari M, Taylor KMG, Malamataris S, Douroumis D, Kachrimanis K (2018) Pharmaceutical nanocrystals: production by wet milling and applications. Drug Discov Today 23(3):534–547. https://doi.org/10.1016/j.drudis.2018.01.016

Malatesta M (2021) Transmission electron microscopy as a powerful tool to investigate the interaction of nanoparticles with subcellular structures. Int J Mol Sci 22(23):12789. https://doi.org/10.3390/ijms222312789

Manca ML, Lai F, Pireddu R, Valenti D, Schlich M, Pini E, Ailuno G, Fadda AM, Sinico C (2020) Impact of nanosizing on dermal delivery and antioxidant activity of quercetin nanocrystals. J Drug Deliv Sci Technol 55:101482

Melian ME, Paredes A, Munguía B, Colobbio M, Ramos JC, Teixeira R, Manta E, Palma S, Faccio R, Domínguez L (2020) Nanocrystals of novel valerolactam-fenbendazole hybrid with improved in vitro dissolution performance. AAPS PharmSciTech 21:1–5

Mengge G, Yazhuo S, Bin L et al (2022) Sustainable preparation of cellulose nanocrystals: state of the art and perspectives. Green Chem 2022(24):9346–9372. https://doi.org/10.1039/D2GC03003A

Meruva S, Thool P, Shah S, Karki S, Bowen W, Ghosh I, Kumar S (2019) Formulation, and performance of Irbesartan nanocrystalline suspension and granulated or bead-layered dried powders—part I. Int J Pharm 568:118189

Miyagi MY, de Oliveira Faria R, de Souza GB, Lameu C, Tagami T, Ozeki T, Bezzon VD, Yukuyama MN, Bou-Chacra NA, de Araujo GL (2023) Optimizing adjuvant inhaled chemotherapy: synergistic enhancement in paclitaxel cytotoxicity by flubendazole nanocrystals in a cycle model approach. Int J Pharm 644:123324

Moroz P, Boddy A, Zamkov M (2018) Challenges and prospects of photocatalytic applications utilizing semiconductor nanocrystals. Front Chem 6:353. https://doi.org/10.3389/fchem.2018.00353

Mu S, Li M, Guo M, Yang W, Wang Y, Li J et al (2016) Spironolactone nanocrystals for oral administration: different pharmacokinetic performances induced by stabilizers. Colloids Surf B Biointerfaces 147:73–80. https://doi.org/10.1016/j.colsurfb.2016.07.0

Müller RH, Jacobs C (2002) Buparvaquone mucoadhesive nanosuspension: preparation, optimisation, and long-term stability. Int J Pharm 237(1–2):151–161

Müller RH, Jacobs C, Kayser O (2001) Nanosuspensions as particulate drug formulations in therapy. Rationale for development and what we can expect for the future. Adv Drug Deliv Rev 47(1):3–19

Müller RH, Gohla S, Keck CM (2011) State of the art of nanocrystals—special features, production, nanotoxicology aspects and intracellular delivery. Eur J Pharm Biopharm 78(1):1–9. https://doi.org/10.1016/j.ejpb.2011.01.007

Murdande SB, Shah DA, Dave RH (2015) Impact of nanosizing on solubility and dissolution rate of poorly soluble pharmaceuticals. J Pharm Sci 104(6):2094–2102

Naduparambath S, Jinitha TV, Shaniba V, Sreejith MP, Balan AK, Purushothaman E (2018) Isolation and characterisation of cellulose nanocrystals from sago seed shells. Carbohydr Polym 180:13–20

Nath D, Singh F, Das R (2020) X-ray diffraction analysis by Williamson-Hall, Halder-Wagner and size-strain plot methods of CdSe nanoparticles-a comparative study. Mater Chem Phys 239:122021

Ndlovu ST, Ullah N, Khan S et al (2019) Domperidone nanocrystals with boosted oral bioavailability: fabrication, evaluation, and molecular insight into the polymer-domperidone nanocrystal interaction. Drug Deliv Transl Res 9:284–297. https://doi.org/10.1007/s13346-018-00596-w

Niculescu AG, Chircov C, Bîrcă AC, Grumezescu AM (2021) Nanomaterials synthesis through microfluidic methods: an updated overview. Nanomaterials (Basel, Switzerland) 11(4):864. https://doi.org/10.3390/nano11040864

Oliveira C, Coelho C, Teixeira J, Santos P, Claudia B (2022) Nanocarriers as active ingredients enhancers in the cosmetic industry. The European and North America Regulation Challenges. Molecules 27:1669. https://doi.org/10.3390/molecules27051669

Ouranidis A, Gkampelis N, Vardaka E, Karagianni A, Tsiptsios D, Nikolakakis I, Kachrimanis K (2020) Overcoming the solubility barrier of ibuprofen by the rational process design of a nanocrystal formulation. Pharmaceutics 12(10):969

Pandi N, Sonawane SH, Kishore KA (2021) Synthesis of cellulose nanocrystals (CNCs) from cotton using ultrasound-assisted acid hydrolysis. Ultrason Sonochem 70:105353

Pardhi VP, Verma T, Flora SJS, Chandasana H, Shukla R (2018) Nanocrystals: an overview of fabrication, characterization and therapeutic Applications in drug delivery. Curr Pharm Des 24(43):5129–5146. https://doi.org/10.2174/1381612825666190215121148

Pardhi VP, Jain K (2021) Impact of binary/ternary solid dispersion utilizing poloxamer 188 and TPGS to improve pharmaceutical attributes of bedaquiline fumarate. J Drug Deliv Sci Technol 62(102349):1–13. https://doi.org/10.1016/j.jddst.2021.102349

Parmar PK, Wadhawan J, Bansal AK (2021) Pharmaceutical nanocrystals: a promising approach for improved topical drug delivery. Drug Discov Today 26(10):2329–2349

Parvizian M, De Roo J (2021) Precursor chemistry of metal nitride nanocrystals. Nanoscale 13(45):18865–18882. https://doi.org/10.1039/d1nr05092c

Patel V, Sharma OP, Mehta T (2018) Nanocrystal: a novel approach to overcome skin barriers for improved topical drug delivery. Expert Opin Drug Deliv 15(4):351–368

Patra JK, Das G, Fraceto LF, Campos EVR, Rodriguez-Torres MDP, Acosta-Torres LS, Diaz-Torres LA, Grillo R, Swamy MK, Sharma S, Habtemariam S, Shin HS (2018) Nano based drug delivery systems: recent developments and prospects. J Nanobiotechnol 16(1):71. https://doi.org/10.1186/s12951-018-0392-8

Patzelt A, Richter H, Knorr F (2011) Selective follicular targeting by modification of the particle sizes. J Control Release 150:45–48. https://doi.org/10.1016/j.jconrel.2010.11.015

Pelikh O, Eckert RW, Pinnapireddy SR, Keck CM (2021) Hair follicle targeting with curcumin nanocrystals: influence of the formulation properties on the penetration efficacy. J Control Release 329:598–613

Permana AD, Paredes AJ, Zanutto FV, Amir MN, Ismail I, Bahar MA, Sudir S, Palma SD, Donnelly RF (2021) Albendazole nanocrystal-based dissolving microneedles with improved pharmacokinetic performance for enhanced treatment of cystic echinococcosis. ACS Appl Mater Interfaces 13(32):38745–38760

Pınar SG, Oktay AN, Karaküçük AE, Çelebi N (2023) Formulation strategies of nanosuspensions for various administration routes. Pharmaceutics 15(5):1520. https://doi.org/10.3390/pharmaceutics15051520

Pireddu R, Caddeo C, Valenti D et al (2016) Diclofenac acid nanocrystals as an effective strategy to reduce in vivo skin inflammation by improving dermal drug bioavailability. Colloids Surf B Biointerfaces 143:64–70

Prasanna NS, Mitra J (2020) Isolation and characterization of cellulose nanocrystals from Cucumis sativus peels. Carbohydr Polym 247:116706

Rahim H, Sadiq A, Khan S, Khan MA, Shah SMH, Hussain Z, Ullah R, Shahat AA, Ibrahim K (2017) Aceclofenac nanocrystals with enhanced in vitro, in vivo performance: formulation optimization, characterization, analgesic, and acute toxicity studies. Drug Des Dev Ther 11:2443–2452. https://doi.org/10.2147/DDDT.S140626

Ran Q, Wang M, Kuang W, Ouyang J, Han D, Gao Z, Gong J (2022) Advances of combinative nanocrystal preparation technology for improving the insoluble drug solubility and bioavailability. Crystals 12(9):1200. https://doi.org/10.3390/cryst12091200

Rana AK, Frollini E, Thakur VK (2021) Cellulose nanocrystals: pretreatments, preparation strategies, and surface functionalization. Int J Biol Macromol 182:1554–1581. https://doi.org/10.1016/j.ijbiomac.2021.05.11

Rana MS, Rahim MA, Mosharraf MP, Tipu MFK, Chowdhury JA, Haque MR, Kabir S, Amran MS, Chowdhury AA (2023) Morphological, spectroscopic and thermal analysis of cellulose

nanocrystals extracted from waste jute fiber by acid hydrolysis. Polymers 15(6):1530. https://doi.org/10.3390/polym15061530
Salazar J, Muller RH, Moschwitzer JP (2013) Performance comparison of two novel combinative particle-size reduction technologies. J Pharm Sci 102(5):1636–1649
Sarnes A, Østergaard J, Jensen SS et al (2013) Dissolution study of nanocrystal powders of a poorly soluble drug by UV imaging and channel flow methods. Eur J Pharm Sci 50(3–4):511–519
Shegokar R, Müller RH (2010) Nanocrystals: industrially feasible multifunctional formulation technology for poorly soluble actives. Int J Pharm 399(1–2):129–139. https://doi.org/10.1016/j.ijpharm.2010.07.04
Sinha B, Müller RH, Möschwitzer JP (2013) Bottom up approaches for preparing drug nanocrystals: formulations and factors affecting particle size. Int J Pharm 453(1):126–141. https://doi.org/10.1016/j.ijpharm.2013.01.019
Sreeharsha N, Naveen NR, Anitha P, Goudanavar PS, Ramkanth S, Fattepur S, Telsang M, Habeebuddin M, Anwer MK (2022) Development of nanocrystal compressed minitablets for chronotherapeutic drug delivery. Pharmaceuticals 15(3):311
Sun B, Yeo Y (2012) Nanocrystals for the parenteral delivery of poorly water-soluble drugs. Curr Opin Solid State Mater Sci 16(6):295–301
Sun L, Hu Y, Zhang L (2018) Recent trends in nanocrystals for pharmaceutical applications. Curr Pharm Des 24(21):2394–2402
Sun L, Xiang H, Ge C, Chen X, Zhang Q, Zhang Y, Miao X (2021) A nanocrystals-based topical drug delivery system with improved dermal penetration and enhanced treatment of skin diseases. J Biomed Nanotechnol 17(12):2319–2337. https://doi.org/10.1166/jbn.2021.3202
Sylvestre J-P, Tang M-C, Furtos A, Leclair G, Meunier M, Leroux J-C (2011) Nanonization of megestrol acetate by laser fragmentation in aqueous milieu. J Control Release 149(3):273–280
Tekko IA, Permana AD, Vora L, Hatahet T, McCarthy HO, Donnelly RF (2020) Localised and sustained intradermal delivery of methotrexate using nanocrystal-loaded microneedle arrays: potential for enhanced treatment of psoriasis. Eur J Pharm Sci 152:105469
Testa-Anta M, Ramos-Docampo MA, Comesaña-Hermo M, Rivas-Murias B, Salgueiriño V (2019) Raman spectroscopy to unravel the magnetic properties of iron oxide nanocrystals for bio-related applications. Nanoscale Adv 1(6):2086–2103. https://doi.org/10.1039/c9na00064j
Thakur M, Sharma A, Ahlawat V, Bhattacharya M, Goswami S (2020) Process optimization for the production of cellulose nanocrystals from rice straw derived α-cellulose. Mater Sci Energy Technol 3:328–334
Tian Z, Mai Y, Meng T et al (2021) Nanocrystals for improving oral bioavailability of drugs: intestinal transport mechanisms and influencing factors. AAPS PharmSciTech 22:179. https://doi.org/10.1208/s12249-021-02041-7
Tran TT, Tran PH, Nguyen KT, Tran VT (2016) Nano-precipitation: preparation and application in the field of pharmacy. Curr Pharm Des 22(20):2997–3006. https://doi.org/10.2174/1381612822666160408151702
Tuomela A, Saarinen J, Strachan CJ, Hirvonen J (2016) Production, applications, and in vivo fate of drug nanocrystals. J Drug Deliv Sci Technol 34:21–31
Uke SJ, Mardikar SP, Bambole DR, Kumar Y, Chaudhari GN (2020) Sol-gel citrate synthesized Zn doped MgFe2O4 nanocrystals: a promising supercapacitor electrode material. Mater Sci Energy Technol 3:446–455
Valle AL, Silva ACA, Dantas NO, Sabino-Silva R, Melo FCC, Moreira CS, Oliveira GS, Rodrigues LP, Goulart LR (2021) Application of ZnO nanocrystals as a surface-enhancer FTIR for glyphosate detection. Nanomaterials (Basel, Switzerland) 11(2):509. https://doi.org/10.3390/nano11020509
Viswanathan P, Muralidaran Y, Ragavan G (2017) Challenges in oral drug delivery: a nano-based strategy to overcome. In: Nanostructures for oral medicine. Elsevier, pp 173–201. https://doi.org/10.1016/b978-0-323-47720-8.00008-0
Waard HH, Frijlink HW, Hinrichs WLJ (2011) Bottom-up preparation techniques for nanocrystals of lipophilic drugs. Pharm Res 28:1220–1223

Wadhawan J, Parmar PK, Bansal AK (2021) Nanocrystals for improved topical delivery of medium soluble drug: a case study of acyclovir. J Drug Deliv Sci Technol 65:102662

Xiong S, Liu W, Zhou Y, Mo Y, Liu Y, Chen X, Pan H, Yuan D, Wang Q, Chen T (2020) Enhancement of oral bioavailability and anti-Parkinsonian efficacy of resveratrol through a nanocrystal formulation. Asian J Pharm Sci 15(4):518–528

Yarbrough R, Davis K, Dawood S, Rathnayake H (2020) A sol-gel synthesis to prepare size and shape-controlled mesoporous nanostructures of binary (II-VI) metal oxides. RSC Adv 10(24):14134–14146. https://doi.org/10.1039/d0ra01778g

Yue P, Zhou W, Huang G, Lei F, Chen Y, Ma Z, Chen L, Yang M (2022) Nanocrystals based pulmonary inhalation delivery system: advance and challenge. Drug Deliv 29(1):637–651. https://doi.org/10.1080/10717544.2022.2039809

Zhou S, Zhao M, Yang TH, Xia Y (2019) Decahedral nanocrystals of noble metals: synthesis, characterization, and applications. Mater Today 22:108. https://doi.org/10.1016/j.mattod.2018.04.003

A Technological Update on Inhalation Drug Delivery Devices

14

Ankaj Kumar, Ishwar Chandra, V. Rajesh, Sourabh Jadhav, Harshita Krishnatreyya, and Arvind Gulbake

Abstract

Inhalation drug delivery (IDD) is an effective approach for treating respiratory diseases, including asthma, cystic fibrosis, lung cancer, and chronic obstructive pulmonary diseases. The advantages of inhalation drug delivery (IDD) include lowering the dose and dosing frequency, overcoming drug resistance, increasing in drug's effectiveness by applying it directly to the expected site of disease, and lowering the systemic toxicities of the drug. Inadequate instruction and training in the use of the inhalation device to control the actuation and aerosol inhalation may lead to the absence of intended clinical outcomes as well as issues with efficacy or any negative drug effects. There have been advances in technology-driven novel strategies in the formulation and development of aerosols, nebulizers, metered dose inhalers, and dry powder inhalers to encounter the problems associated with coordination skills, dosing frequency, and safety. Formulation advancement involves microparticles and nanomedicine-based approaches for IDD. However, the novel advance designs, structures, and functions of IDD devices make them more prominent, precise, and effective. The *in vitro*, *in vivo*, and *ex vivo* models that strengthen and check the possible efficacy of the developed IDD are discussed in the chapter. This chapter gives insight into the advancement of these IDD, novel formulation approaches, their evaluation through various preclinical models, and their benefits over previous IDD.

A. Kumar · I. Chandra · V. Rajesh · S. Jadhav · H. Krishnatreyya · A. Gulbake (✉)
Department of Pharmaceutics, National Institute of Pharmaceutical Education and Research Guwahati, Kamrup, Assam, India
e-mail: arvind@niperguwahati.in

K. Jain, A. K. Yadav (eds.), *Advances in Pharmaceutical Product Development*, https://doi.org/10.1007/978-981-97-9230-6_14

Keywords

Inhalation drug delivery · Microparticles · Inhalation therapy · Drug carriers · Nebulizers

14.1 Introduction

It is very fascinating to study the modern approach of inhalation therapy for drug delivery, especially in the case of lung diseases like chronic obstructive pulmonary disease, lung cancer, asthma, and many more. Inhalation therapy has a long history and hundreds of ingenious devices and medications. The word "inhaler" was first proposed by John Mudge in late 1778 in his book entitled, *A Radical and Expeditious Cure for a Recent Catarrhous Cough* (Mudge 1778). Inhalation therapy was practiced in ancient civilizations such as India, China, and South and Central America. In India, inhaling the fumes of stramonium and hemp was common, whereas in China, opium inhalation and ornate metal inhalers were commonly practiced (Sanders 2007).

The lungs represent the most attractive alternative route for drug delivery, owing to the larger area for the deposition of pharmaceutical agents and high vascularization for the systemic delivery of therapeutic agents. This route of administration prevents the degradation of active components in the gastrointestinal tract and also prevents its first-pass metabolism in the liver (Kuzmov and Minko 2015). The efficiency of IDD mainly depends on lung aerodynamics, the inhaled particle's size, the inhalation method, and its delivery devices (Gagnadoux et al. 2008; Lehofer et al. 2014; Chan et al. 2014). Even though this delivery route has advantages, it is still not used widely due to some limitations, such as high lung toxicity, drug-induced lung injury, and drug degradation by lung macrophages (Labiris and Dolovich 2003). An ideal IDD devices should deliver pharmaceutical active components to the diseased cell without damaging healthy lungs (Kuzmov and Minko 2015). Several conventional IDD devices and drug carriers have been developed, implemented, and studied in various studies. Nebulizers are one of the most modern and developed pulmonary drug delivery systems. Recent advancements in nebulizers are made to optimize the administration of drugs to the lungs by controlling or measuring patient breathing patterns. Jet nebulizers and ultrasonic nebulizers are conventional devices, whereas smart and vibrating mesh nebulizers have recently been developed and improved in the treatment of lung diseases. Metered dose inhalers (MDIs) are the most frequently used device for the treatment of lung diseases such as asthma and chronic obstructive pulmonary disease (Doan et al. 2011). Medihaler Epi, Riker Laboratories, marketed the world's first MDI in late 1956 (Stein et al. 2014). The cost per dose of MDIs is relatively low, which makes it a popular delivery system among all IDD devices (Doan et al. 2011). MDI contains dissolved or suspended drugs in propellant. Though propellant harms the environment, it is still widely used in devices. The hardware of MDI consists of a canister, a metering valve, and an actuator mouthpiece (Stein et al. 2014). Dry powder inhalers (DPIs) are significant segment of the global pulmonary drug delivery systems market

(Shetty and Srinivasan 2017). The DPI is one of the fastest-growing delivery systems, with a market size of USD 19.66 billion in 2022 and expected to reach 27.98 billion in upcoming years. DPIs are breath-actuated devices containing particles with a size of 1-5 µm. The formulation is delivered in the respiratory airways through oral inhalation (Ashurst et al. 2000). DPIs have more advantages than metered dose inhalers as these devices are propellant-free, making them eco-friendly. It also has dose counters that help to calculate the number of doses remaining. It is very user-friendly and easily accessible for patients; hence, it has more patient compliance (Ashurst et al. 2000; O'Connor 2004; Geller 2005). The advances in nanotechnology also enhance effective inhalational therapy by offering sustainability, targetability, dose reduction, and dosing frequency. Polymer-based nanoparticles, dendrimers, lipid-based nanoparticles, nanospheres, complexes with nucleic acid, and magnetic nanoparticles are some examples of nanotechnology-based inhalation drug carriers (Kuzmov and Minko 2015).

The chapter highlights the recent advances in inhalational therapy, with special emphasis on drug carriers and devices. The possible challenges and solutions are addressed, making IDD therapy fruitful with broad applications. The study also included the various preclinical models required to evaluate the performance of such IDD. The wide applications of IDD in various disease treatments explain the prominent role of such therapies.

14.2 Challenges and Their Probable Solutions

Inhalation drug delivery is a very popular route of administration specifically for lung diseases, allowing direct deposition and onset of the drug's action at the disease site with a minimal dose. The choice of drug delivery system mainly depends on drug properties, disease state, patient's condition, and surrounding atmospheric conditions. In the last two decades, traditional inhalation devices have been modified to overcome the problems associated with older inhalers (Sanchis et al. 2013).

14.2.1 Device-Related Challenges

According to Sanchis and colleagues, only 34% of patients adopted adequate inhalation techniques (Sanchis et al. 2013). Giraud and Roche subdivided misuse of inhalers into two categories: either omissions or errors. In omission, improper handling of an inhaler, forced expiration, no expiration, inspiration through the nose, and no inspiration are included (Giraud and Roche 2002; Lavorini et al. 2015). Any misuse during the handling of inhalers leads to serious consequences, such as the risk of hospitalization or emergency department visits (Giraud and Roche 2002). Omission is significantly associated with a lack of patient education (AL-Jahdali et al. 2013). To overcome the misuse of inhalers, the patient should first be educated and trained regarding the use of the device. Whenever needed, a physical demonstration of how to use inhalers is more beneficial than providing just verbal or written instruction. If the patient still misuses the device, an alternative inhaler device should be attempted to overcome patient-specific difficulties. Respimat soft mist inhaler is a new type of device that resembles MDI. Advances in devices and carriers have resolved such improper handling and misuse issues. A detailed description of each advancement and its merits is given in the section below (Giraud and Roche 2002; AL-Jahdali et al. 2013).

The administration of low-potent therapeutics at their higher doses is the main challenge through IDD. Among all IDD, MDI can deliver a maximum of 100–200 μg active pharmaceutical ingredient (API) per shot (Newman 2005), whereas, in the case of nebulizers, it requires a long inhalation time even for a small amount of dose administration (Haque et al. 2016). DPI is the first choice for high-dose administration. Generally, it requires a particle size of formulation below 5 μm to reach the central and peripheral regions of the lungs for therapeutic effect. However, particle agglomeration is a major concern, leading to poor flow properties (Mangal et al. 2017). To encounter such issues, the possible solution is to blend, exploiting the adhesion force between micronized drug particles and suitable larger carriers (Froehlich 2019). A higher carrier amounts can cause serious side effects; thus, safety must also be considered while designing higher excipient formulations (Longest et al. 2019).

14.2.2 Biological Barriers

The mucociliary clearance, phagocytosis, and enzymatic degradation in the lungs are the challenges associated with inhalation drug delivery. There are mucociliary escalators, alveolar macrophages, and enzymes (cytochrome p450, trypsin, antitrypsin, and proteases) that affect the biological performance of the drug carriers. The presence of lung surfactants is also associated with phagocytosis of inhaled particles (Pardhi and Jain 2021; Kumar et al. 2024). To overcome the biological barriers, formulation consideration for the use of biocompatible material generally requires including dipalmitoylphosphatidylcholine (DPPC), 1,2-distearoyl-sn-glycero-3-phosphoglycerol (DSPG), dioleoylphosphatidylethanolamine (DOPE) lipids, and lactose (Haque et al. 2016; Mangal et al. 2017; Froehlich 2019).

14.3 Inhalation Drug Delivery and Devices

14.3.1 Nebulizers

14.3.1.1 Conventional Nebulizers

A nebulizer is a device used to make a poly-disperse aerosol mist suitable through inhalation. It contains liquids and suspensions with a droplet or particle size range of 1–5 μm. It benefits unconscious, accidental, pediatric, and geriatric patients (Longest et al. 2019). The significant and suitable use of nebulizers is in chronic obstructive pulmonary disease (COPD), asthma, and chest diseases (Waldrep and Dhand 2008). Compressed air, connection tubing, and patient interface, including

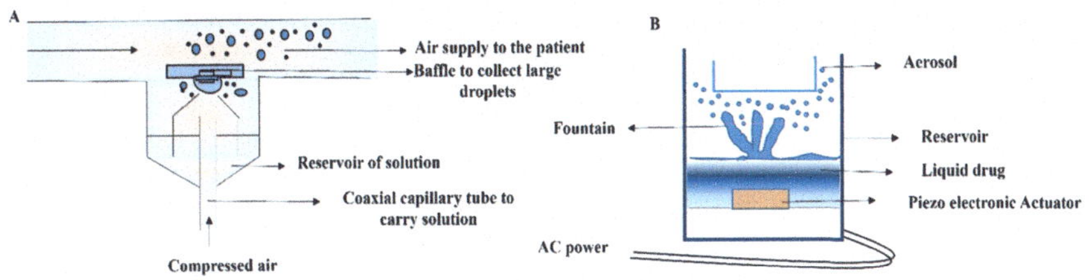

Fig. 14.1 Conventional nebulizers: (**a**) jet nebulizer, (**b**) ultrasonic nebulizer

the mouthpiece or facemask, are the components of the nebulizer system. Such components decide the overall efficiency of aerosol drug delivery. Jet and ultrasonic nebulizers are examples of conventional nebulizers, whereas smart and vibrating mesh nebulizers are categorized as advanced.

14.3.1.1.1 Jet Nebulizers

In a jet nebulizer, negative pressure is created by compressed air (air stream) passing through a small orifice at high speed. It creates low pressure and makes aerosol from liquid or suspension (Ibrahim et al. 2015). The solution form is most suitable for aerosolization owing to higher stability (Dhanani et al. 2016). Overall, nebulization is based on Ventri's principle. The pressure range of 2–10 L/min generates a spacious range of particle sizes that further reduces size through baffles. The design of the jet nebulizer is illustrated in Fig. 14.1a. Examples of products administered through jet nebulizers are antibiotics, mucolytics, liposomal formulations, beta-2-agonists, and recombinant products (Pulmozyme) (Dhanani et al. 2016). There are four categories of jet nebulizers based on the capillary tube, i.e., jet nebulizers with a corrugated tube, jet nebulizers with a collection bag, breath-enhanced jet nebulizers, and breath-activated jet nebulizers (Ari 2014). Each type of nebulizer has pros and cons, as explained in various studies (Ari 2014). The recent modifications to jet nebulizers use a thermostat to overcome the problems with bronchospasm due to a decrease in the temperature of the liquid (Ochowiak et al. 2019). An example of such type is described in Table 14.1.

14.3.1.1.2 Ultrasonic Nebulizer

This nebulizer uses ultrasonic waves to nebulize the liquid formulation into aerosols (Fig. 14.1b). The piezo-electrical crystals are vibrated to produce ultrasonic wave frequency in the range of 1–3 MHz, producing small particles inside the chamber for inhalation and larger particles in the reservoir (Longest et al. 2019). The capillary wave and cavitation theories underlie ultrasonic nebulization's main mechanism (Dhanani et al. 2016). Inhalation therapies include two types of ultrasonic nebulizers: standard nebulizers and water interface nebulizers. In a standard nebulizer, the drug has direct contact with a piezo-electric transducer. However, it is unsuitable for heat-sensitive drugs and products (peptides and proteins) (Longest et al. 2019). In a water interface nebulizer, water is placed in between the drug formulation reservoir and the piezo-electric transducer. The water decreases heat formation during the principal process of nebulization (Waldrep and Dhand 2008).

Table 14.1 Markedly available nebulizer formulations (Waldrep and Dhand 2008)

Types of marketed products	Drug	Aerosol devices	Year approved	Company name	Indications
AeroEclipsell BAN	Methacholine	Breath-actuated jet nebulizer	2020	Trudell Medical International	Asthma, cystic fibrosis
AKITA	Tobramycin	Vibrating mesh	2007	Activaero America, Inc.	Cystic fibrosis
APIXNEB	Tobramycin	Nebulizer	2007	Activaero America, Inc.	Cystic fibrosis
CompAIR	Iloprost	Jet nebulizer	2020	OMRON	Pulmonary arterial hypertension (PAH)
Omron NE C801	Indocyanine green	With virtual value technology	2014	OMRON	Used in lung cancer
I-neb AAD system	Alpha-1 antitrypsin	Vibrating mesh nebulizer	2011	Respironics Respiratory Drug Delivery (UK) Ltd.	Cystic fibrosis
MicroAir NE–U22	Budesonide	Vibrating mesh nebulizer	2020	HC Med Innovations, Co. Ltd.	Persistent asthma

14.3.1.2 Advances in Nebulizers

14.3.1.2.1 Mesh Nebulizer

The nebulizer contains a micropump system to create the aerosols from liquid or suspension (Fig. 14.2a) (Longest et al. 2019). Marketed products *in vitro* studies are proving mesh nebulizers are more efficient than conventional nebulizers, and it takes less time for nebulization (Waldrep and Dhand 2008). The forced formulation, like liquid or suspension, is passed through various apertures in a mesh placed to create the aerosols (Ari 2014). Passive and active vibrating mesh nebulizers are two types of mesh nebulizers. The passive mesh nebulizer includes a piezo-electric element positioned near the mesh and a thin fluid layer to transmit the vibrations to the mesh. An active mesh nebulizer attaches a vibration-guided piezo-electric element nearer the formulation reservoir. Mesh nebulizers have advanced applications in delivering drug nanocarriers, mRNA, and polymyxin B effectively to the lungs (Longest et al. 2019). In contrast to conventional nebulizers, it offers deliverability for higher doses, is suitable for infants and unconscious patients, and does not require device-patient coordination for inhalation. However, bulkiness and lower efficacy are the challenges associated with such nebulizer designs (Labiris and Dolovich 2004).

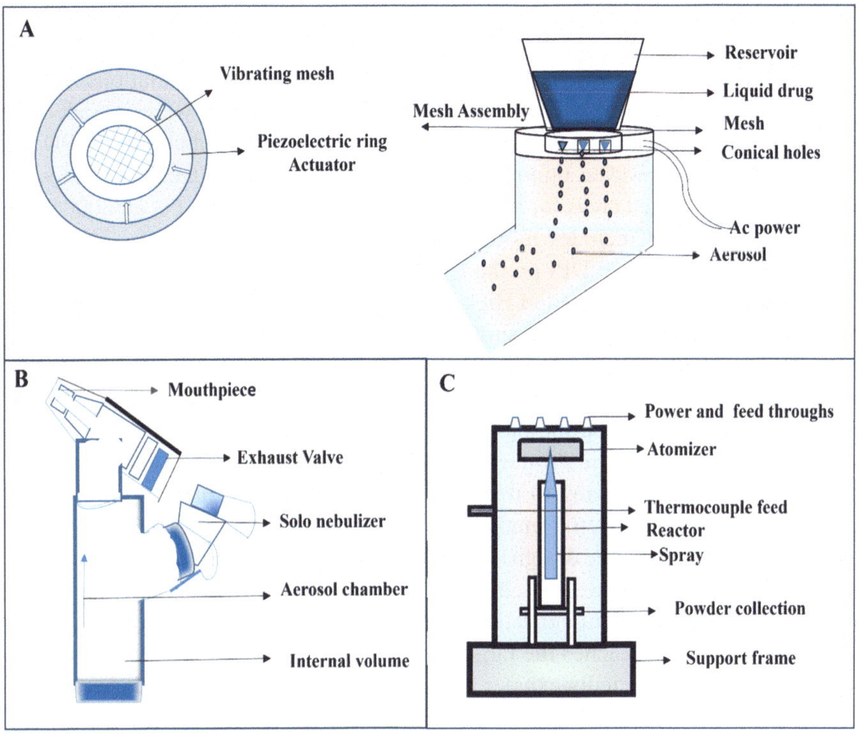

Fig. 14.2 Advancement in nebulizer: (**a**) mesh nebulizer, (**b**, **c**) vibrating mesh nebulizer

14.3.1.2.2 Vibrating Mesh Nebulizer (VMN)

This type of nebulizer includes a perforated membrane attached to a piezo element to assist the vibration mode. Vibrational motion and sound pressure are created nearer to the membrane, pushing the fluid through that membrane to generate the aerosol (AL-Jahdali et al. 2013). The device is best suitable for intensive care unit (ICU) settings, as well as the need for low residual drug volume, and is helpful in attaining deep lung penetration. It can produce the drug particle in a minimal range of 0.5–3 μm, suitable for alveolar deposition. VMN has a horizontal mesh containing 1000 holes. The design of VMN is illustrated in Fig. 14.2b, c. Total emitted dose studies have proved that VMNs would deliver a higher number of aerosols than jet nebulizers. Some examples of VMN are breath-activated VMN systems such as AKITA2, APIXNEB, Aerogen -OnQ®, Aerodose®, Aeroneb Pro® and Solo®, Pari eFlow®, and Philips I-Neb® nebulizer systems (Table 14.1). The novelty here is the personalized therapies achieved by these VMNs (Mccarthy et al. 2020; Sayed et al. 2021).

14.3.2 Dry Powder Inhalers

Dry powder inhalers are devices of medical origin containing formulations like dry powders (Telko and Hickey 2005). Historically, DPIs devices were intended for low-dose medicines to treat lung-related diseases, such as asthma and COPD (de Boer et al. 2016). However, non-respiratory diseases like migraine and diabetes are also the newly targeted areas for recently developed DPIs (Ashurst et al. 2000). The current DPIs market includes more than 20 FDA-approved products. Table 14.2 summarizes the marketed DPI products. The DPI is favored over the nebulizer and the pressurized MDIs (Telko and Hickey 2005; Shetty and Srinivasan 2017). Some MDI devices are breath-actuated single-dose, multidose, and active devices (Cheng et al. 2020). The background of each type is the deliverable dosage quantity, the patient's involvement in forming a colloidal suspension of a powder dispersed in air, and the methods by which the powder is dispersed. Each type is subdivided into single-unit dose inhalers, multi-unit dose inhalers, and multidose reservoir inhalers. The DPI is preferable owing to it offering high stability in comparison to dosage forms consisting of liquid, the smaller size of the device makes it easier to carry and use, and it is propellant-free design for delivery of API to the target site (Rattanupatam and Srichana 2014; Onoue et al. 2008). Regardless of the device's design and formulation, deep and forceful inhalations are required to break the powder into breathable size (5 μm) that reaches the target site (Zheng et al. 2021). The whole dose is lost if a patient unexpectedly exhales into the device. Hence, it is not the best choice in emergencies (Gaikwad et al. 2023). The conventional design of such IDD devices is given in Fig. 14.3a, b.

Table 14.2 Marketed DPI products (Hebbink et al. 2022)

Manufacturer	Device name	API/drug	Device type	Indication	Year of approval
GlaxoSmithKline (GSK)	Ellipta®	Relvar® Incruse®	Blister	Asthma/COPD	2017
Teva	Armonair® Digihaler®	Fluticasone propionate	Reservoir	Asthma	2017
Mylan Pharmaceuticals	Wixela® Inhub®	Fluticasone propionate Salmeterol xinafoate	Blister	Asthma/COPD	2019
Pfizer	Exubera	Insulin	Blisters	Type 2 diabetes mellitus	2006
Novartis	Airmax™	Foradil (formoterol) Miflonide (budesonide)	Reservoir	Asthma, bronchitis Bronchospasm	2007 (delay due to lawsuit)
Orion	Easyhaler®	Budesonide Formoterol	Reservoir	Asthma/COPD	2018

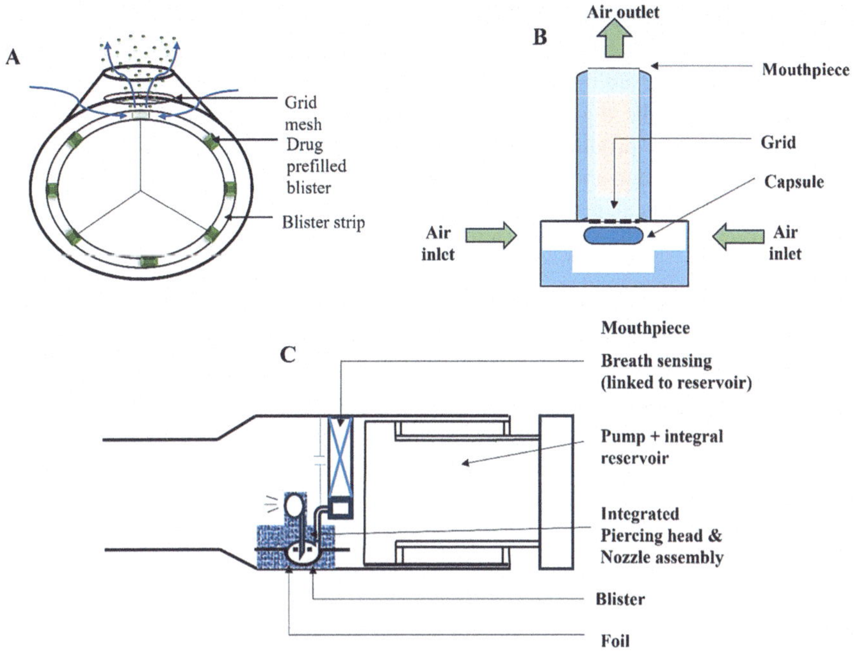

Fig. 14.3 Schematic representation of (**a**, **b**) conventional DPI, (**c**) advancement in DPI (Aspirair device)

14.3.2.1 Challenges of Conventional Dry Powder Inhalers (Need for Advancement)

There are a few challenges for the DPIs in general for delivering the drug to the target site. To achieve effective and reproducible dose delivery, the patient must produce sufficient inspiratory airflow, which is determined by DPI's internal resistance. DPIs are not ideal for patients with considerable airflow blockage since it takes more effort to produce the inspiratory airflow (Colthorpe et al. 2013). The single unit dose requires emptying all the capsule contents into the device before starting the inhalation procedure. Inhalation should be continued until no remaining powder is left in the device to avoid any significant degree of dose variability. Dose variability is also caused by changes in device (brand) by patients, knowingly or unknowingly. The temperature may affect the capsule, making it soft and hard to break in the device before inhalation. Particle agglomeration due to high temperature and humidity causes changes in drug release, leading to dose variability in DPIs. Some of the multidose DPI devices are complex, costly for production, and user-friendly for some patients (O'Connor 2004). The challenges can be encountered by employing advanced IDD devices and formulation.

14.3.2.2 Advancements in DPI Devices

In extension to developments in DPI formulation design, advancements in the design of the devices are ongoing with the introduction of new designs and digital technologies. With the digitalization of many devices, there has been a massive improvement in the product's performance and a decrease in errors. This has improved product usage and execution, along with patient acceptance, compliance, adherence, and clinical results (Xiroudaki et al. 2021). The active and passive devices have been studied more in advancing the DPI device design.

14.3.2.2.1 Active Devices

Most DPIs are referred to as "passive" inhalers since powder aerosolization occurs exclusively because of the energy of air pulled through the device by a patient. The powder dose is preloaded in three different ways: capsules, blisters, and reservoirs. Powder particle deagglomeration in all passive inhalers is often generated by inertial effects caused by airflow patterns and aerosol impaction on the grid and mouthpiece walls. Hence, these passive devices are flow-dependent. To overcome these problems, active DPI devices were introduced, e.g., Exubera (single dose) and Airmax (multidose). Although active devices have a more complicated design and high cost, it is necessary to produce powder aerosolization, which can give the functional advantage of being independent of airflow intensity (Gradoń and Sosnowski 2014). The active device implements energy from additional sources like compressed gas, motor-driven impellers, or electronic vibration for dispersing the API in the formulation, hence making them easy to use for geriatric patients (Young et al. 2004). The active devices are made to deliver drugs without relying on the patient's inspiration; however, the physical or electrical mechanisms are necessary for powder dispersion. Several methods, including priming the device with compressed air, high-frequency piezoelectric vibrators, and battery-powered motors, are used by active DPIs to get around issues caused by dependence on inspiratory airflow (Newman 2004). One example of an active device is Aspirair® (Fig. 14.3c). To optimize drug delivery to the lungs, the Aspirair® device uses a vortex separation chamber and compressed air supply that an airflow sensor activates (Tobyn et al. 2004).

14.3.2.2.2 Digital/Smart Devices

Administration of drugs at the right time, with proper technique, is often a problem for a large population of patients (mostly seen in older people). To overcome such issues, one can arrange a physical timer or diary to keep track of the doses, along with following the proper procedure for drug delivery. This hassle can be overcome by incorporating technology into the device or digitalizing the device to keep all the steps in check. The technique is known as the digital/smart device technique, i.e., add-on devices (the monitor is placed on the exterior part of the device) and integrated devices, where technology is incorporated into the very fabric of the device (Xiroudaki et al. 2021).

14.3.3 Metered Dose Inhaler (MDI)

An MDI is a multidose, compact, and portable device, basically a pressurized reservoir system consisting of a canister (generally made of aluminum) inside a plastic actuator. The drug is released from the chamber at the end of the actuator with the help of an orifice. The canister contains the solution of drug and co-solvent or a suspension of micronized drug crystals within a propellant (becomes gaseous under the atmospheric pressure)/propellant mixture (Janssens et al. 2008). During the 1950s, the first propellant-based, pressurized metered-dose inhalers (pMDIs) came into existence, starting a new age of portable, compact modern inhalers with very effective aerosolization engines (Laube et al. 2011). The pMDIs, estimated to account for more than 80% of the global market, have dominated the pulmonary drug delivery market for the past 30 years (O'Connor 2004). Table 14.3 explained the various marketed formulations for MDI. Respiratory diseases are one of the

Table 14.3 Marketed formulations for MDI devices

Brand name	API	Type	Company	Indication	Year of approval
Atrovent	Ipratropium bromide	Solution	Boehringer Ingelheim Limited	Bronchospasm, emphysema	2004
Serevent	Salmeterol	Suspension	GSK	Long-term airway obstruction in asthma	1998
Flovent HFA	Fluticasone propionate	Suspension	GSK	Prophylactic therapy in asthma	2004
Pulmicort	Budesonide	Suspension	AstraZeneca	Prophylactic therapy in asthma in patients older than 6 years of age	2000
Bevespi aerosphere	Glycopyrrolate	Suspension	AstraZeneca	COPD	2016
Qvar	Dipropionate	Solution	Teva Pharmaceutical Industries Ltd.	Prophylactic therapy in asthma	2000
Ventolin	Albuterol sulfate	Suspension	GSK	Bronchospasm	2001
Primatene mist	Epinephrine	Solution	Amphastar Pharmaceuticals	Intermittent asthma, including wheezing and shortness of breath	2018
Alvesco	Ciclesonide	Solution	Covis Pharma	Prophylactic therapy in asthma in patients older than 12 years of age	2008

significant causes of mortality and morbidity to the health of the general public all over the world. Amid all mentioned devices, the fascinating first choice of the clinician is MDI over the others (Kumar et al. 2022). The formulation or the components of MDI are well protected from any degradation, as well as from environmental factors like moisture, light, oxygen, and microbial contamination due to being packed inside a closed pressurized reservoir within a canister (Janssens et al. 2008). MDIs are relevant options/devices of choice for patients with insufficient negative inspiratory airflow, intubated, or on ventilators, avoid first-pass metabolism, and give quick onset of action. It delivers low concentrations of drugs locally to the lungs for systemic circulation, decreasing the chances of side effects. Despite the positive consequences, various limitations are also associated with MDIs. The inhalers consist of propellants such as chlorofluorocarbons, which are toxic to the environment and should be avoided (Ashurst et al. 2000). The coordination between activation and inhalation is essential while using an MDI. Ineffective therapeutic delivery is the result of poor coordination and is a topic of concern related to the proper use of MDI. The failure to shake before usage will result in non-uniformity in doses and inefficient drug delivery (O'Connor 2004). Keeping track of the daily dose is often a challenge for the patients and decreasing patient compliance (Geller 2005). The MDI can also cause local side effects due to oropharyngeal deposits. In order to overcome these obstacles, various advancements have been made in MDI formulations and devices, as stated below.

14.3.3.1 Recent Advances in MDI Devices

Conventional MDI devices consist of five major components for their successful applications. The actuator/mouthpiece, canister, metering valve, and nozzle are basic device components. The advancement of MDI has been done based on nozzle geometry, extra-fine particle atomization, and the addition of spacers in drug-delivery devices. A brief description of each type is given in Table 14.3.

14.3.3.1.1 Modified Design Based on Nozzle Geometry

Pressurized metered dose inhaler's aerosol properties and output can be impacted by subtle modifications to the actuator design and geometry (Fig. 14.4a) (Dalby et al. 2004). One example is the Respimat® inhaler, which combines the benefits of pMDIs and nebulizers, as shown in Fig. 14.4d. It operates by uniblock, a design nozzle that creates two tiny liquid jets at precisely appropriate angles and this convergence is what causes the soft mist to develop. It can operate at a moderate inhalation flow rate to slowly aerosolize propellant-free drug solutions as a fine mist, similar to nebulizers, reducing the likelihood of oropharyngeal deposition, which is frequently seen with pMDIs (Leach et al. 2005).

14.3.3.1.2 Extra-Fine Particle Atomization

New pMDIs that create aerosols with extra-fine particles have been developed thanks to advancements in drug formulation. With the longer duration of aerosol emission and less impact of inspiratory flow and coordination on lung deposition, these newer formulations are said to be less dependent on inhaler technique than

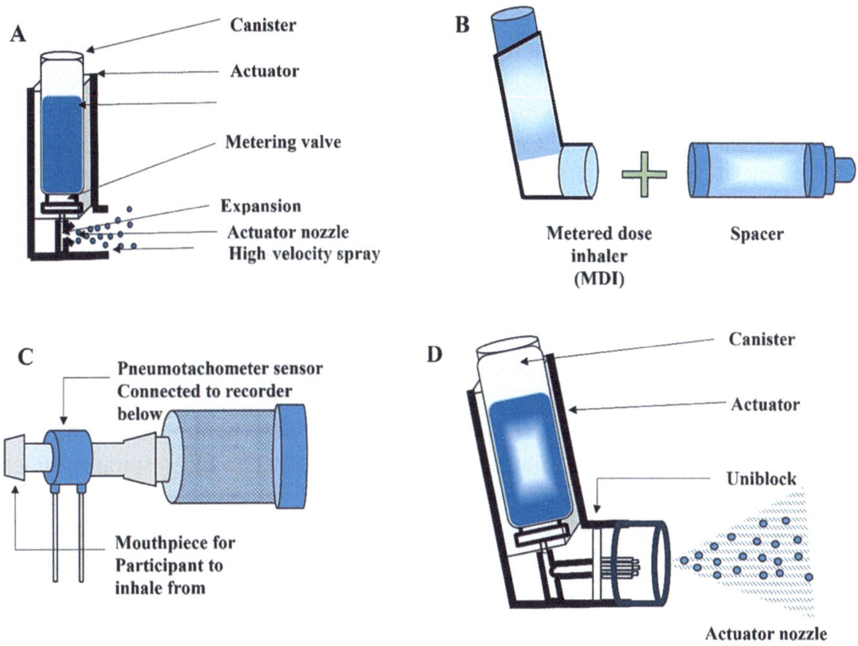

Fig. 14.4 Advancement in MDI: (**a**) Respimat soft mist inhaler, (**b**) spacer added MDI, (**c**) AeroChamber Plus® Flow-Vu®, and (**d**) Respimat® inhaler

other MDIs (Leach et al. 2005). The addition of dissolved CO_2 to a mixture of hydrofluoroalkanes (HFA)-134 and ethanol was suggested as a novel approach to minimize the size of droplets after emission from the pMDIs (Ibrahim et al. 2015).

14.3.3.1.3 Addition of "Spacers"

pMDIs can be equipped with spacers that serve as reservoirs into which drug aerosol can be produced (Fig. 14.4b). This allows the patient to administer the drug without coordinating the inhalation and actuation (Capstick and Clifton 2012). The benefits of utilizing a spacer device include better drug inhalation from pMDIs, especially for people who have trouble using a pMDI by themselves, such as AeroChamber Plus® and Flow-Vu® as shown in Fig. 14.4c (Ho et al. 2004).

14.3.3.2 Advancement in MDI Formulations

Technological advances have driven the use of advanced formulation strategies through MDI. It includes the induction of newer propellants and various micro- and nano-drug carriers. Using newer propellants as substitutes for older ones is an advance in one of the formulation strategies through MDI. The hydrofluoroalkanes (HFA) 134a and 227 were found to be the only HFAs that could effectively replace chlorofluorocarbons (CFC) propellants in pMDIs after significant investigation. HFAs 134a and 227 have numerous traits in common as both propellants exhibit very little impurity, are above 99.9% pure, and are chemically stable under typical storage conditions. HFAs can produce sufficient vapor pressure even at lower temperatures, which helps in the effective delivery of API due to their lower boiling points (BP) (Morton 2001). Salbutamol and fluticasone propionate are the two drugs tested for dosage uniformity and particle size distribution in pMDIs that use HFA 134a as a propellant. Cripps et al. (2000) conducted a study on replacing CFCs with HFAs in pMDI for reproducibility and regulatory requirements. The investigation was conducted on pMDI for dose uniformity, particle size distribution (PDI), single actuation reproducibility, and content uniformity. The findings from the study showed that products (pMDI) provided an emitted dose within ±25% of the mean value, indicating consistent dosing of salbutamol (100 μg per actuation) or fluticasone propionate (125 and 250/μg per actuation) using HFA 134a. The results didn't differ significantly from that of CFCs; hence HFAs were suitable replacements for CFCs without any decrease in therapeutic efficacy (Cripps et al. 2000). Another study by Pickering et al. (2001) uses HFA 227 in pMDI to deliver mometasone furoate (MF). Gamma scintigraphy was used to determine respiratory tract deposition pattern and efficiency of the delivery of formulation containing HFA 227 in a total of 11 patients. It was concluded that lung deposition via HFA 227 pMDI was 10–20% compared to other corticosteroid suspensions, has shown effectiveness in the treatment of asthma, and was a suitable replacement for the CFCs (Pickering et al. 2001).

14.3.3.2.1 Nano- and Microformulations

Despite the availability of pMDIs commercially, deaths due to pulmonary diseases are increasing every year; hence there is a constant search for effective drug delivery systems and advanced pMDIs. Nanotechnology, biotechnology, particle engineering, etc. have shown improvements in MDIs. The different microparticles (MPs) (nanocomposite, porous) and nanoparticles [liposomes, solid lipid nanoparticles (SLN), dendrimers, and polymer hybrid NPs] are tested with MDIs for improvements in the delivery process and therapeutic outcomes for the treatment of pulmonary diseases (Kumar et al. 2017).

14.4 Applications of Inhalational Drug Delivery

There are several advantages of using inhalation medicines over other administration routes for treating pulmonary and systemic diseases. In contrast to oral therapy, this route of administration eliminates the poor absorption of the drug and drug degradation in the GI tract and avoids first-pass metabolism in the liver. However, in contrast to the parenteral route of administration, pain and patient discomfort are overcome by the inhalational route. Systemic toxicities or organ-specific toxicities are also associated with the systemic route of drug administration (Anderson et al. 2022; Okuda and Okamoto 2020). The possible advantages offered by IDDs produce wide applications to deliver drugs, genes, peptides, and vaccines for various diseases (Saleem et al. 2017). Pulmonary drug administration is the prominent route of drug administration to acquire disease specificity by localized targeting. After the publication of the first clinical trial in 1968, the evolution of various IDDs has been raised. Broadly, lung diseases (lung cancer, asthma, COPD, cystic fibrosis, lung injury, etc.) are attractive pathological conditions for the drug-oriented applications of IDD. It offers a non-invasive delivery route for self-medications, can deliver high doses to acquire rapid onset of action, and has low enzymatic degradation, thus increasing effectiveness. Various studies have explained the importance of IDD in pulmonary diseases (Kunde et al. 2022; Chang et al. 2021). The route is suitable for delivering vaccines that induce immunological responses more effectively than the parenteral route (De Castro et al. 2005; Wong-Chew et al. 2006; Henao-Restrepo et al. 2010; Smith et al. 2003; Dilraj et al. 2007). The inhalation route of insulin administration also helped to treat diabetes, as discussed in various studies (Laube et al. 1993, 1998; Cefalu et al. 2001; Beth 2005). The developed IDD devices for insulin Nektar, Aradigam, and Aerogen explained the success of such biopharmaceuticals (Beth 2005). The IDD is also found to be used or delivering genes, proteins, and peptides to treat various diseases (Griesenbach et al. 2006). It has already been explained that proteins or peptides are delicate molecules easily ruptured by gastric juice, high temperature, or enzymatic degradation (Mehta 2016). Most literature favor using DPIs for the pulmonary delivery of protein owing to their inexpensive, propellant-free, and stable formulation properties throughout the shelf life (Hak-Kim and Andy 1997).

14.5 Evaluation of Inhalation Drug Delivery Devices

The fundamental goal of inhalation therapy is to administer medication topically to the lungs. The characteristic features of the device (e.g., the aerosolization system, internal resistance, speed of the aerosol plume, oral/nasal inhalation, and inhaled carrier gas) and formulation (e.g., lipophilicity, particle charge, and hygroscopicity) affect the lung deposition of molecules and their ability to reach the small airways. Each inhaler device includes distinctive instructions for preparing the dose, delivering medication into the airways, and determining the density and size of the aerosolized particulates (Rogliani et al. 2017). Additionally, the qualities and intensity of the respiratory condition and the simplicity with which patients can use the device have a role in the correct delivery of inhalational medication (Hirsh et al. 2008; Laube et al. 2011).

According to Froehlich (2019), the repeatability of aerosol distribution is a crucial factor in determining how well an inhalation device works. Higher doses of ethanol slow the evaporation, causing bigger droplets and oropharyngeal deposition. The possibility of generating turbulences that amplify lung deposition and the interaction between plume velocity and inhalation airflow is complicated. Aerosol deposition for dry powder inhalers (DPIs) is influenced by emitted dose (ED), aerodynamic particle size distribution (APSD), and device resistance (Froehlich 2019). There are various preclinical models (*in vitro*, *ex vivo*, and *in vivo*) to assess the performance of IDD devices.

14.5.1 In Vitro Evaluation of Inhalation Devices

Chrystyn (2007), state that the performance characteristics of inhalers vary considerably based on the wide variation in design characteristics, which impacts their suitability for use in different patient populations. The fine particle fraction (FPF) and total emitted dose (TED) from the IDD devices decide the deposition and actual performance of the therapeutics. Therefore, it is essential to consider its suitability before prescribing a DPI for an individual patient (Chrystyn 2007). The TED is defined as the amount of drug emitted by the inhaler during a single actuation. The quantity of drug particles in the TED that are tiny enough to enter the airways when inhaled provide the therapeutic effect. FPF is the mass of particles produced in an actuation with an aerodynamic diameter of less than 5 μm (Davies 1979). The likelihood of these particles depositing on the airways during inhalation is highest. The oropharynx is where larger aerosol particles usually settle before being swallowed.

In vitro pharmacopoeial procedures are used in the laboratory to measure the TED and FPF delivered by inhalers. Boshra et al. (2018), used a DPI sampling method to determine the TED given by Diskus or Aerolizer inhalers containing 200 mg salbutamol (Boshra et al. 2018). These techniques have shown that the efficacy of the various inhalation systems is highly variable and that they provide helpful information on probable aerosol deposit locations within the airways (Chrystyn 2007). In practice, the airflow capacity of a DPI during use is adaptable and is affected by the patient's inhaling capacity. A patient's inspiratory capacity varies, influenced by parameters including airway obstruction, lung size, and inspiratory musculature (Chrystyn 2007). According to Boehr et al. (2002), a typical consideration during the formulation and characterization of drug inhalation systems is the measurement of the APSD from the aerosol mist as a function of constant inspiratory flow rate. To effectively deposit the dosage at the site of action within the respiratory tract, evaluating the TED and whether it has the necessary size range is essential. The sizes of aerodynamic particles have been measured using a variety of methods. With equipment like cascade impactors, Aerosizers, elutriators, spiral centrifuge aerosol spectrometers, electrical mobility analyzers, etc., direct cloud measurement from an inhaler is possible (Froehlich 2019; de Boer et al. 2003). According to Ho et al. (1986), laser diffraction techniques produce size distributions for nebulized aqueous drug solutions that are more accurate than those produced by inertial impaction methods (Ho et al. 2011). The application of the laser diffraction techniques for particle size characterization in the plumes released from MDIs was described in detail by Ranucci (1992). The apparent permeability (AP) of the inhaled therapeutics is another important property that affects the biological performance of the IDDs. AP is the rate of drug molecule transport through the cell monolayer made up of Calu 3 cells, NCL-H441, HAEVi cells, and human primary cells (Ehrmann et al. 2020; Sakagami 2020; Cidem et al. 2020). The advantages and disadvantages of such are discussed in Table 14.4.

Table 14.4 *In vitro* and *ex vivo* preclinical models to evaluate the IDD

Models (Ehrmann et al. 2020; Sakagami 2020; Cidem et al. 2020; Secher et al. 2020)	Advantages	Disadvantages
Air liquid interface	• Controllable dose • Carry out effectively through impactor machines • Reaction for drugs like a respiratory tract of the lung	• Challenges in holding inserts • It has specificity to particular cell lines • Some restrictions in cell culturing (primary cell lines)
Organoids	• It exactly replicates the heterogenic nature of the lung • Capable of screening personalized medicines in this method • Mimic-like lung micro-level environment	• Not able to predict the bio-pharmacokinetics of inhaled particles in the pathway at the mouth to airways
Lung on chip	• Capability of representing barriers in the lung for drug absorption • Having the capacity to recreate mechanical and shear forces brought on by cyclical breathing • Giving the same heterogenicity and flow rates as the *in vivo* environment of the lung • Capable of creating respiratory airways • It recreates the alveolar-capillary airways • *In situ* aerosol exposure	• Less to optimum throughput • Relatively time-consuming • Moving elements that could be complicated—inconsistent production processes
Precision-cut lung slices (PCLS)	• High platform repeatability • An accurate portrayal of the structural response of the lungs to experimental stimuli is made possible by preserving the lung architecture • Maintain effective cell-cell communication for a multicellular reaction • The experimental design is more controllable than the use of animals in studies • Exact modeling of the upper airway concerning the distal airway	• Lack of clinical inhalation applicability and dosage estimation based on therapy • Failure to replicate muscular stretch, circulation, or airflow seen in the human lung

(continued)

Table 14.4 (continued)

Models (Ehrmann et al. 2020; Sakagami 2020; Cidem et al. 2020; Secher et al. 2020)	Advantages	Disadvantages
Isolated perfused lungs (IPLS)	• It keeps lung tissue functioning • A multicellular response to drug stimuli is made possible by an intact lung • Inhalable delivery devices are suitable for direct administration of inhalation medication therapy • Possibility of carefully regulating pharmacological stimulus • The experimental design is more controllable than the use of animals in studies	• Limitation with lung perfusion • More time-consuming process • Minimal disease models • They are often sourced from rat or rabbit origin, which differs from humans based on tracheobronchial anatomy and structural composition

14.5.2 *Ex Vivo* Evaluation of Inhalation Devices

The flow rates and inhalation profiling of the IDD devices can be assessed through *ex vivo* evaluation techniques. To examine the dosage effectiveness of the fluticasone propionate Diskus and budesonide Turbuhaler inhalers, Burnell et al. (2001), analyzed and compared the inhalation profiles created with patients having severe obstructive pulmonary disease. They used the term *ex vivo* to characterize data produced by combining *in vivo* parameters with *in vitro* analytical techniques while analyzing the performance of the two inhalers. To evaluate the *ex vivo* efficacy of the 250 mg Diskus inhaler and the 200 mg dose Turbuhaler inhaler, they recorded the flow rates and inhalation profiles using an inhalation profile recorder (IPR). Then, they used this data in the electronic lung device (ELD), an inhalation simulator. To validate and feed the data into the ELD, they analyzed these inhalers' pressure drop and flow parameters and compared them with regular batches. A computer-controlled piston in the ELD replicates the IPR's subject profile. After each inhalation, the inhaler was separated from the ELD, and all surfaces that have come into contact with the dosage were cleaned with an appropriate solvent. High-performance liquid chromatography was then used to analyze the liquids from various parts and stages of the Andersen cascade impactor. The TED and FPF of the medication were estimated using approved and validated techniques (Burnell et al. 2001).

The *ex vivo* evaluation of IDDs finds the use of precision-cut lung slices (PCLS) and isolated perfused lungs (IPLS) to study the kinetics of the drugs, toxicity assessment, and therapeutic effect (Gagnadoux et al. 2008). From post-mortem, lung slices are collected in the size range of 100–300 μm. The slices are further introduced in the culture media to maintain the anatomical and functional physiology of the slices. It helps to realize the functional properties of the lung and how it reacts to drug molecules (heterocellular and heterogenicity) of the lungs. PCLS are created from rodents, sheep, and human lung tissue. The isolated perfused lung model (IPLS) is another ex vivo preclinical model to assess the safety and kinetics of the lungs. The whole lungs are isolated from the rodent or rabbit and transferred into an artificial thoracic chamber simply called thorax in physiologically relevant environmental conditions. The advantage and disadvantage of each type are described in Table 14.4 (Ehrmann et al. 2020; Sakagami 2020; Cidem et al. 2020). The advances in preclinical ex vivo models include organs on chips and 3D organoids.

14.5.3 *In Vivo* Evaluation of Inhalation Devices

Various studies reported the use of whole animals for testing inhalational biopharmaceuticals for their pharmacokinetics and pharmacodynamics (PKPD) studies. The use of animals from small rodents (rats, mice, Guinea pigs) to larger animals (rabbits, dogs, and monkeys) has been revealed (Secher et al. 2020; Tandel et al. 2018). Genetically modified animal models are also gaining importance in successfully studying the drug resistance and therapeutic profiling of the drugs. Gamma scintigraphy approaches have evaluated and compared drug delivery from various inhalers depending on the characteristic features of the inhaled drugs (Chrystyn 2007). Borgstrom et al.'s study demonstrates lung deposition of 27% and 28% terbutaline and budesonide, respectively, induced by the Turbohaler, which is increased in individuals with no airway obstruction. These studies establish the importance of *ex vivo* and *in vitro* laboratory techniques in forecasting the consequences that would occur *in vivo* in humans (Borgström et al. 1994).

14.6 Conclusion and Future Perspective

IDD has shown prominent therapeutic results in preventing and mitigating pulmonary diseases over the past years. The prominent IDDs are nebulizers, DPI, and MDI, which provide the disease's applicability. These are well known for delivering drugs and even complex vaccines, insulin, genes, and protein or peptides for pulmonary diseases. However, biological and device-oriented challenges restrict or diminish the therapeutic potential of IDD. Technological advancements have driven the evolution of newly designed devices to address the challenges associated with conventional IDD devices. The advanced technology also includes micro-sized and nano-sized advancements. The evaluation of TED, FPF, aerodynamic particle size, etc. is crucial to reflect the performance of the IDD devices. More physiologically relevant lung models have been designed and considered for lung deposition studies, toxicity assessment, and biological performance. The future direction includes more efficient IDD to overcome the complications for geriatric and pediatric patients.

References

AL-Jahdali H, Ahmed A, AL-Harbi A, Khan M, Baharoon S, Bin Salih S et al (2013) Improper inhaler technique is associated with poor asthma control and frequent emergency department visits. Allergy Asthma Clin Immunol 9(1):8

Anderson S, Atkins P, Bäckman P, Cipolla D, Clark A, Daviskas E et al (2022) Inhaled medicines: past, present, and future. Pharmacol Rev 74:48–118

Ari A (2014) Jet, ultrasonic, and mesh nebulizers: an evaluation of nebulizers for better clinical outcomes. Euras J Pulm 16:1–7

Ashurst I, Malton A, Prime D, Sumby B (2000) Latest advances in the development of dry powder inhalers. Pharm Sci Technol Today 3:246–256

Beth L (2005) The expanding role of aerosols in systemic drug delivery. Respir Care 50:1161–1176

Borgström L, Bondesson E, Morén F, Trofast E, Newman S (1994) Lung deposition of budesonide inhaled via Turbuhaler®: a comparison with terbutaline sulphate in normal subjects. Eur Respir J 7:69–73

Boshra M, Almeldien A, Eldin R, Elberry A, Abdelwahab N, Salem MNS et al (2018) Total emitted dose of salbutamol sulphate at different inhalation flows and inhalation volumes through different types of dry powder inhalers. Exp Lung Res 44:211

Burnell P, Small T, Doig S, Johal B, Jenkins R, Gibson G (2001) Ex-vivo product performance of Diskus™ and Turbuhaler™ inhalers using inhalation profiles from patients with severe chronic obstructive pulmonary disease. Respir Med 95:324–330

Capstick T, Clifton I (2012) Inhaler technique and training in people with chronic obstructive pulmonary disease and asthma. Expert Rev Respir Med 6:91–101; quiz 102

Cefalu WT, Skyler JS, Kourides IA, Landschulz WH, Balagtas CC, Cheng SL et al (2001) Inhaled human insulin treatment in patients with type 2 diabetes mellitus background: despite demonstrated benefits, intensive insulin. Ann Intern Med 134:203. www.annals.org

Chan JGY, Wong J, Zhou QT, Leung SSY, Chan HK (2014) Advances in device and formulation technologies for pulmonary drug delivery. AAPS PharmSciTech 15:882–897

Chang RYK, Chow MYT, Khanal D, Chen D, Chan HK (2021) Dry powder pharmaceutical biologics for inhalation therapy. Adv Drug Deliv Rev 172:64–79

Cheng S, Tan Z, Pandey M, Srichana T, Pichika M, Gorain B et al (2020) A critical review on emerging trends in dry powder inhaler formulation for the treatment of pulmonary aspergillosis. Pharmaceutics 12:1161

Chrystyn H (2007) The Diskus™: a review of its position among dry powder inhaler devices. Int J Clin Pract 61:1022–1036

Cidem A, Bradbury P, Traini D, Ong HX (2020) Modifying and integrating in vitro and ex vivo respiratory models for inhalation drug screening. Front Bioeng Biotechnol 8:581995

Colthorpe P, Voshaar T, Kieckbusch T, Cuoghi E, Jauernig J (2013) Delivery characteristics of a low-resistance dry-powder inhaler used to deliver the long-acting muscarinic antagonist glycopyrronium. J Drug Assess 2:11

Cripps A, Riebe M, Schulze M, Woodhouse R (2000) Pharmaceutical transition to non-CFC pressurized metered dose inhalers. Respir Med 94(Suppl B):S3–S9

Dalby R, Spallek M, Voshaar T (2004) A review of the development of Respimat((R)) soft mist (TM) inhaler. Int J Pharm 283:1–9

Davies CN (1979) The deposition and clearance of inhaled aerosols. J Aerosol Sci 11:227–228

de Boer A, Gjaltema D, Hagedoorn P, Frijlink H (2003) Characterization of inhalation aerosols: a critical evaluation of cascade impactor analysis and laser diffraction technique. Int J Pharm 249:219–231

de Boer A, Hagedoorn P, Hoppentocht M, Buttini F, Grasmeijer F, Frijlink H (2016) Dry powder inhalation: past, present and future. Expert Opin Drug Deliv 14:499

De Castro JF, Bennett JV, Rincon HG, Munoz MTAY, Sanchez LAEP, Santos JI (2005) Evaluation of immunogenicity and side effects of triple viral vaccine (MMR) in adults, given by two routes: subcutaneous and respiratory (aerosol). Vaccine 23(8):1079–1084

Dhanani J, Fraser J, Chan HK, Rello J, Cohen J, Roberts J (2016) Fundamentals of aerosol therapy in critical care. Crit Care 20:269

Dilraj A, Sukhoo R, Cutts FT, Bennett JV (2007) Aerosol and subcutaneous measles vaccine: measles antibody responses 6 years after re-vaccination. Vaccine 25(21):4170–4174

Doan Q, Shefrin A, Johnson D (2011) Cost-effectiveness of metered-dose inhalers for asthma exacerbations in the Pediatric Emergency Department. Pediatrics 127(5):e1105

Ehrmann S, Schmid O, Darquenne C, Rothen-Rutishauser B, Sznitman J, Yang L et al (2020) Innovative preclinical models for pulmonary drug delivery research. Expert Opin Drug Deliv 17:463

Froehlich E (2019) Biological obstacles for identifying in vitro-in vivo correlations of orally inhaled formulations. Pharmaceutics 11:316

Gagnadoux F, Hureaux J, Vecellio L, Urban T, Le Pape A, Valo I et al (2008) Aerosolized chemotherapy. J Aerosol Med Pulm Drug Deliv 21:61–70

Gaikwad S, Pathare S, More M, Waykhinde N, Laddha U, Salunkhe K et al (2023) Dry powder inhaler with the technical and practical obstacles, and forthcoming platform strategies. J Control Release 355:292

Geller D (2005) Comparing clinical features of the nebulizer, metered-dose inhaler, and dry powder inhaler. Respir Care 50:1313–1321; discussion 1321

Giraud V, Roche N (2002) Misuse of corticosteroid metered-dose inhaler is associated with decreased asthma stability. Eur Respir J 19(2):246–251

Gradoń L, Sosnowski T (2014) Formation of particles for dry powder inhalers. Adv Powder Technol 25:43–55

Griesenbach U, Geddes DM, Alton EWFW, Boyd C, Davies J, Gill D et al (2006) Gene therapy progress and prospects: cystic fibrosis. Gene Ther 13:1061–1067

Hak-Kim C, Andy C (1997) Spray dried protein delivery. Pharm Res 14(4):431–437

Haque S, Whittaker M, Mcintosh M, Pouton C, Kaminskas L (2016) Disposition and safety of inhaled biodegradable nanomedicines: opportunities and challenges. Nanomedicine 12:1703

Hebbink G, Jaspers M, Peters H, Dickhoff B (2022) Recent developments in lactose blend formulations for carrier-based dry powder inhalation. Adv Drug Deliv Rev 189:114527

Henao-Restrepo AM, Greco M, Laurie X, John O, Aguado T (2010) Measles aerosol vaccine project. Proc Vaccinol 2:147–150

Hirsh J, Guyatt G, Albers G, Harrington R, Schünemann H, Physician A (2008) Antithrombotic and thrombolytic therapy: American College of Chest Physicians Evidence-Based Clinical Practice Guidelines (8th edition). Chest 133:110S–112S

Ho S, O'Mahony M, Steward J, Breay P, Burr M (2004) Inhaler technique in older people in the community. Age Ageing 33:185–188

Ho K, Kellaway I, Tredree R (2011) Particle size analysis of nebulised aerosols using Fraunhofer laser diffraction and inertial compaction methods. J Pharm Pharmacol 38:26P

Ibrahim M, Verma R, Garcia-Contreras L (2015) Inhalation drug delivery devices: technology update. Med Devices (Auckl) 8:131–139

Janssens W, VandenBrande P, Hardeman E, Langhe E, Philps T, Troosters T et al (2008) Inspiratory flow rates at different levels of resistance in elderly COPD patients. Eur Respir J 31:78–83

Kumar R, Singh A, Garg N, Siril P (2017) Solid lipid nanoparticles for the controlled delivery of poorly water soluble nonsteroidal anti-inflammatory drugs. Ultrason Sonochem 40:686

Kumar R, Mehta P, Shankar K, Rajora M, Mishra Y, Mostafavi E et al (2022) Nanotechnology-assisted metered-dose inhalers (MDIs) for high-performance pulmonary drug delivery applications. Pharm Res 39:25

Kumar A, Vaiphei KK, Gulbake A (2024) A nanotechnology driven effectual localized lung cancer targeting approaches using tyrosine kinases inhibitors: Recent progress, preclinical assessment, challenges, and future perspectives. Int J Pharm 23:124745. (https://doi.org/10.1016/j.ijpharm.124745)

Kunde SS, Ghosh R, Wairkar S (2022) Emerging trends in pulmonary delivery of biopharmaceuticals. Drug Discov Today 27:1474–1482

Kuzmov A, Minko T (2015) Nanotechnology approaches for inhalation treatment of lung diseases. J Control Release 219:500–518
Labiris NR, Dolovich MB (2003) Pulmonary drug delivery. Part I: physiological factors affecting therapeutic effectiveness of aerosolized medications. Br J Clin Pharmacol 56:588–599
Labiris N, Dolovich M (2004) Pulmonary drug delivery. Part II: the role of inhalant delivery devices and drug formulations in therapeutic effectiveness of aerosolized medications. Br J Clin Pharmacol 56:600–612
Laube BL, Georgopoulos A, Adams GK 3rd. (1993) Preliminary study of the efficacy of insulin aerosol delivered by oral inhalation in diabetic patients. JAMA 269:2106. http://jama.jamanetwork.com/
Laube BL, Benedict GW, Dobs AS (1998) Time to peak insulin level, relative bioavailability, and effect of site of deposition of nebulized insulin in patients with noninsulin-dependent diabetes mellitus. J Aerosol Med 11:153
Laube B, Janssens H, De Jongh F, Devadason S, Dhand R, Diot P et al (2011) What the pulmonary specialist should know about the new inhalation therapies. Eur Respir J 37:1308–1331
Lavorini F, Mannini C, Chellini E (2015) Challenges of inhaler use in the treatment of asthma and chronic obstructive pulmonary disease. Eur Med J 3:98
Leach C, Davidson P, Hasselquist B, Boudreau R (2005) Influence of particle size and patient dosing technique on lung deposition of HFA-beclomethasone from a metered dose inhaler. J Aerosol Med 18:379–385
Lehofer B, Bloder F, Jain PP, Marsh LM, Leitinger G, Olschewski H et al (2014) Impact of atomization technique on the stability and transport efficiency of nebulized liposomes harboring different surface characteristics. Eur J Pharm Biopharm 88(3):1076–1085
Longest W, Spence B, Hindle M (2019) Devices for improved delivery of nebulized pharmaceutical aerosols to the lungs. J Aerosol Med Pulm Drug Deliv 32:317
Mangal S, Gao W, Li T, Zhou Q (2017) Pulmonary delivery of nanoparticle chemotherapy for the treatment of lung cancers: challenges and opportunities. Acta Pharmacol Sin 38:782
Mccarthy S, Gonzalez H, Higgins B (2020) Future trends in nebulized therapies for pulmonary disease. J Pers Med 10:37
Mehta P (2016) Dry powder inhalers: a focus on advancements in novel drug delivery systems. J Drug Deliv 2016:1–17
Morton D (2001) The aerosol society drug delivery to the lungs XI—December 11-12, 2000—foreword. J Aerosol Med 14:118
Mudge J (1778) A radical and expeditious cure for a recent catarrhous cough, 2nd edn. Allen E, London, pp 1–252
Newman S (2004) Dry powder inhalers for optimal drug delivery. Expert Opin Biol Ther 4:23–33
Newman S (2005) Principles of metered-dose inhaler design high dose. Respir Care 50:1177–1190
O'Connor BJ (2004) The ideal inhaler: design and characteristics to improve outcomes. Respir Med 98(Suppl A):S10–S16
Ochowiak M, Kasperkowiak A, Doligalski M, Sosnowski T, Matuszak M, Włodarczak S et al (2019) The thermostated medical jet nebulizer: aerosol characteristics. Int J Pharm 567:118475
Okuda T, Okamoto H (2020) Present situation and future progress of inhaled lung cancer therapy: necessity of inhaled formulations with drug delivery functions. Chem Pharm Bull (Tokyo) 68:589–602
Onoue S, Hashimoto N, Yamada S (2008) Dry powder inhalation systems for pulmonary delivery of therapeutic peptides and proteins. Expert Opin Ther Pat 18:429–442
Pardhi VP, Jain K (2021) Impact of Binary/Ternary Solid Dispersion Utilizing Poloxamer 188 and TPGS to Improve Pharmaceutical Attributes of Bedaquiline Fumarate. J Drug Deliv Sci Technol 62(102349):1–13. https://doi.org/10.1016/j.jddst.2021.102349
Pickering H, Pitcairn G, Hirst P, Bacon P, Newman S, Affrime M et al (2001) Regional lung deposition of a technetium 99m-labeled formulation of mometasone furoate administered by hydrofluoroalkane 227 metered-dose inhaler. Clin Ther 22:1483–1493
Ranucci J (1992) Dynamic plume-particle size analysis using laser diffraction. Pharm Technol 16:108–114

Rattanupatam T, Srichana T (2014) Budesonide dry powder for inhalation: effects of leucine and mannitol on the efficiency of delivery. Drug Deliv 21:397

Rogliani P, Calzetta L, Coppola A, Cavalli F, Ora J, Puxeddu E et al (2017) Optimizing drug delivery in COPD: the role of inhaler devices. Respir Med 124:6

Sakagami M (2020) In vitro, ex vivo and in vivo methods of lung absorption for inhaled drugs. Adv Drug Deliv Rev 161–162:63

Saleem I, Petkar K, Somavarapu S (2017) Chapter 19: Rationale for pulmonary vaccine delivery: formulation and device considerations

Sanchis J, Corrigan C, Levy ML, Viejo JL (2013) Inhaler devices-from theory to practice. Respir Med 107:495–502

Sanders M (2007) Inhalation therapy: an historical review. Prim Care Respir J 16:71–81

Sayed N, Abdelrahman M, Abdelrahim M (2021) Effect of functional principle, delivery technique, and connection used on aerosol delivery from different nebulizers: an in-vitro study. Pulm Pharmacol Ther 70:102054

Secher T, Bodier-Montagutelli E, Guillon A, Heuzé-Vourc'h N. (2020) Correlation and clinical relevance of animal models for inhaled pharmaceuticals and biopharmaceuticals. Adv Drug Deliv Rev 167:167

Shetty A, Srinivasan G (2017) Advancements in dry powder inhaler. Asian J Pharm Clin Res 10:8–12

Smith DJ, Bot S, Dellamary L, Bot A (2003) Evaluation of novel aerosol formulations designed for mucosal vaccination against influenza virus. Vaccine 21(21–22):2805–2812

Stein SW, Sheth P, Hodson PD, Myrdal PB (2014) Advances in metered dose inhaler technology: hardware development. AAPS PharmSciTech 15:326–338

Tandel H, Bhatt P, Jain K, Shahiwala A, Misra A (2018) "Chapter 1: In-vitro and in-vivo tools in emerging drug delivery scenario: challenges and updates" in International Book "In-vitro and in-vivo tools in drug delivery research for optimum clinical outcomes". CRC Press, USA. ISBN: 9781351368728, pp 1-24. https://doi.org/10.1201/b22448-1

Telko M, Hickey A (2005) Dry powder inhaler formulation. Respir Care 50:1209–1227

Tobyn M, Staniforth J, Morton D, Harmer Q, Newton M (2004) Active and intelligent inhaler device development. Int J Pharm 277:31–37

Waldrep J, Dhand R (2008) Advanced nebulizer designs employing vibrating mesh/aperture plate Technologies for aerosol generation. Curr Drug Deliv 5:114–119

Wong-Chew RM, Islas-Romero R, García-García MDL, Beeler JA, Audet S, Santos-Preciado JI et al (2006) Immunogenicity of aerosol measles vaccine given as the primary measles immunization to nine-month-old Mexican children. Vaccine 24:683–690

Xiroudaki S, Schoubben A, Giovagnoli S, Rekkas D (2021) Dry powder inhalers in the digitalization era: current status and future perspectives. Pharmaceutics 13:1455

Young P, Thompson J, Woodcock D, Aydin M, Price R (2004) The development of a novel high-dose pressurized aerosol dry-powder device (PADD) for the delivery of pumactant for inhalation therapy. J Aerosol Med 17:123–128

Zheng Z, Leung S, Gupta R (2021) Flow and particle modelling of dry powder inhalers: methodologies, recent development and emerging applications. Pharmaceutics 13:189

Herbal Formulations: Development, Challenges, Testing, Stability, and Regulatory Guidelines

15

Shivam Ramdas Pawar, Parth Patel, and Keerti Jain

Abstract

Herbal formulations consist of herbal medications in the form of plant extracts, enhanced fractions, or biomarker components. Herbal preparations offer distinct benefits due to their enhanced absorption and lower toxicity. There is a scarcity of literature that specifically discuss the typical difficulties encountered in preparation of herbal formulations, such as the process of choosing the appropriate kind or class of formulations for an extract or a phytochemical. This chapter focuses on herbal formulation development from a selection of suitable formulations to getting the regulatory approval of diverse herbal nanoformulations. This chapter intends to present the latest progress in many types of herbal nanomedicines, such as polymeric herbal nanoparticles, solid lipid nanoparticles, micelles, liposomes, dendrimers, nanoemulsions, and nanosuspension. Further, the testing methodologies to maintain the quality of herbal medicine and associated challenges to meet the desired quality are discussed in brief. Additionally, regulatory considerations required to meet the quality and safety standards are also summarized. A combination of herbal medicine with nanotechnology can be an essential instrument to get better therapeutic performance of herbal medicine. Despite the problems associated with the quality, uniformity, and stability of traditional herbal medicine, there is great potential for nanotechnology-based herbal formulations to treat a range of diseases. Overall, herbal nanomedicines have been showing promising therapeutic potential making them a promising alternative of synthetic pharmaceutical and traditional herbal formulations.

S. R. Pawar · P. Patel · K. Jain (✉)
Drug Delivery and Nanomedicine Research Laboratory, Department of Pharmaceutics, National Institute of Pharmaceutical Education and Research (NIPER), Raebareli, Lucknow, Uttar Pradesh, India
e-mail: keertijain.02@niperraebareli.edu.in

K. Jain, A. K. Yadav (eds.), *Advances in Pharmaceutical Product Development*,
https://doi.org/10.1007/978-981-97-9230-6_15

Keywords

Herbal medicine · Nanotechnology · Ayurveda · Regulatory guidelines · Drug delivery

Abbreviations

GCMS	Gas chromatography-mass spectroscopy
HPLC	High-performance liquid chromatography
HPTLC	High-performance thin-layer chromatography
ICPMS	Inductively coupled plasma-mass spectroscopy
LCMS	Liquid chromatography-mass spectroscopy
SFC	Supercritical fluid chromatography
WHO	World Health Organization

15.1 Introduction

Plants have provided food and everyday essentials for humans since immemorial times. They are used in a variety of ways, from papermaking to building making, from spices or fragrances to preventative and therapeutic measures for a range of medical problems (Haslam et al. 1989). The World Health Organization (WHO) has defined herbal medicines as "herbs, herbal materials, herbal preparations and finished herbal products that contain as active components elements of plants, or other plant materials, or combinations" (Sammons et al. 2016). Many civilizations, including India's Ayurvedic medicine, Chinese medicine, Japan's Kampo medicine, Australia's Aboriginal medicine, and New Zealand's TeRongoa, have their distinctive ways of using traditional herbal medicines as a remedy for treatment of many diseases (Barnes et al. 2016).

The years of expertise in designing and developing drugs from medicinal plants have taught us to learn from nature and our predecessors. When Tu Youyou was granted the Nobel Prize in Physiology and Medicine in 2015 for discovering the antimalarial compound artemisinin from *Artemisia annua L.* (Qinghao), it became evident how valuable traditional herbal remedies could turn out to be (Tu 2011). The evolution of herbal medicine has also been closely linked to science, economics, and culture throughout history. There are now many different and meaningful ideas about the current reliance and preference for traditional medicinal products, thanks to the development of modern medicine and industrialization (Leonti and Verpoorte 2017; Sharma et al. 2018). Hence, herbal medicinal products received a significant share in the international pharmaceutical market, and the demand for the same is gradually increasing as people are aware of the benefits of herbal medicines. The size of the global herbal medicine market was projected at USD 151.91 billion in

2021 and is expected to increase at a noteworthy compound annual growth rate (CAGR) of 11.16% from USD 168.86 billion in 2022 to USD 437.59 billion by 2030 (Global Herbal Medicine Market 2024).

Early research on nanotechnologies in medicine mostly focused on applying nanoscale technology to improve the effects of well-known medications. Then, the area grew rapidly in the preparation of multifunctional and customized nanomedicine, which was made possible due to the versatile properties and adaptability of nanomaterials. Multifunctional nanomedicine refers to the integration of various capabilities into a single nanoformulation, such as biomedical imaging, biosensing, diagnostics, and treatment (das Neves et al. 2020; Ahmad et al. 2022). Customization by applying patient- and disease-specific features can enhance the therapeutic benefits of nanomedicines in the treatment of different ailments. A deeper understanding of the critical features that affect the interplay of nanomaterial with mucus and tissues is highly significant for the logical design of more effective nanomedicines and to anticipate any undesirable consequences of acute or chronic exposure to nanoparticulate matter. To logically develop more effective nanomedicines and to foresee any unfavorable effects of acute or chronic exposure to nanoparticulate matter, a greater understanding of the fundamental aspects that govern the interaction of nanomaterial with mucus and tissues is very crucial. The capacity of nanoparticles to enormously enhance the therapeutic value of medication made the best bet for any formulation scientist (Thorley and Tetley 2013; Saini et al. 2022).

The last 10 years have witnessed the use of many herbal drugs in nanoformulations; however, there is still a dearth of knowledge about the best way to formulate nanoparticles for a given class of phytoconstituents or extracts, how to develop herbal nanomedicines systematically, how to characterize formulations, etc. Thus, in the current chapter, we have summarized the development, characterization methods, and regulatory requirements for herbal formulation development. Many of the currently available marketed nanoformulations have provided potential therapeutic benefits in clinical application over traditional medications. Hence, the co-application of nanotechnology and traditional herbal medicine makes sense and may provide suitable results in future. This may broaden the scope of nanotechnology with significant improvement in the therapeutic potential of herbal medicine. Thus, in the following section, we have discussed about some nanoformulations of herbal of medicines.

15.2 Nanoformulations of Herbal Medicines

A wide range of innovative herbal nanoformulations, such as polymeric nanoparticles, nanocapsules, liposomes, phytosomes, nanoemulsions, microspheres, transferosomes, and ethosomes, have been documented by the scientific community. These novel formulations have significant advantages compared to traditional formulations of plant actives and extracts. These advantages include increased solubility, improved bioavailability, protection against toxicity, enhanced pharmacological activity, improved stability, better distribution in tissue macrophages, sustained delivery, and protection against chemical degradation of herbal medicine. The different types of herbal formulations are discussed in the following subsections.

15.2.1 Herbal Nanoemulsion

It is a thermodynamically stable mixture of oil and aqueous phase, in which one phase is homogenously dispersed in another phase and it was stabilized by using proper surfactant. In pharmacy, o/w nanoemulsion is widely used because lipophilic core of emulsion can efficiently deliver the lipophilic drug molecule to the desired site (Suthar et al. 2023; Juneja et al. 2022). Currently, nanoemulsion is emerging as an unprecedented nanocarrier for the delivery of herbal medicine and herbal oils. Instinctually, herbal medicine is comparatively less potent than synthetic active pharmaceutical molecules, which require higher doses to obtain the same therapeutic effect (Pardhi et al. 2022). Further, some volatile herbal oils have therapeutic value, but it became difficult to directly administer the oil due to poor patient compliance. Here, nanoemulsion can be used to solve these problems due to its oil-containing lipophilic core showing high drug loading efficiency and ability to easily accommodate the herbal volatile oil. For instance, Saranya and colleagues formed a eucalyptus oil nanoemulsion and evaluated its antibacterial efficacy against *Staphylococcus aureus* (Suthar et al. 2022). The findings of the microbial study demonstrated that all microbiological viability had been lost within 15 min of treatment. In addition, the ability to heal wounds was evaluated on Wistar rats, which showed a higher rate of wound recovery compared to the neomycin-treated control group, due to significant enhancement in the bioavailability of eucalyptus oil via nanoemulsion (Sugumar et al. 2014). Additionally, Nagi et al. (2017) assessed the pharmacokinetic properties of a silymarin-loaded nanoemulsion formulated using a high-pressure homogenization process. The flavonoid silymarin is derived from the seeds of the milk thistle, known as *Silybum marianum*. Although it has substantial hepatoprotective capabilities, its low bioavailability limits its clinical applications. According to the pharmacokinetic study of silymarin-loaded nanoemulsion, the apparent permeability constant of silymarin rises significantly, which indicates that the silymarin in the nanoemulsion formulation has a higher oral bioavailability (Nagi et al. 2017). These researches showed that the nanoemulsion can significantly improve the bioavailability of herbal oils and herbal medicines.

15.2.2 Herbal Nanoparticles

Nanoparticles include the nanosized particles of polymers, lipids, or a combination of both. Each of them has its advantages and disadvantages. For instance, polymers provide flexibility to formulation scientists to tailor-made drug release, easy processing, and thermostability (Patel et al. 2023; Jain 2018), whereas lipids provide better biocompatibility, biodegradability, and permeation profile than polymers.

The advancement in polymer chemistry has provided large numbers of biocompatible and biodegradable polymers. The lipidic nanoparticles can be further classified as solid lipid nanoparticles, nanostructured lipidic carriers, liposomes, niosomes, etc. Recently, some formulation scientists have also published articles on hybrid nanoparticles prepared using lipids and polymers to improve biocompatibility and permeation and control drug release (Jain and Zhong 2022). Chemically herbal medicines are more prone to degradation by extreme pH and enzymes present in the human body. To protect herbal medicine from the harsh biological environment of the human body, encapsulating them inside the nanocarriers is the best way to improve their bioavailability and therapeutic response (Jain et al. 2014). Multiple articles published in the last decade proved the importance of nanoparticles in the delivery of herbal medicine.

The chief bioactive component of *Curcuma longa* is curcumin, which has been linked to several health benefits, including anti-inflammatory, anticancer, and antioxidant properties. Despite this, it showed significantly low water solubility, which makes it less bioavailable and, thus, limits its use in clinical application. To overcome this, Xie et al. (2011) formulated PLGA nanoparticle formulation of curcumin using the solvent evaporation method to improve its oral bioavailability. The results demonstrated a 5.6-fold increase in oral bioavailability and a roughly 640-fold increase in water solubility with curcumin-loaded nanoparticles in comparison to curcumin powder (Xie et al. 2011). Furthermore, the diterpenoid class of chemicals such as andrographolides discovered from *Andrographis paniculata* is known for its anticancer activity. Andrographolides show poor water solubility, which result in poor bioavailability. To increase its solubility and bioavailability, Praveen et al. (2013) synthesized solid lipid nanoparticles using the solvent injection method and tested for in vitro cell viability against the hormone-positive breast cancer cell line (MCF-7 cells). The results showed the desired safety and activity profile (Parveen et al. 2014). Hence, various types of polymeric and lipidic nanoparticles are the best choice for delivering herbal medicines without any toxicity and stability issues.

15.2.3 Herbal Hydrogels

Hydrogels are three-dimensional polymeric networks of hydrophilic monomers, which have a high water-retention capacity. Hydrogels can be engineered to have mechanical properties that closely resemble various soft biological tissues and often show great biocompatibility (Ho et al. 2022). Moreover, their flexible and soft texture reduces the inflammatory responses of surrounding cells and tissues. Hydrogels can be categorized as natural, synthetic, or hybrid, based on the polymers used in formulating the hydrogel. The size of the holes in the hydrogel network may also be changed based on the need of particular formulation, which is an important factor and directly associated with the drug release from the hydrogel. Hydrogels can also be prepared by conjugating two or more polymers to get the desired properties (Bashir et al. 2020). One such approach includes formation of pH-sensitive bond between polymers, which releases the drug quickly at a particular pH. For instance,

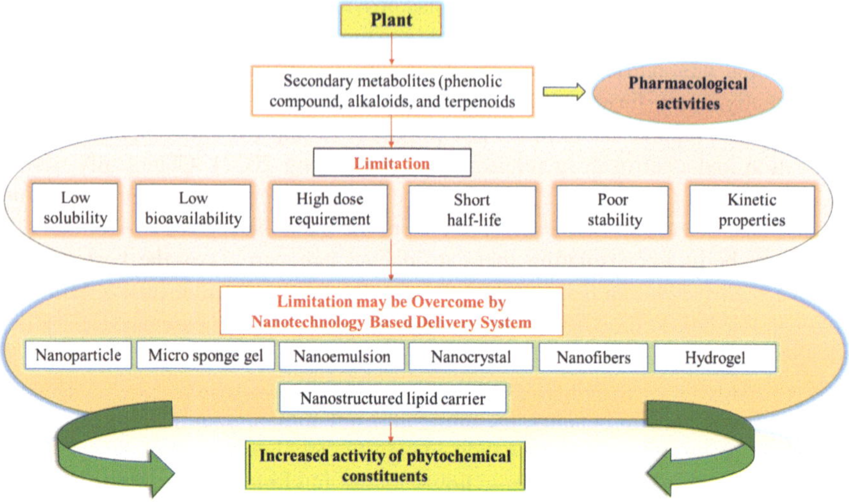

Fig. 15.1 Schematic representation of limitations and strategies to overcome these limitations for development of herbal nanoformulations

a pH-responsive hydrogel for transdermal delivery of naringenin was developed by Park and coworkers using radical polymerization of carboxymethyl cellulose and 2-hydroxyethyl acrylate. This gel showed a significantly higher swelling ratio at the pH of acne-prone skin and atopic skin (pH 7–8) compared to the pH of normal skin (pH 5.5), due to ionization of the carboxyl group at alkaline pH. Further, skin permeation of naringenin was also greatly increased, when topically administered through hydrogel (Park et al. 2018). Thus, hydrogel showed great potential for topical delivery of herbal medicine attributed to its high drug-loading capacity, easy preparation method, and great compatibility with herbal medicines.

The research publications published in the last decade prove that nanotechnology is an efficient way to deliver herbal medicines. The wide range of nanocarriers can solve the problems of herbal medicines like poor solubility, low potency, poor stability, lack of compatibility, etc. Figure 15.1 shows the schematic representation of challenges associated with herbal formulation development and strategies to overcome these challenges for successful preparation of herbal nanoformulations. Overall, nanotechnology has great potential to bypass the pitfalls of herbal medicine. We have discussed the testing methodologies, challenges of nanotechnology, and regulatory requirements for marketing approval of herbal nanoformulations in the subsequent sections.

15.3 Challenges in the Development of Herbal Formulations

The development of herbal formulation is comparatively more challenging than the formulation development with synthetic drugs. Herbal medicine is more prone to degradation by change in temperature, pH, and solvent. Further, herbal medicines are comparatively less explored drug molecules; hence deciding the effective dose, compatibility, stability, degradation profile, etc. demands scientifically sound knowledge to formulate any dosage form containing therapeutically effective dose. Some commonly encountered challenges for formulation scientists during the development of herbal formulations are discussed in the following subsections.

15.3.1 Extraction of Desired Herbal Compound

The herbal extracts obtained immediately after the primary extraction are usually a mixture of several phytoconstituents of varied molecular weight, medicinal potential, solubility, and chemical classes. Few are hydrophilic like glycosides, polyphenolics, etc., others are lipophilic, such as diterpene, tri-terpenoids, monoterpene, sesquiterpene, etc., while some may be median polar with partial aqueous as well as lipid solubility like alkaloids, flavonoids, etc. Altogether, these dissimilarities create challenges for scientists in the characterization of herbal medicines and nanoformulation, which can be solved by bioassay-guided fractionation of these extracts (Bairwa and Jachak 2015). This technique segregates each phytoconstituent, starting from solubility-based differentiation, followed by its separation based on polarity using the chromatographic techniques and screening of molecules.

15.3.2 Difficulties in Crossing Biological Barriers

Any pharmacologically active moiety going inside the body must cross one or more biological barriers before exerting the therapeutic response based on the route of administration and site of action (Wu et al. 2020). Herbal medicines and formulations are comparatively less stable in biological systems than synthetic and semi-synthetic therapeutically active moieties (Banerjee et al. 2018). Hence, delivering herbal medicine at the site of action requires detailed knowledge of formulation development and associated biological interaction. Here, nanotechnology-based approach may provide superior results in protecting herbal medicines.

15.3.3 Higher Cost of Preparation

Nanotechnology is itself a costly technology due to the requirement to maintain very high-quality standards, high-end technical instruments, technically sound support staff, and cleanrooms (ISO 4 or Class 10) to manufacture nanoparticles. Herbal nanomedicine again adds on significant amount of cost due to its additional requirement of extraction, identification, and purification steps (Pathak and Thassu 2009). Purification of herbal extract is itself a humongous task, which requires very high-end instruments like nuclear magnetic resonance spectroscopy, liquid chromatography-mass spectrometry (LCMS), high-performance thin-layer chromatography (HPTLC), supercritical fluid chromatography (SFC), inductively coupled plasma-mass spectroscopy (ICP-MS), and many more (Banerjee et al. 2018). Overall, the extraction of pure herbal medicine and preparation of herbal nanoparticles are high-cost processes.

15.3.4 Physical Stability of Herbal Formulations

Physical stability is the key property of any formulation required to maintain the efficacy and safety profile during its clinical application. Herbal medicine and formulations are very sensitive to changes in temperature, pH, light exposure, and humidity (Jeevanandam et al. 2016). Any alteration in abovementioned parameters can cause the generation of degradation products and may alter the pharmacological activity of herbal medicines (Wu et al. 2011). These changes are most likely to occur during transportation or due to improper storage, although the probability of occurring physical instability are less due to the regulatory guidelines, which made the detailed stability study mandatory for herbal formulations.

15.3.5 Toxicity of Herbal Medicine

Herbal formulations are also associated with different toxicity concerns like non-IgE-mediated hypersensitivity or pseudo-allergic reactions, which may cause anaphylactic shock, face swelling, cardiac distress, flushing, etc. (Hua et al. 2018). Therefore, it is advisable to deeply investigate the safety profile by relevant ex vivo and/or in vivo assays of potential toxicities and adverse effects before clinical approval of herbal formulations (Desai 2012). Further, detailed in vivo toxicity and degradation kinetics studies in relevant animal models should also be carried out to evaluate the short-term and long-term toxicity and biocompatibility of herbal medicines and formulations.

These are some of the challenges that researchers may face during the development of herbal formulations. Further, the testing methods to evaluate the purity and stability of herbal medicines are discussed in the next section.

15.4 Testing of Herbs and Herbal Formulations

The phytochemicals found in herbal plants showed variability based on factors such as climate, soil composition, and geographical condition; these variations impede the process of standardization (Nazim et al. 2018). The increasing growth in deforestation is the cause of the adulteration and substitution of natural medications, which compromise the therapeutic potential of herbal medicines. Thus, it became almost important to ensure the quality of medicinal herbal products by applying the appropriate standards and sophisticated quality control techniques (Marchese et al. 2017). Standardization is important for the identification, evaluation, and purity of herbal medicine as well as the verification of herbal products. The physical, chemical, biological, and preliminary identification characteristics contribute to the purity of herbal medicine and formulations. The quality and freshness of the herbal drugs are determined by their purity (Rashid et al. 2018).

The pharmacognostical scheme for separation, macroscopic evaluation (shape and markings), microscopic evaluation (qualitative and quantitative), evaluation of physicochemical parameters (moisture content, acid insoluble ash, water soluble ash, etc.), identifications (adulterants and genuine drug), and other parameters reported for the first time can all be important tools for the authentication of herbal medicine (Zhang et al. 2012). This process of standardization of formulation under the standard setting of herbal medicines helped to create a monograph in the pharmacopeia and other standard texts (Fernandes and Salgado 2016). The components of herbal medicines are complex; hence certain high-end approaches and tactics are used to verify the quality, purity, and integrity of the herbal formulations, which are discussed in the following subsections.

15.4.1 Thermal Analysis

Thermal analysis is an analytical method widely used for the characterization of herbal medicine such as the determination of purity, thermal stability, compatibility with other phytocompounds and excipients, etc. (Guimaraes et al. 2017). Thermal techniques (such as thermogravimetric analysis and differential thermal analysis) can be used to analyze reaction order (n), activation energy (Ea), frequency factor (A), and degradation constant of the herbal extract. These thermal approaches can also be used to evaluate the absolute water content, crystallization state, and thermal degradation (Liu et al. 2019a). Using thermal analytical techniques like differential thermal gravimetry and differential scanning calorimetry, the interaction between the excipients and therapeutically active herbal medicine can be studied, which is one of the important parameter in marketing approval of herbal medicine and formulations.

15.4.2 High-Performance Thin-Layer Chromatography (HPTLC)

Extraction of herbal medicine is the inevitable step in the development of any herbal formulation. During the process of extraction, active compounds are frequently present in the extraction media with other similar components (Do et al. 2019). Here, HPTLC is used to get a fingerprinting profile as a primary quality control test. Further, it is a widely approved qualitative method and used to match the pharmacopeial standards of diverse medicinal herbal formulations (Patil et al. 2018). Using the HPTLC scientists can determine the efficacy, safety, and quality of divulged phytochemical components.

15.4.3 High-Performance Liquid Chromatography (HPLC)

The quantitative and qualitative separation and identification of different components of phytocompounds are the most stringent step in the extraction of herbal medicine. The sophisticate technologies like preparative and semipreparative HPLC methods are used in the separation of large samples of extraction with utmost accuracy and speed (Hamburger 2019). This technique separates the phytoconstituents of similar chemical groups based on their polarity, hence providing the most accurate separation (Walters et al. 2017). This technique is frequently employed in the purification of herbal extracts.

15.4.4 Liquid Chromatography Mass Spectrometry (LCMS)

In complicated matrix analysis, where prior treatment of purification of the herbal medicine is required to improve the quality and yield, the HPLC method has several limitations when used alone without the assistance of additional methods. This problem is solved by employing LCMS technology, which significantly increases the sensitivity of detection (Marchetti et al. 2019). The capabilities of the LCMS technology include finer separation and a broad range of detection of analytical chemicals, molecular mass, information of fragmentation, retention time, and structure characterization. The LCMS combination technique can be used for the identification, quantification, and quality control of marketed herbal products as well as for the extraction of raw plant material. The ultra-high-performance liquid chromatogram coupled with electrospray ionization tandem quadrupole-time of flight/mass spectroscopy (UHPLC-Q-TOF/MS) (Zhou et al. 2018) is a more powerful tool for analysis that can be used to analyze complex traditional herbal extracts with high resolution, efficiency, and sensitivity.

15.4.5 Supercritical Fluid Chromatography

The SFC method uses compressed carbon dioxide (CO_2) and a small proportion of organic solvents, such as methanol, as the mobile phase. Here, the ratio of carbon dioxide is higher compared to organic solvent; hence it is frequently referred to as an alternative to the chromatographic method. In contrast to traditional organic solvent methods, the SFC technique is used to extract the principal analytes from the extraction media (Liu et al. 2019b). This technique can analyze frequently present compounds in extraction media such as carbohydrates, alkaloids, saponins, flavonoids, phenolics, lipids, etc. The capacity to identify and separate the fat-soluble vitamins is an important advantage of SFC (Murcia-Morales et al. 2019). SFC approach requires a significantly shorter time, utilizes fewer solvents and it is environmentally friendly.

15.4.6 Gas Chromatography-Mass Spectrometry (GCMS)

GCMS is a powerful technique used to separate and subsequently identify the various components of herbal extracts. The examination of volatile derivatives of herbal extract is the most difficult to analyze by another sensitive analytical method. Hence, GCMS is exclusively used to qualitatively and quantitatively analyze the volatile oils present inside the herbal extracts (Zhang et al. 2018). It may also be used to identify the various derivatives and drug metabolites of herbal medicines by separating them from each other. LCMS is relatively more sensitive than GCMS, hence the application of GCMS is limited to the analysis of volatile molecules (Saito et al. 2019). The non-volatile and/or thermolabile components are not suitable for GCMS analysis.

15.4.7 Inductively Coupled Plasma-Mass Spectroscopy

Elemental components present inside the herbal medicines are an essential part of human dietary requirements, although the excess amount of elemental components can cause unwanted side effects to the human body. The quantity of elemental components present inside the plant varies with geological condition, rainfall, quality of soil, and pH (Varhan Oral et al. 2019). Hence, regulatory authorities have decided the certain limits for elemental components, and manufacturers should meet these criteria to get marketing approval for herbal medicine and formulations. Elemental components are generally present either in the ionic form or non-ionic form with or without other phytoconstituents (Tokalıoğlu et al. 2019). To qualitatively and quantitatively identify these elemental components, technologies like ICP-MS and PIXE (partially induced X-ray emission) are used at the industrial level (Kumar et al. 2019). High amount of elemental components in herbal formulations may cause multiple side effects in the human body; hence its control in herbal medicine and formulation is a must for long-term use in humans after marketing approval.

The testing methods discussed in this section are most commonly employed for purity and stability study of herbal medicine. Further, testing methods such as particle size analysis, zeta potential, drug release study, stability study, etc. also need to be performed to get regulatory approval for herbal formulations. In this section we have focused mainly on analytical testing methods for herbal medicines, whereas stability-associated problems and requirements of herbal medicines and formulations are discussed in the next section of this chapter.

15.5 Stability Requirement in Herbal Formulations

Herbal medicines are capturing the attention of scientists of all over the world in treatment of diseases of modern era due to its less toxic potential and multifaceted activity in human body. The clearance of stability testing is a must for any herbal medicine or formulation to get regulatory approval for its clinical application. The stability study is intended to ensure that the herbal medicine and herbal product remain within the prespecified parameters of identification, strength, quality, and purity. The stability of herbal medicine or formulation may be defined as "herbal medicine and excipients should be in a physically, chemically, biologically, and microbiologically stable state for a defined period" (Pingale et al. 2008). Each element, whether medically active or inactive, in a dosage form can affect stability. Environmental conditions such as temperature, light, air (particularly oxygen, carbon dioxide, and water vapors), and humidity can affect stability of herbal medicines and formulations. Further, factors such as particle size, pH, quality of water and other solvents, type of container, and the presence of additional chemicals either from contamination or intentional mixing of different substances can influence the stability (Heigl and Franz 2003). Natural phytochemicals are generally less stable in biological environment; hence detailed stability study is mandatory for such herbal formulations before clinical approval. Multiple reports have been published demonstrating the stability problems of herbal formulations along with its resolution and efficient strategies to mitigate the stability problems (Poetsch et al. 2006).

Herbal medicines can lose their potency or may convert into harmful chemicals upon non-suitable storage or transportation due to degradation. Therefore, it is crucial for the manufacturer of herbal medicine or products to offer sufficient data on its long-term stability testing and degradation kinetic study to ascertain the stability in various temperatures. Nanotechnology-based approaches for formulation development of traditional herbal medicine can primarily address stability-related problems. Generally, herbal medicines degrade in the presence of enzymes, strong antioxidant molecules, or pH changes. Nanocarrier simply encapsulates the herbal medicine inside the core of chemically inert polymeric or lipidic material and protects the sensitive herbal medicines from harsh biological environments (Ahmad et al. 2022). The different types of instability of herbal medicines are discussed below:

15.5.1 Physical Instability

The issue of physical instability in natural medicines is frequently brought on by contaminants and packaging container reactions. The secondary metabolites of plant and chemical composition are impacted by factors such as microbial contamination and moisture content. Further, the herbal formulations containing volatile active ingredients have drawback of becoming less active upon longer storage period (Shinde et al. 2009). To avoid physical instability, quality packaging material should be used with a proper closure system.

15.5.2 Environmental Conditions

Suitable environmental condition is a must for the cultivation of herbal plants. Variations in rainfall, altitude, temperature, and soil in addition to distinct harvesting techniques, collection time, and method can lead to significant variations in product quality, stability, and concentration of desired chemicals in herbal medicines and formulations (Thakur et al. 2011). The frequent change in weather associated with global warming created the need of using innovative cultivation techniques like hydroponics, vertical farming, etc. to improve the yield of herbal crop production.

15.5.3 Chemical Instability

Herbal formulations are frequently degraded by crystallization, hydrolysis, enzymatic deterioration, oxidation, and chemical interactions with the excipients or additives during storage. Two main elements that impact the quality and stability of herbal products are moisture content and temperature. For every 10 °C increase in temperature, a chemical reaction multiplies by two to three folds, which affects the heat-labile phytoconstituents, whereas moisture adsorption by herbal medicines speeds up degradation, as most of these medicines are prone to hydrolysis (Kyriakoudi et al. 2021). The pace of chemical decomposition is also accelerated by the presence of enzymes in the herbal product.

15.5.4 Complex Mixtures

The combinations of various ingredients obtained during the extraction process may impart stability-related problems in herbal medicine or formulations as the shelf-life, therapeutic activity, concentration, and consistency of such components may vary (Wang et al. 2023). Hence, proper purification steps are needed to be performed to protect the final product from such problems. Further, a detailed compatibility study is mandatory, when more than one active ingredients are incorporated in a single herbal formulation.

15.5.5 Lack of Good Manufacturing Practices

The unwanted interactions of active ingredients with the packaging material and the use of poor-quality packaging material are potent causes of instability in herbal medicine or formulations as these increase the moisture content above the critical value and cause mold growth in natural products (Drago et al. 2020). Alkaloids, glycosides, tannins, flavonoids, and other active ingredients are found in large quantities in herbal preparations, and each one has unique stability requirements; hence the stability of herbal formulations differs from that of each of its constituent parts.

Therefore, the primary duty of the maker of herbal drugs is to ensure that the product is safe for patients to use and has sufficient stability for long-term storage. Further, the stability profile of natural medicines should be carefully determined because herbal formulations are generally used in combination of multiple active herbal ingredients. Stability is a must to improve patient compliance, create long-term trust with patients, and get regulatory approval. The distinct regulatory guideline created by various countries for marketing approval of natural medicine and formulations has been discussed in the following section.

15.6 Regulatory Guidelines in Approval of Herbal Formulations

Global concerns about the safety of herbal medicines have led national health authorities to create regulations aimed at ensuring their safe use. WHO launched a worldwide survey in 2001 covering herbal medicine regulations in 191 member countries. The most crucial elements for the regulation of herbal medicines are found to be the availability of research data, associated regulatory mechanisms, and education (World Health Organization 2024a).

Herbal medicine regulations in India are handled by the Department of Ayurveda, Yoga and Naturopathy, Unani, Siddha, and Homoeopathy (AYUSH), which was founded in 1995 under the Ministry of Health and Family Welfare. The 1940 Drugs and Cosmetics Act establishes the regulations governing the manufacturing and distribution of Ayurvedic, Siddha, and Unani (ASU) pharmaceuticals. The good manufacturing practices for ASU medications are particularly addressed in Schedule T of the Drug and Cosmetics Act (US Food and Drug Administration 2024). In the USA herbal remedies are subjected to United States (US) regulations under the botanical drug product (21 CFR sections 331–358). These can be sold in the USA as an authorized NDA or ANDA or as over-the-counter medication. In the USA, to include the botanical component as a new active ingredient in the monograph, the company would have to file a petition in line with 21 CFR 10.30.

Mexico, Brazil, and Argentina are the main competitors in the herbal market in the region of Latin America. Brazil has regulated herbal drugs since 1967, and herbal remedies are available as over-the-counter products and through prescription. The Brazilian Drug Division, which oversees the enforcement of drug regulations, published Directive 6 in 1995, which outlined the legal requirements for phytopharmaceutical drug registration in Brazil. In 2004, the RDC/48/2004 rule was released in its fourth iteration. The directive states that registration of phytopharmaceutical products is granted upon full verification of safety, efficacy, and consistently defined quality. Further, human efficacy and safety test findings are also necessary for the registration of herbal medicine in Brazil (Sahoo et al. 2010).

Globally, there is an urgent need to develop a standard procedure for the development of herbal medicines including interlinking of the monographs for herbal medicines published by various regulatory authorities, such as the Ayurvedic Pharmacopoeia of India, United States Pharmacopoeia, British Herbal Pharmacopoeia, Chinese Pharmacopoeia, Physician's Desk Reference for herbal medicines, and British Herbal Compendium (WHO 2000). The monograph for a certain plant may differ throughout publications, which can create an unnecessary disagreement in the regulatory approval. Further, research units and manufacturing companies involved in herbal formulation development faced various challenges in marketing approval and trading of herbal formulations. To solve these problems and promote the worldwide usage of high-quality and less expensive herbal medicinal products, countries should frame a treaty like "Free Trade Agreement," which may reduce regulatory barriers and facilitate easy commercial transactions by protecting the rights of both countries.

Harmonizing activities have been started with the standardization of pharmacopeial specifications and classification of herbal drugs to guarantee that the same herbal medicines are of the same quality, safety, and efficacy among nations. Similar herbal remedies are included in the pharmacopeias of China, Japan, and Korea, but their details are varied. Although the original plant's family is the same, different species were mentioned, which create a confusion for scientists. Moreover, the name of herbal medicine represented by the regulatory body is in regional character, which may differ for the same plant in different countries. To aid in the promotion of the commercialization of safe and effective herbal medicines across the world, the Western Pacific Regional Forum for the "Harmonization of Herbal Medicine" attempted to harmonize the herbal medicine monographs in the pharmacopeias of six Asian countries, i.e., China, Hong Kong, Japan, Korea, Singapore, and Vietnam (World Health Organization 2024b).

In 2000, American countries started to harmonize their procedural criteria for the registration of herbal medicine and related products. Pan American Network of Drug Regulatory Harmonization (PANDRH) is a working group on medicinal plants, which share advanced knowledge of different categories of herbal products, related terminologies, protocols, and minimal standards for herbal product registration. Another important institute for harmonizing herbal medicine worldwide is WHO's "International Standard Terminologies on Traditional Medicine in the Western Pacific Region" (Working Group on "Access to Health Systems including AYUSH" 2024). Overall, the WHO and governments of some countries have tried to harmonize the regulatory framework to smooth the supply chain of herbal medicines and related products worldwide; however still efforts are required to harmonize the regulatory guidelines for herbal medicines, globally.

15.7 Conclusion

Researchers and scientists are actively focusing on natural compounds due to enormous potential of pharmacologically active herbal medicine in the prevention and treatment of human ailments. Researchers are trying to increase the efficacy and lessen the regulatory restrictions of a therapeutic application of herbal medicine by employing a nanodrug delivery method. Thus far, animal models have demonstrated improved stability and sustained release capabilities of nanopreparations based on natural products, together with optimal therapeutic effects at low dose, which raises the system's long-term safety margin. Some diseases need to be treated over an extended period, which means that specially formulated nanodrugs must satisfy several requirements, including stability in systemic circulation and the ability to release therapeutic concentrations in the appropriate locations without harming healthy tissues. It is important to remember that drug-delivery nanocarriers have the ability to stimulate and suppress the immune system. For this reason, comprehensive immunotoxicology study is necessary before bringing such herbal medications to the clinic.

Simultaneously, researchers also face various issues in the identification of desired species of herbal plants and extractions of herbal medicines. The scientific innovations happened in analytical sciences made the identification and separation of extraction components easy. However, there is an urgent need of streamlining the extraction processes to maintain the uniformity in quality and efficacy of herbal medicines and formulations. Overall, the researches performed in the past proved the potential of herbal medicine in treatment of various diseases; however, there is an immediate need of detailed regulatory guideline for easy clinical translation of such herbal formulations.

Acknowledgments The authors are thankful to the Department of Pharmaceuticals, Ministry of Chemicals and Fertilizers, Government of India, for providing facilities to carry out this research work. The NIPER-R communication number for the manuscript is NIPER-R/Communication/610.

Conflict of Interest The authors declare that they have no known competing financial interests or personal relationships that could have appeared to influence the work reported in this chapter.

References

Ahmad J, Rizwanullah M, Suthar T, Albarqi HA, Ahmad MZ, Vuddanda PR, Khan MA, Jain K (2022) Receptor-targeted surface-engineered nanomaterials for breast cancer imaging and theranostic applications. Crit Rev Ther Drug Carrier Syst 39(6):1–44

Bairwa K, Jachak SM (2015) Anti-inflammatory potential of a lipid-based formulation of a rotenoid-rich fraction prepared from Boerhavia diffusa. Pharm Biol 53(8):1231–1238

Banerjee A, Ibsen K, Brown T, Chen R, Agatemor C, Mitragotri S (2018) Ionic liquids for oral insulin delivery. Proc Natl Acad Sci USA 115(28):7296–7301

Barnes J, McLachlan AJ, Sherwin CM, Enioutina EY (2016) Herbal medicines: challenges in the modern world. Part 1. Australia and New Zealand. Expert Rev Clin Pharmacol 9(7):905–915

Bashir S, Hina M, Iqbal J, Rajpar AH, Mujtaba MA, Alghamdi NA et al (2020) Fundamental concepts of hydrogels: synthesis, properties, and their applications. Polymers (Basel) 12(11):2702

das Neves J, Sverdlov Arzi R, Sosnik A (2020) Molecular and cellular cues governing nanomaterial–mucosae interactions: from nanomedicine to nanotoxicology. Chem Soc Rev 49(14):5058–5100

Desai N (2012) Challenges in development of nanoparticle-based therapeutics. AAPS J 14:282–295

Do T, Santi I, Reich E (2019) A harmonized HPTLC method for identification of various caffeine containing herbal drugs, extracts, and products, and quantitative estimation of their caffeine content. J Liq Chromatogr Relat Technol 42(9–10):274–281

Drago E, Campardelli R, Pettinato M, Perego P (2020) Innovations in smart packaging concepts for food: an extensive review. Food Secur 9(11):1628

Fernandes FHA, Salgado HRN (2016) Gallic acid: review of the methods of determination and quantification. Crit Rev Anal Chem 46(3):257–265

Global Herbal Medicine Market (2024). https://www.skyquestt.com/report/herbal-medicine-market

Guimaraes GP, Santos RLBD, Cartaxo-Furtado NA, Cavalcanti AL, Macedo RO (2017) Thermoanalytical characterization of herbal drugs from Poincianella pyramidalis in different particle sizes. J Therm Anal Colorim 131:661–670

Hamburger M (2019) HPLC-based activity profiling for pharmacologically and toxicologically relevant natural products—principles and recent examples. Pharm Biol 57(1):328–334

Haslam E, Lilley T, Cai Y, Martin R, Mangnolato D (1989) Traditional herbal medicines—the role of polyphenols. Planta Med 55(01):1–8

Heigl D, Franz G (2003) Stability testing on typical flavonoid containing herbal drugs. Int J Pharm Sci 58(12):881–885

Ho TC, Chang CC, Chan HP, Chung TW, Shu CW, Chuang KP et al (2022) Hydrogels: properties and applications in biomedicine. Molecules 27(9):2902

Hua S, de Matos MBC, Metselaar JM, Storm G (2018) Current trends and challenges in the clinical translation of nanoparticulate nanomedicines: pathways for translational development and commercialization. Front Pharmacol 9:9

Jain K (2018) Nanohybrids of dendrimers and carbon nanotubes: a benefaction or forfeit in drug delivery? Nanosci Nanotechnol Asia 9(1):21–29

Jain K, Zhong J (2022) Theranostic applications of nanomaterials. Curr Pharm Des 28(2):77–77

Jain K, Mehra NK, Jain NK (2014) Potentials and emerging trends in nanopharmacology. Curr Opin Pharmacol 15:97–106

Jeevanandam J, Chan YS, Danquah MK (2016) Nano-formulations of drugs: recent developments, impact and challenges. Biochimie 128–129:99–112

Juneja M, Suthar T, Pardhi VP, Ahmad J, Jain K (2022) Emerging trends and promises of Nanoemulsions in therapeutics of infectious diseases. Nanomedicine 17(11):793–812

Kumar V, Singh S, Singh A, Subhose V, Prakash O (2019) Assessment of heavy metal ions, essential metal ions, and antioxidant properties of the most common herbal drugs in Indian Ayurvedic hospital: for ensuring quality assurance of certain Ayurvedic drugs. Biocatal Agric Biotechnol 18:101018

Kyriakoudi A, Spanidi E, Mourtzinos I, Gardikis K (2021) Innovative delivery systems loaded with plant bioactive ingredients: formulation approaches and applications. Plan Theory 10(6):1238

Leonti M, Verpoorte R (2017) Traditional Mediterranean and European herbal medicines. J Ethnopharmacol 199:161–167

Liu Y, Yang L, Ma C, Zhang Y (2019a) Thermal behavior of sweet potato starch by non-isothermal thermogravimetric analysis. Materials 12(5):699

Liu TT, Cheong LZ, Man QQ, Zheng X, Zhang J, Song S (2019b) Simultaneous profiling of vitamin D metabolites in serum by supercritical fluid chromatography-tandem mass spectrometry (SFC-MS/MS). J Chromatogr B 1120:16–23

Marchese A, Barbieri R, Coppo E, Orhan IE, Daglia M, Nabavi SF et al (2017) Antimicrobial activity of eugenol and essential oils containing eugenol: a mechanistic viewpoint. Crit Rev Microbiol 43(6):668–689

Marchetti L, Pellati F, Graziosi R, Brighenti V, Pinetti D, Bertelli D (2019) Identification and determination of bioactive phenylpropanoid glycosides of Aloysia polystachya (Griseb. et Moldenke) by HPLC-MS. J Pharm Biomed Anal 166:364–370

Murcia-Morales M, Cutillas V, Fernández-Alba AR (2019) Supercritical fluid chromatography and gas chromatography coupled to tandem mass spectrometry for the analysis of pyrethroids in vegetable matrices: a comparative study. J Agric Food Chem 67(46):12626–12632

Nagi A, Iqbal B, Kumar S, Sharma S, Ali J, Baboota S (2017) Quality by design based silymarin nanoemulsion for enhancement of oral bioavailability. J Drug Deliv Sci Technol. 40:35–44

Nazim M, DMohd A, Khatoon R, DrMohd A, Chaudhary SS (2018) Physico-chemical standardization of Hansraj (Adiantum capillus-Veneris). J Drug Deliv Therap 8(6-s):195–203

Pardhi VP, Suthar T, Sharma A, Jain K (2022) Bedaquiline fumarate microemulsion: formulation optimization, rheological characterization and *in vitro* studies. Nanomedicine 17(21):1529–1546

Park SH, Shin HS, Park SN (2018) A novel pH-responsive hydrogel based on carboxymethyl cellulose/2-hydroxyethyl acrylate for transdermal delivery of naringenin. Carbohydr Polym 200:341–352

Parveen R, Ahmad FJ, Iqbal Z, Samim M, Ahmad S (2014) Solid lipid nanoparticles of anticancer drug andrographolide: formulation, *in vitro* and *in vivo* studies. Drug Dev Ind Pharm 40(9):1206–1212

Patel P, Kumar K, Jain VK, Popli H, Yadav AK, Jain K (2023) Nanotheranostics for diagnosis and treatment of breast cancer. Curr Pharm Des 29(10):732–747

Pathak Y, Thassu D (2009) Drug delivery nanoparticles formulation and characterization. Drugs and pharmaceutical science series. Inform Healthcare USA 67:1–30

Patil PR, Bhurat MR, Sanghavi RS, Nagdev SA, Patil PR (2018) HPTLC fingerprinting and quantification of shatavarin IV in extract and polyherbal formulations. World J Pharm Res 7:442

Pingale SS, Pokharkar RD, Pingale MS (2008) Stability study of a herbal drug. Pharmacology online. Univ Salerno 1:20–23

Poetsch F, Asche S, Steinhoff B (2006) Stability testing of herbal medicinal products. Pharm Indust 68(4):476–483

Rashid S, Zafar M, Ahmad M, Lone FA, Shaheen S, Sultana S et al (2018) Microscopic investigations and pharmacognostic techniques used for the standardization of herbal drug *Nigella sativa* L. Microsc Res Tech 81(12):1443–1450

Sahoo N, Manchikanti P, Dey S (2010) Herbal drugs: standards and regulation. Fitoterapia 81(6):462–471

Saini V, Singh A, Shukla R, Jain K, Yadav AK (2022) Silymarin-encapsulated xanthan gum–stabilized selenium nanocarriers for enhanced activity against amyloid fibril cytotoxicity. AAPS PharmSciTech 23(5):1–14

Saito K, Kaneko S, Furuya Y, Asada Y, Ito R, Sugie K et al (2019) Confirmation of synthetic cannabinoids in herb and blood by HS-SPME-GC/MS. Forensic Chem 13:100156

Sammons HM, Gubarev MI, Krepkova LV, Bortnikova VV, Corrick F, Job KM et al (2016) Herbal medicines: challenges in the modern world. Part 2. European Union and Russia. Expert Rev Clin Pharmacol 9(8):1117–1127

Sharma A, Jain K, Flora SJS (2018) Vitamins based novel target pathways/molecules as possible emerging drug targets for the Management of Tuberculosis. Med Chem (Los Angeles) 14(3):212–224

Shinde VM, Dhalwal K, Potdar M, Mahadik KR (2009) Application of quality control principles to herbal drugs. Int J Phytomed 1(1):4–8

Sugumar S, Ghosh V, Nirmala MJ, Mukherjee A, Chandrasekaran N (2014) Ultrasonic emulsification of eucalyptus oil nanoemulsion: antibacterial activity against Staphylococcus aureus and wound healing activity in Wistar rats. Ultrason Sonochem 21(3):1044–1049

Suthar T, Jain VK, Popli H, Jain K (2022) Nanoemulsions as effective carriers for targeting brain tumors. In: Nanocarriers for drug-targeting brain tumors. Elsevier, pp 347–363

Suthar T, Patel P, Singh P, Datusalia AK, Yadav AK, Jain K (2023) Hesperidin microemulsion: formulation optimization, characterization, and in vitro evaluation. J Drug Deliv Sci Technol 80:104166

Thakur L, Ghodasra U, Patel N, Dabhi M (2011) Novel approaches for stability improvement in natural medicines. Pharmacogn Rev 5(9):48

Thorley AJ, Tetley TD (2013) New perspectives in nanomedicine. Pharmacol Ther 140(2):176–185

Tokalıoğlu Ş, Dokan FK, Köprü S (2019) ICP-MS multi-element analysis for determining the origin by multivariate analysis of red pepper flakes from three different regions of Turkey. LWT 103:301–307

Tu Y (2011) The discovery of artemisinin (qinghaosu) and gifts from Chinese medicine. Nat Med 17(10):1217–1220

US Food and Drug Administration (2024) Guidance for industry: botanical drug products (the botanical guidance). https://www.fda.gov/regulatory-information/search-fda-guidance-documents/botanical-drug-development-guidance-industry

Varhan Oral E, Tokul-Ölmez Ö, Yener İ, Firat M, Tunay Z, Terzioğlu P et al (2019) Trace elemental analysis of *allium* species by inductively coupled plasma-mass spectrometry (ICP-MS) with multivariate chemometrics. Anal Lett 52(2):320–336

Walters NA, de Villiers A, Joubert E, de Beer D (2017) Improved HPLC method for rooibos phenolics targeting changes due to fermentation. J Food Compos Anal 55:20–29

Wang H, Chen Y, Wang L, Liu Q, Yang S, Wang C (2023) Advancing herbal medicine: enhancing product quality and safety through robust quality control practices. Front Pharmacol 14:14

WHO (2000) Regional meeting on regulatory aspects of herbal products, Final report. Division of Health Systems and Services Development, Pan American Health Organization. https://applications.emro.who.int/dsaf/dsa545.pdf

Working Group on "Access to Health Systems including AYUSH" (2024) Eleventh five year plan (2007–2012) Government of India Planning Commission. https://www.niti.gov.in/sites/default/files/2023-08/11th_vol1.pdf

World Health Organization (2024a) The declaration of Alma-Ata. https://www.who.int/teams/social-determinants-of-health/declaration-of-alma-ata#:~:text=The%20Alma%2DAta%20Declaration%20of,goal%20of%20Health%20for%20All

World Health Organization (2024b) International standard terminologies on traditional medicine in the western pacific region. https://www.who.int/westernpacific/health-topics/traditional-complementary-and-integrative-medicine#tab=tab_1

Wu L, Zhang J, Watanabe W (2011) Physical and chemical stability of drug nanoparticles. Adv Drug Deliv Rev 63(6):456–469

Wu LP, Wang D, Li Z (2020) Grand challenges in nanomedicine. Mater Sci Eng C 106:110302

Xie X, Tao Q, Zou Y, Zhang F, Guo M, Wang Y et al (2011) PLGA nanoparticles improve the Oral bioavailability of curcumin in rats: characterizations and mechanisms. J Agric Food Chem 59(17):9280–9289

Zhang J, Wider B, Shang H, Li X, Ernst E (2012) Quality of herbal medicines: challenges and solutions. Complement Ther Med 20(1–2):100–106

Zhang A, Sun H, Wang X (2018) Mass spectrometry-driven drug discovery for development of herbal medicine. Mass Spectrom Rev 37(3):307–320

Zhou J, Cai H, Tu S, Duan Y, Pei K, Xu Y et al (2018) Identification and analysis of compound profiles of sinisan based on 'individual herb, herb-pair, herbal formula' before and after processing using UHPLC-Q-TOF/MS coupled with multiple statistical strategy. Molecules 23(12):3128

Drug Repurposing and Virtual Screening 16

Ruchika Sharma, Sweta Roy, and Anoop Kumar

Abstract

Drug repurposing is a promising strategy in which already approved drugs are explored for their alternate indications. It saves time and money which is involved in the development of new chemical entities as behaviour of a molecule in a large number of populations is already known. Virtual screening could play a significant role in the identification of potential repurposing candidates using high-performance computational techniques. A number of virtual methods like molecular docking, molecular dynamics, quantitative structure activity relationship (QSAR), etc. are available for the identification. Recently, researchers have also used pharmacovigilance data to identify promising drug repurposing candidates. This chapter starts with introduction to drug repurposing and its importance along with various approaches which are being used. The commonly used databases/software along with steps involved are also discussed.

Keywords

Virtual screening · Drug repurposing · In silico · High-throughput screening · Protein database · Algorithms

R. Sharma
Centre for Precision Medicine and Pharmacy, Delhi Pharmaceutical Sciences and Research University, New Delhi, India

S. Roy · A. Kumar (✉)
Department of Pharmacology, Delhi Institute of Pharmaceutical Sciences and Research (DIPSAR), Delhi Pharmaceutical Sciences and Research University (DPSRU), New Delhi, India

K. Jain, A. K. Yadav (eds.), *Advances in Pharmaceutical Product Development*, https://doi.org/10.1007/978-981-97-9230-6_16

Abbreviations

ACE2	Angiotensin-converting enzyme 2
BBB	Blood-brain barrier
CCLE	Cancer Cell Line Encyclopaedia
CML	Chronic myeloid leukaemia
CNS	Central nervous system
COVID-19	Coronavirus disease 2019
DARS	Division of Applied Regulatory Science
DR	Drug repurposing
FDA	Food and Drug Administration
FFT	Fast Fourier transform
GEO	Gene Expression Omnibus
GIST	Gastrointestinal stromal tumours
GOLD	Genomes Online Database
HTS/HCS	High-throughput or high-content screening
IRAK-4	Interleukin-1 receptor-associated kinase
LBVS	Ligand-based virtual screening
MD	Molecular dynamics
MDDR	MDL Drug Data Report
ML	Machine learning
NCBI-GEO	National Centre for Biotechnology Information-Gene Expression Omnibus
PD	Protein database
PHASE	Public health assessment via structural evaluation
QSAR	Quantitative structure-activity relationship
SARS	Severe acute respiratory syndrome
SBVS	Structure-based virtual screening
TBVS	Target-based virtual screening
UCSF	University of California, San Francisco
VS	Virtual screening

16.1 Introduction

In the previous decade, the research and manufacturing methods for new pharmaceuticals have been constantly evolving, with benefits and results directly tied to improving the global population's life quality (Sahoo et al. 2021). Production and manufacturing tactics are becoming more effective, but the costs remain a hurdle. The average cost of developing a new medicine is between \$1 and \$2 billion USD, and the entire process can take 10–17 years, from target selection to registration. These costs indicate an expensive, time-consuming, and laborious process that involves drugs with a high added value that are frequently out of reach for at least one-third of the global population. Therefore, drug repurposing (DR), often referred to as drug recycling, drug rescue, drug repositioning, drug reprofiling, and therapeutic flipping, is a ground-breaking strategy for resolving this problem (Sahoo et al. 2021). Drug repurposing (DR) includes the exploration of novel applications or indications for existing drugs, extending beyond their initial medical uses. Utilizing drug repurposing has the potential to reduce the average time required for drug development by 5–7 years. Numerous instances of successful drug repurposing exist, such as imatinib, initially designed for chronic myeloid leukaemia (CML) and later found effective in treating malignant gastrointestinal stromal tumours (GIST). Another illustration is aspirin, a non-steroidal anti-inflammatory drug taken orally that is also used to prevent colorectal cancer and cardiovascular disease (Gan et al. 2023). It is predicted that the drug repositioning procedure will take approximately 5 years only.

The traditional drug development process incurs an average cost of USD 1.24 billion to bring a new medicine for public use, while drug repurposing involves approximately ≤60% of the expenditure associated with traditional methods (Napolitano et al. 2013). The accessibility of earlier gathered data on structural optimization, pharmacokinetics, toxicology, clinical effectiveness, and safety of drugs in the traditional approach contributes to a saving in the time and cost of medication development, along with a decreased risk of failure or a higher success level in medicine repurposing (Wu et al. 2013).

Importance

- It has the potential to enter directly into the phase II clinical trials.
- Reduce the attrition rate.
- Pharmacokinetic profile is already known.
- Safety profile is already known.

The extensive studies are not required for the repurposed candidates; however few studies need to be conducted for the following reasons: need to make sure the drug works in alternate indication, biology of diseases is different, required dose may be different from the approved dose, and can interfere with current treatment regimen. Therefore, few studies need to be done for the repurposing candidates.

16.2 A Prime Candidate for Repurposing

A pharmaceutical product, having completed clinical medicine development and positioned as an excellent applicant for repurposing, may bypass additional clinical trials if it possesses robust safety and toxicity data from prior clinical studies and if it has received regulatory approval. The selected drug's mechanism of action must be clearly defined (Dinić et al. 2020).

Medicines that have progressed through various levels of clinical development but were failed due to reasons unrelated to safety are well-suited for repurposing.

Some instances of DR have occurred even within the course of clinical trials, exemplified by the renowned case of viagra (sildenafil). Originally designed for hypertension and angina, its novel application for treating erectile dysfunction was discovered during clinical trials (Kulkarni et al. 2023).

16.3 Approaches

The primary challenge in drug repositioning involves identifying new associations between drugs and diseases. Two general categories for systematic drug repurposing are in silico methods and experimental screening techniques. While the latter strategy employs current data and new computational techniques that are substantially less expensive than the former, the former approach depends on specially developed high-throughput trials. Virtual screening is one such technology that can be used in drug development to quickly search compound libraries using computer-aided drug design (CADD). To tackle this challenge, various strategies have been devised, encompassing computational methods, biological experiments, and hybrid approaches (Fig. 16.1).

A number of databases including information on medicine and diseases have emerged as a result of the rapid development of biological microarray techniques, including DrugBank (Wishart et al. 2018), ChemBank (Seiler et al. 2008), OMIM (Hamosh et al. 2005), KEGG (Ogata et al. 1999), and PubMed (2024). Large-scale genomic databases have also been constructed at the same time, including MIPS (Mewes et al. 1998), PDB (Bank RPD 2024), GEO (Barrett et al. 2013), and GenBank (Benson et al. 2013). The expansion of various computing techniques has been greatly expedited by this abundance of information and data. Computational methodologies are characterized by cheaper costs and fewer restrictions than biological experimental methods (Oprea and Overington 2015). A number of methods are available for the identification of repurposing drugs. The details about the available methods are explained below.

16.3.1 Phenotypic Screening

The approach of phenotypic screening was employed in the identification of compounds and biologics sanctioned by regulatory authorities. These techniques lack pharmaceutical or biological data, making it less probable for them to unveil the mechanisms of drug action. Typically, they rely on fortuitous identification through tests targeting specific disease and medications. The notable benefit of these approaches, specifically off-label use and phenotypic screening, lies in their heightened potential for applicability across various drugs or conditions (Jin and Wong 2014).

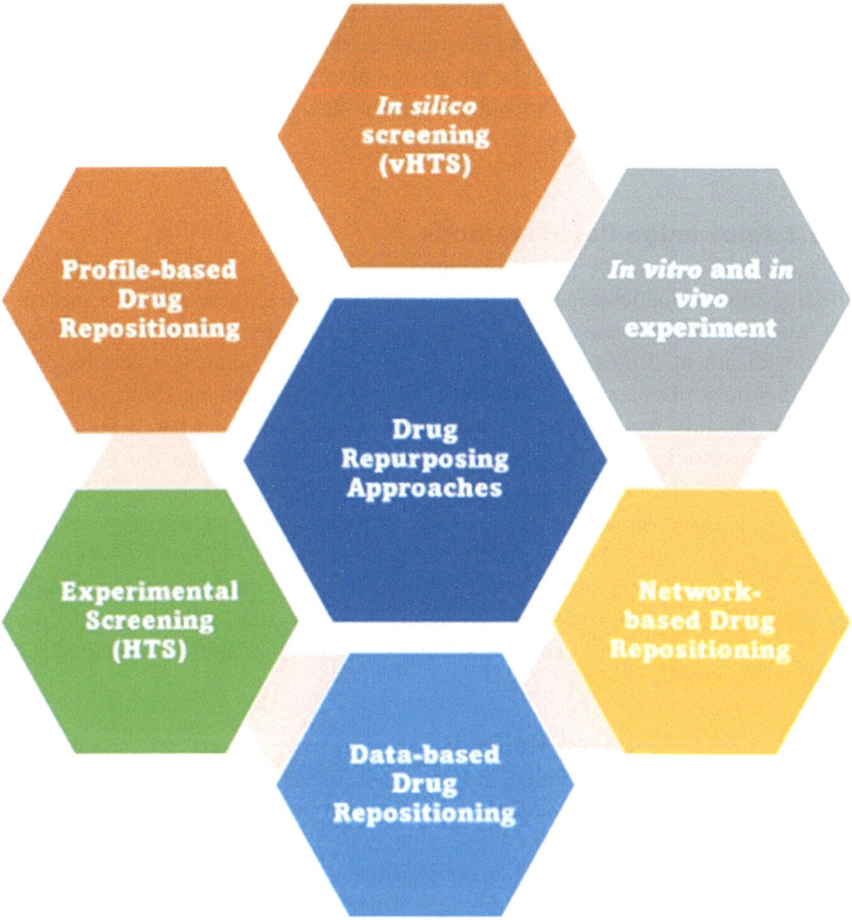

Fig. 16.1 The computational approaches for drug repositioning

16.3.2 Target-Based Methods

This approach necessitates specialized information concerning targets, like 3D protein type of structures. Knowledge-driven approaches rely on information about medicine or diseases, encompassing adverse effects, governing approval labels, clinical trial documents, and available biomarkers (potential targets) or pathways associated with diseases. Scientists can quickly screen a large number of pharmacological compounds with known chemical structures using these techniques (e.g. simplified molecular response line-entry system (SMILES)). These strategies fall under the category of target-based DR techniques.

Conducting high-throughput or high-content screening (HTS/HCS) of drugs in vitro and animal studies for a specific protein/biomarker, along with in silico selection using methods like ligand-based screening or docking, constitutes the exploration process. In contrast to blinded approaches, target-based methods enhance the likelihood of medicine discovery by directly linking targets to the disease/syndrome mode of action. The integration of target evidence and DR amplifies the potential for identifying therapeutically advantageous compounds (Benson et al. 2013; March-Vila et al. 2017).

16.3.3 Knowledge-Based Methods

Utilizing bioinformatics/cheminformatics methodologies, these methods incorporate various data sources into drug repositioning studies, encompassing details about medicament, drug-target networks, target and drug chemical structures, clinical studies data, FDA approvals, as well as their signalling and metabolic pathways. However, the informational content of both blinded and target-based approaches might be insufficient to uncover novel mechanisms away from the established targets.

In contrast, knowledge-based methods integrate existing data to predict unknown mechanisms, like identifying new drug targets, revealing hidden drug-drug likenesses, and pinpointing novel disease biomarkers. By incorporating substantial known information, knowledge-based methods enhance the DR process, thereby improving the accuracy of predictions (Benson et al. 2013; March-Vila et al. 2017).

16.3.4 Signature-Based Methods

Methods for DR based on signatures involve the utilization of gene signatures extracted by using disease omics data, either with the treatment or without treatment, to unveil undiscovered off-target or disease mode of action. The sequence read archive (http://www.ncbi.nlm.nih.gov/Traces/sra/), NCBI-GEO (http://www.ncbi.nlm.nih.gov/geo/), SRA, CMAP Connectivity Map, and CCLE, Cancer Cell Line Encyclopaedia, are among the publicly accessible databases from which the genomic data used in these analyses is sourced. Signature-based approaches play a vital role in unravelling previously unidentified mechanisms for compound and drugs. Computational techniques employed in this context often focus on molecular-level mode of action, such as genes that undergo significant changes (Benson et al. 2013; March-Vila et al. 2017).

16.3.5 Pathway or Network-Based Methods

This method reconstructs pathways specific to a given disease by using genetic disease information, available signalling or metabolic pathways, and protein interaction networks. Through this process, broad signalling networks that previously involved many proteins are refined to more particular networks with a smaller number of proteins or targets, which helps identify important targets for drug repurposing. Utilizing pathway exploration or network biology approaches, these techniques explore into genetic, metabolic information, proteomic, and genomic related to diseases to pinpoint essential pathways, uncovering new targets suitable for drug repositioning. For example, exploring the signalling mode of action for metastatic subtypes in breast cancer poses a challenge due to the difficulty in clarifying these subtype-specific signalling mechanisms through already known breast cancer pathways or gene signatures (Benson et al. 2013; March-Vila et al. 2017).

16.3.6 Targeted Mechanism-Based Methods

To clarify the unknown mechanism of drug action, this method combines treatment omics (genetic) data, pre-existing signalling pathway data, and protein interaction networks. The objective is to uncover the mechanisms of drug mechanism by identifying off-target or targeted pathways affected by drug treatment, utilizing drug omics data collected earlier and afterward the administration of the drug. For instance, addressing drug resistance in cancer therapy is critical. Patients often have resistance to a medication after a few months of treatment, even with initial positive responses. In order to find more effective therapeutic targets, further understanding of the mechanisms underlying drug action is necessary for successful drug treatment. However, there are a very few or limited studies available on mechanism-based methodology that focus on targets and employ sophisticated in silico study to predict drug effects and their associated pathways. This scarcity is attributed to the challenges involved in developing actual computational models (Benson et al. 2013; March-Vila et al. 2017).

16.3.7 Pharmacovigilance-Based Drug Repurposing

Pharmacovigilance-based drug repurposing is a recent concept in which repurposed candidates are identified from pharmacovigilance data using data mining approaches. Recently, few studies have used pharmacovigilance-based approach in the identification of potential repurposing drugs against various diseases (Table 16.1). Recently, in our lab, we have used this approach for the identification of potential drugs against bacterial infections.

16.4 Virtual Screening (VS)

Virtual screening (VS) is a useful in silico technique used in the search for new medicines (Fig. 16.2). A crucial prerequisite for engaging in virtual screening involves having access to the 3D type of structure of every target protein (Miró-Canturri et al. 2019; de Mello et al. 2020). Consequently, several databases have been established to house 3D molecular structures. The application of VS has nowadays become widespread in the identification of novel drug molecules and has

Table 16.1 Few examples of repurposed drugs

S. no	Drug name	Drug known for	Repurposed for	Type of study
1.	Auranofin	Rheumatoid arthritis	Antibacterial and antifungal (Miró-Canturri et al. 2019)	In vitro
2.	Aripiprazole	Antipsychotic/ anti-depressant	Active against fungal biofilms (Rajasekharan et al. 2019)	In vitro
3.	Aspirin and ibuprofen	NSAIDS	Antibacterial and antifungal (Ogundeji et al. 2016)	In vitro
4.	Dacarbazine	Anti-tumour	Active against fungal biofilms (Wakharde et al. 2018)	In vitro
5.	Niclosamide	Helminthiasis/ anti-infective	Antibacterial and antifungal and active against fungal biofilms (Hamdoun et al. 2017)	In vitro and in silico
6.	Artesunate	Anti-infective	Active against fungal biofilm (De Cremer et al. 2015)	In vitro
7.	Auranofin	Rheumatoid arthritis	Anti-bacterial and antifungal (Siles et al. 2013)	In vitro
8.	Bithionate disodium	Anti-infective	Active against fungal biofilms (De Cremer et al. 2015)	In vitro
9.	Bleomycin	Anti-tumour	Active against fungal biofilms (Wakharde et al. 2018)	In vitro
10.	Avermectin B1a	Anti-infective	Active against fungal biofilms (Siles et al. 2013)	In vitro
11.	Bromperidol	Anti-psychotic/ anti-depressant	Active against fungal biofilms (Holbrook et al. 2017)	In vitro
12.	Benzbromarone	Vasodilator	Active against fungal biofilms (Siles et al. 2013)	In vitro
13.	Carboplatin	Anti-tumour	Active against fungal biofilms (Wakharde et al. 2018)	In vitro
14.	Broxyquinoline	Anti-infective	Active against fungal biofilms (De Cremer et al. 2015)	In vitro

(continued)

Table 16.1 (continued)

S. no	Drug name	Drug known for	Repurposed for	Type of study
15.	Celecoxib	Anti-inflammatory/ immunomodulatory	Active against fungal biofilms (Alem and Douglas 2004)	In vitro
16.	Clarithromycin	Anti-infective	Active against fungal biofilms (Fernández-Rivero et al. 2017)	In vitro
17.	Chloroquine	Anti-malarial	Active against fungal biofilms (Shinde et al. 2013)	In vitro
18.	Cyclophosphamide	As immune-modulator in autoimmune diseases	Breast cancer (Aggarwal et al. 2021)	
19.	Doxepin	Anti-psychotic/ anti-depressant	Active against fungal biofilms (Caldara and Marmiroli 2018)	In vitro and in vivo
20.	Disulfiram (Antabuse)	Reduces ethanol tolerance in alcoholism	Metastatic breast cancer (The Institute of Molecular and Translational Medicine, Czech Republic 2023)	Phase II clinical trial
21.	Diazepam	Anti-psychotic/ anti-depressant	Active against fungal biofilms (Kathwate et al. 2015)	In vitro
22.	Dexpramipexole	ALS and other neurological diseases: phase 3 trials did not meet the endpoint	Hypereosinophilic syndrome (National Institute of Allergy and Infectious Diseases (NIAID) 2017)	Clinical trial
23.	Ebastine	Anti-inflammatory/ immunomodulatory	Active against fungal biofilms (Dennis and Garneau-Tsodikova 2019)	In vitro
24.	Etodolac	Anti-inflammatory/ immunomodulatory	Active against fungal biofilms (Alem and Douglas 2004)	In vitro
25.	Edaravone	Neuroprotective agent in acute ischemic stroke and ALS	Multiple sclerosis (Eleuteri et al. 2017)	Clinical studies
26.	Exemestane	Ovulation induction	Breast cancer (Zucchini et al. 2015)	Clinical studies
27.	Eltrombopag	Anti-inflammatory/ immunomodulatory	Active against fungal biofilms (Ko et al. 2020)	In vitro
28.	Everolimus (Votubia, Evertor)	Immunosuppressant during organ transplants, wound healing	Metastatic breast cancer (Royce and Osman 2015)	Clinical studies
29.	Favipiravir	Inhibitors of RNA-dependent RNA polymerase of virus (anti-viral drug)	Effective against SARS-CoV-2 (COVID-19) (Wang et al. 2020)	Clinical studies

(continued)

Table 16.1 (continued)

S. no	Drug name	Drug known for	Repurposed for	Type of study
30.	Finasteride	Benign prostatic hyperplasia	Anti-bacterial and antifungal (Chavez-Dozal et al. 2014)	In vitro
31.	Fulvestrant	Anti-oestrogen	Breast cancer (Turner et al. 2018)	Clinical studies
32.	Fluvastatin	Lipid lowering	Active against fungal biofilms (Rana et al. 2019)	In vitro
33.	Gemcitabine	Anti-viral drug	Breast cancer (Stemmler et al. 2011)	
34.	Goserelin	Prostate cancer, uterine fibroids, assisted reproduction	Breast cancer (Noguchi et al. 2016)	Clinical studies
35.	Imipramine	Anti-psychotic/ anti-depressant	Active against fungal biofilms (Caldara and Marmiroli 2018)	In vitro
36.	γ-Secretase inhibitors (GSI)	Alzheimer disease: prevent amyloid precursor cleavage	Several inhibitors are being tested against a variety of cancers (Schein 2020)	Clinical trails
37.	Ivermectin	Anti-parasitic	Effective against SARS-CoV-2 (COVID-19) (Caly et al. 2020)	Clinical trails
38.	Iodoquinol	Anti-infective	Active against fungal biofilms (Wall et al. 2019)	In vitro
39.	Letrozole	Ovulation induction	Breast cancer (Finn et al. 2016)	Clinical trails
40.	Ketoprofen	Anti-inflammatory/ immunomodulatory	Active against fungal biofilms (Abdelmegeed and Shaaban 2013)	In vitro
41.	Itraconazole	Antifungal	Anti-cancer (Kim et al. 2010)	In vitro
42.	Ketorolac	Anti-inflammatory/ immunomodulatory	Active against fungal biofilms (Kim et al. 2010)	In vitro
43.	Lopinavir/ritonavir	Anti-viral drug	Effective against SARS-CoV-2 (COVID-19) (Cao et al. 2020)	Clinical studies
44.	Lorazepam	Anti-psychotic/ anti-depressant	Active against fungal biofilms (Domhan et al. 2008)	In vitro
45.	Losartan	Blood pressure reduction	Alzheimer disease (Schein 2020)	Pre-clinical trials
46.	Mycophenolic acid	Immunosuppressant	Anti-cancer (Domhan et al. 2008)	In vitro and in vivo

(continued)

Table 16.1 (continued)

S. no	Drug name	Drug known for	Repurposed for	Type of study
47.	Verapamil	Anti-arrhythmic	Active against fungal biofilms (Yu et al. 2013)	In vitro
48.	Tocilizumab	Immunosuppressive drug for cytokine release syndrome	Severe COVID-19 infection (Toniati et al. 2020)	Clinical studies
49.	Ribavirin	Anti-viral drug	Effective against SARS-CoV-2 (COVID-19) (Elfiky 2020)	In silico
50.	Rifampicin	Anti-infective	Active against fungal biofilms (Fernández-Rivero et al. 2017)	In vitro

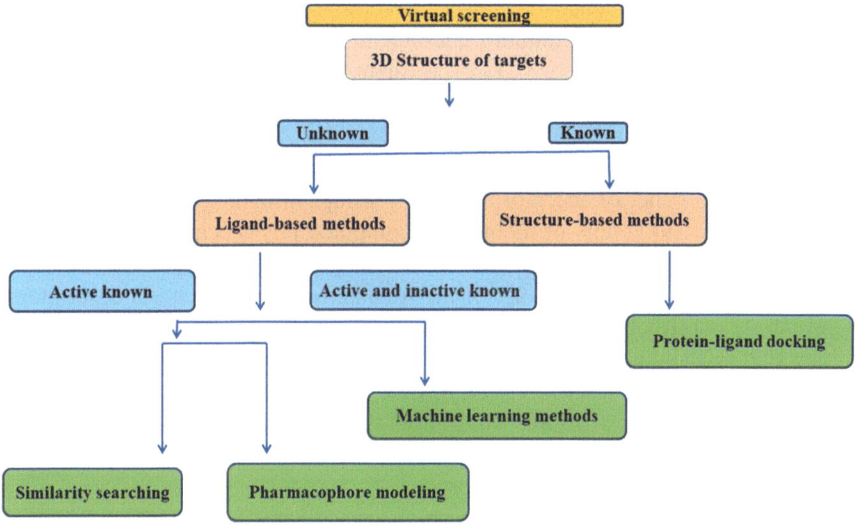

Fig. 16.2 Methods of virtual screening

already played a significant role in bringing compounds to market. Some notable examples of medications that are sold through virtual screening are zanamivir, captopril, ritonavir, and indinavir, as well as saquinavir, tirofiban, dorzolamide, and saquinavir (Maia et al. 2020).

16.4.1 Molecular Docking

It is a flexible method for estimating the structure and evaluating the intensity of the interaction between a protein and a minor chemical in a complex. Extensive sets of ligands and targets can be quickly and effectively screened via molecular docking, which provides a large number of sample options (Kitchen et al. 2004; Sgobba et al. 2012).

16.4.2 Ligand-Based Virtual Screening (LBVS)

LBVS utilizes ligands through established biological action to uncover compound sharing analogous structures within virtual compound public library. In this method, the structure of the molecular target is disregarded, with LBVS relying on the premise that molecules possessing structural resemblance might demonstrate comparable biological activity. Thus, the goal of LBVS is to identify molecules with similar molecular scaffolds or pharmacophore components in order to increase the likelihood of finding compounds that are biologically active. In LBVS, compounds stored in databanks are equated with the descriptors or qualities of recognized molecules derived from reference molecules. This comparison is conducted using similarity measures, with the Tanimoto coefficient being widely employed for this purpose (de Oliveira et al. 2023). One prevalent LBVS classification technique is based on the number of dimensions represented by the 1D, 2D, or 3D descriptor:

1D descriptors: 1D descriptors include molecular characteristics such as molecular weight, the number of acceptor and donor groups in hydrogen bonds, rotatable bonds, types and counts of atoms, and computed physicochemical parameters such as log*P* and water solubility (log*D*), among others.

2D descriptors: Molecular linkage is the foundation upon which 2D descriptors, which are constructed, are built. In the form of topological indices and structural descriptors, these appear (Siles et al. 2013). Binary vectors or 2D charts are two ways that structural descriptors can be used to illustrate a substance by describing its chemical substructure. Descriptors that are used to calculate molecular similarity include the carbo index, connection index, and count of aromatic rings. These are examples of 2D descriptors.

3D descriptors: When it comes to the spatial arrangement of atoms, their characteristics, or chemical groups, 3D descriptors provide molecular insights. These descriptors take into account the conformations of molecules, necessitating the examination of chemical groups, particle distribution in space, and molecular characteristics (Holbrook et al. 2017). Studies have revealed that when it comes to virtual screening, no particular descriptor performs better than others overall (Alem and Douglas 2004). Therefore, a variety of descriptors are usually used to describe molecules. Van der Waals forces and electrostatic potential are two examples of 3D descriptors (de Oliveira et al. 2023).

16.4.3 Pharmacophore Modelling

Pharmacophore modelling is a firmly established approach utilized for screening vast databases in the initial stages of medicament development. Pharmacophore modelling VS is an in silico method which used identification of new drug by matching the structural and chemical features of a small compound (pharmacophore) with the target site on a biological macromolecule, like a protein receptor. This technique helps in the efficient identification of large chemical public library to predict compound that are likely to bind and interact with the target, thus aiding in the discovery of a new drug (Giordano et al. 2022).

16.4.4 Similarity Searching

Similarity searching has become a straightforward and cost-efficient method for comparing information across different chemical databases to discover connections between active structures in the database. With this method, it's now simpler to track down the original active aspects based on how much the structures resemble each other. Because of its ease and effectiveness, much chemo informatics software relies on likeness searching with single target structure methods. To conduct several searches or investigate target structures that aren't structurally associated, similarity searching is carried out using chemical databases such as a MDL Drug Data Report (MDDR) (Bastikar et al. 2022).

16.4.5 Machine Learning (ML)

Machine learning (ML) is a computational approach used in drug development and materials science to expedite the identification of promising compounds with desired properties. It influences machine learning algorithms to analyse large datasets of chemical structures and their associated properties, predicting which compounds are most expected to exhibit the desired biological actions or material characteristics (Carpenter and Huang 2018).

16.4.6 Structure Based

Target-based virtual screening (TBVS), another name for structure-based virtual screening (SBVS), is a technique that predicts the ideal combination of two chemicals to create a stable molecule. This technique involves various approaches to investigate the 3D structure of the molecular target. SBVS is the favoured approach when there is a 3D structure of the molecular target which has been empirically determined. It endeavours to anticipate the probability of interaction among applicant ligands and the target protein, taking into consideration the binding strength of the resulting complex. Among SBVS techniques, molecular anchoring is commonly utilized due to its cost-effectiveness and reliable outcomes (Maia et al. 2020).

In the 1980s, molecular docking emerged with Kuntz et al.'s development of an algorithm for investigating the geometrically plausible docking of a compound and target (Pinzi and Rastelli 2019). Their work demonstrated the ability to achieve structures closely resembling the correct ones post-docking. This method aims to forecast the optimal positioning and alignment of a ligand within a molecular target's binding site to facilitate the formation of a stability of complex (Shah et al. 2024). SBVS is commonly employed with proteins and enzymes like receptors, but it can also extend its application to other structures like carbohydrates and DNA. Despite its promising potential, it wasn't until the 1990s that SBVS saw widespread adoption, facilitated by advancements in techniques, increased computational capabilities, and enhanced accessibility to structural data of target molecules (Smith 2016).

16.4.7 Molecular Dynamics Studies

Molecular dynamics studies are usually done after the molecular docking studies to check the stability of drugs in the active site of the target. These simulations depend upon the Newton second law, i.e. $F = m*a$ where F is force, m is mass, and a is acceleration. Further, change in velocity with respect to time is acceleration. Overall, simulation is provided using high-performance computational techniques to analyse the physical movements of atoms and molecules. The stability of ligand in the active site of the target is analysed by calculating various parameters like root mean square deviation (RMSD), root mean square fluctuations (RMSF), radius of gyration, potential energy, principal component analysis (PCA), free energy landscape (FEL), and so on.

16.4.8 Quantitative Structure-Activity Relationship (QSAR)

Quantitative structure-activity relationship (QSAR) is a mathematical as well as computational approach aimed at establishing a statistically meaningful connection between the structure and function of molecules, employing chemometric techniques. The method of ligand-based drug design known as QSAR analysis was originated by Hansch and Fujita back in 1964, marking over 50 years since its inception (Bastikar et al. 2022).

QSAR assists in identifying compounds possessing specific characteristics by analysing their chemical details and their impact on biological functions. It considers factors such as how they interact with their environment and the presence or absence of certain chemical traits. QSAR aims to establish connections between a compound's structure, chemistry, and physical attributes with its biological effects, employing diverse mathematical techniques. The resulting QSAR models are utilized to anticipate and categorize the biological behaviours of novel chemical molecules (Bastikar et al. 2022). QSAR models are beneficial for regulatory purposes when there is limited or no empirical data available, and there is a need to assess toxicological or pharmacological potential at regulatory endpoints. The division of Applied Regulatory Science (DARS) focuses its research efforts in computational pharmacology and toxicology on developing QSAR models and carefully selected datasets for endpoints that are pertinent to regulatory matters. Two statistical QSAR models were recently developed to predict drug permeability across the blood-brain barrier (BBB). These models are designed to help regulators assess if drug metabolites or related chemicals have the potential to be abused. They can also forecast whether an unknown substance of abuse will be able to cross the blood-brain barrier and cause the effects that the public health assessment via structural evaluation (PHASE) approach predicted. A sophisticated computer technique called PHASE is employed to determine the possible threat that recently developed street drugs represent to public safety. It has been used in the past to evaluate kratom alkaloids and derivatives of fentanyl. Pharmaceutical companies can use BBB models to identify novel treatments for illnesses of the central nervous system (CNS) (Research C for DE 2023).

The reason for creating in silico QSAR models delves into and encompasses the following factors:

- Using mathematical techniques to predict the biological action of molecules and comprehend their physical-chemical properties. QSAR models can be developed for various drug classes to study and forecast the activity of compounds.
- Understanding and explaining the mechanisms across a set of chemicals. By constructing a QSAR model based on this established mode of action for a series of molecules, it becomes possible to predict the action of unidentified molecules. Typically, molecules that is similar in structure exhibit similar types of activity within a certain range. Therefore, when a novel compound is discovering belonging to a similar type of class, its action can be also predicted using a QSAR model. This aids in enhancing the activity of the molecule and designing new ones.
- Cost savings in compound development (e.g. drugs, pesticides) through reduced expenses in molecule synthesis, manufacturing, and in vitro/in vivo testing. Mathematical validation of improved activity among newly designed molecules allows for selective synthesis, saving time and resources compared to traditional drug design methods.
- Predictions could reduce the need for costly animal testing, addressing ethical concerns and sacrificing animals for every new molecule, in terms of price, time, and ethics. QSAR helps in side-step unnecessary animal tests for new compound (Bastikar et al. 2022).

QSAR Application

- Identifying unidentified leads with medicinal, biocidal, or pesticide actions is dependable when using the QSAR model.
- The QSAR makes it possible to find harmful substances in the early phases of ligand creation or when scanning various databases of already-existing compounds.

- The toxicity to native species is predicted by the QSAR model. It directs the choice of mixes, whether synthetic or naturally occurring, with the best pharmacokinetic qualities.
- Predicting a range of physical and chemical properties of molecules, whether they are medicament, pesticides, consumer products, or specialty chemicals (Bastikar et al. 2022).

16.4.9 Commonly Used Databases/Software in Virtual Screening

The most commonly used databases/software in virtual screening are explained below:

16.4.9.1 Available Software for Docking Studies

Molecular docking uses computers to forecast how molecules fit together. It has two main stages: the first one is sampling the ligand and the second is scoring function. Sampling contains algorithms which assist in recognizing the most energetically advantageous arrangements of the ligand inside the protein's active site, considering how they bind together, and then those positions are ranked based on how well they fit on the basis of scoring functions (Kitchen et al. 2004; Das et al. 2020). An overview of various molecular docking software options highlights the most commonly used software package.

16.4.9.1.1 AutoDock

The widely used AutoDock programme from Scripps Research Institute allows the general public to freely and conveniently use both rigid and flexible docking. It provides several scoring systems for determining affinity and maximizes ligand employment within receptor binding sites through the use of a Lamarckian genetic algorithm. AutoDock (2024) can be accessed via http://autodock.scripps.edu and supports a variety of input file formats, including PDB, MOL2, and SDF.

16.4.9.1.2 Chimera

The University of California, San Francisco (UCSF), created Chimaera, a flexible programme for modelling, analysis, and visualization of molecular structures. It provides resources for exhibiting tiny molecules, proteins, and nucleic acids in three dimensions. The "Dock Prep" module adds hydrogens, charges, and creates molecular surfaces to direct the location of ligands in target proteins. Additionally, Chimera makes it easier to analyse docking results by visualizing binding postures and calculating binding energies (UCSF Chimera 2024). At https://www.cgl.ucsf.edu/chimera/, it is accessible.

16.4.9.1.3 Discovery Studio

Discovery Studio, established by Dassault Systèmes BIOVIA, is a comprehensive molecular modelling and simulation software. It offers tools for molecular docking, VS, protein modelling, and molecular dynamics examination. Its docking

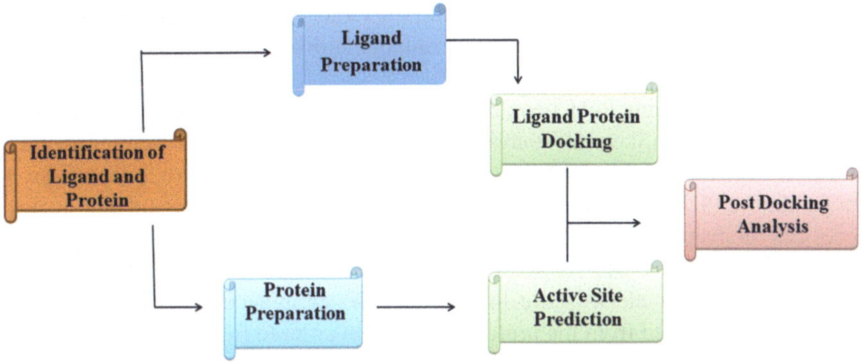

Fig. 16.3 Flowchart for molecular docking

component predicts ligand binding modes and interaction strength with target proteins using various algorithms like CDOCKER, GOLD, and LibDock. The software enables visualization, analysis, and comparison of docking results (Dassault Systèmes 2023). It is available at https://discover.3ds.com/discovery-studio-visualizer-download.

16.4.9.1.4 Dock

Dock is created by the UCSF Chimera team, and it's a user-friendly molecular docking software for positioning small molecules within receptor-binding sites. It utilizes a grid-based approach to assess ligand-receptor binding affinity and offers scoring functions for pose ranking. Supporting various input file formats such as PDB, MOL2, and SDF, Dock (UCSF Dock 2024) can be accessed at http://dock.compbio.ucsf.edu/.

16.4.9.1.5 MolDock

MolDock, created by MolSoft LLC, is a rapid and effective molecular docking software for positioning small molecules within receptor-binding sites. Utilizing a fast Fourier transform (FFT) algorithm, it assesses ligand-receptor binding affinity (Fig. 16.3). The software incorporates a scoring function. Supporting multiple input file formats such as PDB, MOL2, and SDF, MolDock (Molsoft L.L.C. 2024) is accessible at https://www.molsoft.com/about.html.

16.4.9.1.6 Argus Lab

Argus Lab, created by Mark Thomson at Pacific Northwest National Laboratory for the Department of Energy, simulates solvent effects using a combination of quantum mechanics and classical mechanics algorithms. It manages jobs including visual design, medicinal design, and molecular modelling (ArgusLab 2024). You can visit Argus Lab at http://www.arguslab.com.

The steps which need to perform in molecular docking study are as follows (Fig. 16.3): identifying essential target proteins and ligands is an essential component for conducting docking. Ensure the target protein is accessible through Swiss UniProt or PDB databases. If not available, utilize Swiss model repository or modeller programs.

Find the ligand using PubChem, Zinc, or ChmBl databases, or synthesize it using ChemDraw or ChemSketch if unavailable. Protein preparation is essential for precise docking simulations. It involves refining the protein structure, either from a databank like PDB or by utilization of tools, i.e. SWISS-MODEL, and adding any necessary atoms or residues. The protein undergoes energy minimization to relax its structure and remove any crowding. Ionizable residues' protonation states are adjusted for correct electrostatic interactions. Water molecules and unnecessary ligands are eliminated from the protein structure in order to further simplify it (Agu et al. 2023). In molecular docking, initially the binding sites can be determined. The ligand undergoes docking with the protein, and the resulting interactions are examined. A counting function assigns a score to the most favourable docking complex identified (The Institute of Molecular and Translational Medicine, Czech Republic 2023; Torres et al. 2019). After the docking of ligands to the protein, an analysis is conducted to pinpoint the most viable candidates for subsequent investigation. Each ligand's binding affinity is determined by predicting interaction energy, leading to the ranking of ligands based on their affinity scores. Furthermore, the docked ligand and protein complex structures are examined in detail to analyse important interactions between the ligands and the protein, such as electrostatic, hydrophobic, and hydrogen bonding interactions. These interactions give insights into the mechanism of the ligands and pointers for further structural optimization (The Institute of Molecular and Translational Medicine, Czech Republic 2023; Pinzi and Rastelli 2019).

Few examples of virtual screening are the following: interleukin-1 receptor-associated kinase 4 (IRAK-4) is an attractive target for treatment because of its critical role in immunological and inflammatory disease pathways, as demonstrated by a study conducted in 2016 by Zhong et al. Research has utilized ligand-based pharmacophore modelling and three-dimensional QSAR analysis to investigate ATP competitive inhibitors of IRAK-4, leading to the discovery of 12 strong and unique inhibitors. These results were further confirmed by virtual screening and molecular dynamics (MD) simulation analyses, which showed encouraging IRAK-4 inhibitory potential, especially in anthraquinones ZINC09047206 and ZINC09477176 (Zhong et al. 2016).

Jawarkar et al. (2023) reported a study in which they used QSAR modelling to a diverse dataset of 657 compounds to uncover structural features essential for ACE2 inhibitory activity, aiming to identify novel hit molecules. High predictivity (R2tr = 0.84, R2ex = 0.79) was demonstrated by the created QSAR model, which unveiled previously undiscovered characteristics and novel mechanistic insights. This model was used to estimate the ACE2 inhibitory activity (PIC50) of 1615 ZINC FDA compound. This allowed for the identification of a hit molecule (ZINC000027990463) with a docking score of −9.67 kcal/mol (RMSD 1.4) and a

PIC50 of 8.604 M. Molecular docking highlighted 25 connections with the residue ASP40, while MD simulation affirmed the stability of the hit compound-ACE2 receptor complex over 400 ns, supporting its potential as a viable ACE2 inhibitor (Jawarkar et al. 2024).

16.4.9.2 Available Software for Molecular Dynamics Studies

The most commonly used is GROMACS which is freely available; however, graphical processing unit with good configuration is a must to run the simulations for longer duration. Desmond, AMBER, CHARMM, Discovery Studio, etc. are also available. The steps need to follow for molecular dynamics are as follows: step 1, preparation of molecules; step 2, define the box; step 3, solvation (implicit, GBMB, PBSA, etc.; explicit, SPC, TIP3P, etc.); step 4, ionization, neutralization, and minimization; step 5, equilibration (NVT, NPT); step 6, production and simulation; step 7, analysis (Jacob et al. 2017).

16.4.9.3 Databases/Software for Network

Various databases are available to perform the network pharmacological studies to identify the potential repurposing drugs. The databases available for the prediction of targets of drugs are SwissTargetPrediction (http://swisstargetprediction.ch/), SEA database (https://sea.bkslab.org/), and so on. The targets involved in the particular diseases can be extracted from GeneCard (https://www.genecards.org/) and DisGeNET (https://www.disgenet.org/). Venny tool (https://csbg.cnb.csic.es/BioinfoGP/venny.html) is helpful to identify the common targets between the drugs and diseases. The interaction among the common targets is usually checked by STRING server. Most of the researchers have used Cytoscape for the creation and analysis of network. Briefly the steps need to follow for network pharmacology are as follows: identification of targets of drugs, identification of targets involved in disease, identification of common targets, interactions among common targets, network construction and analysis, and enrichment analysis (Saima et al. 2024; Bhati et al. 2024).

16.5 Conclusion

Virtual methods are playing a significant role in the identification of repurposed candidates against various diseases. There are number of methods to identify potential repurposing drugs; however, each method has its own merits and demerits. The access to considerable volumes of data, including compound libraries, patent data, pharmacological data, and published scientific literature, is one of the biggest challenges in the field of drug repurposing. Further, experimental studies are required to confirm the potential of identified drug in a particular disease. The extensive safety studies are required if repurposed dose is found higher than the approved dose.

References

Abdelmegeed E, Shaaban MI (2013) Cyclooxygenase inhibitors reduce biofilm formation and yeast-hypha conversion of fluconazole resistant Candida albicans. J Microbiol 51(5):598–604

Aggarwal S, Verma SS, Aggarwal S, Gupta SC (2021) Drug repurposing for breast cancer therapy: old weapon for new battle. Semin Cancer Biol 68:8–20

Agu PC, Afiukwa CA, Orji OU, Ezeh EM, Ofoke IH, Ogbu CO et al (2023) Molecular docking as a tool for the discovery of molecular targets of nutraceuticals in diseases management. Sci Rep 13(1):13398

Alem MAS, Douglas LJ (2004) Effects of aspirin and other nonsteroidal anti-inflammatory drugs on biofilms and planktonic cells of Candida albicans. Antimicrob Agents Chemother 48(1):41–47

ArgusLab. http://www.arguslab.com/arguslab.com/ArgusLab.html. Accessed 10 Mar 2024

AutoDock. Homepage. https://autodock.scripps.edu/. Accessed 10 Mar 2024

Bank RPD. RCSB PDB: Homepage. https://www.rcsb.org/. Accessed 10 Mar 2024

Barrett T, Wilhite SE, Ledoux P, Evangelista C, Kim IF, Tomashevsky M et al (2013) NCBI GEO: archive for functional genomics data sets—update. Nucleic Acids Res 41(Database issue):D991–D995

Bastikar V, Bastikar A, Gupta P (2022) Quantitative structure–activity relationship-based computational approaches. In: Computational approaches for novel therapeutic and diagnostic designing to mitigate SARS-CoV-2 infection. Academic Press, p 191

Benson DA, Cavanaugh M, Clark K, Karsch-Mizrachi I, Lipman DJ, Ostell J et al (2013) GenBank. Nucleic Acids Res 41(Database issue):D36–D42

Bhati V, Kumar A, Lather V, Sharma R, Pandita D (2024) Association of temozolomide with progressive multifocal leukoencephalopathy: a disproportionality analysis integrated with network pharmacology. Expert Opin Drug Saf 23(5):649–658

Caldara M, Marmiroli N (2018) Tricyclic antidepressants inhibit Candida albicans growth and biofilm formation. Int J Antimicrob Agents 52(4):500–505

Caly L, Druce JD, Catton MG, Jans DA, Wagstaff KM (2020) The FDA-approved drug ivermectin inhibits the replication of SARS-CoV-2 in vitro. Antivir Res 178:104787

Cao B, Wang Y, Wen D, Liu W, Wang J, Fan G et al (2020) A trial of lopinavir–ritonavir in adults hospitalized with severe Covid-19. N Engl J Med 382(19):1787–1799

Carpenter KA, Huang X (2018) Machine learning-based virtual screening and its applications to Alzheimer's drug discovery: a review. Curr Pharm Des 24(28):3347–3358

Chavez-Dozal AA, Lown L, Jahng M, Walraven CJ, Lee SA (2014) In vitro analysis of finasteride activity against Candida albicans urinary biofilm formation and filamentation. Antimicrob Agents Chemother 58(10):5855–5862

Das DR, Kumar D, Kumar P, Dash BP (2020) Molecular docking and its application in search of antisickling agent from Carica papaya. J Appl Biol Biotechnol 8(1):105–116

Dassault Systèmes (2023) Visualization [cited 2024 Mar 10]. https://www.3ds.com/products/biovia/discovery-studio/visualization

De Cremer K, Lanckacker E, Cools TL, Bax M, De Brucker K, Cos P et al (2015) Artemisinins, new miconazole potentiators resulting in increased activity against Candida albicans biofilms. Antimicrob Agents Chemother 59(1):421–426

de Mello TP, Silva LN, de Souza Ramos L, Frota HF, Branquinha MH, dos Santos ALS (2020) Drug repurposing strategy against fungal biofilms. Curr Topics Med Chem 20(7):509–516

de Oliveira TA, da Silva MP, Maia EHB, da Silva AM, Taranto AG (2023) Virtual screening algorithms in drug discovery: a review focused on machine and deep learning methods. Drugs Drug Candidates 2(2):311–334

Dennis EK, Garneau-Tsodikova S (2019) Synergistic combinations of azoles and antihistamines against Candida species in vitro. Med Mycol 57(7):874–884

Dinić J, Efferth T, García-Sosa AT, Grahovac J, Padrón JM, Pajeva I et al (2020) Repurposing old drugs to fight multidrug resistant cancers. Drug Resist Updat 52:100713

Domhan S, Muschal S, Schwager C, Morath C, Wirkner U, Ansorge W et al (2008) Molecular mechanisms of the antiangiogenic and antitumor effects of mycophenolic acid. Mol Cancer Ther 7(6):1656–1668

Eleuteri C, Olla S, Veroni C, Umeton R, Mechelli R, Romano S et al (2017) A staged screening of registered drugs highlights remyelinating drug candidates for clinical trials. Sci Rep 7:45780

Elfiky AA (2020) Ribavirin, Remdesivir, Sofosbuvir, Galidesivir, and Tenofovir against SARS-CoV-2 RNA dependent RNA polymerase (RdRp): a molecular docking study. Life Sci 253:117592

Fernández-Rivero ME, Del Pozo JL, Valentín A, de Diego AM, Pemán J, Cantón E (2017) Activity of amphotericin B and anidulafungin combined with rifampicin, clarithromycin, ethylenediaminetetraacetic acid, N-acetylcysteine, and farnesol against Candida tropicalis biofilms. J Fungi (Basel) 3(1):E16

Finn RS, Martin M, Rugo HS, Jones S, Im SA, Gelmon K et al (2016) Palbociclib and letrozole in advanced breast cancer. N Engl J Med 375(20):1925–1936

Gan J-H, Liu J-X, Liu Y, Chen S-W, Dai W-T, Xiao ZX et al (2023) DrugRep: an automatic virtual screening server for drug repurposing. Acta Pharmacol Sin 44(4):888–896

Giordano D, Biancaniello C, Argenio MA, Facchiano A (2022) Drug design by pharmacophore and virtual screening approach. Pharmaceuticals (Basel) 15(5):646

Hamdoun S, Jung P, Efferth T (2017) Drug repurposing of the anthelmintic niclosamide to treat multidrug-resistant leukemia. Front Pharmacol 8:110. https://doi.org/10.3389/fphar.2017.00110. https://www.frontiersin.org/journals/pharmacology/articles/. Accessed 16 Mar 2024

Hamosh A, Scott AF, Amberger JS, Bocchini CA, McKusick VA (2005) Online Mendelian Inheritance in Man (OMIM), a knowledgebase of human genes and genetic disorders. Nucleic Acids Res 33(Database issue):D514–D517

Holbrook SYL, Garzan A, Dennis EK, Shrestha SK, Garneau-Tsodikova S (2017) Repurposing antipsychotic drugs into antifungal agents: synergistic combinations of azoles and bromperidol derivatives in the treatment of various fungal infections. Eur J Med Chem 139:12–21

Jacob KS, Ganguly S, Kumar P, Poddar R, Kumar A (2017) Homology model, molecular dynamics simulation and novel pyrazole analogs design of Candida albicans CYP450 lanosterol 14 α-demethylase, a target enzyme for antifungal therapy. J Biomol Struct Dyn 35(7):1446–1463

Jawarkar RD, Zaki MEA, Al-Hussain SA, Al-Mutairi AA, Samad A, Mukerjee N et al (2024) QSAR modeling approaches to identify a novel ACE2 inhibitor that selectively bind with the C and N terminals of the ectodomain. J Biomol Struct Dyn 42(5):2550–2569

Jin G, Wong STC (2014) Toward better drug repositioning: prioritizing and integrating existing methods into efficient pipelines. Drug Discov Today 19(5):637–644

Kathwate GH, Shinde RB, Karuppayil SM (2015) Antiepileptic drugs inhibit growth, dimorphism, and biofilm mode of growth in human pathogen Candida albicans. Assay Drug Dev Technol 13(6):307–312

Kim J, Tang JY, Gong R, Kim J, Lee JJ, Clemons KV et al (2010) Itraconazole, a commonly used antifungal that inhibits hedgehog pathway activity and cancer growth. Cancer Cell 17(4):388–399

Kitchen DB, Decornez H, Furr JR, Bajorath J (2004) Docking and scoring in virtual screening for drug discovery: methods and applications. Nat Rev Drug Discov 3(11):935–949

Ko HT, Hsu LH, Yang SY, Chen YL (2020) Repurposing the thrombopoietin receptor agonist eltrombopag as an anticryptococcal agent. Med Mycol 58(4):493–504

Kulkarni VS, Alagarsamy V, Solomon VR, Jose PA, Murugesan S (2023) Drug repurposing: an effective tool in modern drug discovery. Russ J Bioorg Chem 49(2):157–166

Maia EHB, Assis LC, de Oliveira TA, da Silva AM, Taranto AG (2020) Structure-based virtual screening: from classical to artificial intelligence. Front Chem 8:343. https://doi.org/10.3389/fchem.2020.00343. https://www.frontiersin.org/articles/. Accessed on 10 Mar 2024

March-Vila E, Pinzi L, Sturm N, Tinivella A, Engkvist O, Chen H et al (2017) On the integration of in silico drug design methods for drug repurposing. Front Pharmacol 8:298

Mewes HW, Hani J, Pfeiffer F, Frishman D (1998) MIPS: a database for protein sequences and complete genomes. Nucleic Acids Res 26(1):33–37

Miró-Canturri A, Ayerbe-Algaba R, Smani Y (2019) Drug repurposing for the treatment of bacterial and fungal infections. Front Microbiol 10:41
Molsoft L.L.C. About Molsoft L.L.C. https://www.molsoft.com/about.html. Accessed 10 Mar 2024
Napolitano F, Zhao Y, Moreira VM, Tagliaferri R, Kere J, D'Amato M et al (2013) Drug repositioning: a machine-learning approach through data integration. J Cheminform 5(1):30
National Institute of Allergy and Infectious Diseases (NIAID) (2017) An open-label, proof-of-concept study to evaluate the safety and efficacy of dexpramipexole (KNS-760704) in subjects with hypereosinophilic syndrome. clinicaltrials.gov. Report No.: NCT02101138. https://clinicaltrials.gov/study/NCT02101138. Accessed 1 Jan 2024
Noguchi S, Kim HJ, Jesena A, Parmar V, Sato N, Wang HC et al (2016) Phase 3, open-label, randomized study comparing 3-monthly with monthly goserelin in pre-menopausal women with estrogen receptor-positive advanced breast cancer. Breast Cancer 23(5):771–779
Ogata H, Goto S, Sato K, Fujibuchi W, Bono H, Kanehisa M (1999) KEGG: Kyoto encyclopedia of genes and genomes. Nucleic Acids Res 27(1):29–34
Ogundeji AO, Pohl CH, Sebolai OM (2016) Repurposing of aspirin and ibuprofen as candidate anti-Cryptococcus drugs. Antimicrob Agents Chemother 60(8):4799–4808
Oprea TI, Overington JP (2015) Computational and practical aspects of drug repositioning. Assay Drug Dev Technol 13(6):299–306
Pinzi L, Rastelli G (2019) Molecular docking: shifting paradigms in drug discovery. Int J Mol Sci 20(18):4331
PubMed. https://pubmed.ncbi.nlm.nih.gov/. Accessed 10 Mar 2024
Rajasekharan SK, Lee JH, Lee J (2019) Aripiprazole repurposed as an inhibitor of biofilm formation and sterol biosynthesis in multidrug-resistant Candida albicans. Int J Antimicrob Agents 54(4):518–523
Rana R, Sharma R, Kumar A (2019) Repurposing of fluvastatin against Candida albicans CYP450 lanosterol 14 α-demethylase, a target enzyme for antifungal therapy: an in silico and in vitro study. Curr Mol Med 19(7):506–524
Research C for DE (2023) New developments in regulatory QSAR modeling: a new QSAR model for predicting blood brain barrier permeability. FDA [cited 2024 Mar 10]. https://www.fda.gov/drugs/regulatory-science-action/new-developments-regulatory-qsar-modeling-new-qsar-model-predicting-blood-brain-barrier-permeability
Royce ME, Osman D (2015) Everolimus in the treatment of metastatic breast cancer. Breast Cancer (Auckl) 9:BCBCR.S29268
Sahoo BM, Ravi Kumar BVV, Sruti J, Mahapatra MK, Banik BK, Borah P (2021) Drug repurposing strategy (DRS): emerging approach to identify potential therapeutics for treatment of novel coronavirus infection. Front Mol Biosci. https://www.frontiersin.org/articles/10.3389/fmolb.2021.628144
Saima LS, Sharma R, Kumar A (2024) Role of network pharmacology in prediction of mechanism of neuroprotective compounds. In: Neuroprotection: method and protocols. Springer US, New York, pp 159–179
Schein CH (2020) Repurposing approved drugs on the pathway to novel therapies. Med Res Rev 40(2):586–605
Seiler KP, George GA, Happ MP, Bodycombe NE, Carrinski HA, Norton S et al (2008) ChemBank: a small-molecule screening and cheminformatics resource database. Nucleic Acids Res 36(Database issue):D351–D359
Sgobba M, Caporuscio F, Anighoro A, Portioli C, Rastelli G (2012) Application of a post-docking procedure based on MM-PBSA and MM-GBSA on single and multiple protein conformations. Eur J Med Chem 58:431–440
Shah M, Patel M, Shah M, Patel M, Prajapati M (2024) Computational transformation in drug discovery: a comprehensive study on molecular docking and quantitative structure activity relationship (QSAR). Intell Pharm 2:589. https://www.sciencedirect.com/science/article/pii/S2949866X24000340. Accessed 10 Mar 2024
Shinde RB, Raut JS, Chauhan NM, Karuppayil SM (2013) Chloroquine sensitizes biofilms of Candida albicans to antifungal azoles. Braz J Infect Dis 17(4):395–400

Siles SA, Srinivasan A, Pierce CG, Lopez-Ribot JL, Ramasubramanian AK (2013) High-throughput screening of a collection of known pharmacologically active small compounds for identification of Candida albicans biofilm inhibitors. Antimicrob Agents Chemother 57(8):3681–3687

Smith KJ (2016) Computational approaches to fragment based screening. PhD thesis. University of Essex. https://repository.essex.ac.uk/17574/. Accessed 10 Mar 2024

Stemmler HJ, diGioia D, Freier W, Tessen HW, Gitsch G, Jonat W et al (2011) Randomised phase II trial of gemcitabine plus vinorelbine vs gemcitabine plus cisplatin vs gemcitabine plus capecitabine in patients with pretreated metastatic breast cancer. Br J Cancer 104(7):1071–1078

The Institute of Molecular and Translational Medicine, Czech Republic (2023) Phase II open labeled trial of disulfiram with copper in metastatic breast cancer. clinicaltrials.gov. Report No.: NCT03323346. https://clinicaltrials.gov/study/NCT03323346. Accessed 1 Jan 2024

Toniati P, Piva S, Cattalini M, Garrafa E, Regola F, Castelli F et al (2020) Tocilizumab for the treatment of severe COVID-19 pneumonia with hyperinflammatory syndrome and acute respiratory failure: a single center study of 100 patients in Brescia, Italy. Autoimmun Rev 19(7):102568

Torres PHM, Sodero ACR, Jofily P, Silva FP Jr (2019) Key topics in molecular docking for drug design. Int J Mol Sci 20(18):4574

Turner NC, Slamon DJ, Ro J, Bondarenko I, Im SA, Masuda N et al (2018) Overall survival with palbociclib and fulvestrant in advanced breast cancer. N Engl J Med 379(20):1926–1936

UCSF Chimera. Home Page. https://www.cgl.ucsf.edu/chimera/. Accessed 10 Mar 2024

UCSF Dock. Google Search. https://dock.compbio.ucsf.edu/. Accessed 10 Mar 2024

Wakharde AA, Halbandge SD, Phule DB, Karuppayil SM (2018) Anticancer drugs as antibiofilm agents in Candida albicans: potential targets. Assay Drug Dev Technol 16(5):232–246

Wall G, Herrera N, Lopez-Ribot JL (2019) Repositionable compounds with antifungal activity against multidrug resistant Candida auris identified in the medicines for malaria venture's pathogen box. J Fungi (Basel) 5(4):92

Wang Y, Fan G, Salam A, Horby P, Hayden FG, Chen C et al (2020) Comparative effectiveness of combined Favipiravir and oseltamivir therapy versus oseltamivir monotherapy in critically ill patients with influenza virus infection. J Infect Dis 221(10):1688–1698

Wishart DS, Feunang YD, Guo AC, Lo EJ, Marcu A, Grant JR et al (2018) DrugBank 5.0: a major update to the DrugBank database for 2018. Nucleic Acids Res 46(D1):D1074–D1082

Wu C, Gudivada RC, Aronow BJ, Jegga AG (2013) Computational drug repositioning through heterogeneous network clustering. BMC Syst Biol 7(Suppl 5):S6

Yu Q et al (2013) In vitro activity of verapamil alone and in combination with fluconazole or tunicamycin against Candida albicans biofilms. Int J Antimicrob Agents 41(2):179–182

Zhong L, Zhou L, Tian Y, You R (2016) Identification of novel IRAK-4 inhibitors through pharmacophore modeling, atom-based 3D-QSAR, docking strategies and molecular dynamics simulation. Lett Drug Des Discov 13(9):879–887

Zucchini G, Geuna E, Milani A, Aversa C, Martinello R, Montemurro F (2015) Clinical utility of exemestane in the treatment of breast cancer. Int J Womens Health 7:551–563

Pre-clinical and Clinical Studies, Pharmacovigilance, Pharmacogenomics, and Commercialization of Pharmaceutical Products

17

Mit Joshi and Bhoomika M. Patel

Abstract

Even with the development of new technologies and understanding of pathophysiology of diseases, drug discovery is still considered as a long and tedious process comprised of different stages which eventually leads to the availability of new drugs to the entire population. Development of new drug starts with research on target identification which generally takes place during academic research. Target identification and validation lead to the synthesis of new molecules through different in silico techniques such as high-throughput screening and other molecular modeling. New molecules are then examined for possible interaction with the target molecule which leads to identification and validation of new molecule. New molecules then undergo pre-clinical and clinical studies. Pre-clinical evaluations mainly involve the assessment of the safety and efficacy of novel molecules in vitro and in vivo setup before going into the clinical phase. In the clinical phase, new drug molecules are scrutinized for their safety and efficacy in human patients. After successful clinical trials, the new drug is allowed for commercialization in the market. Clinical trials assess the safety and efficacy of new drugs in a relatively small population compared to the entire population. When a drug enters the market, a pharmacovigilance program is initiated to track and report any adverse drug reaction associated with the drug. Sometimes, certain drug does not follow the expected ADME due to variation in the genetic build of a particular population. Due to that, pharmacogenomic assessment of each patient is important to personalized medication to certain

M. Joshi
Department of Pharmacology, Institute of Pharmacy, Nirma University, Ahmedabad, India

B. M. Patel (✉)
National Forensic Sciences University, Gandhinagar, Gujarat, India
e-mail: bhoomika.patel@nfsu.ac.in

K. Jain, A. K. Yadav (eds.), *Advances in Pharmaceutical Product Development*, https://doi.org/10.1007/978-981-97-9230-6_17

patient populations. Drug discovery is a vast process and each step has its advantages and disadvantages. Here we discuss pre-clinical and clinical studies, along with pharmacovigilance and pharmacogenomic programs. In the end, we discuss the commercialization of pharmaceutical products which is different from any other product available in the market.

Keywords

Clinical trials · Pre-clinical · Pharmacovigilance · Pharmacogenomics · Toxicity · Efficacy

17.1 Introduction

The need for new drug candidates arises in two scenarios: First is a medical condition or disease or the emergence of a new disease such as COVID-19 where no suitable treatment or drug is available. Second, there is an unmet medical need, or currently available intervention is not enough to mitigate the disease (Hughes et al. 2011).

The primary or basic research, most carried out in academia, provides hypotheses regarding the potential pathophysiology and identifying particular targets. The result of this research further requires proper validation before beginning the next phase of lead discovery to support the drug discovery effort (Mohs and Greig 2017).

Despite significant achievements made in unraveling the pathophysiology of biological systems at the molecular level and substantial progress in the innovation of novel technologies, the journey of drug discovery remains extensive and resource-intensive. The process of drug development is characterized by prolonged duration, substantial expenses, and a notable rate of setbacks. To address these challenges, innovative methodologies, such as artificial intelligence (Paul et al. 2021) and groundbreaking in vitro technologies, are being harnessed to accelerate research and development efforts, aiming to deliver novel medications to patients in a more efficient manner.

Developing a new drug is time-consuming and tedious and divided into several segments. Drug development starts with genomic and proteomic studies with target identification and validation (Schenone et al. 2013). This data leads to the synthesis of a drug or a molecule and its optimization (Bruno et al. 2019). The discovery of the molecule was further investigated through pre-clinical (in silico studies, toxicity studies, efficacy studies) and clinical trials (Steinmetz and Spack 2009). When the drug is approved for commercial use, it is studied and tracked in a large population through a pharmacovigilance program (Kalaiselvan et al. 2019). The entire process takes around 10–15 years (Fig. 17.1).

Every category is further divided into many subcategories like drug discovery divided into basic research, target identification and validation, discovery and identification of hit and lead, optimization of lead, pre-clinical studies, and clinical studies.

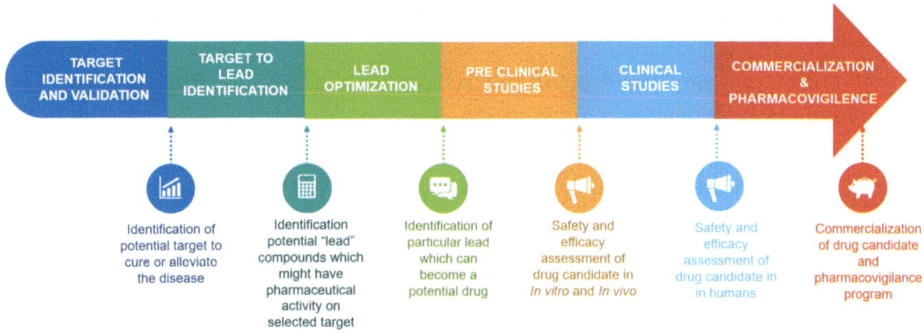

Fig. 17.1 Process of drug discovery

In this chapter, we will discuss pre-clinical and clinical studies, pharmacovigilance, pharmacogenomics, and commercialization of pharmaceutical products.

17.2 Pre-clinical Evaluations

During the beginning stage of drug discovery, the activity and specificity of the candidate molecule were evaluated first followed by pharmacokinetics and toxicity studies, although many lead molecules failed at the pre-clinical evaluation due to failure in achieving safety and efficacy of lead possibly due to failure in drug absorption, distribution, metabolism, and excretion (ADME).

Pre-clinical studies can be divided into two parts: exploratory and confirmatory. Exploratory studies mainly involve the exploration of potential hypotheses that are scientifically driven and explore either the pathophysiology of the disease or identification of potential drug targets. Exploratory studies are often not regulated as their sole purpose is to explore the disease progression and identify the potential target, although computational technologies and in silico studies were developed to avoid pre-clinical in vivo exploratory studies to minimize research waste. Confirmatory studies mainly involved confirmation of efficacy and safety of potential drug candidates. These studies must be properly designed and follow different regulatory guidelines. Data achieved from confirmatory studies must be properly analyzed, stored, and reported to the highest regulatory authorities with transparency (Huang et al. 2020).

With the rapid progression in the advancement of techniques such as high-throughput screening, computational biology, and combinatorial chemistry, the synthesis of different molecules has become easy and fast. Different methods to evaluate the ADME of drug candidates increased rapidly. Many in silico methods successfully predicted the ADMS profile of the drug, and due to that, in silico methods replaced in vivo exploratory and ADME methods.

17.2.1 In Vitro Pharmacological Studies

Over the past few years, the drug discovery approach of the new drug candidate has been largely dependent on evaluating the compounds for their biological activity and establishing structure-activity relationships, with assessment for better drug efficacy, based on different experiments in moralized and immortalized cell lines as the first pre-clinical step which is easy and easily replicable (Perego et al. 2018; Ziegler et al. 2000). Using monolayer culture cells which are the easiest and replicable in vitro models, the mechanism of action of novel drugs and their toxicity can be easily measured (Phillips 2016). Advancements in high-throughput screening methods and incorporating genomics and proteomics approaches are improving the pace of pre-clinical drug evaluation from conventional to target-based approaches, focusing on targeting gene as well as protein-level biomarkers for the sensitivity of drugs and their resistance (Roma-Rodrigues et al. 2019). In the context of cancer drug discovery, new insights into cellular and molecular mechanisms of different cancers have led researchers to use different kinds of normal as well as tumor cell models to efficiently screen new potential anticancer drug molecules, including cell lines having particular mutations, stem cells, and endothelial cells (Hamburger and Salmon 1977; Rhim et al. 1998). Pathophysiology and molecular understanding between complex interaction of cancer cells and the surrounding microenvironment are not well understood. Interaction between cancer cell and microenvironment gave rise to metastatic events. How much tumor cells influence their microenvironment is an important factor in developing new drug strategies, and for that development of a new kind of in vitro system is required (Whiteside 2008).

Two pivotal pharmacological determinants governing cellular responses encompass the drug's concentration (C) and the duration of its exposure (T). In the context of traditional cytotoxic agents, cell proliferation and survival within two-dimensional (2D) monolayer cultures ordinarily correlate with the product of $C \times T$. However, this relationship may deviate for drugs exhibiting cell-cycle phase specificity, wherein the cellular response beyond a certain concentration threshold tends to be directly proportional to the exposure duration. To capture temporal alterations in response and suit the context of plate-based experiments, an array of time-dependent measurement techniques can be employed. The attainment of consistent, pertinent, and replicable outcomes hinges on the discerning selection of a suitable assay and experimental framework (Kavallaris 2010; Capula et al. 2019).

17.2.2 In Vivo Toxicity Studies

The main goals of initial in vivo preclinical toxicity assessments for drug candidates are to identify possible untoward outcomes and provide preliminary safety margin assessments. These evaluations guide the suitability of a compound for human application or the necessity to exclude it from subsequent developmental stages. Furthermore, additional investigations can be carried out to find possible molecular mechanisms behind observed toxic effects, thereby changing the strategies in the preclinical drug development phase and evaluating possibilities for human hazard evaluation (Gross and Kramer 2003; Greaves et al. 2004). Based on the industry inputs, toxicity data from rodents anticipate only 43% of human toxic reactions, whereas non-rodent inquiries foresee 63%. Interestingly, the synergy arising from considering both these species escalates the predictability to approximately 71% for human toxicities. Interpretation of toxicity data is a straightforward task, predicting human response based on these data remains controversial (Olson et al. 2000). This difference in interpretation arises mainly due to biological as well as pathological differences between species along with intrinsic variations in pharmacokinetic aspects like absorption, distribution, metabolism, and excretion properties. Many studies have documented variations in toxicity profiles where toxicity was present exclusively in rodents in drug metabolism (Mutlib et al. 2000), thyroid hormone regulation (Wu and Farrelly 2006), or the formation of bladder tumors (Cohen and Lawson 1995). Enhancing human risk assessment and the judicious selection of viable drug candidates for developmental pursuits goes beyond simply advancing regulatory toxicity investigations during the early stages of discovery or conducting tests across a wider array of models. Instead, it hinges upon cultivating a deeper understanding of the intricacies of toxicity mechanisms. This proactive investment should yield substantial returns by facilitating a more astute utilization and interpretation of both in vivo and in vitro models (Fielden and Kolaja 2008).

As of now, the Organization for Economic Co-operation and Development (OECD) provides guidelines to conduct toxicity studies of different drug candidates. It includes Acute Oral Toxicity, Acute Dermal Toxicity, Acute Inhalation Toxicity, Acute Dermal Irritation/Corrosion, Acute Eye Irritation/Corrosion, Skin Sensitization, Repeated Dose 28-day Oral Toxicity Study in Rodents, Repeated Dose 90-day Oral Toxicity Study in Rodents, Repeated Dose 90-Day Oral Toxicity Study in Non-Rodents, Repeated Dose Dermal Toxicity: 21/28-day Study, Subchronic Dermal Toxicity: 90-day Study, Subacute Inhalation Toxicity: 28-Day Study, Subchronic Inhalation Toxicity: 90-day Study, Prenatal Development Toxicity Study, One-Generation Reproduction Toxicity Study, Two-Generation Reproduction Toxicity, Toxicokinetics, Acute Oral Toxicity: Up-and-Down Procedure, and many more (Buschmann 2013).

17.2.3 In Vivo Efficacy Studies

In vivo, efficacy studies mainly involve the assessment of pharmaceutical drugs in biological systems except humans. In vivo, efficacy studies mainly involve the induction of a particular disease into the animals and then treating the disease with investigational drugs. As of now, animals such as mice, rats, rabbits, dogs, zebrafish, and fruit flies are the most commonly used animals to evaluate toxicity and effectiveness of drugs due to their somewhat resemblance to human anatomy and physiology. However, it remains controversial regarding the translation of animal data into human data. Regarding conducting the efficacy studies, the Institutional Animal Ethics Committee (IAEC) is formed which provides approval and guidelines for public participation in the regulation of animal research. IAEC decides whether the study requires in vivo testing or not and if required then how many numbers of animals are required. The three Rs (3Rs) are another ethical principle coined by W.M.S. Russell and R.L. Burch for the utilization of animals in experiments. The 3Rs are replacement, development of alternative procedures to replace the use of animals in experiments; reduction, use of protocols that allow the generation of more meaningful data with the use of fewer animals; and refinement, use of protocols that alleviate or minimize distress, pain, and discomfort to animals while enhancing the comfort and welfare of animals used in experiments.

The 3Rs have a broad value and meaning to improve animal care and quality of research and encourage researchers to use alternatives to avoid excessive use of animals. In many countries, these 3Rs are now introduced into the governing body for animal use (Lewis 2019).

Many different animal models have been developed over the years for almost all diseases. With the progression of genetic and molecular insights into each disease, different gene knock-in and knock-down models were developed to understand the pathophysiology in a better way (Robinson et al. 2019).

17.3 Clinical Evaluations

Clinical research is a study of investigational drugs. In some cases, it includes both healthy volunteers and patients. Types of clinical trial can broadly divide to observational trial and clinical trials.

Observational inquiries meticulously observe individuals within their natural circumstances. Investigators gather insights from individuals and scrutinize shifts that unfold over time. To illustrate, a cohort of seniors might be queried about their physical activity routines, followed by a year of monthly memory assessments. This endeavor aims to unveil the correlation between physical engagement and cognitive well-being. While observational studies abstain from testing medical interventions like pharmaceuticals or devices, their outcomes can potentially unearth innovative treatments or prevention tactics warranting assessment in clinical trials.

17.3.1 Clinical Trial Phases

17.3.1.1 Phase 0

The main difference between phase 0 and another phase trial in the regard of traditional Investigational New Drug (IND) application lies in the absence of therapeutic intent within phase 0 trials. Both types of subjects, healthy individuals and patients, are involved in the study and exposed to doses of an intervention that are subtherapeutic yet still pharmacologically active. The primary objective of these trials is not to evaluate the toxicity of drug, unlike other traditional trials. In phase 0 trials, the focus remains on administering doses that are below the level of expected therapeutic effects. While the exposure of the drug remains limited to subjects, flexibility exists for dose escalation. The anticipation of significant adverse events related to the drug is low due to the use of lower doses. This specific advantage was recognized by many regulatory bodies, allowing for a more regulated approach to preclinical toxicology studies. Instead of necessitating extensive dose-limiting toxicity studies, phase 0 trials remain dependent on single dose or short-term treatment or short-term toxicity assessment. Moreover, the reduced quantity of the investigational drug required for phase 0 trials does not require the need for full-scale production and clinical-grade manufacturing that adheres to good manufacturing practices (GMP) before the commencement of the trial. Consequently, this advantage allows the initiation of phase 0 trials at a much earlier stage compared to the traditional phase I studies. This early initiation allows researchers to understand the pharmacokinetics (PK) and the effects on drug targets in human participants at an earlier developmental stage of the therapeutic agent. In summary, the requirement of phase 0 trials arises from their non-therapeutic intent, employing subtherapeutic yet pharmacologically active doses in a limited exposure setting, accompanied by the graded dose escalation. Lower chances of significant adverse events allow the utilization of condensed pre-clinical toxicology studies. Furthermore, the minimal quantity of study drugs needed for these trials allows for their initiation ahead of traditional phase I studies. This, in turn, grants researchers an invaluable opportunity to delve into the pharmacokinetic and drug target effects in human subjects much earlier in the trajectory of clinical development (Kummar et al. 2008).

17.3.1.2 Phase I

Phase I clinical trial is carried out primarily to identify the recommended dose for phase II. It is commonly carried out by dose escalation and achieves the maximum tolerated dose (MTD) using different methods such as 3 + 3 algorithms and continuous evaluations. Phase I trials are very small, and a maximum of around ten patients are treated with MTD. To gather more data on the toxicity and efficacy of the intervention, phase I trials consider the inclusion of "expansion cohorts" that allow for the treatment of additional participants at the MTD. In terms of efficacy, meaning undesirable clinical effects, several factors are taken into consideration after administration of the drug. The conventional approach in the phase I trial is to achieve a minimum recommended dose (P2RD) by focusing only on toxicity outcomes from a limited sample size, while not focusing on the integration of efficacy within the dose-finding algorithm, leading to certain unfavorable ramifications. In a recent advancement, the number of patients in expansion cohorts has dramatically increased with 100–200 in some trials. A large "phase I expansion cohort" can be considered a phase II study without any particular design and comparison with a placebo or any other intervention. This approach can eliminate the problems associated with small cohorts. Phase I can be considered as a phase II trial, but carried out without having any particular plan and design other than a fixed sample size. This method mitigates the problem associated with small expansion cohorts. The MTD easily may turn out to be either toxic or ineffective.

The main focus of this clinical trial and drug discovery is to select the right dose for the large population. To achieve this, modifications are required from time to time in terms of patient selection, trial design, cautiously selected objectives, endpoints, and inclusion and exclusion criteria. Future efforts should be made in the direction of improving patient selection, along with identifying the mechanism of action of intervention and minimizing the adverse effects of the drug (Muglia and Digiovanna 1998).

17.3.1.3 Phase II

The objective of the phase II clinical trial is to identify the efficacy of the intervention that can then proceed to further scrutiny in the conclusive phase III trials.

Randomized arrangements in phase II trials are classified within three distinct frameworks:

Parallel noncomparative exploration: This category involves the random assignment of participants to separate experimental regimens, each with its independent decision-making criteria.

Randomized selection: Within this design, various experimental regimens are presented, and the selection of the most auspicious among them is randomized.

Randomized screening design: This type facilitates the comparison between an experimental regimen and the established standard of care through randomized allocation.

In a current scenario with limited availability of patients and resources with several potential therapies for various diseases, the utilization of randomized phase II designs is surging in popularity. This mainly has a role in the field of cancer drug development, where the abundance of prospects calls for astute clinical trial strategies. By employing randomized phase II designs in a manner consistent with their foundational inference principles, the judicious allocation of phase III financial and patient resources can be ensured (Van Norman 2019).

17.3.1.4 Phase III

Based on the previous findings that have demonstrated the safety and efficacy of the intervention, a phase III trial is sometimes referred to as a "therapeutic confirmatory," "comparative efficacy," or "pivotal" trial carried out. This particular phase of drug evaluation occurs within a broader and often more diverse population. Its primary objectives are to validate safety and effectiveness, as well as the determination and estimation of possible adverse reactions.

It is noteworthy that phase III trials while encompassing a range of 300–5000 participants have more statistical power compared to phase II studies. This statistical power allows for the establishment of adverse event rates not exceeding 1 in 100 individuals, in alignment with Hanley's "Rule of 3." This prominently proves the importance of phase IV trials in the detection of infrequent adverse drug reactions (Eypasch et al. 1995). Consequently, it's apparent why the FDA commonly requires multiple phase III trials as a prerequisite for the establishment of both the safety and efficacy of a drug.

The different types and designs of phase III trials, often recognized as "superiority" or "placebo-controlled trials," focused on comparative efficacy outcomes. These trials compare the new intervention of interest with either an established standard therapy or a placebo. A distinctive feature of the phase III trial framework lies in the balance achieved in the allocation of treatments through randomization to evaluate the treatment effectiveness unbiasedly (Cahana and Romagnioli 2007; Foddy 2009; Wilcox 2008).

The preferred method to evaluate the phase III trials is the intention-to-treat analysis. This analytical approach evaluates participants based on the treatment arm to which they were initially randomized, regardless of the actual treatment they eventually received. This principle, often referred to as the "analyzed as randomized" rule, serves to uphold the integrity of randomization and eradicate the influence of selection bias. Consequently, any observed disparities in outcomes can be directly correlated to the treatment itself rather than any confounding factor.

Another type of analysis is termed "as-treated" or "per-protocol." Here, participants are assessed according to the treatment they received, irrespective of their initial randomization. This approach, while offering valuable insights, diverges from the intent of randomization and can, in essence, approximate the conditions of an interventional cohort study, which introduces the potential for treatment selection biases (Tsiatis 1990).

To further minimize the intended biases, the phase III trial can be designed and designated as "blinded" (or masked) where the investigator or patient does not know about the intervention arm. Specific blinding strategies such as "single blinding" (subject only), "double blinding" (both subject and investigator), or "triple blinding" (data analyst, subject, and investigator) can be applied based on the study requirement and study design. Unfortunately, it is not possible to blind all the trials. To design a phase III trial, phase III study, the Consolidated Standards of Reporting Trials (CONSORT) guideline was created to improve the design of the trial while minimizing intended and unintended biases (Umscheid et al. 2011) (Fig. 17.2).

17.4 Pharmacovigilance

The World Health Organization (WHO) has defined pharmacovigilance as the "science and activities relating to the detection assessment, understanding, and prevention of adverse effects or any other possible drug-related problems." The responsibility to establish proper guidelines was taken up by the Council for International Organization of Medical Sciences (CIOMS). Initially founded in 1949 as an independent body for the worldwide communication of medical science knowledge, CIOMS underwent reconfiguration in 1966 to create fundamental concepts, regulations, and protocols concerning the safe utilization of pharmaceutical products (medications and vaccines). Employing a collaborative network involving a worldwide working group comprised of medical and paramedical professionals from governmental, academic, and industry, CIOMS undertook the resolution of expanding the scope of pharmacovigilance. Pharmacovigilance mainly includes

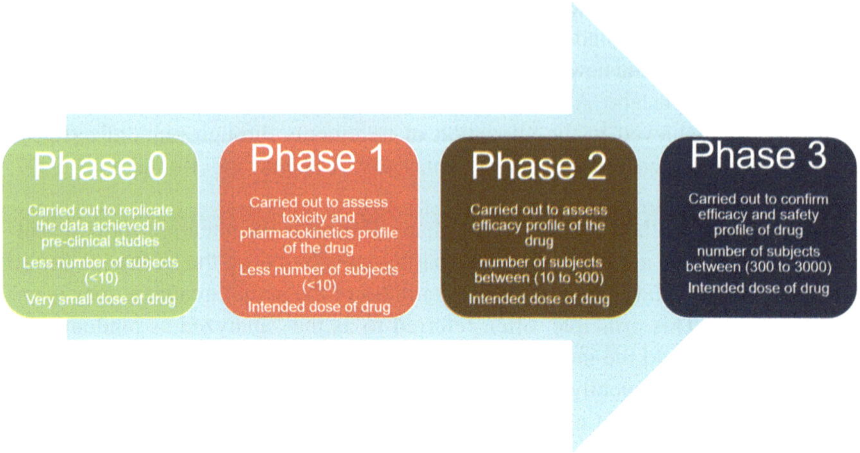

Fig. 17.2 Different phases of clinical trials

adverse drug reactions, intermittent reporting, risk-benefit evaluations, safety documentation during clinical trials, signal recognition, and risk mitigation (Tsintis and La Mache 2004).

Thus, the three main objectives of PV are case management, signal management, and benefit management.

Activities and task currently included in the PV domain are monitoring clinical trials to ensure the safety of the patient, identifying the safety profile of the drug in regard to selecting first safe dose for humans, communication regarding safety and progress in clinical trials to stakeholders, determining the casual relationship between drug and adverse effects, monitoring issues related to drug safety and quality of manufactured drugs, and preparing and carrying out inspection in various departments.

It is important to keep in mind that pharmacovigilance has broadened the data and knowledge of pharmacological drugs. It is an important addition to the lack of knowledge of adverse drug reactions during clinical trials. Reporting and analyzing those adverse drug reactions increase our understanding of the mechanism of drugs, their efficacy on different populations, and their impact on genetics. Pharmacovigilance remains in a state of constant modifications and is adapted finely in the context of drug safety and also in personalized medicine.

Pharmacovigilance mainly deals with detecting and reporting adverse drug reactions (ADRs). Adverse drug reactions can be defined as unintended or obnoxious drug responses that take place at the normal dose levels intended to be used as therapy, diagnosis, or prophylaxis.

ADR can be divided into six different categories:

Type A: Dose-dependent effect mainly due to exaggerated pharmacological response
Type B: Not dose-dependent on dose and occurs in a small number of populations

Type C: Chronic effect associated with the development of spontaneous disease
Type D: Carcinogenicity, delayed effect, and teratogenicity
Type E: ADR reported after the end of treatment
Type F: Failure of treatment (Zazzara et al. 2021)

17.4.1 Role of PV in the Drug Development Process

Direct involvement of PV professionals in the clinical trials generally fell out scope of PV activities although based on their field experience, they could be strategically positioned to assume a vital role in comprehending the broader context and ensuring effective communication with all stakeholders regarding ongoing advancements.

The process of gathering essential data to establish and uphold drug safety profiles throughout a product's life cycle is generally consistent on a global scale. The uncertainties inherent in post-marketing encounters serve as cues that underscore the overarching realization that safety issues linked to drugs can manifest unexpectedly and from diverse sources. PV professionals are tasked with upholding a state of constant "risk-awareness" concerning both the intended and unintended impacts of their drugs.

The procedural aspects associated with these documents are typically, though not universally, overseen by a company's regulatory affairs department. However, the accountability for the substance of these documents lies with the departments responsible for pre-clinical animal studies, clinical indications, adverse events, and product packaging. Within this framework, PV assumes a proactive role in suggesting modifications to the product's safety profile. This necessitates the establishment of robust governance structures and processes to effectively advocate for changes in the content, format, and placement of such proposals within the relevant user documents.

The contrasting regulatory strategies employed by the FDA and the EMA necessitate PV professionals to be exceptionally vigilant and perceptive in the face of emerging signals. Each new signal's implications may diverge depending on the regulatory jurisdiction, demanding distinct approaches to management (Beninger 2018).

17.4.2 Clinical Trial Designs

During the era of evidence-grounded medicine, randomized controlled clinical trials hold a paramount status in enlightening clinicians regarding the relative effectiveness and safety of treatments. The insights emanating from a clinical trial hold merit only when the trial itself is meticulously crafted. This entails an explication of the research queries and a precise delineation of fitting criteria for participants, outcomes, and study frameworks. These measures serve to mitigate distortions and extraneous influences. It's important to note that no solitary trial blueprint reigns supreme across all conceivable scenarios. Hence, a robust comprehension of the merits and demerits intrinsic to diverse clinical trial architectures becomes imperative.

17.4.3 Randomized Controlled Trials

Randomized controlled trials (RCTs) are type of a study design that evaluates the safety and efficacy of novel interventions or treatments. While individual studies might not independently establish causation, randomization of subjects in the trial can effectively mitigate biases, offering a robust mechanism to scrutinize the causal links between interventions and outcomes. This arises from the fact that the process of randomization can divide the characteristics of subjects in a balanced way among different groups.

During the process of designing the RCTs, the type of subjects, different interventions, and desired outcomes should be taken into consideration carefully. After considering these factors, the next step should be a selection of the total number of required participants which can robustly establish the presence of any potential connections (conducted through power calculation). Subsequently, participants are enlisted and arbitrarily divided into either the intervention cohort or the comparator group (Torres-Saavedra and Winter 2022).

17.4.3.1 Parallel Arm Design

In the parallel arm design, participants are assigned randomly to one or more study arms, with each branch being linked to a unique intervention. After the randomization procedure, each participant remains exclusively associated with their designated treatment arm throughout the entire study period. This parallel group design can be used in any disease condition, and many studies can be carried out simultaneously. This study design requires strong vigilance to avoid unintended co-interventions or cross-overs.

17.4.3.2 Cross-Over Design

In the cross-over design, each subject receives both treatments at different time intervals. For example, subjects receiving treatment A will be receiving treatment B after the washout period, and subjects receiving treatment B will be receiving treatment A after the washout period. A washout period must be present to eliminate any effect of the first intervention and avoid synergic and antagonist interaction between the two interventions. In this method, the outcome can be compared in the same subject as well as between the groups. One of the important requirements to design cross-over RCT is that the disease must be chronic (e.g., diabetes, hypertension) and the effect of the drug should be washed out before starting the next intervention.

17.4.3.3 Randomized Withdrawal Design

In this design, all the subjects will receive the intervention, and no randomization was carried out. Subjects who responded to therapy will be considered in the study, while non-responders were left out of the study. Responders were randomized in different arms and a study was carried out. Analysis of the data mainly depends upon the withdrawal phase, wherein the anticipated result commonly encompasses the resurgence of symptoms. The withdrawal phase serves to enhance the statistical analysis within a predetermined sample size. In a randomized withdrawal design, the primary objective is to evaluate the effect of an intervention for a particular duration. An important benefit of this approach is to minimize the use of placebo interventions, as only responders are enrolled in the study. This design ensures more acceptability among trial participants, thus increasing subject recruitment. Ultimately, this design stands poised to ascertain whether the ongoing administration of treatment is imperative or if cessation can be contemplated.

17.4.3.4 Factorial Design

By employing this design, multiple objectives can be achieved in a single trial with efficacy using minimum and constant sample size. In a 2 × 2 factorial design, involving a placebo, patients are subjected to randomization across four distinct groups: (1) treatment A alongside placebo; (2) treatment B alongside placebo; (3) a combination of treatments A and B; or (4) a placebo-only scenario without either treatment. Outcomes were analyzed using a two-way analysis of variance (ANOVA), pitting patients who receive treatment A (groups 1 and 3) against those who do not (groups 2 and 4). Likewise, a comparison is made between patients receiving treatment B (groups 2 and 3) and those not receiving it (groups 1 and 4) (Nair 2019).

17.4.4 Types of Randomization in RCTs

17.4.4.1 Stratified Randomization

Stratified randomization is a two-step process in which subjects entering clinical trials are first assigned to strata based on their clinical features that may influence the outcome of the intervention. Within each stratum, subjects are assigned to intervention based on the randomization. For example, patients over the age of 50 years may not be randomized with patients under 50. Although stratification is mostly used for RCTs, their significance in trial and outcome is not fully understood (Kernan et al. 1999).

17.4.4.2 Block Randomization

Complete randomization may produce uneven sample allocation, for example, randomizing all subjects receiving intervention in a single batch and all the subjects receiving placebo in another batch. Due to this, both intervention and batch are confounded, and it became very difficult to analyze the data concerning the treatment. In such kind of scenario, it is not possible to simply adjust the batches and subjects until they become balanced. This method will remain unreproducible and contain unintended biases. A structured way is to tackle the problem using block randomization. The method of block randomization involves different groups of different sizes exhibiting very minor differences. The initial step involves the creation of blocks of samples, ensuring proportional representation within each group.

For instance, in cases where one group outnumbers the other by a factor of two, every block would encompass three subjects: one from the smaller group and two from the larger group (Yang et al. 2019).

17.4.4.3 Cluster Randomization

The typical approach in most clinical trials involves the random assignment of participants as individuals to various treatment arms. Yet, there are situations where individual allocation isn't feasible or preferable. In these cases, the randomization occurs at a group level, which is referred to as cluster or group randomization. The primary units of randomization are groups of participants, which could encompass entities like general practices or hospital wards. Additionally, periods can also constitute clusters. For instance, segments of time, like weeks within a year, might be subjected to randomization. This allows for the introduction of an intervention during 1 randomly selected week, followed by its withdrawal during another week designated as the control (Hauck et al. 1991).

17.5 Pharmacogenomics

The concept of pharmacogenomics is based on the fact that different types of populations have diverse genetic variations across the genome and play an important role in influencing individual reactions to medications. The International Conference on Harmonization has officially defined pharmacogenomics as the investigation into differences within DNA and RNA attributes concerning response to pharmaceutical drugs. Additionally, pharmacogenetics, within the same framework, is recognized as the exploration of discrepancies within DNA sequencing in connection to drug responsiveness (Burt and Dhillon 2013).

The efficacy of drug dosages experienced by certain patients might not translate uniformly, inevitably leading to either ineffectiveness or the emergence of adverse drug reactions (ADRs) in others. These ADRs have been identified as a significant factor contributing to hospital admissions, as one study demonstrated that they accounted for 6.5% of all hospitalizations across two major UK medical facilities. Genomic variation leads to different drug responses and can be divided into pharmacokinetic and pharmacodynamic variations (Roden et al. 2018).

17.5.1 Pharmacokinetic Gene Variation

Two different scenarios have shown how gene variation between different populations can affect the efficacy of the drug and ADR. The first scenario is a prodrug. Prodrugs must be metabolized into the body and converted into the active form which exerts pharmacological effects. For example, codeine must be converted into the morphine by CYP2D6 enzyme, while clopidogrel must be converted into its bioactive form by CYP2C19 enzymes. It was found that in certain populations, deficiency of CYP2D6 is present as an autosomal recessive trait and classified as poor metabolizers. Due to a lack of CYP2D6, codeine cannot convert into its active form, and therefore, no analgesic activity was achieved. Many other drugs such as β-blockers (metoprolol, timolol, propranolol), antiarrhythmic drugs (flecainide, propafenone), antidepressants (amitriptyline, clomipramine, fluoxetine, nortriptyline), and neuroleptics (perphenazine, haloperidol) cannot be metabolized due to CYP2D6 deficiency, leading to unwanted drug response (Bertilsson et al. 2002).

In the second scenario, gene variation in DNA or RNA may have resulted in an extremely high effect of the drug which has a narrow therapeutic window. For example, 6-mercaptopurine, an anticancer drug, is inactivated by two enzymes TPMT and xanthine oxidase. Changes in TPMT variants may result in the inhibition of the inactivation of the drug which can cause higher drug concentration in blood and higher levels of cytotoxic thioguanine nucleotide metabolites resulting in the incorporation of these metabolites into DNA and association with a drug response (Relling et al. 2019). Similarly, the toxic effects of 5-fluorouracil and other fluoropyrimidines such as capecitabine were associated with the loss-of-function of DYPD variants (Henricks et al. 2018).

17.5.2 Pharmacodynamics Gene Variation

Variations in the pharmacodynamics of the drug can also influence the drug response. One study identified that overexpression of wild-type VKORC1 is involved in increased levels of VKOR activity which causes inhibition of warfarin activity (Rost et al. 2004). A retrospective study identified at least five major haplotypes and ten common noncoding single-nucleotide polymorphisms in VKORC1. VKORC1 haplotypes were used to stratify the patients in different groups (low-, intermediate-, and high-dose groups) and showed different dose requirements in patients with different ancestries (Rieder et al. 2005). Another study revealed a high prevalence of CYP2C92 (p.R144C), CYP2C93 (p.I359L), and VKORC1 promoter (g.-1639G/A) polymorphisms in patients classified as "sensitive to warfarin" necessitating less dosages. Conversely, patients exhibiting VKORC1 missense mutations demonstrated resistance to warfarin, demanding higher doses for efficacy. To compare allele and genotype frequencies of CYP2C9 and VKORC1 in a cohort of 260 Ashkenazi (AJ) and 80 Sephardi Jewish (SJ) individuals, Tag-It Mutation Detection Kit and PCR-RFLP assays were carried out to determine the presence of six CYP2C9 alleles and eight VKORC1 alleles. The results showed that around 85% of individuals belonging to the AJ group and approximately 90% of those in the SJ group possess at least one "sensitive" allele (CYP2C9*2, *3, VKORC1 g.-1639G/A) or a "resistant" allele (VKORC1 p.D36Y). This suggests a clear divergence in the warfarin pharmacogenetics between these two groups, highlighting the potential advantages of dosage predictions based on genotype information for each group (Rieder et al. 2005).

17.5.3 Pharmacogenomics and Diagnosis

Before applying pharmacogenomics to hospital setup, genetic testing must comply with certain parameters encompassing their analytical and clinical efficacy and accuracy. Establishing accurate genetic tests presents a complex challenge, particularly when applied to pharmacogenes such as CYP2D6. This complexity arises from the susceptibility of the gene to inherent variations in copy number within the germline, along with the intricate emergence of hybrid gene fusions, both of which pose difficulties in terms of accurate measurement and accurate interpretation.

Dimension of clinical effectiveness encompasses a twofold assessment: firstly, the determination of whether the employment of the genetic test directly contributes to enhanced health outcomes in subjects undergoing testing and, secondly, an evaluation of the potential risks arising from the testing process. The matter of precisely which outcomes signify clinical efficacy is an area marked by significant divergence in viewpoints. Certain investigations have expanded their purview to encompass the broader impact of testing on the entire healthcare ecosystem. This includes a comparative analysis of the expenses linked with genetic testing against alternative healthcare interventions, alongside an exploration of how such testing can influence the conduct of healthcare.

Multiple factors are taken into consideration regarding the analytical and clinical effectiveness and validation of a pharmacogenomic test. This substantiation further justifies its implementation in writing medication prescriptions. The determination of analytical validity highly depends upon the quality of genetic test data and the operational performance attributes of the test, containing both positive and negative predictive values. Data received from various sources can be employed to appraise both validity and clinical use. This contains the extent to which genetic variation impacts the effectiveness of drugs, deduced from retrospective studies. Furthermore, the pathways through which this genetic phenotype affects the efficacy of drugs, or a pertinent endophenotype (an intermediary phenotype like drug-metabolizing enzyme activity), can also be enlisted for assessment.

A pivotal aspect to consider in evaluating the practicality of interaction between drug and gene lies in the presence of an alternative therapeutic option, which, to some extent, hinges on the underlying mechanism of the gene-drug linkage. In instances where the gene's influence pertains to a drug's pharmacokinetics (as seen in the case of CYP3A5-mediated breakdown of the immunosuppressive drug tacrolimus), substantial documented evidence might endorse dose adjustments analogous to those based on factors like age or kidney/liver function. This is especially justifiable when the drug's blood concentration can be conveniently monitored, as is the case with therapeutic drug monitoring (Crews et al. 2014).

Decisions to adjust dosages become particularly defensible if genetic tests suggest ineffectiveness or potential adverse reactions specific to a certain genotype. Recommendations for alternative therapies in such situations depend on weighing the evidence for both the efficacy and potential toxicity of the substitute. For instance, individuals homozygous for nonfunctional CYP2D6 alleles cannot convert the analgesic drug codeine into its active form, morphine, yet they can respond to various other opiate analgesics. In scenarios where genetic tests show a notably elevated risk of a severe adverse event—such as carriers of the HLA-B*57:01 allele being highly susceptible to hypersensitivity reactions with the antiretroviral drug abacavir—the alternate therapy must be equally effective while carrying an acceptable risk of adverse effects, which might be influenced by other genetic variations or not (Relling and Evans 2015; Mallal et al. 2008).

17.5.4 Implementation in a Clinical Setup

The primary reason for difficulties in adopting pharmacogenomics in a clinical setup is the lack of proper guidelines regarding the practicality of gene variations in a clinical setup. Many healthcare professionals, regulatory bodies, and healthcare societies remain in dilemma regarding the use of pharmacogenetic testing and, if so, how. Examples include clinical use of warfarin and clopidogrel.

17.6 Commercialization of Pharmaceutical Products

The final part of drug development is the commercialization of drugs into the market. Regarding the target audience of pharmaceutical drugs, the company targets medical professionals who can prescribe drugs to patients and the targeted patient group who can choose to take drugs. To achieve success with a new drug, companies have to make strategies to target potential medical professionals and patients. This could encompass a diverse range of considerations, spanning from strategizing product launches and ensuring logistical and operational preparedness to navigating the impact of payer preferences. While the specifics might differ among pharmaceutical companies, the general drug development and commercialization journey usually adheres to two main phases: the pre-launch strategies and the post-launch strategies.

During the pre-launch strategy, the main focus for commercialization of the drug revolves around securing market entry. Following the FDA's approval, manufacturers must start shaping their brand image and communication approaches, pinpointing the distinctive value their product offers. This phase typically spans 1–2 years and necessitates harmonizing a multitude of viewpoints from diverse stakeholders. Naturally, a brand must formulate distinct strategies to engage payers, healthcare providers, and patients.

Once the product is introduced, building awareness becomes paramount. The effort initiated through patient support initiatives must persist, carried forward by continual education for healthcare providers and patients, coupled with strategic marketing campaigns. This endeavor gains significant traction for pharmaceutical firms that can assemble a versatile team comprising indispensable stakeholders. Presently, much of this endeavor is centered around digital avenues. The majority of pharmaceutical enterprises are transitioning to a digital-first approach to introducing new drugs. This approach capitalizes on insights gleaned from valuable data, facilitating well-informed decision-making. However, launch events have made a resurgence. Meticulously planned events offer additional avenues to establish and nurture connections with field representatives, who in turn offer valuable insights regarding messaging strategies.

The progression of pharmaceutical innovation plays a vital role in ushering novel therapies to individuals in need. This journey comprises numerous phases, each demanding meticulous forethought and implementation, coupled with adherence to stipulated regulatory prerequisites and benchmarks for quality control. Despite the intricate and protracted nature of this expedition, the potential dividends of introducing a fresh pharmaceutical to the market are momentous. These rewards encompass enhanced patient prognoses, an elevated standard of living, and a curtailment of medical expenditure adopting a methodical sequence for pharmaceutical innovation; drug manufacturers can be instrumental in guaranteeing that patients gain access to treatments that are both secure and efficacious, tailor-made to address their medical requisites. As the pipeline of groundbreaking pharmaceuticals continues to burgeon, the process of pharmaceutical innovation will persist as a cornerstone of the healthcare domain, playing a pivotal role in furnishing novel remedies to patients and augmenting global health outcomes (Calfee 2002).

17.7 Conclusions

The process of uncovering, crafting, and producing effective molecules targeted at established objectives, guided by thorough clinical and nonclinical investigations, will ultimately culminate in a submission package seeking regulatory endorsement. When strategizing a developmental regimen, it is of utmost significance to ascertain whether the planned studies will fulfill the stipulated regulatory prerequisites for assessing both safety and efficacy. This approach will facilitate the creation of an illuminating description for the new medicinal product.

While only a handful of treatments have gained approval for AD, regulatory bodies in the United States and worldwide have been notably proactive in devising guidelines governing novel pharmaceuticals in this domain. Nonetheless, there exists the potential scenario wherein a proposed medication operates in a manner not encompassed by established regulatory directives. In such instances, developers must engage in collaborative discourse with regulatory authorities to devise a comprehensive strategy that honors the proposed mechanism while meeting regulatory imperatives.

References

Beninger P (2018) Pharmacovigilance: an overview. Clin Ther 40:1991–2004

Bertilsson L, Dahl ML, Dalen P, Al-Shurbaji A (2002) Molecular genetics of CYP2D6: clinical relevance with focus on psychotropic drugs. Br J Clin Pharmacol 53:111–122

Bruno A, Costantino G, Sartori L, Radi M (2019) The in silico drug discovery toolbox: applications in lead discovery and optimization. Curr Med Chem 26:3838–3873

Burt T, Dhillon S (2013) Pharmacogenomics in early-phase clinical development. Pharmacogenomics 14:1085–1097
Buschmann J (2013) The OECD guidelines for the testing of chemicals and pesticides. Methods Mol Biol 947:37–56
Cahana A, Romagnioli S (2007) Not all placebos are the same: a debate on the ethics of placebo use in clinical trials versus clinical practice. J Anesth 21:102–105
Calfee JE (2002) The role of marketing in pharmaceutical research and development. PharmacoEconomics 20(Suppl 3):77–85
Capula M, Corno C, El Hassouni B, Li Petri G, Arandelovic S, EORTC PAMM Group (2019) A brief guide to performing pharmacological studies in vitro: reflections from the EORTC-PAMM course “preclinical and early-phase clinical pharmacology”. Anticancer Res 39:3413–3418
Cohen SM, Lawson TA (1995) Rodent bladder tumors do not always predict for humans. Cancer Lett 93:9–16
Crews KR, Gaedigk A, Dunnenberger HM, Leeder JS, Klein TE, Caudle KE, Haidar CE, Shen DD, Callaghan JT, Sadhasivam S, Prows CA, Kharasch ED, Skaar TC, Clinical Pharmacogenetics Implementation Consortium (2014) Clinical Pharmacogenetics Implementation Consortium guidelines for cytochrome P450 2D6 genotype and codeine therapy: 2014 update. Clin Pharmacol Ther 95:376–382
Eypasch E, Lefering R, Kum CK, Troidl H (1995) Probability of adverse events that have not yet occurred: a statistical reminder. BMJ 311:619–620
Fielden MR, Kolaja KL (2008) The role of early in vivo toxicity testing in drug discovery toxicology. Expert Opin Drug Saf 7:107–110
Foddy B (2009) A duty to deceive: placebos in clinical practice. Am J Bioeth 9:4–12
Greaves P, Williams A, Eve M (2004) First dose of potential new medicines to humans: how animals help. Nat Rev Drug Discov 3:226–236
Gross CJ, Kramer JA (2003) The role of investigative molecular toxicology in early stage drug development. Expert Opin Drug Saf 2:147–159
Hamburger AW, Salmon SE (1977) Primary bioassay of human tumor stem cells. Science 197:461–463
Hauck WW, Gilliss CL, Donner A, Gortner S (1991) Randomization by cluster. Nurs Res 40:356–358
Henricks LM, Lunenburg C, De Man FM, Meulendijks D, Frederix GWJ, Kienhuis E, Creemers GJ, Baars A, Dezentje VO, Imholz ALT, Jeurissen FJF, Portielje JEA, Jansen RLH, Hamberg P, Ten Tije AJ, Droogendijk HJ, Koopman M, Nieboer P, Van De Poel MHW, Mandigers C, Rosing H, Beijnen JH, Werkhoven EV, Van Kuilenburg ABP, Van Schaik RHN, Mathijssen RHJ, Swen JJ, Gelderblom H, Cats A, Guchelaar HJ, Schellens JHM (2018) DPYD genotype-guided dose individualisation of fluoropyrimidine therapy in patients with cancer: a prospective safety analysis. Lancet Oncol 19:1459–1467
Huang W, Percie Du Sert N, Vollert J, Rice ASC (2020) General principles of preclinical study design. Handb Exp Pharmacol 257:55–69
Hughes JP, Rees S, Kalindjian SB, Philpott KL (2011) Principles of early drug discovery. Br J Pharmacol 162:1239–1249
Kalaiselvan V, Srivastava S, Singh A, Gupta SK (2019) Pharmacovigilance in India: present scenario and future challenges. Drug Saf 42:339–346
Kavallaris M (2010) Microtubules and resistance to tubulin-binding agents. Nat Rev Cancer 10:194–204
Kernan WN, Viscoli CM, Makuch RW, Brass LM, Horwitz RI (1999) Stratified randomization for clinical trials. J Clin Epidemiol 52:19–26
Kummar S, Rubinstein L, Kinders R, Parchment RE, Gutierrez ME, Murgo AJ, Ji J, Mroczkowski B, Pickeral OK, Simpson M, Hollingshead M, Yang SX, Helman L, Wiltrout R, Collins J, Tomaszewski JE, Doroshow JH (2008) Phase 0 clinical trials: conceptions and misconceptions. Cancer J 14:133–137
Lewis DI (2019) Animal experimentation: implementation and application of the 3Rs. Emerg Top Life Sci 3:675–679

Mallal S, Phillips E, Carosi G, Molina JM, Workman C, Tomazic J, Jagel-Guedes E, Rugina S, Kozyrev O, Cid JF, Hay P, Nolan D, Hughes S, Hughes A, Ryan S, Fitch N, Thorborn D, Benbow A, PREDICT-1 Study Team (2008) HLA-B*5701 screening for hypersensitivity to abacavir. N Engl J Med 358:568–579

Mohs RC, Greig NH (2017) Drug discovery and development: role of basic biological research. Alzheimers Dement (N Y) 3:651–657

Muglia JJ, Digiovanna JJ (1998) Phase 1 clinical trials. J Cutan Med Surg 2:236–241

Mutlib AE, Gerson RJ, Meunier PC, Haley PJ, Chen H, Gan LS, Davies MH, Gemzik B, Christ DD, Krahn DF, Markwalder JA, Seitz SP, Robertson RT, Miwa GT (2000) The species-dependent metabolism of efavirenz produces a nephrotoxic glutathione conjugate in rats. Toxicol Appl Pharmacol 169:102–113

Nair B (2019) Clinical trial designs. Indian Dermatol Online J 10:193–201

Olson H, Betton G, Robinson D, Thomas K, Monro A, Kolaja G, Lilly P, Sanders J, Sipes G, Bracken W, Dorato M, Van Deun K, Smith P, Berger B, Heller A (2000) Concordance of the toxicity of pharmaceuticals in humans and in animals. Regul Toxicol Pharmacol 32:56–67

Paul D, Sanap G, Shenoy S, Kalyane D, Kalia K, Tekade RK (2021) Artificial intelligence in drug discovery and development. Drug Discov Today 26:80–93

Perego P, Hempel G, Linder S, Bradshaw TD, Larsen AK, Peters GJ, Phillips RM, EORTC PAMM Group (2018) Cellular pharmacology studies of anticancer agents: recommendations from the EORTC-PAMM group. Cancer Chemother Pharmacol 81:427–441

Phillips RM (2016) Targeting the hypoxic fraction of tumours using hypoxia-activated prodrugs. Cancer Chemother Pharmacol 77:441–457

Relling MV, Evans WE (2015) Pharmacogenomics in the clinic. Nature 526:343–350

Relling MV, Schwab M, Whirl-Carrillo M, Suarez-Kurtz G, Pui CH, Stein CM, Moyer AM, Evans WE, Klein TE, Antillon-Klussmann FG, Caudle KE, Kato M, Yeoh AEJ, Schmiegelow K, Yang JJ (2019) Clinical Pharmacogenetics Implementation Consortium guideline for thiopurine dosing based on TPMT and NUDT15 genotypes: 2018 update. Clin Pharmacol Ther 105:1095–1105

Rhim JS, Tsai WP, Chen ZQ, Chen Z, Van Waes C, Burger AM, Lautenberger JA (1998) A human vascular endothelial cell model to study angiogenesis and tumorigenesis. Carcinogenesis 19:673–681

Rieder MJ, Reiner AP, Gage BF, Nickerson DA, Eby CS, Mcleod HL, Blough DK, Thummel KE, Veenstra DL, Rettie AE (2005) Effect of VKORC1 haplotypes on transcriptional regulation and warfarin dose. N Engl J Med 352:2285–2293

Robinson NB, Krieger K, Khan FM, Huffman W, Chang M, Naik A, Yongle R, Hameed I, Krieger K, Girardi LN, Gaudino M (2019) The current state of animal models in research: a review. Int J Surg 72:9–13

Roden DM, Van Driest SL, Wells QS, Mosley JD, Denny JC, Peterson JF (2018) Opportunities and challenges in cardiovascular pharmacogenomics: from discovery to implementation. Circ Res 122:1176–1190

Roma-Rodrigues C, Mendes R, Baptista PV, Fernandes AR (2019) Targeting tumor microenvironment for cancer therapy. Int J Mol Sci 20:840

Rost S, Fregin A, Ivaskevicius V, Conzelmann E, Hortnagel K, Pelz HJ, Lappegard K, Seifried E, Scharrer I, Tuddenham EG, Muller CR, Strom TM, Oldenburg J (2004) Mutations in VKORC1 cause warfarin resistance and multiple coagulation factor deficiency type 2. Nature 427:537–541

Schenone M, Dancik V, Wagner BK, Clemons PA (2013) Target identification and mechanism of action in chemical biology and drug discovery. Nat Chem Biol 9:232–240

Steinmetz KL, Spack EG (2009) The basics of preclinical drug development for neurodegenerative disease indications. BMC Neurol 9(Suppl 1):S2

Torres-Saavedra PA, Winter KA (2022) An overview of phase 2 clinical trial designs. Int J Radiat Oncol Biol Phys 112:22–29

Tsiatis A (1990) Methodological issues in AIDS clinical trials. Intent-to-treat analysis. J Acquir Immune Defic Syndr (1988) 3(Suppl 2):S120–S123

Tsintis P, La Mache E (2004) CIOMS and ICH initiatives in pharmacovigilance and risk management: overview and implications. Drug Saf 27:509–517

Umscheid CA, Margolis DJ, Grossman CE (2011) Key concepts of clinical trials: a narrative review. Postgrad Med 123:194–204

Van Norman GA (2019) Phase II trials in drug development and adaptive trial design. JACC Basic Transl Sci 4:428–437

Whiteside TL (2008) The tumor microenvironment and its role in promoting tumor growth. Oncogene 27:5904–5912

Wilcox CM (2008) Exploring the use of the sham design for interventional trials: implications for endoscopic research. Gastrointest Endosc 67:123–127

Wu KM, Farrelly JG (2006) Preclinical development of new drugs that enhance thyroid hormone metabolism and clearance: inadequacy of using rats as an animal model for predicting human risks in an IND and NDA. Am J Ther 13:141–144

Yang HL, Wu XB, Mao C (2019) Block randomization in clinical trials. Zhonghua Yu Fang Yi Xue Za Zhi 53:437–440

Zazzara MB, Palmer K, Vetrano DL, Carfi A, Onder G (2021) Adverse drug reactions in older adults: a narrative review of the literature. Eur Geriatr Med 12:463–473

Ziegler CJ, Silverman AP, Lippard SJ (2000) High-throughput synthesis and screening of platinum drug candidates. J Biol Inorg Chem 5:774–783

The manufacturer's authorised representative in the EU is Springer
Nature Customer Service Centre GmbH, Europaplatz 3, 69115 Heidelberg,
Germany. If you have any concerns regarding our products, please
contact ProductSafety@springernature.com

Printed and bound by CPI Group (UK) Ltd, Croydon, CR0 4YY
01/10/2026
02225956-0014